D0079026

Invertebrate Zoology

THIRD EDITION

Invertebrate Zoology

Joseph G. Engemann
Professor of Biology, Western Michigan University

Robert W. Hegner
Late Professor of Protozoology, The Johns Hopkins University

Macmillan Publishing Co., Inc.
New York
Collier Macmillan Publishers
London

Macmillan Publishing Co., Inc.
866 Third Avenue, New York, New York 10022
Collier Macmillan Canada, Ltd.

Library of Congress Cataloging in Publication Data

Engemann, Joseph G
  Invertebrate zoology.
  In the 2d ed. Hegner's name appeared first on T. P.
  Bibliography: p.
  Includes index.
  1. Invertebrates. I. Hegner, Robert William,
1880–1942, joint author. II. Title.
QL362.H4 1981   592   80-12063
ISBN 0-02-333780-X

Printing: 1 2 3 4 5 6 7 8   Year: 1 2 3 4 5 6 7 8

to my wife, Nancy,
*whom I have learned to like even
better than invertebrates*

Preface

Twenty-five printings of the first edition of this text in a period of three decades attest to the value of Hegner's approach to invertebrate zoology. I was one of many satisfied students who received their initiation into the world of the invertebrates through the pages of the late author's book.

The philosophy of the first edition is retained: the *type approach* is used to give specific examples as a base for understanding adaptation and variation within the phyla. This approach also provides a more flexible text with greater usefulness in both lectures and laboratory. The topical treatment used permits the combination of theory and application to specific groups, so that the student is not overwhelmed with conceptual knowledge before he or she knows about the organisms to which it is pertinent.

Many topics included in this edition are new to invertebrate zoology since the appearance of the first edition in 1933. Details of surface ultrastructure have greatly increased knowledge since the wide availability of scanning electron microscopy in the late 1960s made study of surface detail convenient. "Killer bees," as well as the destruction of tropical reefs by starfish, have evoked public concern and investigation by zoologists. Those findings and many other new facets of classical biology revitalized by the methods and findings of molecular biology find expression in this edition.

Some references from the previous editions are retained, and new references have been added that purposely emphasize recent articles in *Science* and *Transactions of the American Microscopical Society.* A small college library can readily provide these journals in a minimal core of science references.

The elimination of references to the human species by a convenient three-letter word may seem awkward in places to some but should make growth and development of half of the users free of those stifling stereotypes of language usage.

Chapter 14 is entirely new. A theory about a major portion of our biosphere is introduced. My hope is that beginning zoologists will find it an exciting and stimulating introduction to thinking about invertebrate zoology beyond the often rote memorization of factual information frequently emphasized. Unresolved questions in other chapters provide similar opportunities for thought and discussion.

Although the recent progress made in the field of invertebrate zoology is enormous and wide ranging, I hope that the sample presented in this book will convey some idea of the range and excitement of recent research trends as well as most of the more important concepts of invertebrate zoology.

I am grateful to the many zoologists upon whose work textbooks such as this one are dependent. I am especially grateful to those who so generously provided illustrations or gave their permission for reproducing such material. Mr. and Mrs. Sam Rizzetta receive my thanks for copying and improving many of the drawings retained from the first edition as well as for executing a number of original ones for the second edition which were retained in this edition. The aid of editors Woodrow Chapman and David Garrison of Macmillan were essential to the completion of this text. Macmillan production supervisor J. Edward Neve, designer William Gray, and anonymous copy editors have added much to the quality of this book. The review efforts of Drs. Dean Dillery, William Pohley, Frank Rokop, and David Rubin were most helpful in eliminating errors and improving content. I wish that time and space had permitted inclusion of more of their excellent suggestions. The interest, encouragement, and help of colleagues at Western Michigan University and elsewhere, including Drs. R. Brewer, D. Buthala, R. Husband, R. Pippen, T. W. Porter, and E. B. Steen, are valued. The stimulation and help of former students and graduate students have been important, and I especially acknowledge J. Amato, W. Andre, J. Badgerow, M. Campbell, N. Elliott, D. Flanagan, W. Harvey, R. Hollinger, J. Medlin, S. Miller, G. Mimidis, D. Oosting, T. Petersen, J. Price, W. Rosenbaum, W. Schroeter, A. Swehli, R. Ulrich, J. Wenger, and R. Wolf.

For errors and omissions I bear sole responsibility. For the prayers and help of friends, for the neglect accepted and help rendered by my family—Nancy, along with Jennifer, Molly, and John—I will be eternally grateful.

J. G. E.
Kalamazoo

Contents

Introduction

Most animals are invertebrates. However, most of the time spent on the study of animals in introductory courses is on the study of vertebrates. Many fundamental principles of physiology, genetics, ecology, and taxonomy apply equally well to invertebrates and vertebrates. Biochemical pathways common to all living things can frequently be studied with greater ease in invertebrates. A knowledge of the invertebrates can simplify the selection of an organism that can be used to greater advantage in molecular biology. Invertebrates, with small size and short life cycles, are excellent organisms for investigations of population dynamics and other ecological phenomena. Perhaps more important are the direct effects of invertebrates on humans and their possessions. Of growing importance is the use of invertebrates in the biological control of pests now being held in check with poisons which may accumulate in our soils and waters.

Beneficial and detrimental effects of invertebrates are an important factor in modern living. Of utmost importance is the pollination of many crops by insects. Other benefits include the production of commercially valuable substances (e.g., waxes, honey, and silk), the use of some invertebrates as food animals, service as scavengers and for aeration of the soil, and an important role as intermediate forms in the food chain. Their value more than offsets the harm done as disease organisms, carriers of disease, and destroyers of food and other valuable substances.

As our utilization of space on earth increases, knowledge of the exact role of the various invertebrates must increase if we are to manage our resources in a way that will not be disastrous.

Molecular biology is important to—and has benefitted from—invertebrate zoology. An attempt has been made in this book to show the relationship between the invertebrates and the advances in molecular biology with which they are most closely associated. For the complete story of macromolecules, their formation, and degradation, replete with alphabetic abbreviations such as DNA, RNA, and ATP, one must turn to genetics, physiology, and biochemistry. This is not because molecular biology is an unimportant part of invertebrate zoology, but because the vast numbers and varieties of invertebrates make it impossible to adequately cover both the classical aspects of invertebrate zoology and the molecular aspects. Thus the emphasis here is on a balanced approach with as many guideposts as possible added to connect this subject with other areas of biology.

Since all biologists are ultimately involved with organisms, and organisms that are closely related may differ markedly in their molecular, ecological, and genetic aspects, it is important to be able to distinguish organisms in a systematic way. That is one reason why invertebrate zoology is an important subject early in the career of a biologist.

The next to the last chapter, "Invertebrates—An Overview", is intended to summarize and connect some unifying threads of invertebrate zoology; it may also be read as an introduction, along with the first chapter, for those who have not had a recent course in general zoology or general biology. The last chapter is an application of many diverse principles toward a suggested explanation of a question difficult to investigate directly. Those interested in the Piagetian theory of intellectual development will find the chapter a logical step into somewhat abstract thinking from the more concrete material of earlier chapters.

The information and theories accumulated about invertebrates are too extensive to be contained in one text. Morphological, ecological, and to a lesser extent taxonomic and physiological studies reported in the older literature retain much of their usefulness today. Thus the person who would become an invertebrate zoologist must become familiar with the accumulated literature or know how to tap the wealth of information there.

References from this and other texts will provide one entry into the literature. Three of the most important bibliographic tools are *Biological Abstracts, Science Citation Index,* and the *Zoological Record.* Each provides an index of most pertinent information published before or during the year that they are dated. Through them one can learn of useful literature found in an increasing array of journals, museum reports, and government publications. Use of literature cited in such papers is an efficient way of locating additional information on the subject. Of equal importance to the researcher is communication with his or her colleagues, either informally or at society meetings.

Two important recent treatises on the invertebrates are the still incompleted works of Hyman (1940–) and Grasse (1948–). Among recent texts that students will find useful are Russell-Hunter (1979) with stimulating and sometimes unusual viewpoints, Gardiner (1972) with strong physiology and systems information, and Beklemishev (1969) for comparative anatomy and development. Books especially useful for identification include Pennak (1978; also containing many valuable general discussions), Smith and Carlton (1975), Gosner (1971), Pratt (1935), and the *How to Know* series of the Brown Publishing Company. Useful laboratory aids include Brown (1950), Dales (1969), Bullough (1960), Beck and Braithwaite (1961), and Sherman and Sherman (1976). Titles of the general references listed below indicate their content.

General References

ALLEE, W. C., A. E. EMERSON, O. PARK, AND K. P. SCHMIDT. 1949. *Principles of Animal Ecology*. Saunders, Philadelphia.

BARNES, R. D. 1974. *Invertebrate Zoology*. Saunders, Philadelphia.

BAYER, F. M., AND H. B. OWRE. 1968. *The Free-Living Lower Invertebrates*. Macmillan, New York.

BECK, D. E., AND L. F. BRAITHWAITE. 1961. *Invertebrate Zoology Laboratory Workbook*. Burgess, Minneapolis.

BECKLEMISHEV, W. N. 1969. *Principles of Comparative Anatomy of Invertebrates*. Translated by J. M. MacLennan; edited by Z. Kabata. Univ. of Chicago Press, Chicago.

BLACKWELDER, R. E. 1963. *Classification of the Animal Kingdom*. Southern Illinois Univ. Press, Carbondale.

BORRADAILE, L. A., F. A. POTTS, L. E. S. EASTHAM, J. T. SAUNDERS, AND G. A. KERKUT. 1959. *The Invertebrata*. Cambridge Univ. Press, New York.

BREWER, R. 1979. *Principles of Ecology*. Saunders, Philadelphia.

BRONN, H. G. (Ed.). 1880– . *Klassen und Ordnungen des Tier-reichs*. Leipzig.

BROWN, F. A. (Ed.). 1950. *Selected Invertebrate Types*. Wiley, New York.

BRUSCA, G. J. 1975. *General Patterns of Invertebrate Development*. Mad River Press, Eureka, Calif.

BULLOCK, T. H., AND G. A. HORRIDGE. 1965. *Structure and Function in the Nervous System of Invertebrates*. Freeman, San Francisco.

BULLOUGH, W. S. 1960. *Practical Invertebrate Anatomy*. Macmillan, London.

BYKHOVSKAYA-PAVLOVSKAYA, I. E., et al. 1962. *Key to Parasites of Freshwater Fish of the U.S.S.R.* Translated by A. Birron and Z. S. Cole, 1964. U.S. Dept. Comm., Washington.

CARTHY, J. D. 1958. *An Introduction to the Behavior of Invertebrates*. Macmillan, New York.

CHENG, T. C. 1973. *General Parasitology*. Academic Press, New York.

CORNING, W. C., J. A. DYAL, AND A. O. D. WILLOWS (Eds.). 1975. *Invertebrate Learning*. Vol. **3**. *Cephalopods and Echinoderms*. Plenum Press, New York.

DALES, R. P. (Ed.). 1969. *Practical Invertebrate Zoology*. Univ. of Washington Press, Seattle.

DAWYDOFF, C. 1928. *Embryologie comparée des invertébrés*. Masson et Cie, Paris.

DICKINSON, C. H., AND G. J. F. PUGH (Eds.). 1974. *Biology of Plant Litter Decomposition*. Vol. 2. Academic Press, London.

DOUGHERTY, E. (Ed.). 1963. *The Lower Metazoa*. Univ. of California Press, Berkeley.

EDMONDSON, W. T. (Ed.). 1959. *Fresh-Water Biology*. 2nd Ed. Wiley, New York.

FLORKIN, M., AND B. T. SCHEER (Eds.). 1969. *Chemical Zoology*. Vol. **IV**. *Annelida, Echiura, and Sipuncula*. Academic Press, New York.

FRETTER, V., AND A. GRAHAM. 1976. *A Functional Anatomy of Invertebrates*. Academic Press, London.

GARDINER, M. S. 1972. *The Biology of Invertebrates*. McGraw-Hill, New York.

GIESE, A. C., AND J. S. PEARSE. 1975. *Reproduction of Marine Invertebrates. Vol. **II**. Entoprocts and Lesser Coelomates*. Academic Press, New York.

GOSNER, K. L. 1971. *Guide to Identification of Marine and Estuarine Invertebrates*. Wiley-Interscience, New York.

GRASSE, P. (Ed.). 1948– . *Traité de zoologie*. (Tomes **I–XI** on invertebrates.) Masson et Cie, Paris.

HALSTEAD, B. W. 1965. *Poisonous and Venomous Marine Animals of the World*. G.P.O., Washington.

HESSE, R., W. C. ALLEE, AND K. P. SCHMIDT. 1937. *Ecological Animal Geography*. Wiley, New York.

HICKMAN, C. P. 1967. *Biology of the Invertebrates*. Mosby, Saint Louis.

HYMAN, L. H. 1940– . *The Invertebrates*. McGraw-Hill, New York.

HYNES, H. B. N. 1960. *The Biology of Polluted Waters*. Liverpool Univ. Press, Liverpool.

KEVAN, D. K. 1962. *Soil Animals*. Philosophical Library, New York.

KOZLOFF, E. N. 1973. *Seashore Life of Puget Sound, the Strait of Georgia and the San Juan Archipelago*. Univ. of Washington Press, Seattle.

KUKENTHAL, W., AND T. KRUMBACH (Eds.). 1923. *Handbuch der Zoologie*. Berlin.

KUME, M., AND D. KATSUMA. 1968. *Invertebrate Embryology*. Translated by J. C. Dan. Nat. Library of Med., PHS, Washington.

LANKESTER, R. (Ed.). 1909– . *A Treatise on Zoology*. Black, London.

MACGINITIE, G. E., AND N. MACGINITIE. 1949. *Natural History of Marine Animals*. McGraw-Hill, New York.

MALLIS. 1960. *Handbook of Pest Control*. MacNair Dorland, New York.

McCONNAUGHEY, B. H. 1978. *Introduction to Marine Biology*. Mosby, Saint Louis.

MEGLITSCH, P. A. 1972. *Invertebrate Zoology*. Oxford Univ. Press, New York.

MOORE, R. C. (Ed.) 1953– . *Treatise on Invertebrate Paleontology*. Geol. Soc. Am., New York.

MORGAN, A. H. 1930. *Field Book of Ponds and Streams*. Putnam, New York.

NOBLE, E. R., AND G. A. NOBLE. 1976. *Parasitology*. Lea & Febiger, Philadelphia.

PENNAK, R. W., 1978. *Fresh-Water Invertebrates of the United States*. 2nd Ed. Wiley, New York.

PRATT, H. S. 1935. *A Manual of the Common Invertebrate Animals*. McGraw-Hill, New York.

RICKETTS, E. F., AND J. CALVIN. (Rev. by J. W. Hedgpeth.) 1968. *Between Pacific Tides*. Stanford Univ. Press, Stanford.

RUSSELL-HUNTER, W. D. 1979. *A Life of Invertebrates.* Macmillan, New York.

SALANKI, J. (Ed.). 1973. *Neurobiology of Invertebrates Mechanisms of Rhythm Regulation.* Akademiai Kiado, Budapest.

SCHMIDT, G. D., AND L. S. ROBERTS. 1977. *Foundations of Parasitology.* Mosby, Saint Louis.

SHERMAN, I. W., AND V. G. SHERMAN. 1976. *The Invertebrates: Function and Form.* 2nd Ed. Macmillan, New York.

SHROCK, R. R., AND W. H. TWENHOFEL. 1953. *Principles of Invertebrate Paleontology.* McGraw-Hill, New York.

SMITH, R. I., AND J. T. CARLTON (Eds.). 1975. *Light's Manual: Intertidal Invertebrates of the Central California Coast.* Univ. of California Press, Berkeley.

STEEN, E. B. 1970. *Dictionary of Biology.* Barnes and Noble, New York.

SWEETMAN, H. L. 1958. *The Principles of Biological Control.* Brown, Dubuque.

WELSCH, U., AND V. STORCH. 1976. *Comparative Animal Cytology & Histology.* Univ. of Washington Press, Seattle.

WHIPPLE, H. E. (Ed.). 1963. Some Biochemical and Immunological Aspects of Host–Parasite Relationships. *Ann. N. Y. Acad. Sci.* **113** (1).

WILSON, E. O. 1975. *Sociobiology.* Belknap Press of Harvard Univ. Press, Cambridge.

ZWEIFEL, F. W. 1961. *A Handbook of Biological Illustration.* Univ. of Chicago Press, Chicago.

Basic Principles of Invertebrate Zoology

Invertebrates are more numerous in kinds, variations, and functions in the environment than any other group of organisms. Their extreme variety and importance has led to a variety of approaches, all attempting to improve our understanding of them.

The "father of invertebrate zoology," Aristotle, although emphasizing structure (morphology), recognized that all aspects of animals' biology are of use in classifying them. He stated 22 centuries ago that (Mayr, Lindsley, and Usinger, 1953) "Animals may be characterized according to their way of living, their actions, their habits, and their bodily parts."

Morphology has stood the test of time as the backbone of classification and a basis for study of invertebrates, because structure is intimately related to all other aspects. Although the genotype constitutes the ultimate definition of an organism, it is impossible to use as the basis for an effective classification scheme because of the tremendous amount of information involved and the specialized capabilities required to process it. Structure is determined by the genetics of the individual as modified by environmental and developmental variables. Morphology determines in a major way the functions of parts involved and the ecological roles of the whole organisms. In their development, the biological sciences seem to progress through descriptive phases beginning with classification, followed by greater attention to structure, then function. Then an experimental phase expands in various directions to provide greater understanding, generally of the more important aspects, but sometimes just aspects determined by the expertise and hunches of the investigator. Because invertebrates are so varied, they provide excellent opportunities for investigation of many important biological topics from cellular biology through ecology. That is why it is important for a scientist in any of the many burgeoning experimental disciplines of biology to have invertebrate zoology training.

As an alert student you may find ways to utilize the invertebrates to clarify biological principles that will become outstanding examples along with the fruit fly in genetics, the squid giant axon in nervous system depolarization phenomena, the fiddler crab in biological rhythms, protozoan communities or insect communities in species interactions, and the many other ways that will be indicated as we go along.

If Aristotle was the "father of invertebrate zoology", then Libbie Hyman deserves the title "mother of invertebrate zoology" for what may be the last great individual attempt to comprehensively monograph the invertebrates in a work that saw six volumes completed before her death.

Many individuals deserve recognition as outstanding scholars of particular groups and many will be recognized in pertinent chapters. It is more difficult to assess the impact of modern workers such as Beklemishev, Brooks, Huxley, and Pennak or the value of editorial services provided by Bronn, Grasse, or Moore than it is to recognize the impact of Linnaeus or Darwin from past centuries. Some of the more notable people and events or contributions are noted in Table 1-1. It is of interest to note that the fields of current high interdisciplinary interest in biology are less easily credited to one individual and are thus omitted from the table, although invertebrate zoologists often make major contributions, as did the student of insect societies, E. O. Wilson, with his book, *Sociobiology* (1975).

European males dominated the early development of invertebrate zoology, possibly because wealth from the industrial revolution may have given them the time and opportunity. Contributions today are not restricted by geographic, racial, or sexual boundaries. Names like Barnard or Marcus do not automatically reveal their African or South American locations. The black heritage of Finley is not apparent from his name, although the oriental ancestry of Kudo may be. Libbie Hyman and many of her sex, such as Deichman, Manton, and Rogick, were established leaders well before the recent emphasis on equality and made major contributions to various invertebrate studies. This universality of appeal and freedom of access is apparent among the outstanding contributors of today.

The remainder of this chapter is a rapid review of basic principles

Table 1-1 Chronology of Some Major Contributions to Invertebrate Zoology

Person	Years	Country	Contribution
Aristotle	384–322 B.C.	Greece	"Father of invertebrate zoology," text on natural history of animals.
Ray, John	1627–1705	England	Distinction between genus and species and higher categories of classification.
Linnaeus, Carolus	1707–1778	Sweden	Consistent application of binomial system in classification in his *Systema Naturae*, the tenth edition of which is the official starting point for modern zoological taxonomy (1758).
Darwin, Charles	1809–1882	England	Evolution by means of natural selection, wrote *The Origin of Species* (1859).
Wallace, Alfred	1823–1913	England	Independently developed concept of natural selection and thus stimulated publication by Darwin.
van Leeuwenhoek, Anthony	1632–1723	Holland	Invention of microscope, "father of protozoology."
Schleiden and Schwann	1830s	Germany	Clarification of cell theory.
Liebig and Blackman	1840s	Germany	Minimum amounts of essential substances needed for growth.
Haekel, Ernst	1834–1919	Germany	Clarified invertebrate relationships
Hyman, Libbie	1888–1969	United States	Modern statement of invertebrate relationships.

and common themes that have broad application throughout most major groups of invertebrates.

THE EVOLUTIONARY KEY

The basic principle to consider in evaluating biological phenomena is that the characteristics of any group have had some value, during the evolutionary history of the group, in terms of reproductive success at leaving reproductively successful offspring. An example of the speed with which useless characteristics are lost is the relatively few generations in which pigmentation is lost by forms that move into subterranean habitats. The value of a characteristic is determined by how it functions in relationship to the composite of characteristics of the individual or group and its contribution to its or their survival and reproductive success. The characteristics of concern are inheritable organization and range from molecular through physical and social characteristics. Most consider mechanistic explanations adequate to explain the ongoing processes of their production. Many find theistic explanations needed to explain ultimate origins and causes, while a few still view the *Book of Genesis* as a statement of scientific truths in addition to a guide for religious and historical values. If a person believed in the special creation of each individual form, it would be reasonable to expect that a Creator would have endowed similar forms with similar attributes, so that person could infer relationships similar to those arrived at with an evolutionary model, assuming that the fossil record is ignored or interpreted as an aberrant additional form of creation. The purpose of features in an evolutionary sense is in terms of past and present value. The purpose of features in a teleological, or goal-seeking, sense is in terms of present and future value and not accepted as valid in science, although valid conclusions are often reached when present values are considered because of partial overlap with the more accurate evolutionary model.

Our evolutionary key has three premises, listed in descending order of probability: (1) an existing feature is of present value to the organism or its group; (2) an existing feature was of value to the ancestors of the organism; and (3) the feature was never of value but was intimately linked since its development to a valuable characteristic of the group. Examples of the last type are unknown and perhaps nonexistent. Excluded from consideration are those characteristics of no value that are rapidly eliminated from the group and thus never characteristic of a species for long. Premise 1 usually applies and yields the best understanding. Premise 2 may be true simultaneously with premise 1.

SOME MOLECULAR PRINCIPLES

Protoplasm, the living substance of cells and organisms, is a variable assemblage of chemical substances. Quantities vary with different functional states, but water, proteins, carbohydrates, lipids, nucleic acids, minerals, and dissolved gases are always present. Some of these, the macromolecules (protein, carbohydrate, or nucleic acid molecules), have a special importance. Their large size may be of value in living systems for one or more of the following reasons.

1. Localization. They may keep their functional parts in position. For example, functions that are related may be tied together for increased efficiency. This may be continued on an even larger scale by the many membranes that occur in protoplasm.

2. Retention. Permeability results in loss of small molecules more rapidly than loss of macromolecules. The energy used in producing macromolecules is not wasted by loss.

3. Storage.

4. Reduced osmotic impact. Since osmotic impact is proportional to the number of separate particles, a macromolecule has only the amount of effect of one ion or one small molecule. Energy stores accumulated as macromolecules do not decrease the osmotic concentration of protoplasm and disturb the physiological balance as much as if they were retained as many small molecules. Thus glucose can be stored as glycogen, fatty acids in triglycerides, and amino acids in proteins.

5. Reduced chemical reactivity. Chemical activity is also reduced or modified by such storage. This permits self-regulation without interference but allows the ready availability of needed supplies within the cell.

6. Diversity. The basic chemical subunits of protoplasm can be combined into a greater variety of materials to meet the diverse functional needs of the cells. This is especially true of proteins and nucleic acids, where substitution of one amino acid or nucleotide at one position in a long chain can have a major functional effect.

7. Reproduction. In sexual reproduction the macromolecule, DNA, forming part of a chromosome, enables functional combinations to be maintained in offspring. Some characteristics are of value only in association with other particular variants. When hereditary characteristics are linked in a valuable combination of high survival and reproductive value, they will tend to become the most common forms found as a result of natural selection.

Molecules of all sizes are subject to selective influences for their utility to the cell or organism. Information molecules include the large nucleic acids and the much smaller hormones. These may convey information due to their chemical uniqueness or physical characteristics. Lipid materials with complex ring structures having minor variations in rings and side chains have provided variety with underlying uniformity. Because such information molecules often function at very

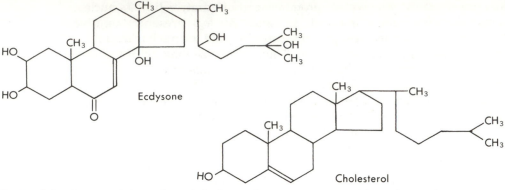

Figure 1-1 Structural formulas of cholesterol, a common steroid, and one derivative, ecdysone, an insect molting hormone.

low concentrations, they are difficult to isolate from natural sources for study. Over the course of time the same molecule may be utilized for different purposes in different groups. Thus ecdysone (Fig. 1-1), the molting hormone of insects, occurs in bracken fern, where its function is unknown. Undoubtedly, many important substances and functions remain to be discovered for many organisms in which low concentrations are involved in the natural state.

Changes can sometimes occur in molecules without causing a change in function. Change may modify function slightly by changing the rate of enzymatic function, or changing pH, temperature, or other optima. Such changes have enabled invertebrates to speciate into groups specialized for remarkably slight environmental variations.

Energy storage molecules have involved lipids most frequently because high concentrations can be reached in droplets that are virtually insoluble in the aqueous phase of protoplasm and very rich in stored energy. Glucose is important fuel for all organisms, but its chemical reactivity prohibits the use of high concentrations. Glucose is stored as trehalose (Fig. 1-2) in many insects, as glycogen in perhaps all animals, or is converted even more radically into two carbon units for addition to the carbon chains of lipids or proteins, or for breakdown during the formation of ATP for the immediate energy needs of the cell.

Working molecules such as enzymes and muscle protein are often recognized or named for their function. Because the protein portion can be changed slightly in ways that make a difference identifiable by density or electrophoresis without changing its chemical function in the cell, accumulated changes or mutations in such proteins can be used as a measure of evolutionary difference between groups. This can be done with any cellular chemical that is recognized as derived from the same ancestral substance. Because structural proteins and many enzymes change readily without necessarily losing functional capacity, they are excellent for determining close relationships. Some materials can tolerate relatively little change before losing functional capacity, so they are useful for determining more distant relationships—examples are cytochrome *c* and hemoglobin. Other substances have not exhibited the ability to be replaced by related substances. Examples are glucose, ATP, and other subunits of the macromolecules.

Waste molecules from protein metabolism have evolved in most groups because the ammonia formed as the immediate end product cannot be tolerated in high concentrations. Common solutions to the problem for those animals unable to diffuse ammonia rapidly into an aqueous environment have been the development of metabolic ma-

Figure 1-2 Glucose and related polymers. Chitin is a polymer of acetyl glucosamine.

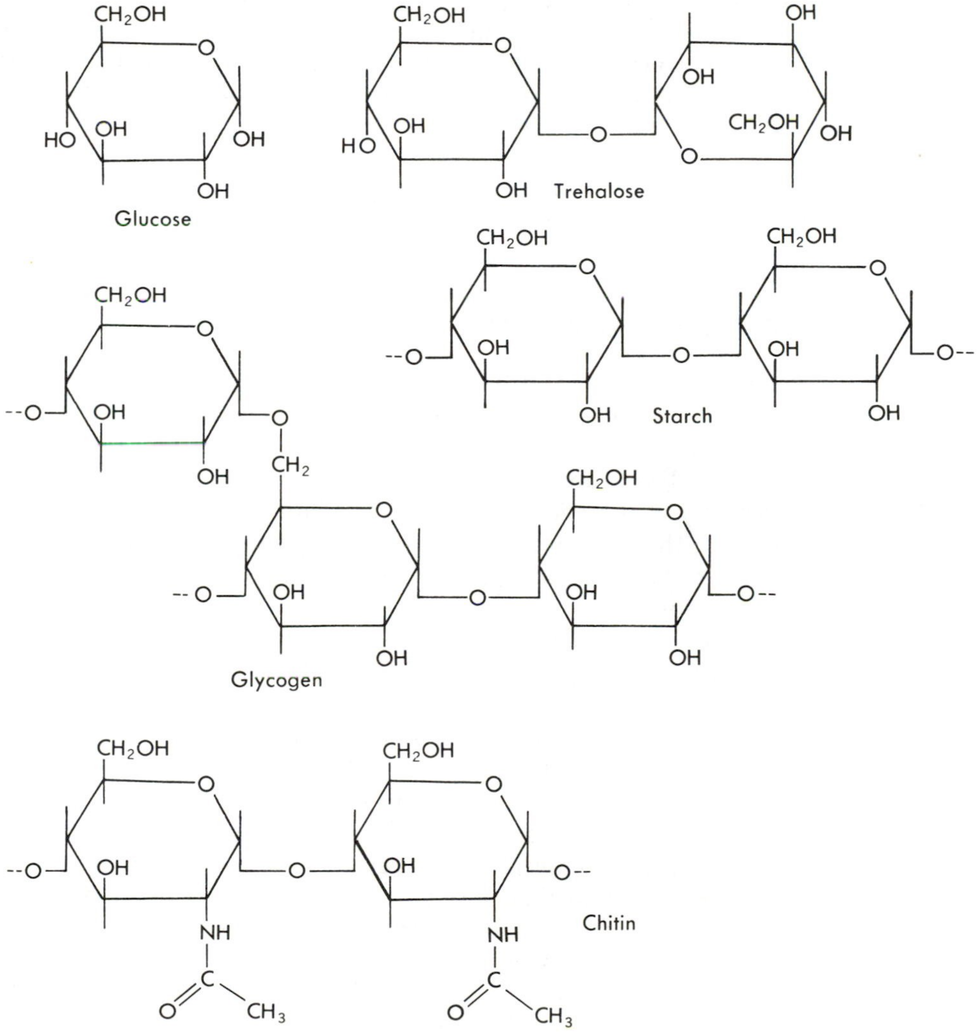

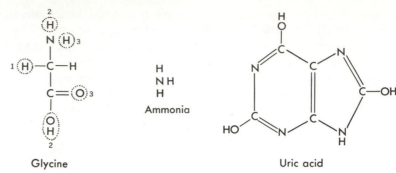

Glycine   Uric acid

Figure 1-3 Nitrogenous compounds. The amino acid, glycine, and related amino acids form other amino acids, peptides, and proteins by substitutions or combinations indicated: 1, hydrogen replaced by various carbon side chains in other amino acids; 2, sites of peptide bonding and release of water molecule during formation of a peptide chain; 3, site of hydrogen bonding with another peptide chain. Ammonia, the toxic nitrogen-containing compound remaining when amino acids are used for energy. Uric acid, a compound of low solubility often used for disposal of ammonia.

chinery to convert the ammonia to more complex substances, such as the less toxic urea or the insoluble uric acid (Fig. 1-3). This allows temporary storage or excretion with less water loss for terrestrial animals.

SOME CYTOLOGICAL AND HISTOLOGICAL PRINCIPLES

Cells are the structural and functional units of organisms that may be specialized for different functions. The cells are often aggregated in groups of like cells as tissues. Different tissues are assembled to form organs. Specialized functions of cells in these organs are thought to be improved by the aggregation, making regulation of the immediate environment of the cells possible. Complex organisms have organs connected into organ systems, but functioning is still heavily dependent on the component cells.

Cells have a plasma membrane surrounding the cytoplasm which usually contains one or more nuclei as well as mitochondria, membranes, ribosomes, Golgi apparatus, microtubules, microfibrils, and other inclusions. The constituents and amounts vary with function. Functional and structural similarities of cells from different groups of complex invertebrates demonstrate an underlying relationship. Recognizable cells and tissues (Fig. 1-4) include epithelial, glandular, muscular, nervous, connective, and reproductive types. Flagellated or ciliated, and amoeboid types occur as organisms in the Protozoa and constituent cells in individuals of most phyla.

The protozoans are discussed in Chapter 2 as a logical introduction to the invertebrates at the cellular level and as representatives of the most direct descendents of the ancestral type of all animals.

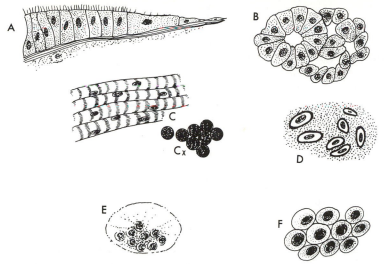

Figure 1-4 Invertebrate tissues. (A) Epithelial tissues cover surfaces or line cavities; nuclei are often prominent; typically one cell-layer thick; often ciliated, or with a secreted covering; appearance in cross section often varies with degree of stretching. (B) Glandular tissues often have vacuolated cells clustered around ducts. (C) Muscle tissues have a linear orientation and fewer nuclei are evident than in the preceding tissues; often cross-striations are present; frequently in sheets or bundles; they appear very different when the section (C_x) is across the bundle. (D) Connective tissue has few cells in a uniform ground substance as in cartilage of the squid; other forms often appear similar to epithelial, nerve, and muscle tissues, but with very few cells or nuclei evident. (E) nerve tissue often has erratic aggregations of cells mixed with fibrous structures; often poorly stained in cross sections because of fatty substances in fiber sheaths. (F) Reproductive tissue is often an aggregation of cells with prominent nuclei; ovaries usually have much more cytoplasm per cell than testes; thus, testes usually appear more darkly stained then ovaries.

STRUCTURE AND FUNCTION

Structure and function are generally so intimately related that one is a good predictor of the other in the analysis of an organism. Some functions are so basic to the integrity of the cell that they are not assigned to specialized systems in any animals, and are collectively called cellular metabolism in a broad sense, including growth and division.

Functions assigned to different tissues, organs, or systems, in higher invertebrates form four functional clusters.

1. Structure, protection from the environment, and locomotion form a functional cluster, including the *integumentary system, skeletal system,* and *muscular system* that collectively have a great impact on the appearance of an organism.

2. Maintenance of the organism is largely relegated to the *digestive system, circulatory system, respiratory system,* and *excretory system.*

3. Integration and control is largely dependent on the *nervous system* and *endocrine system,* aided by the organs of special sense and the previously mentioned circulatory system.

4. Perpetuation of the species is made possible by the *reproductive system,* coordinated with developmental processes that often include different stages in the life cycle that may have differing functional arrangements of the systems.

The latter system's functions or processes are the major focus of genetics and embryology, whereas physiology would be concerned with all functional aspects and morphology or anatomy with all structural aspects. When we consider the organism in relation to its environment, the area of ethology or behavior is concerned with activities and ecology with the total relationship. The distributional aspects of animals in space is considered in zoogeography and the distribution in time in paleontology and evolution. When we consider all aspects of the study of organisms or life the field is called biology, the equivalent term for animals being zoology.

The *integumentary system* protects the organism from the environment. For many invertebrates it consists of a single layer of cells resting on a basement membrane with the opposite and outer surface often covered by secreted material. Those animals in contact with solid surfaces often have mucus secretions that lubricate and also keep the surface moistened when they leave aqueous environments. Many of the most successful invertebrates have the secretion forming a toughened cuticle that may be hardened to provide skeletal functions. Pigmentation is usually in close association with this system. The epidermal layer of many aquatic forms is ciliated.

The *skeletal system* may be internal, external, or both. Besides the rigid skeletal elements evident in some dead organisms, the live organism often uses a hydraulic skeletal mechanism in which pressurized body fluids maintain the containing portion in appropriate shape. Support, protection, and muscle attachment functions are present but no hemocytopoeitic function is known as is characteristic of many vertebrates. Chemical composition varies but shells and endoskeletons are often calcareous, whereas exoskeletons of other types are often chitinous. Some protozoans and sponges also have glassy (siliceous) materials in their skeletal parts.

The *muscular system* is composed of smooth or striated muscles. They do their work by, when they are shortened by interdigitating actin and myosin filaments, deriving their energy from adenosine triphosphate. Numerous physiological variations of muscles exist. The most conspicuous muscles of many invertebrates are the outer circular layer and inner longitudinal layer of the body wall of species utilizing a hydraulic skeletal mechanism. Muscles occur as antagonistic groups, or unopposed where some force is able to lengthen the muscle following the work of contraction.

The *digestive system* shows great variation in structure, with many accessory food-gathering features. Structures for mechanical breakdown of food may be external or internal. Digestion is enzymatic in most cases, although hosts of symbionts may provide some of these functions. Digestion may occur extracellularly or intracellularly. Few invertebrates have the ability to utilize cellulose or chitin that are often abundant and available in their food materials. Storage of nutrient materials may be a generalized function of most tissues, may be in localized energy storage tissues such as the fat bodies of insects, or may even be external, as in the honey stores of bees.

The *circulatory system* achieves its function of internal transport by moving internal fluids. The movement may be due to body movements, ciliary propulsion, or muscular hearts. Direct diffusion or protoplasmic movements may suffice in small organisms where systems are not recognizable for circulation, respiration, or excretion.

The *respiratory system* may consist of external gills, or equivalents, or internal passageways for transport of oxygen and carbon dioxide. Respiratory pigments in blood or blood cells of the circulatory system may increase this transport ability. Blood cells may also function as plungers or pistons to improve flow through small blood vessels. Cellular respiration shows great uniformity of the anaerobic process of glycolysis and the aerobic processes through the citric acid cycle and mitochondrial respiration. Uniformity is so great that it is more common among researchers to study and note the variations.

The *excretory system* has the responsibility of getting rid of most of the nitrogenous metabolic wastes and maintaining salt and water balance of body fluids by secretion or filtration and selective reabsorption. The task is assigned to body surfaces, nephridia, or kidney-type structures and becomes progressively more challenging as organisms go from marine to freshwater and terrestrial environments.

The *nervous system* of brain, ganglia, and nerves may have evolved from cells having an endocrine function by providing a structural method for rapid delivery of regulatory substances to specific sites. Neurosecretions with endocrine functions have been identified in many groups of invertebrates. The propagation of waves of depolarization is not restricted to the outer membranes of the nervous system neurons but has been demonstrated on a variety of other non-nervous-system locations. Rapid regulation and control by nerve depolarization phenomena is often replaced by slower, longer-lasting endocrine substances when sustained effects are required. Some of the more advanced invertebrates show the beginnings of an *endocrine system* arising from the nervous system to provide this function.

A *reproductive system* of testes, and/or ovaries, and related structures becomes important in large and complex organisms that cannot conveniently reproduce by fission or budding and need a mechanism for (1) delivering progeny or reproductive products to the outside, (2) internal fertilization to avoid waste or environmental destruction of gametes, (3) protection and support during part of development, and/or (4) packaging the zygote with sufficient resources to support development. The provision of the ovum or sperm with the appropriate genetic material can be accomplished by the most primitive and undifferentiated of tissues; the complex portions of a reproductive system fulfill other needs. Differentiation of cells during development often results in the loss of reproductive and regenerative abilities.

The complex life cycles with various stages associated with the reproduction and development of many species may have various selective reasons. The large adult may not function well in scaled-down models and require alternate stages of functional value to achieve adult size. Developmental stages may be of use for dispersal or survival during unfavorable times. Different stages with different requirements reduce intraspecific competition of the stages and may enable a species to survive in a resource-poor environment. When an entire population goes through the life cycle in synchrony it may prevent predation or parasitism from developing to restrictive levels.

All features have some functional significance to the organism relative to one or more of the foregoing systems. Interesting research topics are generated when the relationship is not readily apparent.

Over a million distinct kinds of invertebrates exist. To list or describe all aspects of the biology of each is beyond the current capability of the largest existing computer or library. The factor that makes possible an understanding of the biology of this vast diversity is the fact that they occur in natural groupings. The study of these natural groupings with emphasis on the naming of the groups is called taxonomy or systematics.

TAXONOMIC PRINCIPLES

The outline used to organize the many species into understandable groupings is called a taxonomic hierarchy. The taxonomic hierarchy ranges from the most inclusive group, the kingdom, to the least inclusive group, the species, which consists of all those organisms considered to be potentially interbreeding or sharing a common gene pool. Of course, further subdivision is possible below the species level into subspecies and populations and even the individual. These lower levels are of less utility for conveying general information, although they are essential for studying evolution, ecology, and population biology. Subspecific identification is important to the experimental biologist in that the same genetic stock of a species should be used

The Taxonomic Hierarchy

Taxon	Example	Usual Ending
Kingdom	Animalia	
Phylum	Arthropoda	-a
Class	Insecta	-a
Order	Diptera	-ida
Family	Muscidae	-idae
Genus	*Musca*	
Species	*Musca domestica*	

when possible for duplicating experimental results or making some comparative studies.

The classification may be expanded by using prefixes, as illustrated below, or by the insertion of new categories, such as division, cohort, or tribe.

Class
Subclass
Superorder
Order

Zoological family names always end in -idae. Other taxonomic levels lack uniform endings, although they are often uniform within certain groups. The genus (plural = genera) and the species (singular and plural = species) names are italicized or underlined to show that they are written in Latin.

Species names, since the time of Linnaeus, conform to the binomial system with the generic part capitalized and the specific trivial portion of the name not capitalized even when based on a proper noun (in contrast to the policy of botanists). Proper taxonomic works include additional information with the name to show the name of the describer and the year of the description.

The purpose of a classification for most biologists is for communication of accurate designation of the organisms. Most active taxonomists give a high priority to having their classification also show the best assessment of the evolutionary relationships of the groups being discussed. Since relationships are imperfectly known, the assessment often changes with new information and new publications tend to reflect the changes. The most frequent changes encountered are often merely changes of level. Consequently, the information most important is the relationship, not the level. For example, the ciliates may be considered to be a phylum, subphylum, superclass, or class. They may be recognized as probably constituting the same group of organisms by the same root in all cases even though the endings may vary, as in Ciliatea, Ciliophora, and Ciliata.

The *law of priority* is a principle of classification intended to provide stability of classification of species names. By this principle the first validly applied species name is the only valid name. Any species names applied after the original descriptions become synonyms and are dropped when the original names are found. Because a prior name is often found after a synonym has become widely used, the changes under this principle are often disruptive to good communication. A new principle has been established by taxonomists to avoid these disruptive changes. The *law of conservation* allows the retention of a synonym as the valid name if it has been in use for 50 years before the prior name is discovered. Exceptions to these and other voluntarily followed rules are sometimes approved by the International Commission of Zoological Nomenclature.

A phenetic classification gives primary consideration to the resemblances of groups; a phyletic classification gives primary consideration to the evolutionary relationships of the groups. Because both approaches require evaluation and the possible introduction of personal biases, many objective taxonomists have incorporated numerical approaches made popular by Sneath and Sokal (1973). Numerical taxonomy may require some weighting of data and departure from its original totally objective thrust, but it has been an important factor in the introduction of greater objectivity into classification.

Not all new phyla and higher taxonomic categories are the result of shifting, splitting, or lumping older taxonomic categories. Two new or proposed phyla are based on organisms of major types unknown before this century, the Pogonophora and the Gnathostomulida. New species are occasionally found in most groups. The frequency is dependent on how well known the group is, where it occurs, and how many people are studying it. The discovery of the potential for disease transmission by mosquitoes resulted in such intensive study of mosquitoes that known species increased dramatically following that discovery (Fig. 1-5).

Many species of lesser known groups remain to be discovered. Such discoveries are thrilling to contemplate, but the new species discovered in most instances are very similar in their function to the most closely related known species. The greater the distance in the taxonomic hierarchy of the relationship of two species, the greater the probable differences in feeding habits, life cycle, and other aspects of their biology. Animals from the same family generally occupy similar ecological niches or trophic positions. The distribution or range is somewhat related to taxonomic rank—thus species occupy smaller ranges in most instances than the higher taxonomic categories to which they belong since the range of the higher taxonomic category must include the ranges of all species in the group. Willis proposed,

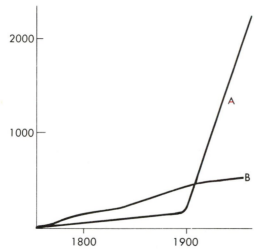

Figure 1-5 Trend curves for rates of species description. Ordinate, number of described species; abscissa, year. (A) The family Culicidae (mosquitoes). (B) Four families of Rhopalocera (butterflies and skippers) from North America. (*Adapted from Steyskal, 1965.*)

many years ago, that the range of a group was related to its age since it takes time to expand the range and adapt to new habitats. By this reasoning we would also expect that higher categories would have broader distributions since it also takes time to differentiate into the many types representative of most higher categories.

Our purpose is to study invertebrates as a whole. This will help to clarify the significance of differences in physiology, morphology, and ecology. The fundamental tool for developing this appreciation is understanding the basic assumptions of classification and phylogeny. Phylogeny is the study of evolutionary relationships or relatedness by a philosophical system.

Phylogeny

The many classification schemes proposed in the past usually have tried to reflect the groupings that were thought to have occurred phylogenetically. Because the development of groups is hidden in the past, major divisions of the animal kingdom have been variously based on morphological, physiological, and developmental features. Symmetry, metamerism, coelom development, and type of cleavage in the egg are some of the more commonly used criteria for establishing relationships. Although none of the many schemes warrant complete acceptance at this time, the student should be familiar with the considerations involved.

SYMMETRY

A major dichotomy of the animal kingdom into the Radiata and Bilateria is sometimes proposed. Symmetry is a major feature of considerable stability, but it also tends to modification related and adaptive to the environment. Thus, sessile animals with radial symmetry seem to have a selective advantage over sessile bilateral forms. A basic bilateral symmetry appears to occur with a motile way of life. Spherical symmetry is almost restricted to planktonic forms. Thus, symmetry is a risky base for a truly phylogenetic classification.

METAMERISM

The serial repetition of similar structures is used for major groupings into those with no segments, few segments, and many segments, respectively, the Ameria, Oligomeria, and Polymeria, which are the invertebrate metazoan phyla of Hadzi (1963). Although his views merit further investigation, they have not had wide acceptance, probably because of his unpopular attempt to derive the Coelenterata from the Turbellaria. One problem in the use of metamerism as the criterion is the presence of metamerism in one group of mollusks (*Neopilina*) only. A reduction in segmentation has occurred in numerous arthropods. Such reduction could well have occurred in other groups in separate lines, and thus the Ameria and Oligomeria are likely polyphyletic groups.

COELOM (BODY CAVITY)

Absence of a coelom, presence of a pseudocoel, or presence of a true coelom are features given considerable weight in most phylogenetic schemes. The method of coelom formation is also of importance. The fact that a coelom may form by different methods, such as budding off from the enteron or developing from a split in the mesoderm, is an indication that a coelom may have arisen in different lines and just be an adaptation to increased size.

STRUCTURAL LEVEL

Division of the Metazoa into diploblastic and triploblastic grades of organization is not considered of great phylogenetic utility because the diploblastic phylum, Coelenterata, has only two cell layers in species that are thought to have undergone a reduction in the level of development. The vast majority of metazoans are clearly triploblastic.

EARLY EMBRYOLOGY

Early cleavage divisions, with either a radial, indeterminate type of development or a spiral, determinate type of development, provide a major dichotomy in the metazoans. One objection is the variation in embryology that can occur within a single phylum. As previously mentioned, the method of formation of the coelom in the embryo is given considerable significance. Similarity of larval types is often thought to be significant.

COMPARATIVE BIOCHEMISTRY

Insufficient data have been accumulated, tabulated, and analyzed to make comparative biochemistry contribute significantly to an understanding of phylogeny as it may well do in the future. Chitin is generally present in major groups of invertebrates other than the deuterostome phyla. Cytochrome c has the potential for revealing phylogeny from amino acid sequencing studies if methods are ever developed to make inexpensive determinations from minute amounts of cytochrome c. Studies of cytochrome c are practical at present if organisms are very large or abundant. Isozyme studies have proven a useful procedure for estimating relationships on the population level because the number of enzymes showing electrophoretic variants generally increases in proportion to the number of events contributing to incipient speciation of the populations compared.

COMPARATIVE MORPHOLOGY

Structural similarities have always been, and will continue to be, important criteria for determining relationships. Gross appearance can often be a reflection of the ecology of the organism as indicated under symmetry. The lophophorate invertebrates may have lophophores as a consequence of their sessile, filter-feeding habits. Internal structures such as flame cells and their modification are more likely to be indicative of a relationship. Absence of a structure may not indicate a wide phylogenetic separation from other organisms having the structure. For example, subterranean species often lack eyes and pigmentation found in very closely related surface-dwelling species. The relationship of form and function has brought about many cases of convergent evolution. Bivalved shells have developed independently in at least three phyla, the Brachiopoda, Mollusca, and Arthropoda.

In general, gross types of similarities and differences in the categories discussed are of use in determining relationships at the phylum level. Groups of characteristics are likely to have greater validity than single features as indices of relationships. As we proceed to lower and closer taxonomic relationships, the characteristics of use become more specific until at the species level even structure size may be significant.

The sequence of phyla in the chapters of this book approximates one of increasing complexity. Such a linear sequence does not show a phylogenetic pattern, because such a pattern would be branching and not linear. Thus, the arrangement in the sequence is arbitrary, as must all alleged phylogenetic schemes be with presently available knowledge. All phylogenetic relationships are not equally speculative. The relationship of annelids and arthropods is close and well documented. Others are highly enigmatic; for example, the chaetognaths are often put on a branch with the echinoderms and chordates, but there are some grounds for putting them near nematodes. Perhaps it only means that the phyla have diverged less than we think. The incompleteness of the fossil record means that phylogeny is likely to remain one of the most fascinating puzzles of nature.

The arrangement of phyla in the accompanying Outline of the Invertebrate Phyla incorporates some of the groupings popular among American zoologists. Other arrangements are proposed by some zoologists. Although this grouping of phyla does not account for all relationships satisfactorily, it is useful as a base for discussion.

Outline of the Invertebrate Phyla

I. Unicellular, or all somatic cells alike	Protozoa
II. Multicellular (Metazoa)	
A. Stereoblastula grade construction	Mesozoa
B. Cellular grade construction (Parazoa)	Porifera
C. Tissue grade construction (Eumetazoa, in part)	Coelenterata, Ctenophora
D. Organ-system grade construction (Eumetazoa, in part)	
1. Acoelomate	Platyhelminthes, Rhynchocoela
2. Pseudocoelomate	Aschelminthes, Entoprocta, Acanthocephala
3. Coelomate	
a. Protostomes (spiral, determinate cleavage)	
(1) Lophophorate protostomes	Ectoprocta, Phoronidea, Brachiopoda
(2) Schizocoelous protostomes	Mollusca, Priapulida, Sipunculida, Annelida, Arthropoda
b. Deuterostomes (radial, indeterminate cleavage; enterocoelous)	Chaetognatha, Pogonophora, Hemichordata, Echinodermata, Chordata

Because there is a high degree of subjectivity in weighting factors that did not leave a fossil record of clarity, an Alternative Outline of the Invertebrate Phyla is presented (modified from Beklemishev, 1969) to illustrate another equally reasonable approach. A distinctive feature of the alternative outline is the isolated position of two groups, the Brachiopoda and Chaetognatha.

Alternative Outline of the Invertebrate Phyla

I. Unicellular, or all somatic cells alike	Protozoa
II. Multicellular (Metazoa)	
A. Cellular grade construction, no digestive tract	Porifera
B. Tissue or organ grade construction, usually with a digestive tract (Enterozoa)	
1. Basic radial symmetry	Coelenterata, Ctenophora
2. Basic bilateral symmetry	
a. Protostomes (spiral, determinate cleavage)	
(1) Protrochula-type larva	Platyhelminthes, Acantho-cephala, Aschelminthes
(2) Trochophore larval type	Mollusca, Annelida, Arthropoda, Sipunculida
(3) Actinotroch larval type, lophophorate	Phoronidea, Entoprocta, Ectoprocta
b. Bivalved lophophorates	Brachiopoda
c. Deuterostomes (radial, indeterminate cleavage)	Pogonophora, Hemichordata, Chordata, Echinodermata
d. *Sagitta*-type development	Chaetognatha

The outline does not contain the details of larval types and development that are weighted heavily in Beklemishev's scheme. It does rely on basic similarities of general features when some features are missing, as in the absence of a trochophore larva in arthropods that otherwise are clearly related to those types having such a larva.

STORAGE, RETRIEVAL, AND DISSEMINATION OF INFORMATION

A stable classification of organisms can be the basis for a filing system for information on invertebrates. Because emphasis of interests may change with time, many people find that a filing system based on authors' names is more versatile because it does not require total reorganization as interest shifts from taxonomic to morphological to physiological topics.

Communication begins with the original observation. Note-taking that is clear, complete, and adequately labeled in a form intelligible to others will also be intelligible to the notetaker when memory fades or confusion arises. A book such as this is primarily dependent on the information contained in journals and other books (Table 1-2). The immense amount of information people in science deal with precludes direct communication with all sources of the information, although direct communication is still an important part of science.

Direct communication with others studying the same subjects is in the form of conversation or letters. Because many scientists often share a common interest and want to join the conversation, meeting of professional societies and journals of professional societies have evolved. Peer and editorial review of the published articles is intended to ensure that the material is clear, accurate, and significant for the presentation of either new information or a needed update reviewing and collating scattered information on a subject.

Table 1-2 Library Call Numbers Illustrating Some Subject Areas That May Have Information on Invertebrates

Subject	Library of Congress System	Dewey System
Societies, academies	AS	
Oceanography	GC	
Official documents	J	
Science (general)	Q	500
Natural history (general)	QH	
Ecology	QH 540-549	
Botany	QK	580
Zoology	QL	590
Human anatomy	QM	
Medical sciences		610
Physiology	QP	
Medicine	R	
Agriculture (general)	S	630
Diseases, pests, and control	SB 599-999	
Animal culture, insects	SF 521-561	
Aquaculture (including lobster)	SH	
Technology	T	600
Environmental pollution	TD 172-195	
Specific Examples		
Cheng (1973)	QL 757 .C49	591.5 24
Hegner and Engemann (1968)	QL 362 .H4	592
Moore, Lalicker, and Fischer (1952)	QE 770 .M6	562
Pennak (1978)	QL 141 .P45	592 .09 2973
Sneath and Sokal (1973)	QH 83 .S58	574 .01 2
Tombes (1970)	QP 187 .T64	592 .01 4
Treat (1975)	QL 458 .T73	595 .42
Wetzel (1975)	QH 96 .W47	551.4 82

Knowledge of the style of research reports presented in such journals can enable readers to use their time economically when seeking specific information from them. Several thousand different journals each publish articles of potential interest to an invertebrate zoologist. Most follow a common style consisting of (1) an informative title that identifies the topic, and then the body of the report follows with (2) an introduction that may be equivalent to an expanded title or else a rationale for the report intended to sell readers on its importance. The introduction may be combined with or be followed by (3) a review of the literature pertinent to the article. The next section, (4) materials and methods, is detailed or refers to detailed information that would enable the reader to repeat the research or evaluate the suitability of the procedures used. (5) A results section may include the actual data or data summarization and statistical evaluation. A wide variety of tabular and graphic methods are used for presentation of results. (6) The discussion section is an evaluation of the results and their integration with prior research. (7) An abstract or summary provides a brief view of the new information contained. Reading the abstract saves time if the article does not cover the topics you are searching for. (8) The literature cited portion is a useful guide to additional information, and the inclusion of appropriate citations gives evidence that the author was informed of the state of the art.

The symposium volume is a popular type of book in science because symposia usually deal with specialized topics and the resulting articles are a convenient source of current information on many aspects of the subject. Reprint volumes on specialized topics that include the classic papers in a field also provide convenience for the student or newcomer to a field. Large libraries have bibliographic tools in addition to the card catalog and *Reader's Guide to the Periodical Literature* that are of special value to scientists.

Three of the most valuable are *Zoological Record, Biological Abstracts,* and *Science Citation Index. Zoological Record* is essential for taxonomic work and is especially useful when an animal is the main reference point. *Biological Abstracts* is especially useful for the brief summaries that it provides as well as its detailed indexes. *Science Citation Index* is an effective tool for locating the most recent work on a subject because it has an index based on the references included in articles. Some references are a more specific indication of content than are key words from titles. Author indexes are found in all three tools and are useful because of specialization among researchers.

Because such tools are so voluminous and some of them are constructed from computerized data bases it is possible to have computers list literature meeting specifications such as subject or author. The output from such a search is limited to data stored in the system and must be based on careful search criteria to get the needed information uncluttered by irrelevant information. The computer principle of GIGO (garbage in, garbage out) applies to the search request as well and must be heeded if an excellent data base is to provide useful information at reasonable expense.

The student wishing to enter the field of research may develop an interest and understanding of a specific area of study by reading journals such as *Science, Nature, Biological Bulletin, Transactions of the American Microscopical Society,* or *Ecology.* Progression to more specialized journals dealing with the taxonomic group or biological subdiscipline should follow.

Attendance at scientific meetings, symposia, and seminars as available will give an opportunity for insights and views of interest that do not appear in print, because authors and editors are frequently reluctant to publish information that is either not thoroughly substantiated or else presumed to be overly published or obvious.

The organisms themselves are the ultimate reference. Invertebrates are found in all but the most extreme habitats. Patient observation of organisms in their natural environment or the laboratory can yield an abundance of information and increase understanding of biological principles taught in texts and classrooms. Morphology becomes meaningful when parts are seen in use. The role of the organism may be understood when it is seen feeding and being fed upon.

ECOLOGICAL IMPORTANCE

Invertebrates can be found in all the trophic roles (Figs. 1-6 to 1-11) enumerated for organisms cycling energy in the biosphere. Relatively few are producers outside of the photosynthetic plantlike protozoa, but many plants depend on invertebrates for their success—from the pollination dependence on insects of most flowering plants to the nutritional benefits for plants provided by earthworms and soil invertebrates. All types of plant materials have specialized invertebrate primary consumers or herbivores feeding upon them. All other roles are filled by some species of invertebrate, with the possible exception of the top carnivore. The largest carnivore, the sperm whale, normally feeds upon squid, but perhaps the outcome is uncertain when that squid is a giant squid, the largest of invertebrates. Invertebrates almost monopolize the role of parasite if kinds of species are considered. Only lower bacteria and fungi rival invertebrates in importance as reducers in disposing of dead organisms and organic waste.

Their numbers, size, and important roles, together with abundance and ubiquity, make invertebrates of special value in evaluating aquatic environments via aquatic communities.

The stunning diversity of invertebrates should not divert us from the many common principles that they can illustrate. For example, three principles seem to determine evolutionary success for living organisms—self-duplication, self-regulation, and cooperation. The

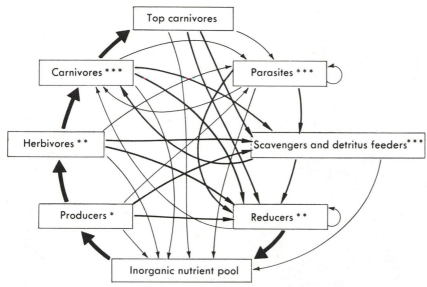

Figure 1-6 Trophic importance of invertebrates. *Phytomastigophorea make an important contribution. **Invertebrates common and important with numerous species. ***Invertebrates dominate in biomass, numbers, or species.

Figure 1-7 Producers, Carnivore. *Fungia* (Madreporarida, Fungidae), one of the few corals with a non-colonial adult stage. The juvenile skeleton (upper right) still shows the scar on its aboral surface where it broke from a stalk produced by the larval coral. Like other reef-forming corals, *Fungia* participates in the producer trophic level through a mutualistic association with intracellular photosynthetic symbionts. Corals also function at the carnivore level when they feed on planktonic animals. (*Photo by author.*)

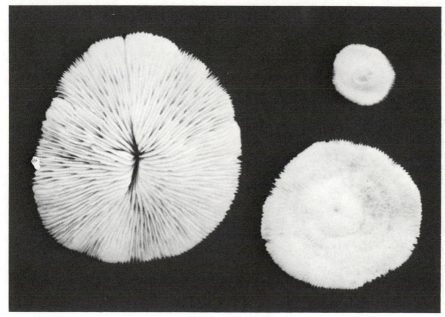

23

Figure 1-8 Herbivore. *Malacosoma americanum* (Lepidoptera, Lasiocampidae) larvae. These herbivores cooperatively build tents as larvae, but have no social structure as adult moths. (*Photo by author.*)

Figure 1-9 Carnivore. *Cheumatopsyche* (Trichoptera, Hydropsychidae) larvae in their self-made, silken nets which are used for filtering food from streams. Gregarious behavior of these plankton predators often results in shared supports for their nets. (*Photo by author.*)

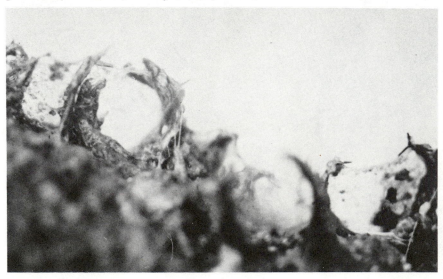

Figure 1-10 Parasite. A flabelliferan isopod parasitic upon a small coral reef fish. (*Photo by author.*)

Figure 1-11 Scavengers and detritus feeders. *Phreatoicopsis terricola*, phreatoicid isopods from tunnels in the rain forests of the Ottway Mountains of Australia. This species contains the largest terrestrial isopods and they show the reduced eyes and reduced pigmentation characteristic of subterranean animals. (*Photo by author.*)

principle of self-duplication requires some inexactitude to allow adaptation to change. Invertebrates have excelled at using two types of reproductive processes to take advantage of the need for uniformity when conditions are stable and to experiment with variation when conditions are changing due to time or location. Many have an alternation of vegetative or asexual reproduction with periods of sexual reproduction, the latter often at times of stress when variation may be useful.

Self-regulation was made easier by the cell membrane and later by increased complexity and size that sometimes reduces the impact of the immediate environment.

Cooperation is found in a staggering variety of symbiotic and other relationships.

REFERENCES

BEKLEMISHEV, W. N. 1969. *Principles of Comparative Anatomy of Invertebrates*. Translated by J. M. MacLennan; edited by Z. Kabata. Univ. of Chicago Press, Chicago. 1019 pp.

CHENG, T. C. 1973. *General Parasitology*. Academic Press, New York. 965 pp.

FITCH, W. M., and E. MARGOLIASH. 1967. Construction of phylogenetic trees. *Science,* **155**:279–284.

HADZI, J. 1963. *The Evolution of the Metazoa*. Macmillan, New York. 499 pp.

HEGNER, R. W., and J. G. ENGEMANN. 1968. *Invertebrate Zoology*. 2nd Ed. Macmillan, New York. 619 pp.

MAYR, E., E. G. LINSLEY, and R. L. USINGER. 1953. *Methods and Principles of Systematic Zoology*. McGraw-Hill, New York. 328 pp.

MOORE, R. C., C. G. LALICKER, and A. G. FISCHER. 1952. *Invertebrate Fossils*. McGraw-Hill, New York. 766 pp.

PENNAK, R. W. 1978. *Fresh-Water Invertebrates of the United States*. 2nd Ed. Wiley, New York. 803 pp.

SNEATH, P. H. A., and R. R. SOKAL. 1973. *Numerical Taxonomy*. Freeman, San Francisco. 573 pp.

STEYSKAL, G. C. 1965. Trend curves of the rate of species description in zoology. *Science,* **149**:880–882.

TOMBES, A. S. 1970. *An Introduction to Invertebrate Endocrinology*. Academic Press, New York. 217 pp.

TREAT, A. E. 1975. *Mites of Moths and Butterflies*. Comstock, Ithaca. 362 pp.

WETZEL, R. G. 1975. *Limnology*. Saunders, Philadelphia. 743 pp.

WILSON, E. O. 1975. *Sociobiology*. Belknap Press of Harvard University Press, Cambridge. 697 pp.

2

Protozoa

Within the Protozoa are animals with a great diversity of complexity, size, and behavior. Although not subdivided into cells of differing functions, as the other phyla are, the protozoan carries on all the life processes with one cell. They range from simple cells, containing little more than cytoplasm and nucleus, to complex cells with variation in vacuoles, fibrils, and other organelles. Some are so small as to be barely visible with the light microscope. Others are large enough to be seen with the unaided eye. All require the microscope as an aid to study of their structure. In recent years the electron microscope has revealed much about their ultrastructure to many researchers. Many Protozoa are economically important parasites, which live in or upon the bodies of other animals or plants. Free-living Protozoa are found occupying a variety of niches in fresh water, salt water, and soil communities. Whether parasitic or free-living, all protozoans require the presence of sufficient moisture to prevent drying for activity and growth to occur. Only the resistant spores or cysts of some species of Protozoa are able to withstand desiccation.

Protozoa are of importance in the study of molecular biology for the following reasons. Many of the biochemical pathways used by protozoans are used by all animals, including humans. Some protozoans can be cultured in chemically defined media, in which no other species are present, to rapidly produce large populations of essentially identical cells for use as experimental and control organisms. They are sometimes used for bioassay because of their sensitive growth responses to various nutrients.

Free-living Protozoa are not distributed among bodies of water in a haphazard fashion; each species is more or less restricted to a definite type of habitat just as are higher animals. Some species live in fresh water; others, only in salt water; some live in contact with the bottom; others, float about suspended in the water; some are known to live only in the soil, and others, only in sphagnum swamps. Free-living Protozoa are to be found almost everywhere on the surface of the earth where moisture exists.

Parasitic Protozoa are likewise rather definitely restricted in their habitats. Every higher animal that has been carefully studied has been found to harbor parasitic Protozoa. Each species of animal, as a rule, is parasitized by its own peculiar species of protozoa. For example, 28 different species of Protozoa are known to live in humans. Many of these appear to occur in monkeys also, but only a few of them have been recorded from other animals. The parasitic species are usually separated into two groups; those that live in the digestive tract are known as intestinal Protozoa and those that live principally in the blood, as blood-inhabiting Protozoa.

The number and variety of Protozoa is very great. About 46,000 species have been named, although over 20,000 of them are fossil forms. Thousands of species of free-living Protozoa have been described, and each kind of higher animal seems to possess one or more types of Protozoa peculiar to itself, hence the number of parasitic Protozoa is probably at least as great as that of all other animals combined. The number of individuals is also enormous. Billions of a free-living species may exist in a single pond. Among parasitic varieties we need only refer to the millions of ciliates that occur in the stomach of cattle and the millions of flagellates that live in the cecum of almost every rat.

These immense numbers of Protozoa are very diverse in their characteristics and have been assigned to three kingdoms and a dozen different phyla in a reasonable attempt by Whittaker (1969) to provide a classification reflecting their diversity. Their assignment to a single phylum in another kingdom, the Animal Kingdom, was accepted by many in the past, including previous editions of this book and by Blackwelder (1963) in his useful classification. Many of the groups have fundamental biological differences that justify phylum ranking, especially the ciliates (Corliss, 1974) with their great diversity and complex ultrastructure.

The superclasses of the classification used here could be considered phyla or subphyla to satisfy current concepts of differences resulting from polyphyletic origin of groups collectively referred to as Protozoa. The diverse origin with undoubted relationships in some cases has made it reasonable to omit a phylum, superphylum, or subkingdom designation until the relationships are clarified. Bovee and Jahn (1973) credit Kerkut with the concept of "the name Protozoa denoting a state of organization and not a natural taxonomic group." The sporozoa, all parasites that typically form spores and lack organelles for locomotion are split into three distinct groups. Sporozoa is omitted as a taxon. All but the Ciliata of the following superclasses are put in the subphylum Plasmodroma.

Protozoa

Unicellular and closely related animals.
    Subphylum **Plasmodroma.** One type of nucleus, often with pseudopodial activity.
        SUPERCLASS MASTIGOPHORA. Flagella are predominant locomotor organelles.
        SUPERCLASS SARCODINA. Pseudopodia are predominant locomotor organelles.
        SUPERCLASS APICOMPLEXA. Sporozoa with an apical complex.
        SUPERCLASS MYXOSPORA. Sporozoa with complex valved spores having polar filaments.
        SUPERCLASS MICROSPORA. Sporozoa with vesicular nuclei and simple spores with one or no polar filaments.
    Subphylum **Ciliophora.** Usually with two types of nuclei, cilia, and a complex infraciliature.
        SUPERCLASS CILIATA.

Superclass Mastigophora: The Flagellated Protozoa

EUGLENA

The genus *Euglena* contains over a dozen species. Two commonly found in algae-choked ponds are *E. viridis* and *E. gracilis*. They are green, and—although a single animal cannot be seen with the naked eye—a great many massed together impart a green tint to the water.

Cytology

Euglena (Fig.2-1) is a single elongated cell pointed at the posterior, and blunt at the anterior end. Two kinds of cytoplasm can be distinguished in *Euglena* as in *Amoeba* and *Paramecium,* a dense outer layer, the *ectoplasm,* and a central mass, the *endoplasm,* which is more fluid. A thin cuticle is present, as in *Paramecium,* covering the entire surface of the body. Parallel ridges of this cuticle spiral around the animal, making it appear striated. A little to one side of the center of the anterior blunt end of the body is a funnel-shaped depression known as the *cytostome.* At the bottom of this depression is an opening that leads into a short duct called the *gullet.* This, in turn, enters a large spherical vesicle, the *reservoir,* into which several minute *contractile vacuoles* discharge their contents. The mouth (cytostome) and gullet

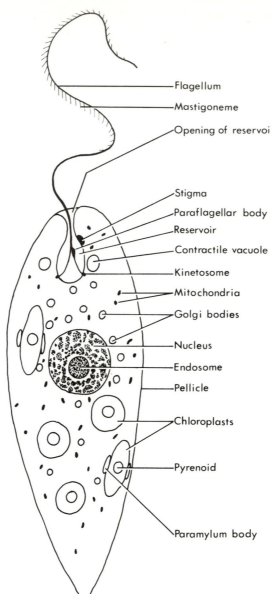

Flagellum
Mastigoneme
Opening of reservoi
Stigma
Paraflagellar body
Reservoir
Contractile vacuole
Kinetosome
Mitochondria
Golgi bodies
Nucleus
Endosome
Pellicle
Chloroplasts
Pyrenoid
Paramylum body

FIGURE 2-1 *Euglena*. Diagram of main cytological features. (*Modified after several sources.*)

are not used for the ingestion of food but as a canal for the escape of fluid from the reservoir.

A conspicuous structure in *Euglena* is the red eye spot or *stigma*. This is placed near the inner end of the gullet close to the reservoir. It consists of protoplasm in which are embedded a number of granules of red pigment. It is thought that the *paraflagellar body* is a light-sensitive structure, which is enabled to give a directional light response because of the one-sided light filtering action of the stigma.

Euglena has a single oval *nucleus* lying in a definite position near the center of the body. There is a distinct nuclear membrane, which electron microscope studies have shown to consist of a double layer penetrated by numerous pores. The nucleus contains a central body known as the *endosome* (syn. *nucleolus, karyosome*).

Euglena derives its green color from a number of oval disks suspended in the cytoplasm. These are known as *chromatophores*. Located in the center of each chromatophore is a collection of granules, the *pyrenoid*. The chromatophore contains chlorophyll, which enables *Euglena* to produce food by the process of photosynthesis. This plantlike characteristic has caused botanists to classify *Euglena* as an alga.

Other structures within *Euglena* appear when properly stained. Numerous small granules, termed *mitochondria*, are scattered throughout the cell. The mitochondria are important as a location of many of the biochemical reactions of the cell during respiration. *Golgi bodies* are abundant but of unknown function.

A long, whiplike *flagellum* extends from the opening of the reservoir. It comes from one of the two flagella that originate from *kinetosomes* located in the wall of the reservoir. The electron-microscopic structure of a flagellum is similar to that of a cilium (Fig. 2-2) in that each has 11 electron dense fibers arranged in a pattern of two central fibers and nine peripheral fibers. Only the peripheral fibers extend into the kinetosome. The distal portion of the flagellum contains numerous minute fibers, known as *mastigonemes,* which project laterally.

Figure 2-2 Some fibril patterns demonstrated by electron micrographs of cilia cross sections. (A) *Didinium* cilium. (B) *Opalina* flagellum. (C) Generalized ciliate cilium. (D) Generalized subpellicular pattern of ciliate cilium. (*A, after Pitelka; B, after Noirot-Timothee in Grell; C and D, adapted from Corliss.*)

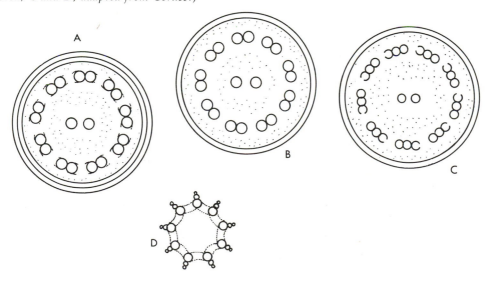

Nutrition

Although *Euglena* has a mouth (cytostome) and gullet, it is doubtful, as noted above, if any food is ingested. Food is manufactured photosynthetically, as in plants, with the aid of light and the chlorophyll in the chromatophores. This mode of nutrition is known as *holophytic*, or *autotrophic*. Photosynthesis enables the oxygen of a water molecule to be released, while the hydrogen of the water molecule reduces the carbon dioxide present through a series of steps to form carbohydrates and more water. Light of blue and red wavelengths is utilized with greatest efficiency. The carbohydrates are stored as a starchlike substance called *paramylum*. The conversion of the simple sugars produced by photosynthesis to paramylum is presumed to occur due to the action of the pyrenoid. Both pyrenoids and chromatophores are permanent cell structures and increase in number by division and not by the origin of new ones from other parts of the body. All the food necessary for the life of *Euglena* need not be produced by photosynthesis. When the chlorophyll is destroyed by treatment with certain drugs, the colorless *Euglena* resulting can continue to live if provided with a medium rich in organic nutrients. This seems to indicate that organic substances in solution are absorbed through the surface of the body, that is, *saprophytic* nutrition can occur and may supplement the normal holophytic nutrition. The nutrition of *Euglena* differs from that of the majority of animals, since the latter live by ingesting solid particles of food and are said to be *holozoic*, or *heterotrophic*.

Biochemical similarities with bacteria and blue green algae indicate that mitochondria and chloroplasts have had a symbiotic origin in plants and photosynthetic protozoa. *Euglena* may pass on the benefit; Willey, Bowen, and Durban (1970) have found that a *Euglena* sp. may live symbiotically part of the year in the rectum of a damselfly nymph.

Locomotion

Euglena changes its shape frequently, becoming shorter and thicker, and shows certain squirming movements. These prove that it possesses considerable elasticity, because the normal shape is regained if enough water is present. Often in a favorable specimen, the threadlike flagellum may be seen projecting from the anterior end of the body and bending to and fro, drawing the animal after it. Pitelka and Schooley (1955) have found that when the flagellum is fixed at a pH near that of the culture medium, the mastigonemes are closely applied to the flagellum. Under other conditions, the mastigonemes (Fig. 2-3) may extend laterally from one side of the flagellum. Jahn and co-workers (1964) have shown that the presence of a sufficient number of laterally projecting mastigonemes can reverse the direction of the propulsive force (Fig. 2-4) provided by the flagellum of *Ochromonas*, although the undulations of the flagellum are the same. Some workers

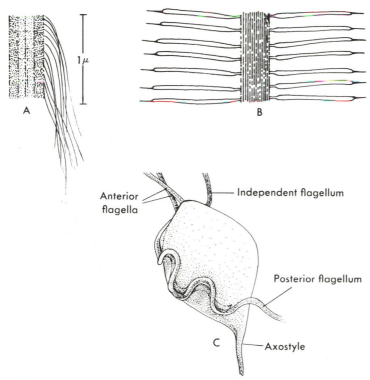

Figure 2-3 (A) Lateral view of an *Euglena* flagellum. (*Adapted from Wolken.*) (B) Lateral view of an *Ochromonas malhamensis* flagellum with pantoneme mastigonemes. (*Adapted from Pitelka and Schooley.*) (C) *Pentatrichomonas hominis*, showing the relative position of the posterior flagellum, independent flagellum, and four anterior flagella. (*Drawn from a scanning electron micrograph in Andrzej Warton and B. M. Honigberg. Structure of trichomonads as revealed by scanning electron microscopy. Copyright 1979 by* The Journal of Protozoology. J. Protozool., **26**:56–62.)

think that the propulsive force comes from the sculling action of the tapered *Euglena* as it vibrates due to the movement of the flagellum.

Nichols and Rikmenspoel (1978) found that divalent calcium ions inhibit flagellar motility, whereas internal divalent magnesium ions may restore flagellar motility. Negative direct electric current also was inhibitory to flagellar motility, whereas positive direct electric current was stimulatory.

Behavior

Euglena swims through the water in a spiral path. The effect of this is the production of a perfectly straight course through the trackless water. When stimulated by a change in the intensity of the light. *Euglena*, in the majority of cases, stops or moves backward, turns strongly toward the dorsal surface, but continues to revolve on its long

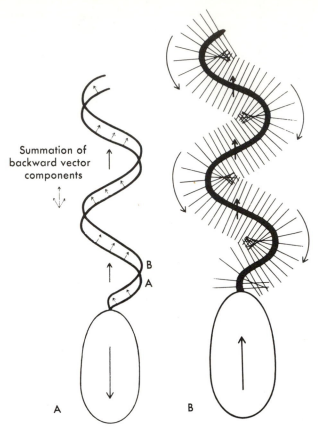

Figure 2-4 Flagellar movement. Diagram to indicate how mastigonemes may reverse the force exerted by a flagellum. (A) Distally directed sine wave produces the force indicated by small arrows as the wave progresses from *A* to *B*. Organism moves in opposite direction. (B) Action of mastigonemes on a similar wave reverses effective force. (*From T. L. Jahn, M. D. Landman, and J. R. Fonseca. The mechanism of locomotion of flagellates. II. Function of the mastigonemes of* Ochromonas. *J. Protozool.,* **11**(3):291–296. *Copyright 1964 by the Society of Protozoologists.*)

axis. The posterior end then acts as a pivot, while the anterior end traces a circle of wide diameter in the water. The animal may swim forward in a new direction from any point in this circle. This is the *avoiding reaction.*

Observations by Diehn, Fonseca, and Jahn (1975) indicate that this may be based on the posterior end of the body shading the photosensitive paraflagellar body more effectively than the stigma at high light intensity.

Euglena is very sensitive to light. It swims toward an ordinary light such as that from a window, and if a culture containing euglenae is examined, most of the animals will be found on the side toward the brightest light. This is of distinct advantage to the animal, because light is necessary for the assimilation of carbon dioxide by means of its chlorophyll. *Euglena* will swim away from the direct rays of the sun. Direct sunlight will kill the organism if allowed to act for a long time. The ultraviolet portion of the solar spectrum is probably the most injurious portion, because large amounts of ultraviolet light are lethal for many organisms. If a drop of water containing euglenae is placed in the direct sunlight and then one half of it is shaded, the animals will avoid the shady part and also the direct sunlight, both of which are unfavorable to them, and will remain in a small band between the two in the light best suited for them, that is, their optimum. By shading various portions of the body of a *Euglena,* it has been found that the region in front of the eye spot is more sensitive than any other part. It should be noted that when *Euglena* is swimming through the water it is this anterior end which first reaches an injurious environment; the animals give the avoiding reaction at once, and are thus carried out of danger.

Reproduction

Reproduction in *Euglena* takes place by binary longitudinal fission. Nuclear division takes place within the nuclear membrane. The chromatin, which is in the form of paired strands of chromomeres in the vegetative stage, forms pairs of chromosomes each of which divides longitudinally into two (Fig. 2-5). The endosome becomes constricted into two approximately equal parts. The intranuclear body also divides into two. The body begins to divide at the anterior end. The old flagellum is retained by one half, whereas a new flagellum is developed by the other. Often division takes place while the animals are in the encysted condition. Occasionally, euglenae are found which have become almost spherical and are surrounded by a rather thick gelatinous covering which they have secreted. Such an animal is said to be encysted. In this condition periods of drought are successfully passed; the animals become active when water is again encountered. In cultures brought into the laboratory many cysts are usually found on the sides of the dish. Encystment frequently takes place without any apparent cause; the animal rests in this condition for a time and then emerges to its free-swimming habit. Before encystment the flagellum is thrown off, a new one being produced when activity is again resumed. One cyst usually produces two euglenae, although these may divide while still within the old cyst wall to make four in all. Certain observers have recorded as many as 32 young flagellated euglenae that escaped from a single cyst. Fission occurs only during the hours of darkness in ordinary circumstances; however, synthesis of protoplasmic constituents occurs during hours of light with replication of DNA limited to the latter part of those hours.

Jarrett and Edmunds (1970) found a temperature-dependent cir-

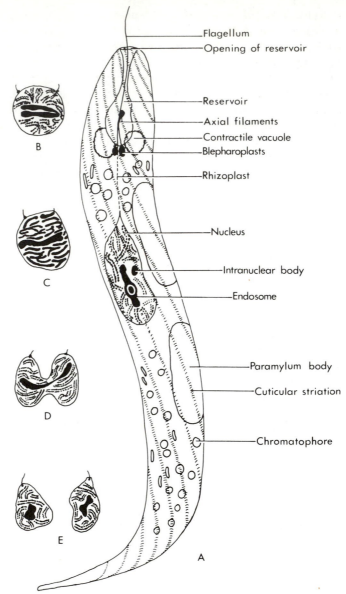

Figure 2-5 *Euglena spirogyra:* nuclear division. (A) *Vegetative form* with nuclear chromatin in the form of paired rows of chromomeres. Only the proximal portion of the flagellum is included. (B) *Prophase nucleus.* The chromosomes are in pairs; the endosome has elongated and divided into two; the axial filaments are attached to two blepharoplasts on the nuclear membrane that have arisen from the intranuclear body. (C) *Metaphase nucleus.* The chromosomes separate and the endosome elongates. (D) *Anaphase nucleus.* The nuclear membrane constricts; the chromosomes divide longitudinally and become granular; and the endosome constricts. (E) *Nuclear division completed.* The nuclei next assume the appearance indicated in A. (*After Ratcliffe.*)

cadian rhythm of cell division in a heterotrophically grown mutant *Euglena*. Light–dark-cycle-induced synchrony of cell division approximated 24 hours at 19°C during continuous dark, but generation time was only about 10 hours at 25°C.

OTHER MASTIGOPHORA

The superclass Mastigophora includes a very large number of minute Protozoa that are rather difficult to classify. It is convenient to separate them into two classes: (1) plantlike species or Phytomastigophorea, and (2) animal-like species or Zoomastigophorea. Certain of the more common and interesting species will be referred to under some of the orders recognized in these two subclasses.

Phytomastigophorea

Order Chrysomonadida

This order includes a number of small forms that contain typically one or two brownish chromatophores and one to three flagella. The body is often amoeboid and food may even be taken in by pseudopodia. Nutrition is either holophytic, holozoic, or both.

Protozoa of Drinking Water. Several species of Chrysomonadida are of particular interest because they sometimes become very numerous in water confined in reservoirs and render it unfit for drinking purposes. They are all colonial in habit. *Uroglenopsis americana* (Fig. 2-6, A) forms spherical colonies the individuals of which are embedded in the periphery of a gelatinous matrix. A stigma and a platelike chroma-

Figure 2-6 Chrysomonadida. Colonial flagellates sometimes obnoxious in drinking water. (A) *Uroglenopsis americana*. (B) *Dinobryon sertularia*. (C) *Synura uvella*. (*After various authors.*)

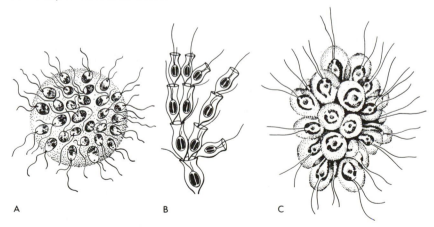

A   B   C

tophore are present. *Dinobryon sertularia* (Fig. 2-6, B) has a vaselike, hyaline, or yellowish cellulose test, usually a stigma, and one or two brownish chromatophores. *Synura uvella* (Fig. 2-6, C) consists of from 2 to 50 individuals arranged in radial fashion; each has a stigma and two brown chromatophores. *Uroglenopsis* is said to be the worst of the species, giving rise to a fishy odor; when large numbers are present an odor resembling cod liver oil results. *Dinobryon* also produces a fishy odor. *Synura* is responsible for an odor resembling ripe cucumbers and a bitter and spicy taste. These odors are due to aromatic oils elaborated by the organisms during growth and liberated when they die and disintegrate. They are recognizable when present in minute amounts; for example, the oil of *Synura* produces a perceptible odor in dilution of 1 part oil to 25 million parts water. Treating water with copper sulfate kills these Protozoa, so that the odors eventually disappear.

Order Cryptomonadida

Chilomonas paramecium (Fig. 2-7, B) is a common species belonging to this order, one that frequently occurs in laboratory cultures. It is about 35 μm in length, without chromatophores and hence colorless, with two anteriorly directed flagella, a contractile vacuole on one side near the anterior end, a spherical nucleus posterior to the center, and a mass of endoplasm that is distinctly alveolar in structure. No solid food particles are ingested, but nutritive substances are absorbed through the surface of the body.

Order Dinoflagellida

(Figs. 2-7, A and D). The most characteristic features of the Dinoflagellida are their shell and flagellar apparatus. The shell or *lorica* is a rigid structure, sometimes very bizarre in form, composed of cellulose, or an allied substance, which usually has a longitudinal groove, the *sulcus*, and a circular groove, the *girdle* (annulus). Two flagella are present which issue from pores in the lorica and lie within these grooves. The flagellum in the sulcus extends beyond it and provides much of the motive force. The undulations of the flagellum in the girdle rise enough above the surface of the organism to provide part of the forward thrust.

The majority of the Dinoflagellida are holophytic and possess chromatophores, which may be brown, pale green, or yellow. A few species, however, have taken on a holozoic method of nutrition and may even ingest their food by pseudopodia. Frequently, a stigma is present. Asexual reproduction by fission takes place.

Ecology of the Dinoflagellates. Most of the dinoflagellates are saltwater species; a large number are parasites, and many live in fresh water. *Ceratium hirundinella* (Fig. 2-7, A) has a distinctive appearance

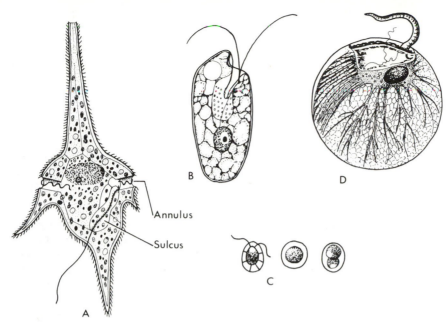

Figure 2-7 Phytomastigophorea. (A) Dinoflagellida; *Ceratium macroceras*. (B) Cryptomonadida; *Chilomonas paramecium*. (C) Volvocida; *Haematococcus lacustris*. Causes red rain and red snow when present in large numbers. Active, resting, and division stages. (D) Dinoflagellida; *Noctiluca scintillans*. (*After various authors*.)

and is found in both fresh water and salt water. An interesting marine species is *Noctiluca scintillans* (Fig. 2-7, D). It is spherical in shape, bilaterally symmetrical, and usually from 500 to 1,000 μm in diameter. It possesses a small flagellum, a tentacle, and cytoplasm that is much vacuolated. Sometimes noctilucas are so abundant in the Atlantic and Pacific as to color the water red by day and render it phosphorescent by night. The "red tide" coincides with the abundance of some species of dinoflagellate, often *Gymnodinium brevis*. The oils, which are stored as food by the dinoflagellates causing the "red tide," are thought to be toxic to many other marine animals when they are liberated in large amounts by the natural death of many of the dinoflagellates. The Gulf of Mexico has been the site of many of the "red tide" fish kills.

The effect can be produced along an extended portion of the coast by a drifting "bloom" of dinoflagellates. Cobalt or iron may be involved in such population outbreaks. They may leave resting stages on the bottom and become an annual problem, contaminating filter-feeding shellfish when favorable conditions for the dinoflagellates reoccur at some East Coast locations. Their close relationship to plants is indicated by the growth-stimulating effects of a plant hormone (Paster and Abbott, 1970).

Dinoflagellates are the symbiotic protozoan present in the cells of many marine invertebrates, such as corals and mollusks. Such symbiotic dinoflagellates, which are presumed to provide a nutritional advantage to their host, are known by the term *Zooxanthellae*. In such a mutualistic association, the dinoflagellates are not recognizable as such, but they regain their dinoflagellate characteristics when removed from the host.

Order Euglenida

Euglenida contains those flagellates that store their food reserve as paramylum. All Euglenida can undergo the squirming undulations known as *metaboly;* such metabolic movements are not amoeboid movements. *Euglena* is a rather typical genus and presents the principal characteristics of the group. A number of interesting species are common in fresh water. *Euglena pisciformis* is a highly active spindle-shaped species. *E. spirogyra* is large and sluggish. *E. sanguinea* has a hematochrome. *E. deses* has a very short flagellum and is highly metabolic. The genus *Phacus* (Fig. 2-8, A) is represented by several fresh-water species. The body is much flattened, and the cuticle has conspicuous striations. Green chromatophores are present. *Peranema trichophorum* (Fig. 2-8, C) is a common species. The colorless body is broad and truncated at the posterior end when in locomotion, and a long flagellum with vibrating tip extends out from the pointed anterior end. The body is very metabolic when stationary. *Heteronema acus* (Fig. 2-8, B) occurs in fresh water and in the soil. Two flagella arise from the anterior end: one directed forward and the other backward. The body is colorless, metabolic, and pointed at both ends.

Figure 2-8 Euglenida. (A) *Phacus longicaudus.* (B) *Heteronema acus.* (C) *Peranema trichophorum.* (D) *Copromonas subtilis.* (E) *Euglenamorpha hegneri.* (*After various authors.*)

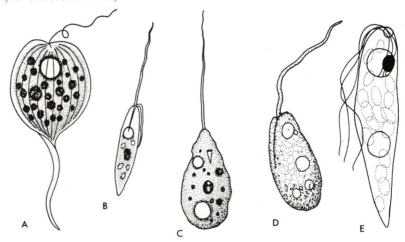

Copromonas subtilis (Fig. 2-8, D) is an interesting species because it is coprozoic, that is, it occurs in the feces of higher animals, which it finds a favorable environment. It has been recorded from the feces of frogs, toads, pigs, and man. A parasitic species, *Euglenamorpha hegneri* (Fig. 2-8, E), that occurs commonly in the intestine of frog and toad tadpoles probably belongs to this group, and, if so, is the only representative that possess three flagella. It lives between the food mass and the wall of the intestine or rectum of tadpoles and can be transferred from one tadpole to another of the same or different species either by association of infected tadpoles with clean tadpoles or by feeding clean tadpoles with the rectum from infected animals.

Order Volvocida

The Volvocida are undoubtedly closely allied to the algae. The body is covered with a rigid cellulose membrane. Biflagellate forms are typical, the flagella being inserted through pores in the cellulose wall. Most species are exclusively holophytic and possess a large cup-shaped green chromatophore. A few colorless forms that live saprophytically are referred to this order, but in no case do any of the forms ingest solid food. A red stigma or eye spot is also present. Asexual reproduction takes place giving rise to numbers of small, flagellated swarmers. Sexual phenomena are definitely known for a large number of species. Colony formation is quite frequent. An interesting group of genera belonging mostly to the family Volvocidae can be arranged to illustrate the evolution of multicellular from unicellular organisms and also the evolution of sex, as indicated in the following paragraphs and in Figures 2-9 and 2-10.

Spondylomorum (Fig. 2-9, A) is colonial in habit with 16 cells in each colony. These cells are practically independent. Each of them reproduces by fission to form a colony like the parent colony. No gamete formation is known.

Chlamydomonas (Fig. 2-9, B) is a unicellular type that reproduces vegetatively by fission into two, four, or eight daughter cells and also produces gametes, all of one size, that fuse together in pairs forming zygotes. The zygotes undergo fission, resulting in a number of vegetative unicellular individuals. After vegetative fission, cells may be retained in irregularly shaped groups by an abundance of gelatinous material that ordinarily separates as the coating of the individual cells. Such a grouping of cells, which is not truly colonial, is known as a *palmella* formation.

Pandorina (Fig. 2-9, C) is a colonial form of 16 cells embedded in a gelatinous matrix, each independent of the others. New colonies are produced by each cell by fission. Gametes are also formed, some being larger than others. The larger gametes appear to fuse with the smaller ones to form zygotes. Because sexual reproduction usually includes the fusion of a larger female gamete with a smaller male gamete, the pro-

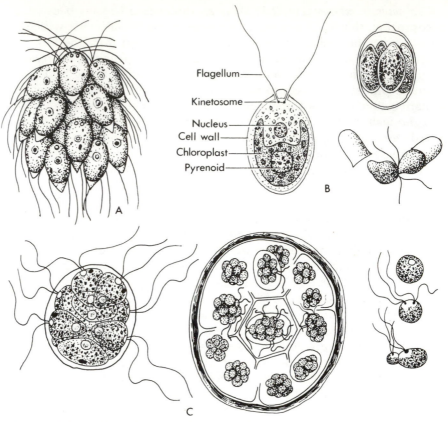

Flagellum

Kinetosome

Nucleus
Cell wall
Chloroplast
Pyrenoid

B

C

Figure 2-9 Colonial Volvocida. (A) *Spondylomorum quaternarium.* (B) *Chlamydomonas monadina;* active, dividing, and fusion of gametes. (C) *Pandorina morum;* active, dividing, and fusion of gametes into a zygote. (*After Oltmanns and others.*)

cess in *Pandorina* seems to illustrate an early stage in the evolution of sex.

Eudorina is a colonial type with 32 cells. Each cell may reproduce a colony vegetatively by fission. At times the cells of certain colonies become large; this is characteristic of female colonies and the large cells are macrogametes. In male colonies, each cell divides to form 16 or 32 microgametes. The microgametes fuse with the macrogametes in pairs to form zygotes. In this genus there is no doubt about the difference in size between the conjugating gametes.

Pleodorina is a somewhat spherical colony, like *Eudorina,* with the cells arranged at the surface with their flagella out. However, some of the cells are small *somatic* cells that do not have a reproductive capacity. The somatic cells are located at one end; the larger generative cells complete the majority of the spherical arrangement.

Volvox globator (Fig. 2-10) represents the final stages in the series.

The thousands of vegetative cells in a colony are united by protoplasmic strands; physiological continuity is thus established between the cells, a condition not found in the colonies previously described. Most of the cells contain an eye spot, chlorophyll, a contractile vacuole, and two flagella. These are the somatic cells. The production of daughter colonies is accomplished by special reproductive cells that are set aside for this purpose. The asexual method proceeds as follows. Certain cells of the colony are larger than others and lack flagella; a number of these in one colony increase in size and divide by simple fission into a great number of cells, producing new colonies without being fertilized. The cells that act in this way are named parthenogonidia. The sexual method of reproduction may be observed in colonies that contain as many as 50 of the larger nonflagellated cells. Some of these grow larger and may be recognized as female cells or macrogametes; others produce by simple division a flat plate, containing about 128 spindle-shaped male cells or microgametes. These fuse with the macrogametes. The zygote thus formed secretes a surrounding wall and in this condition the winter is passed. The following spring the zygote breaks out of the wall and by division produces a new colony. The smaller

Figure 2-10 *Volvox globator:* a large colonial flagellate. (A) A sexually ripe colony, showing microgametes and macrogametes in various stages of development; surface detail is omitted over the internal gametes. (B) A portion of the edge of the colony highly magnified, showing three flagellate cells united by protoplasmic strands, and a single reproductive cell. (C) A single cell connected with surrounding cells by protoplasmic strands. (*After Kölliker, from Bourne.*)

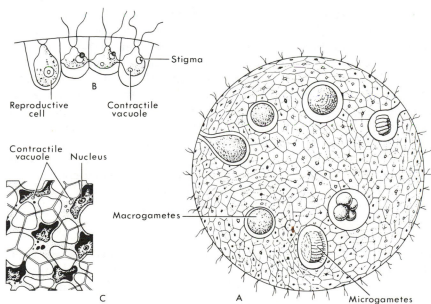

somatic cells contained in the mother colony fall to the bottom and disintegrate as soon as the new colonies produced by the fertilized germ cells have escaped.

In *Pleodorina* and *Volvox,* true somatic cells are encountered for the first time, that is, cells that function only vegetatively and are unable to reproduce the colony. In the other forms described every cell is capable of reproducing the whole. *Volvox* also contains true germ cells, that is, cells that have given up nutritive functions to carry on reproduction. Furthermore, a clear case of natural death occurs in the somatic cells when they fall to the bottom of the pond and disintegrate. The bodies of higher animals consist of many cells that may be separated into somatic and germ cells. The latter are either male or female. In most cases the fusion of a male cell with a female cell is necessary before a new animal can be reproduced. At any rate, some of these germ cells maintain the continuation of the species by producing new individuals, whereas the somatic cells perish when the animal dies.

Zoomastigophorea

Order Choanoflagellida

The order Choanoflagellida contains a small number of free-living species having a single flagellum surrounded at the base with a delicate collar. A freshwater species is *Monosiga brevipes* (Fig. 2-11, B). Collared, flagellated cells occur in this order of Protozoa, in the sponges, and in some echinoderms.

Order Rhizomastigida

The order Rhizomastigida is characterized by the fact that the organisms ingest their food at any point on the body by means of pseudopodia. They exhibit so many transitional characters between the true Sarcodina on the one hand and the Mastigophora on the other that it is exceedingly difficult to define their exact relationships. A common species in fresh water and soil is *Mastigamoeba aspera.* This species has a single long flagellum and pushes out fingerlike pseudopodia. *Mastigamoeba hylae* is a parasitic species that occurs in the large intestine of frog tadpoles. A short, inactive flagellum arises from the nucleus at the anterior end. *Histomonas meleagridis* causes a liver infection in turkeys and other fowl. No cyst stage is known, but the amoeboid flagellate is able to get from one host to the next in the eggs of a different parasite of fowl, the nematode *Heterakis gallinae.*

Order Kinetoplastida

The order Kinetoplastida comprises a number of flagellates generally of small size, and only one to four flagella. Most species are parasitic. The order is of particular interest because it includes the flagellates living in the blood and tissues of humans.

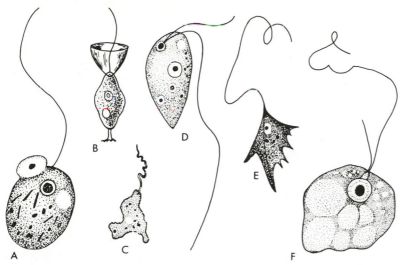

Figure 2-11 Zoomastigophorea. (A) *Oikomonas termo.* (B) *Monosiga brevipes.* (C) *Mastigamoeba invertens.* (D) *Bodo saltans.* (E) *Cercomonas longicauda.* (F) *Monas vulgaris.* (*After various authors.*)

Of the free-living Kinetoplastida, *Oikomonas termo* (Fig. 2-11, A) is a simple type that is often common in fresh water and soil. It is very small (4 to 5 microns in diameter), spherical in shape, and has a single long flagellum. *Bodo caudatus, Cercomonas longicauda* (Fig. 2-11, E), and *Monas vulgaris* (Fig. 2-11, F) are very common in stagnant water and infusions, and the first two are often coprozoic. *Bodo* (Fig. 2-11, D) is ovoid and plastic with two anterior flagella, one directed forward, the other trailing behind, and a spherical nucleus near or anterior to the center of the body. *Cercomonas* is similar to *Bodo,* but the trailing flagellum is attached to the side of the body and the nucleus is pyriform. *Monas* has two anterior flagella directed forward, one a longer primary flagellum and the other a shorter secondary flagellum.

The six genera of parasitic species described in the following paragraphs are obviously closely related. They include those usually grouped together as blood-inhabiting flagellates to distinguish them from another group that live in the digestive tract, called intestinal flagellates. During its life cycle *Trypanosoma* may assume trypanosome, crithidial, herpetomonad, and leishmania stages (see Fig. 2-12); *Herpetomonas* may appear in trypanosome, crithidial, herpetomonad, and leishmania stages; *Crithidia* in crithidial, herpetomonad, and leishmania stages; and *Leishmania* in herpetomonad and leishmania stages. This phenomena of a single species having different shapes at different times is known as *polymorphism.*

The Genus *Trypanosoma.* The trypanosomes are widely spread among animals. They occur in humans, many other mammals, birds,

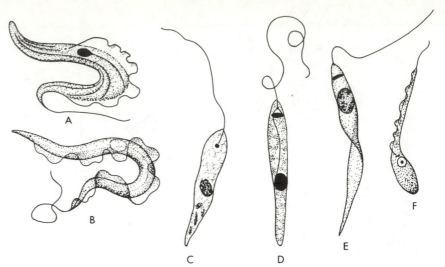

Figure 2-12   Kinetoplastida. (A) *Trypanosoma rotatorium*, from the frog. (B) *Trypanosoma diemyctyli*, from the newt. (C) *Leishmania donovani*, from humans. (D) *Herpetomonas muscarum*, from the house fly. (E) *Phytomonas elmassiani*, from the milkweed plant. (F) *Crithidia gerridis*, from the water strider. Some Kinetoplastida have several stages: the trypanosome stage has an undulating membrane extending past the nucleus; the crithidial stage has a shorter undulating membrane; the herpetomonad stage has no undulating membrane; and the leishmania stage has neither undulating membrane nor external flagellum. (*After various authors.*)

reptiles, amphibians, and fish. The type species, *T. rotatorium* (Fig. 2-12, A), occurs in the blood of frogs and is not uncommon in this country. Another species, *T. diemyctyli* (Fig. 2-12, B), is present in the blood of the common crimson-spotted newt, *Triturus viridescens*, in certain localities. Three species are known from humans: *Trypanosoma gambiense* the organism of Gambian sleeping sickness and *T. rhodesiense* of Rhodesian sleeping sickness occur in Africa; and *T. cruzi* of Chagas' disease is localized in South and Central America. The African types live also in wild game, such as antelope, and are transmitted by tsetse flies; the American type occurs especially in armadillos, which serve as reservoirs from which the transmitting agents, the triatoma kissing bugs, may acquire their infection. All human trypanosomes are pathogenic and frequently bring about the death of the host.

Animal trypanosomes are mostly nonpathogenic. *T. lewisi* of the wild rat is a good example of this type. The rat is thought to produce a substance, called *ablastin*, by some unknown method. Ablastin inhibits division of *T. lewisi* but does not destroy them. Ablastin may achieve this by stimulating increased concentrations of cyclic AMP in the trypanosomes (Strickler and Patton, 1975). Among the more important trypanosomes that cause disease in lower mammals are *T. equiperdum,* the organism of the disease of horses known as dourine; *T. brucei,* the organism of nagana in various domesticated animals; *T. evansi,* of surra in domesticated mammals; *T. equinum,* of mal-de-caderas in horses and mules; and *T. hippicum,* of murrina in horses and mules. Trypanosomes are often local in their distribution, that is, many species are limited to the animals in certain definite geographical areas. This is due primarily to the fact that the transmitting agents are geographically restricted. For the most part, the transmitting agents of trypanosomes of terrestial animals are blood-sucking insects and those of aquatic animals are blood-sucking leeches. The transmitting agents are usually active carriers; that is, the trypanosomes pass through part of their life cycle within their bodies and are not merely transferred mechanically by them as typhoid fever germs are spread by house flies, which are passive carriers.

The Genus *Leishmania.* The leishmanias are probably transmitted from person to person by sand flies. Kala-azar, a disease widely distributed in Asia, is due to *Leishmania donovani* (Fig. 2-12, C), a species that attacks especially the endothelial cells and macrophages. Also localized in the Far and Near East is *L. tropica,* the organism of oriental sore or cutaneous leishmaniasis. One attack by this species gives immunity to further infection. The type of cutaneous leishmaniasis that occurs in South and Central America is due to *L. braziliensis,* a species that resembles *L. tropica* morphologically but differs from it in its serological reactions.

The genus *Crithidia* is parasitic in arthropods and other invertebrates. A species that may easily be obtained for study is *C. gerridis* (Fig. 2-12, F) from the intestine of water striders of the genus *Gerris.*

The genus *Herpetomonas* also lives in invertebrates. The species that can most readily be secured for examination is *H. muscarum* (Fig. 2-12, D), which lives in the intestine of the common house fly and other flies.

The genus *Leptomonas* is similar to *Herpetomonas* in structure and also only parasitizes invertebrates. A rare species, *Leptomonas karyophilus,* is the only known zooflagellate parasitic in another protozoan. It is found in the macronucleus of *Paramecium.*

The genus *Phytomonas* (Fig. 2-12, E) is of peculiar interest because it occurs in the latex of plants, such as milkweeds and euphorbias, and is transmitted by hemipterous insects.

Order Diplomonadida

Diplomonadida are protozoans with a definite bilateral symmetry. They include one of the more common intestinal flagellates of humans, *Giardia lamblia* (Fig. 2-13, E). *G. lamblia* lives in the duodenum, where it maintains itself against the force of peristalsis by clinging to the in-

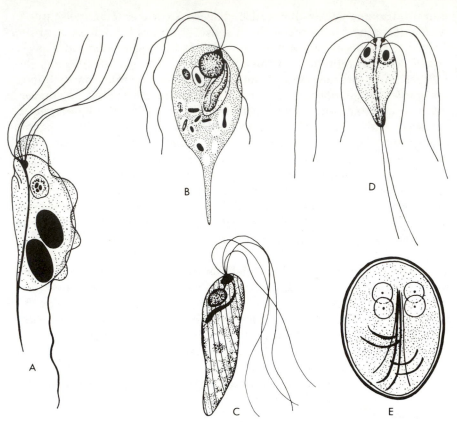

Figure 2-13 Intestinal flagellates. (A) *Pentatrichomonas hominis,* from humans. (B) *Chilomastix mesnili,* from humans. (C) *Polymastix bufonis,* from the toad. (D) *Hexamita salmonis,* from the salmon. (E) *Giardia lamblia, cyst,* from humans. (*After various authors.*)

testinal wall by means of its sucking disk. It multiplies by binary fission and forms cysts that pass out of the body in the feces and are infective to other humans whose food or drink may become contaminated by them. Giardias occur in many other vertebrates and can best be obtained for study from the duodenum of laboratory rats and frog tadpoles. Dogs, cats, mice, rabbits, guinea-pigs, herons, and many other animals are parasitized by their own peculiar species of giardias.

Hexamita (Fig. 2-13, D) is a diplomonad with six anterior and two posterior flagella, with species living in a wide variety of animals, including the laboratory rat, turkeys, salmon, and cockroaches.

Order Trichomonadida

The order Trichomonadida contains several intestinal flagellates of humans. Trichomonads have three to five anteriorly directed flagella and one recurrent flagellum often associated with an undulating membrane. They also have a longitudinal supporting structure known as an *axostyle*. Trichomonads live in the mouth, large intestine, and vagina and a large percentage of the general population is infected by them. They represent three species that are more alike morphologically than physiologically (Cheng, 1973). *Trichomonas tenax* lives in the mouth, *T. vaginalis* lives in the vagina, and *Pentatrichomonas hominis* lives in the intestine. *Pentatrichomonas* is distinguished from *Trichomonas* by having one of five anterior flagella arising independently (Fig. 2-3, C). No cysts are formed by trichomonads; hence, they must be transmitted in the active, trophozoite stage. Trichomonads are abundant in lower vertebrates.

Many other flagellates that live in the digestive tract of various animals are in the Trichomonadida or closely related orders. Among them are species of the genera *Chilomastix* and *Polymastix*. *Chilomastix mesnili* (Fig. 2-13, B) lives in the large intestine of humans. Other species occur in rats, rabbits, and other common mammals. *Polymastix* lives in many lower animals but not in humans. *P. melolonthae* is found in the cockchafer. *P. bufonis* (Fig. 2-13, C) lives in frogs and toads.

The intestinal flagellates of humans are thought to be nonpathogenic in most cases of infection.

Order Hypermastigida

The order Hypermastigida includes a number of peculiar flagellates that live in the gut of certain insects and possess some very complex structures. All of the forms bear many flagella, which may be arranged in bunches or may be distributed over the entire body. This is the only order of zooflagellates in which sexual reproduction has been demonstrated. In addition to finding sexuality in them, Cleveland has demonstrated that a correlation exists between reproductive processes of some of these protozoans and the hormonal cycles of their wood roach host.

Intestinal Flagellates of Termites. The flagellate inhabitants of the intestine of termites are of special interest because of the number of species, the number of individuals in a single termite, their complexity, the large number of flagella possessed by many of them, and the symbiotic relations between some of them and their hosts. Species from at least 11 families and 40 genera of flagellates have been reported from termites. A species that well illustrates the enormous development of flagella and the complexity of the body is *Trichonympha campanula* (Fig. 2-14). An excellent example of symbiotic relations is furnished by certain of these flagellates and their hosts. The flagellates render the cellulose (in wood) eaten by termites digestible by the insects; without the aid of the flagellates the wood eaten is not digested

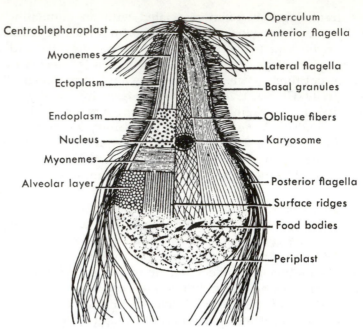

Labels on figure:
Centroblepharoplast — Operculum
Myonemes — Anterior flagella
Ectoplasm — Lateral flagella
Endoplasm — Basal granules
Nucleus — Oblique fibers
Myonemes — Karyosome
Alveolar layer — Posterior flagella
— Surface ridges
— Food bodies
— Periplast

Figure 2-14 Hypermastigida. *Trichonympha campanula*. Sections of the body show the structures found at different levels. Surface ridges form the outer layer with their rows of flagella; beneath are successively oblique fibers, alveolar layer, and transverse myonemes. In the endoplasm are longitudinal myonemes. (*From Kofoid and Swezy.*)

Figure 2-15 Intestinal flagellates of termites. (A) *Sunderella tabogae*, from *Lobitermes longicollis*. (B) *Spirotrichonympha flagellata*, from *Reticulitermes lucifugus*. (C) *Proboscidiella kofoidi*, from *Cryptotermes dudleyi*. (D) *Streblomastix strix*, from *Termopsis laticeps*. (*A, C, and D, after Kirby; B, after Grassi.*)

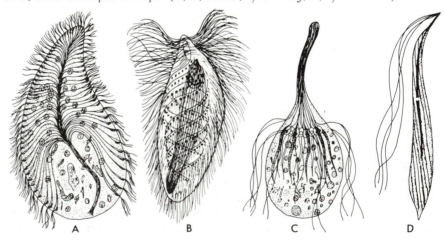

A   B   C   D

and the termites starve to death. Some of the characteristics of these termite protozoa are illustrated in Figure 2-15.

The genus *Lophomonas* contains species that are common in the intestine of cockroaches. *L. blattarum* has a pyriform body and bears a tuft of flagella at the anterior end. *L. striata* is spindle-shaped and is characterized by rodlike bodies obliquely arranged at the periphery of the body.

Opalinata

The Opalinata were once thought to be primitive ciliates and called Protociliata to indicate the relationship. More recently they have been considered intermediate between ciliates and flagellates. The fact that (1) all are symbiotic in the rectum of insect predators, (2) they are covered with numerous cilia or flagella (Fig. 2-16), (3) all nuclei are similar in an individual, and (4) they have sexuality, with hormones of the host appearing to influence development, suggests a close relationship with the Hypermastigida, in which they possibly should be considered a suborder. In recognition of the distinctive status accorded by recent workers, the group is here treated as a class of the

Figure 2-16 Opalinidae. (A) *Protoopalina*, a binucleate species; nuclei in anaphase of mitosis. (B) *Cepedea lanceolata*, a quadrinucleate species. (C) *Zelleriella macronucleata*, a binucleate species; nuclei in prophase of mitosis. (D) *Opalina ranarum*, a multinucleate species. (E) Nucleus of *Opalina* in anaphase of mitosis, showing large, flat, "trophic" chromosomes, and small, linear, "reproductive" chromosomes. (*After Metcalf.*)

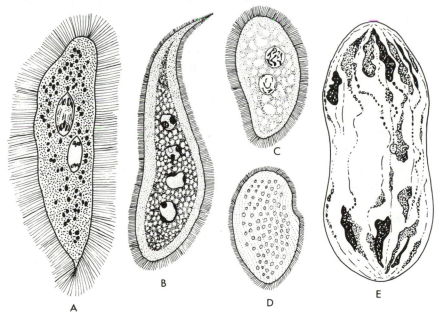

Mastigophora. The Opalinata contains a single order, Opalinida, and a single family, Opalinidae, with species that are common in the rectum of frogs, toads, and tadpoles. There are four genera: *Protoopalina* (Fig. 2-16, A), with a cylindrical binucleate form; *Zelleriella* (Fig. 2-16, C), with a flattened binucleate form; *Cepedea* (Fig. 2-16, B), with a cylindrical multinucleate form; and *Opalina* (Fig. 2-16, D), with a flattened multinucleate form. These genera are all characterized by a covering of cilia of equal length arranged in parallel rows, by the absence of a cytostome, and by the presence of only one type of nucleus. The nuclei, however, are of considerable interest because in some species, during mitosis, two types of chromosomes have been observed. Large chromosomes lie near the surface of the mitotic spindle and smaller chromosomes, in the form of slender rows of granules, are in the center of the mitotic figure (Fig. 2-16, E). These two types of chromosomes may represent the material of the macronucleus and micronucleus, respectively, of the ciliates. Another interesting characteristic of the Opalinidae is the fact that in many species the nuclei come to rest in some stage of mitosis. The number of nuclei in the individuals of various species ranges from two to several thousand.

Opalina ranarum (Fig. 2-16, D), which lives in the rectum of the frog, is the commonest and best known species. When alive, it is opalescent in appearance. Some specimens reach a length of about a millimeter and may be seen with the naked eye. The body contains a large number of spherical nuclei evenly distributed throughout the cytoplasm. Many small spindle-shaped bodies of unknown character are present in the endoplasm. Reproduction is by binary division during most of the year, but in the spring rapid division results in the production of many small individuals, which encyst and pass out in the feces of the frog. When these cysts are ingested by tadpoles, they hatch in the rectum and give rise either to macrogametocytes or microgametocytes. The gametocytes, which contain from three to six nuclei, divide into uninucleate macrogametes or microgametes. These conjugate, forming zygotes from each of which a young opalina develops. *Opalina ranarum* may be kept alive outside the body of the frog for as long as 3 days in Locke's solution.

Superclass Sarcodina

AMOEBA PROTEUS

Amoeba proteus (Fig. 2-17) is a one-celled animal about 250 μm (0.25 mm) in diameter and so transparent that it is invisible to the naked eye. Under the compound microscope, it appears as an irregular colorless or light blue particle of animated jelly that is constantly changing its shape by thrusting out fingerlike processes.

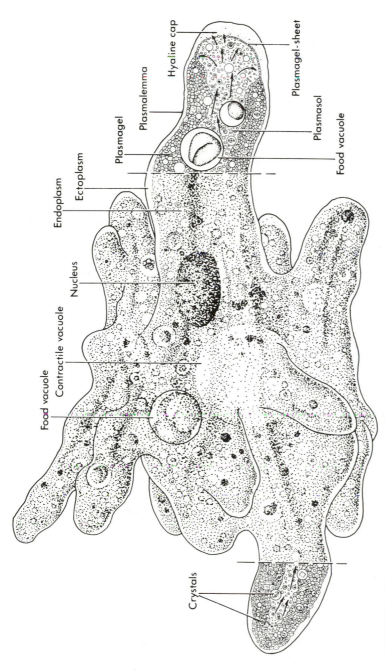

FIGURE 2-17 *Amoeba proteus*: cytology. Horizontal optical section through portions of an advancing pseudopodium and a retreating pseudopodium. (*Modified from Mast and Woodruff.*)

Hyaline cap

Plasmalemma

Plasmagel-sheet

Plasmagel

Plasmasol

Ectoplasm

Food vacuole

Endoplasm

Nucleus

Contractile vacuole

Food vacuole

Crystals

Habitat

Amoeba proteus lives in freshwater ponds and streams. It may be obtained for laboratory use from a variety of places, such as the organic ooze from decaying vegetation or the lower surface of lily pads. About 2 weeks before the specimens are needed, a mass of pond weed (*Ceratophyllum* is the best) should be gathered, placed in flat dishes, and immersed in water. The vegetation soon decays, and a brown scum appears on the surface. In this scum amoeba may be found.

The fact that amoebae appear in large number in cultures, such as just described, indicates that decaying pond weed furnishes a good habitat for them. Here they find their food, which consists of algae, other protozoans, bacteria, and other animal and vegetable matter.

General Cytology

(Fig. 2-17). Two regions are distinguishable in the body of amoeba: an outer colorless layer of clear cytoplasm, the *ectoplasm,* and a comparatively large central mass of granular cytoplasm, the *endoplasm.* A single clear spherical body, usually lying near the end of the animal away from the direction of motion, and disappearing at more or less regular intervals, is the *contractile vacuole.* Suspended in the endoplasm is a nucleus, usually one or more *food vacuoles,* material ready for excretion, foreign substances such as grains of sand, and undigested particles, the amount of the latter depending upon the feeding activity of the specimen at the time of examination.

The *nucleus* is not easily seen in living specimens. In animals that have been properly killed and stained, it appears as a biconcave disk in young specimens but it is often folded and convoluted in older specimens. Electron microscope studies show a honeycomb layer under a double-layered nuclear envelope. Nuclei of some genera of amoebae are distinct enough to be used for identification of the genus. The position of the nucleus in the endoplasm is not definite, but changes during the movements of the amoeba. Within the nucleus are many small, spherical chromosomes said to number over 500. During the life of an amoeba, before the period of reproduction, the nucleus plays an important role in the metabolic activity of the cell. This has been proved by experiments in which the animal was cut in two. The streaming of the cytoplasm ceases within a few minutes in the piece without a nucleus but is resumed after a few hours. The enucleated amoeba may attach itself to the substratum and exhibits irritability although its responses to stimuli are modified. Food bodies are engulfed and digested in an apparently normal manner but death finally ensues. The part with the nucleus continues its life as a normal amoeba. Other studies have shown that nuclei transferred to the cytoplasm of another amoeba from which the nucleus has been removed will produce characteristics in the hybrid amoeba resembling those of the cell donating the nucleus. The characteristics of the cell donating the cytoplasm are retained to varying degrees in different experiments. Because such blends of characteristics are maintained through successive generations, it is clear that both nucleus and cytoplasm have an important role in determining cell characteristics and inheritance.

Jeon, Lorch, and Danielli (1970) were able to separate nucleus, cytoplasm, and membrane and reassemble them, with 80 percent successfully dividing to produce new clones if the components were from the same individual. The rate of success was very low when components were from different strains. Division of the nucleus during reproduction will be described later.

The *contractile vacuole* is thought to be a water-regulatory organelle because its rate of activity is inversely proportional to the osmotic concentration of the media. Further evidence comes from a recent study that found the fluid in the vacuole is hypoosmotic to the protoplasm. It derives its name from the fact that at more or less regular intervals it suddenly disappears, its walls having collapsed or contracted, its contents leaving the cell. That energy is required for its activity is indicated by the ring of mitochondria surrounding it. Because no contractile elements have been demonstrated, it is thought that the energy is used in filling it against a diffusion gradient. It is located posterior to the nucleus (i.e., away from the direction of the advancing pseudopodia). The water that diffuses into the cell and is expelled by the contractile vacuole may have an important function in transporting oxygen in and waste products out of the cell, although diffusion through the cell membrane is adequate in marine and parasitic forms that lack contractile vacuoles.

Food vacuoles, mitochondria, oil droplets, various crystals, and small granules are also found in the endoplasm.

Physiology

Although extremely simple in structure, amoeba carries on practically all of the vital activities characteristic of the higher animals. It is capable of automatic movement, of reacting to various stimuli, of carrying on metabolic processes, of growth, and of reproduction. These are all fundamental properties of protoplasm and are here exhibited in simple form.

Locomotion. Amoeba moves from place to place by means of fingerlike protrusions of the body, known as *pseudopodia*. Four basic types of pseudopodia exist based on their structure (Fig. 2-18). (1) *Lobopodia* (Fig. 2-18, A) are broad and blunt and contain both an outer layer of ectoplasm and an inner portion of endoplasm. They are the type of pseudopodia found in *Amoeba proteus*. (2) *Filopodia* (Fig. 2-18, B) are narrow and contain only ectoplasm. (3) *Axopodia* (Fig. 2-18, C) have a fine coating of cytoplasm over an axial filament. Although the coating

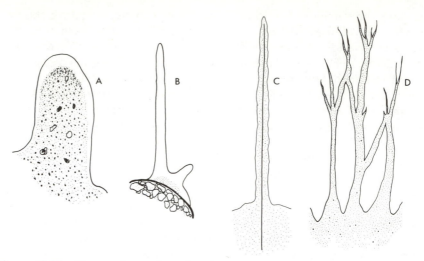

Figure 2-18 Types of pseudopodia. (A) Lobopodium of *Amoeba proteus*. (B) Filopodium of *Difflugia oblonga*. (C) Axopodium of *Actinosphaerium eichhorni*. (D) Rhizopodia of *Gromia*.

is of ectoplasm, it is more granular than the ectoplasm of the previous types. Axopodia are capable of retraction and bending even though they have an axial structure. (4) *Rhizopodia* (Fig. 2-18, D) are thin, branching pseudopodia that produce a network by fusion of the branches. Axopodia and rhizopodia may have the cytoplasm streaming in opposite directions on opposite sides of the same pseudopodium, a condition never observed in lobopodia. Thus, there may be more than one explanation for pseudopodial movement. Even in lobopodia, the most intensively studied type, protozoologists do not agree on the explanation for their method of formation. All have observed a rather fluid endoplasm flowing into or against the enlarging ectoplasmic front of the pseudopodium. Allen (1962) proposes that the explanation involves a contraction of contractile elements at the anterior, thus drawing the endoplasm into the advancing pseudopodium. Mast (1931), Goldacre and Lorch (1950), and Rinaldi and Jahn (1963) have proposed that the contraction occurs at the posterior in the gelated endoplasm, and that at the same time solation is occurring where it is in contact with the solated endoplasm and flows up with the endoplasm to be again gelated at the anterior of the advancing pseudopodium. Most protozoologists have supported the latter view. Unfortunately, electron microscopy has been unable to contribute to the solution with present techniques. Rinaldi and Jahn (1963) have analyzed motion pictures of granule movements in advancing pseudopodia as well as the various concepts and observations given in support of the two theories and concluded that the concept of amoeboid movement as given by Mast is essentially correct.

A recent volume on amoeba (Joen, 1973) includes a chapter by Allen (Chapter 7) that discusses the evidence for the frontal-contraction theory; a chapter by Bovee and Jahn (Chapter 8) is also on amoeboid movement. The controversy does not seem to be resolved. Allen, Francis, and Zeh (1971) interpret the amoeba's ability to form pseudopodia while one is sucked into a capillary tube as negating the contraction-hydraulic theory. That does not really negate the contraction-hydraulic theory since amoebae can form numerous pseudopodia simultaneously and their photographs provide little evidence of reduced pressure beyond the capillary tip.

According to Mast, the body of an amoeba is divided into four parts (Fig. 2-19). (1) The *plasmasol* is a central elongated portion of colloidal substances in the sol phase. These substances move toward the direction of lowest pressure. (2) The *plasmagel* is a colloidal substance of protoplasm in the gel phase that forms a tube around the plasmasol. The plasmagel contracts at the posterior and undergoes a change to the sol phase at its junction with the plasmasol. (3) The *plasmalemma* is a very thin elastic surface layer or membrane (in some amoebas it is actually a thick and rigid structure). (4) A *hyaline layer* exists between the plasmagel and the plasmalemma. The plasmasol and plasmagel constitute the endoplasm, and the hyaline layer and the plasmalemma constitute the ectoplasm. The contraction of the posterior plasmagel produces a hydraulic pressure on the plasmasol. The resistance to the pressure is least at the anterior where the thin sheet of plasmagel is continually stretched and occasionally broken, but continually reformed by gelation of plasmasol. The plasmagel is strongest at the sides and of intermediate strength at the posterior. Thus, a pressure at the anterior causes a reversal of flow, or a pressure at the side of the anterior causes a deflection away from the pressure. The plasmagel is under continuous tension and any localized reduction in elastic strength will cause the formation of a pseudopodium.

Whether or not the theory of Mast is correct, actin and actomyosin are present in amoebas with ATP to provide the energy, and divalent

Figure 2-19 Diagram of regions involved in amoeboid movement according to the contraction-hydraulic theory. (*Modified from Mast.*)

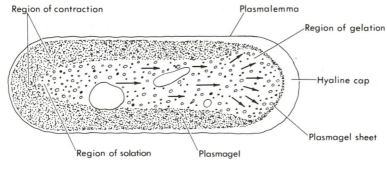

Figure 2-20 *Difflugia oblonga* in locomotion. Pseudopodia, numbered in the sequence formed, are extended, attached, and contracted to pull the amoeba forward.

Ca and Mg ions are involved in regulation. The basic similarity with other systems of motility is great on the molecular level.

Pseudopodial movement in *Difflugia* (Fig. 2-20) a shelled amoeba, is interesting when compared with that of *Amoeba*. The pseudopodia are extended through the opening of the shell one after another and attached to the substratum at the tip. The pseudopodium then contracts and the shell containing the body is drawn forward. The action is thought to be due to the same contraction-hydraulic system that functions in the previously described lobopodia.

Forms having axopodia and rhizopodia (or reticulopodia) have a flow of protoplasm in both directions on opposite sides of narrow pseudopodia. Jahn and Bovee (1965) have proposed that a different mechanism is involved with an active shearing or sliding that warrants revision of the classification of the Sarcodina. Their classification is followed here (Jahn and Bovee, 1965) because the differences in pseudopodial movement seem to provide a better basis for reflecting a major phylogenetic difference than the presence or absence of axial structures in the pseudopodia used in earlier systems. A genetic change resulting in the loss of the ability to produce a structure such as an axial filament is not considered to be a sufficient reason to place the shorted relative in a different class if other characteristics indicate a close relationship.

Metabolism. Growth in *Amoeba*, as in any living thing, involves a complex series of changes. The chemical compounds that make up the bodies of animals are extremely labile; they are constantly breaking down into simpler substances or becoming more complex by the addition of new materials. The changes are accomplished through the aid of enzymes, which are thought to be produced by an interaction of substrate, nucleic acids, proteins, and other substances. *Metabolism* is the term used to include this great complex of incessant changes. Some processes use energy, releasing it from high-energy phosphate compounds such as adenosine triphosphate (ATP), to build up compounds; others break down substances to produce energy, usually in the form of high-energy phosphate compounds. These changes occur within the cell, for the most part, so that the term *cellular metabolism* is sometimes applied. They also occur after substances such as oxygen and amino acids have been received in the cell and before the ultimate breakdown products of carbon dioxide, water, and nitrogenous wastes are eliminated so that the term *intermediate metabolism* is applied also. Because some substances are produced from other substances via a long sequence of particular changes, such sequences are referred to as *metabolic pathways*.

One of the principal sources of energy for cellular activity comes from the *cellular respiration* of glucose. A reversible series of reactions result in the breakdown of glucose and the net production of ATP. The first part of the series of reactions is known as *glycolysis*. The last part of the series is known as the *citric acid cycle*, because the oxaloacetate that combines with the acetyl group left from glycolysis to form citric acid is eventually reformed by the degradation of citric acid. The citric acid cycle has been localized in the mitochondria. Although the metabolic pathways have not been demonstrated in their entirety in any protozoan, enough has been demonstrated in many to make their occurrence in most very likely.

A portion of the energy stored in fats and proteins can also be released by some of the same pathways by which glucose and other carbohydrates are metabolized. However, protein metabolism results in the formation of ammonia, which may diffuse out of a simple aquatic animal like *Amoeba*. Ammonia is toxic in large concentrations and does not readily diffuse from water to air, so that terrestrial animals tend to have a large part of their nitrogenous wastes accumulated as compounds of low solubility or low toxicity such as in uric acid and urea. Urea has been identified in the contractile vacuoles of *Amoeba*.

Ingestion. The ingestion, or taking in of food (Fig. 2-21), occurs without the aid of a mouth. Other protozoans, algae, or other particulate organic matter serving as food may be engulfed at any point on the surface of the body, but is usually taken in at what may be called the temporary anterior end, that is, the part of the body toward the direction of locomotion. Jennings (1906) describes ingestion as follows. The amoeba flows against the food particle, which does not adhere but tends to be pushed forward away from the animal. That part of the body directly back of the food ceases its forward movements while, on either side and above, pseudopodia are extended which gradually form a concavity in which the food lies, and finally bend around the particle until their ends meet and fuse. A small amount of water is taken in with the food, so that there is formed a vacuole whose sides were formerly the outside of the body, and whose contents consist of a particle of nutritive material suspended in water. The whole process of food-taking occupies 1 or more minutes, depending on the character of the food. Amoeba is not always successful in accomplishing what it undertakes, but when it does not capture its prey at once, it seems to show a persistence usually only attributed to higher organisms. No doubt the reactions in food-taking depend upon both mechanical and chemical stimuli.

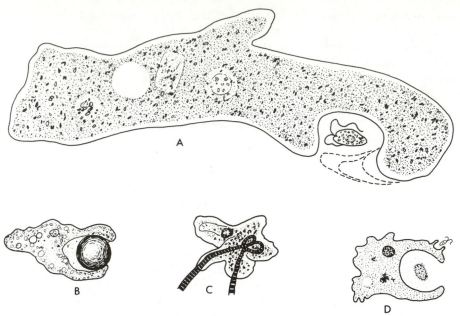

Figure 2-21 *Amoeba*. Ingestion of food. (A) Successive positions of a pseudopodium of an amoeba capturing a flagellate, *Chilomonas*. (B) Ingesting a cyst of a flagellate, *Euglena*. (C) Ingesting a filament of *Oscillatoria*. (D) A food cup for ingesting a flagellate superimposed on a food cup containing a ciliate. (*A, after Kepner and Taliaferro; B, after Jennings; C, after Rhumbler; D, after Becker.*)

A somewhat similar process occurs on a smaller scale with no particles actually ingested. Small pockets may form on the surface, their openings then close to pinch off a small vacuole of water, and the enclosed water vacuole is said to have been formed by the process of *pinocytosis*. Such a process may be important in nutrition of higher forms by providing a way in which large molecules can pass the barrier of the cell membrane.

Digestion. Food vacuoles serve as temporary digestive organelles. After a food vacuole has become embedded in the endoplasm, its walls pour into it a secretion of some acid, probably HCl. Various enzymes are also part of the digestive fluid, which breaks down protein substances, and perhaps also fats and sugars, but not starch.

Egestion. Indigestible particles are egested at any point on the surface of the amoeba, there being no opening to the exterior for this waste matter. Usually, such particles are heavier than the protoplasm, and as the animal moves forward they lag behind, finally passing out at the end away from the direction of movement; that is, amoeba flows away, leaving the indigestible solids behind. This process is not so simple in a species such as *Amoeba verrucosa*, which possesses an ec-

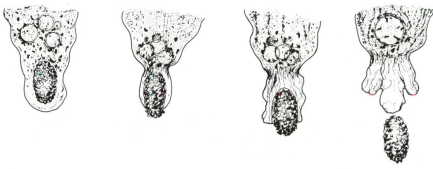

Figure 2-22 *Amoeba verrucosa.* Four stages in the ejection of a waste pellet. Note the rigidity of the pellicle. (*After Howland.*)

toplasmic pellicle that is a thick, tough membrane. Waste pellets are extruded as shown in Figure 2-22; a new pellicle, which prevents the outflow of the endoplasm, is formed at the point of exit.

Assimilation. The peptones and amino acids, derived from the digestion of protein substances, together with the water and mineral matter taken in when the gastric vacuole was formed, are absorbed by the surrounding protoplasm and pass directly into the body substance of the animal. These particles of organic and inorganic matter are then assimilated; that is, they are rearranged to form new particles of living protoplasm that are deposited among the previously existing particles. The ability to thus manufacture protoplasm from unorganized matter is one of the fundamental properties of living substance. Nucleic acids are thought to have an important role in the process.

Dissimilation. The energy for the work done by amoeba comes from the breaking down of complex molecules by oxidation or "physiological burning." This is the cellular respiration previously mentioned. The products of this slow combustion are the energy of movement, heat, and residual matter. Ordinarily, the residual matter consists of solids and fluids, mainly water, some mineral substances, urea, and CO_2. Secretions, excretions, and the products of respiration are included in this list.

Secretion. We have already noted that an acid is poured into the gastric vacuoles by the surrounding protoplasm. Such a product of dissimilation, which is of use in the economy of the animal, is known as a secretion.

Excretion. Materials representing the final reduction of substances in the process of metabolism are called excretions. These are deposited either within or outside of the body. A large part of the excretory matter, including urea and CO_2, passes through the general surface of the body. The fluid content of the contractile vacuole is known to contain urea; therefore, this organ has an excretory function in addition to its osmoregulatory action.

Respiration. The contractile vacuole is also respiratory, because CO_2 probably makes its way to the exterior by way of this organ. Oxygen dissolved in water is taken in through the surface of the body. This gas is necessary for the life of the animal; if replaced by hydrogen, movements cease after 24 hours; if air is then introduced, movements begin again; if not, death ensues.

Growth. If food is plentiful, more substance is added to the living protoplasm of the amoeba than is used up in its various physical activities. The result is an increase in the volume of the animal. This is growth, and, as in all other living organisms, growth by the addition of new particles among the preexisting particles.

Reproduction (Fig. 2-23). There is, however, a limit with regard to the size that may be attained by *Amoeba proteus,* as it rarely exceeds 0.25 mm in diameter. When this limit is reached, the animal divides into two parts. It is supposed that this division is inaugurated through some unknown change in the relations between the nucleus and the cytoplasm.

Figure 2-23 *Amoeba proteus.* Binary fission. Graph showing pseudopodial width during phases of mitosis. Drawings show the appearance during late prophase, early anaphase, and late telophase. (*Modified from Chalkley and Daniel.*)

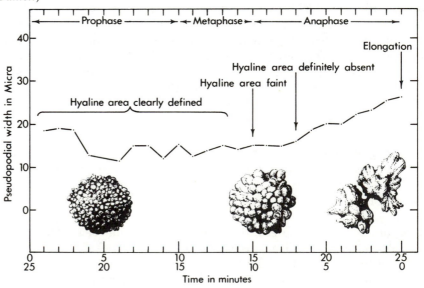

We are certain of only one type of reproduction in *Amoeba proteus:* binary fission. Sporulation and other types of reproduction have been described by investigators but have never been satisfactorily established. A description of the binary fission of *Amoeba proteus* (Fig. 2-23) is as follows. Nuclear division is mitotic, with the spindle forming while the nuclear membrane is disappearing at metaphase.

The average time for the division process at a temperature of 24°C is 33 minutes, with an average deviation of about 2 minutes. The *prophase* lasts about 10 minutes, the *metaphase* not over 5 minutes, probably less, the *anaphase* about 10 minutes, and the *telophase* about 8 minutes. The nuclear phases are accompanied by typical changes in cell form that are readily distinguishable under a magnification of 30 diameters. During prophase the cell is spherical, studded with fine pseudopodia and exhibits under reflected light a clearly defined hyaline area at its center. During metaphase the picture is similar, except that the hyaline area is very faint. During anaphase the pseudopodia become rapidly coarser, and the hyaline area is no longer visible. The telophase is accompanied by elongation of the cell and cleft formation.

Under the cultural conditions used, the average measured width of the pseudopodia in early prophase was 18 micra; from mid prophase to metaphase it was 13 to 14 micra. At anaphase the average width began to increase and at the beginning of telophase attained a value of 26 micra. It is possible merely from observation of the external appearance of the living cell to select from culture under a magnification as low as 30 diameters cells showing any desired stage of *nuclear* division except actual metaphase. Such selection can readily be made with 97 percent accuracy, thus making possible the study of the effect of reagents upon the separate nuclear phases in cell division. The modal volume of *Amoeba proteus* at division under the cultural conditions employed is 0.0027 cubic millimeter. Binucleate forms are produced in culture by failure of cytoplasmic fission. Evidence is presented that the contractile vacuole ceases to function during division from the time that the nucleus enters the metaphase stage until cell division is completed. The close correlation of the disappearance of the nuclear membrane with the cessation of the activity of the contractile vacuole and the beginning of the rapid increase in average width of the pseudopodia during the nuclear anaphase, suggests a causative relation between these phenomena (Chalkley and Daniel, 1933).

The *development* of amoeba is simply a matter of growth; the daughter cells resulting from binary division become full-grown specimens by means of a gradual increase in volume.

Behavior

The sum total of all the various movements of an animal constitute what is known as its behavior. In amoeba these movements may be separated into those connected with locomotion and those resulting from external stimuli. We have already given an account of the loco-

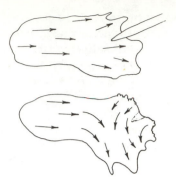

Figure 2-24 *Amoeba:* reaction to contact. A specimen moves away from a mechanical stimulus. (*From Jennings.*)

motion of amoeba, and so shall confine ourselves now to a discussion of its responses to different kinds of stimuli. The reactions of amoeba to stimuli have been grouped by Jennings into positive, negative, and food-taking. The last named were discussed under ingestion. The following account, then, will deal with positive and negative reactions of *Amoeba proteus* to external stimulation.

Reactions to Stimuli. Amoebae react to various kinds of stimuli, including contact, chemicals, heat, light, and electricity.

If the animal reacts by a movement toward the stimulus, such as light, it is said to react *positively;* if away from the stimulus, *negatively.*

Amoeba reacts negatively when touched at any point with a solid object; the part affected contracts and the animal moves away (Fig. 2-24). When, however, an amoeba is floating freely in the water and a pseudopodium comes in contact with the substratum, the animal moves in the direction of that pseudopodium until the normal creeping position has been attained (Fig. 2-25). Contact with food also results in positive reactions. Amoeba, therefore, reacts negatively to a strong mechanical stimulus and positively to a weak one.

Reactions to chemicals prove that amoeba is sensitive to changes in the chemical composition of the water surrounding it. *Amoeba* reacts negatively to many chemicals (Fig. 2-26) and changes in culture water. Experimental studies indicate that amoebae are able to "sense" food some distance away; this is probably the result of chemical stimuli.

Changes in the environment have a profound influence on the shape of amoeba. When an organism is transferred from culture fluid

Figure 2-25 *Amoeba:* reaction to contact. Changes in shape when a floating amoeba encounters a solid. (*From Jennings.*)

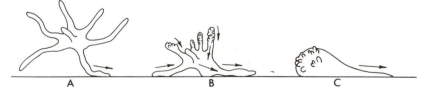

Figure 2-26 *Amoeba:* reaction to a chemical. Methyl green results in a negative reaction. *(From Jennings.)*

to pure water it becomes radiate in form. If it is then placed in the proper concentration of any one of a number of salts it changes its form as indicated in Figure 2-27. The form assumed by the organism appears to depend directly upon its water content.

Negative reactions result if amoeba is locally affected by heat, since the animal will move away from heat stimuli. Cold and excessive heat retard its activities, which cease altogether between 30 and 35°C.

Amoeba will orient itself in the direction of the rays of a strong light and move away from it (Fig. 2-28), but may react positively to a very weak light. The light causes gelation of the plasmasol adjoining the plasmagel, making it thicker and increasing the elastic strength of the portion illuminated. The response to light is thus due to the contraction of the plasmagel in the region stimulated, owing to the increase in the elastic strength of the plasmagel in this region. The evidence at present indicates that, while gelation or solation produced in a pseudopod of an amoeba by localized illumination is not followed by gelation or solation in other pseudopods, the effect of localized gelation and solation is transmitted, that *Amoeba* acts as an organized unit, and that its actions are, on the whole, adaptive. *Amoeba* responds to an electric current by moving toward the cathode; this results from the solation of the plasmagel on the cathodal side of the organism.

In amoeba there are no organs that can be compared with what we call sense organs in higher animals, and we must attribute its reactions to stimuli to that fundamental property of protoplasm called irritability. The superficial layer of cytoplasm receives the stimulus and transfers the effects to some other part of the body; thus may be shown the phenomenon of *internal irritability* or *conductivity*. The stim-

Figure 2-27 Changes in the shape of an amoeba when transferred from pure water to an N/1,000 solution of NaCl. The numbers indicate the time of day and the arrows the direction of protoplasmic streaming in the pseudopodia. *(After Mast.)*

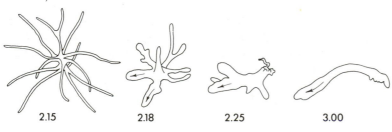

2.15   2.18   2.25   3.00

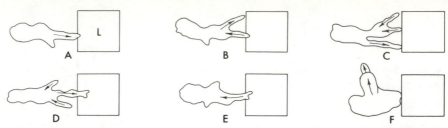

Figure 2-28 Reaction of an amoeba toward an area of intense light (L), the light rays of which were perpendicular to the slide. The successive positions of the amoeba are indicated at intervals of about ½ minute. The arrows show the direction of protoplasmic streaming in the pseudopodia. (*After Mast.*)

ulus causing a reaction seems to be in most cases *a change in the environment.* The behavior of amoeba in the absence of external stimuli, for example when it is suspended freely in the water, shows that some of its activities are initiated by internal causes.

The reactions of amoeba to stimuli are of undoubted value to the individual and to the preservation of the race, for the negative reaction is in most cases produced by injurious agents such as strong chemicals, heat, and mechanical impacts, whereas positive reactions are produced usually by beneficial agents. The responses, therefore, in the former cases carry the animal out of danger; in the latter, to safety.

Amoeba is of interest to animal psychologists because it represents animal protoplasm at a very low level of structural organization that exhibits behavior. The potential of protoplasm lacking an organized nervous system can be studied here.

A review of the facts obtained thus far seems to show that factors are present in the behavior of amoeba "comparable to the habits, reflexes, and automatic activities of higher organisms," and "if amoeba were a large animal, so as to come within the everyday experience of human beings, its behavior would at once call forth the attribution to it of states of pleasure and pain, of hunger, desire, and the like, on precisely the same basis as we attribute these things to the dog" (Jennings, 1906).

OTHER SARCODINA

The Sarcodina are divided into two classes, Hydraulea and Autotractea. In the Hydraulea all sides of a pseudopodium have the same direction of movement with respect to the axis of the pseudopodium. In the Autotractea the pseudopodia are slender and movement may occur in opposite directions on opposite sides of the same pseudopodium; the pseudopodia frequently have axial structures.

Order Amoebida

The order Amoebida contains amoeboid Sarcodina with lobose pseudopodia; but there is neither skeleton nor shell.

Amoeba proteus is the type of the superclass Sarcodina usually selected for detailed study and probably no other member of this superclass is as satisfactory for this purpose. However, many species of both free-living and parasitic Sarcodina that differ in various respects from the type may easily be secured for observation. Some of these are described in the following paragraphs.

Free-Living Amoebae (Fig. 2-29). Several different species of amoeba have been grouped together in the past under the name *Amoeba proteus*. *Amoeba proteus* was described by Leidy in 1879. It is characterized by the possession, in the younger forms, of a nucleus in the shape of a biconcave disk, which may become folded and convoluted in the older forms. It also has definite ectoplasmic ridges that are constant cell structures. A second form is much like the true *Amoeba pro-*

Figure 2-29 Amoebida. (A) *Amoeba dubia;* nucleus ovoidal. (B) *Amoeba discoides;* nucleus discoidal. (C) *Pelomyxa palustris;* many nuclei. (D) *Chaos carolinensis;* many nuclei. (E) *Amoeba verrucosa;* the tough pellicle results in many pseudopodia. (*A and B, after Schaeffer; C, after Doflein; D, after Kepner and Edwards; E, after Wang.*)

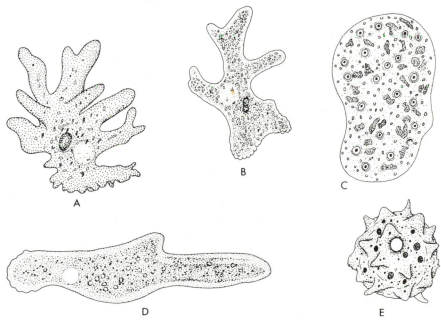

teus except that it possesses a disklike nucleus that never becomes folded and its ectoplasm never shows the ridges constantly present in *Amoeba proteus.* The name *Amoeba discoides* has been given to this species (Fig. 2-29, B). A third form possesses an ovoid nucleus instead of a discoidal one and does not show any cytoplasmic ridges. Furthermore, it is characterized by a type of locomotion different from that of the other species. To this form the name *Amoeba dubia* (Fig. 2-29, A) has been given. By means of carefully pedigreed cultures, Schaeffer (1920) has been able to show that these three forms breed true and probably represent true species. Another fairly common species is *Amoeba verrucosa,* a sluggish form that has a tough pellicle and many short pseudopodia (Fig. 2-29, E).

Chaos and *Pelomyxa* (Fig. 2-29, C and D) are genera of very large freshwater amoebae with many nuclei. *Pelomyxa* differs from *Amoeba* and *Chaos* in that its nuclei are spherical with a dispersed nucleolus and it typically has only a single pseudopodium but many particulate inclusions in its cytoplasm. Encystment has been reported for both genera.

Protein Differences of Species. The availability of techniques for comparison of proteins by chromatographic, electrophoretic, and antigenic characteristics has made it possible to compare species for similarities of their protein composition. In general, species in the same genus have a greater similarity of proteins than species in different genera. Although these techniques have not enabled biologists to determine phylogenetic relationships as well as was hoped, they have provided much interesting information. Only slight differences, if any, have been detected in the spectrum of proteins of *Amoeba discoides* and two different strains of *Amoeba proteus* by electrophoretograms. However, the three differ markedly in comparison with proteins of a species of *Chaos* (Fig. 2-30). The fact that the same bands of protein or proteinlike compounds appear in electrophoretograms does not prove that the amoebae are identical, but it does indicate that they have a grouping of compounds with approximately the same properties of migration in electric fields. When one considers that a single gene may often have a profound effect on an organism, it seems reasonable to assume that organisms alike enough to be classified in the same species should have great similarity of protein composition. Considerable changes could probably be made in some protein molecules without markedly changing the size, weight, and charge, which would determine the electrophoretic properties. Similarly, changes may not affect the properties by which they are recognized chromatographically or antigenically. When more information on differences within the better known species and genera has been accumulated, taxonomists may be able to use such techniques for establishing generic boundaries.

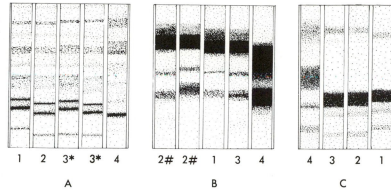

1 2 3* 3* 4   2# 2# 1 3 4   4 3 2 1

A   B   C

Figure 2-30 Drawings of electrophoretograms of amoebae. (A) Stained to show some water-soluble proteins. (B) Pattern of acid phosphatase activity. (C) Pattern of esterase activity. (1) *Amoeba proteus,* strain Bk. (2) *A. proteus,* strain T. (3) *A. discoides.* (4) *Chaos chaos.* Those marked with an * were run at the same time and those marked with a # were run at different times to show the reproducibility of the method. (*From J. R. Kates and L. Goldstein. A comparison of protein composition of three species of amoebae.* J. Protozool. **11**(1):30–35. *Copyright 1964 by the Society of Protozoologists.*)

Parasitic Amoebae. Of special interest in the Superclass Sarcodina are the parasitic amoebae, particularly those that occur in people. Humans may be infected by as many as six species. One, *Entamoeba gingivalis,* lives in the mouth in the tartar around the base of the teeth. Probably 50 percent of the general population are infected. This amoeba feeds principally on bacteria and leucocytes and is probably harmless, although it is known to be present more frequently in diseased than in healthy mouths. Transmission occurs by contact during kissing.

The other five parasitic amoebae of humans live in the large intestine. One of them, *Entamoeba histolytica* (Fig. 2-31, C) is pathogenic. The infective stage is a spherical cyst containing four nuclei. Encystment takes place in the large intestine and the cysts pass out in the feces. Cysts that are swallowed in contaminated food or drink hatch in the small intestine giving rise to amoebae with four nuclei, which subsequently divide into uninucleate amoebae. This species infects about 10 percent of the general population, but most of the infected persons are carriers; that is, the amoebae live and multiply in their large intestine and produce cysts but do not injure them. In a few persons, however, the amoebae attack the wall of the intestine, producing ulcers and giving rise to amoebic dysentery. Several races exist which vary in size and pathogenicity. They seem to be particularly fond of red blood cells as a food supply. From the intestinal wall, the amoebae may be carried in the blood stream to the liver, lungs, brain, etc., where they sometimes produce abscesses. Compounds containing iodine are used successfully to cure amoebiasis.

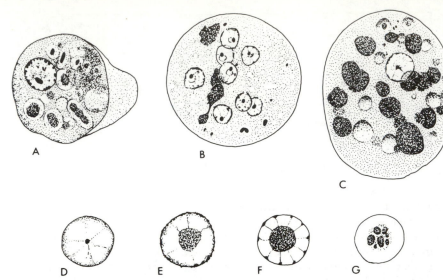

Figure 2-31 Amoebae parasitic in the large intestine of humans. (A) *Entamoeba coli*, a harmless species showing pseudopodium, nucleus and food vacuoles. (B) *Entamoeba coli*, cyst containing eight nuclei and a few chromatoid bodies (in black). (C) *Entamoeba histolytica*, the organism of amoebic dysentery, containing a nucleus and many red blood corpuscles (in black). (D) Nucleus of *Entamoeba*. (E) Nucleus of *Endolimax*. (F) Nucleus of *Iodamoeba*. (G) Nucleus of *Dientamoeba*. (*A–C, after Stabler; D–G, after various authors.*)

The remaining four species of human amoebae are supposed to be harmless. They are *Entamoeba coli, Endolimax nana, Iodamoeba bütschlii,* and *Dientamoeba fragilis*. These amoebae differ from one another especially in their nuclear structure (Fig. 2-31). The last named frequently possesses two nuclei. The parasitic amoebae resemble in general the free-living amoebae but do not possess contractile vacuoles.

Practically every lower animal that has been carefully examined has been found to be infected with parasitic amoebae. For example, *Hydramoeba hydroxena* (Fig. 2-32) is an amoeba that occurs on the external surface and in the gastrovascular cavity of *Hydra*. It dies in from 4 to 10 days if removed from its host and is hence an obligate parasite. The hydras are injured by the amoebae and usually are killed by them in about a week. Perhaps the parasitic amoebae most easily obtained for study are those that occur in the intestine of the frog, frog tadpole, cockroach, and rat.

Naegleria fowleri is one of several amoebas that may also exist in a flagellate stage. They are generally free-living but sometimes enter the body through the olfactory epithelium and gain access to the brain to produce a rapidly fatal meningoencephalitis. Griffin (1972) found that virulent strains had a high temperature tolerance. The high temperature tolerance may be considered a preadaptation to parasitism.

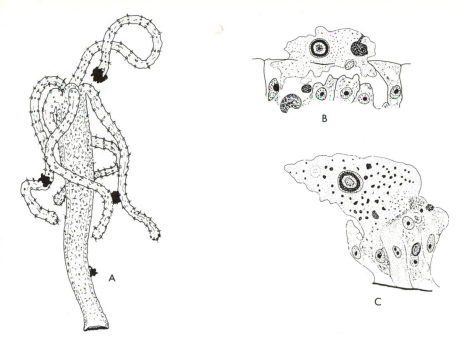

Figure 2-32 *Hydramoeba hydroxena:* a parasite of *Hydra.* (A) Four specimens (represented in black) on the surface of a hydra. (B) An amoeba destroying ectodermal cells. (C) An amoeba ingesting endodermal cells. (*After Reynolds and Looper.*)

Order Arcellinida

A number of common freshwater genera belong to this order, including *Arcella, Difflugia, Centropyxis, Lecquereusia,* and *Euglypha.*

The arcellas possess a shell secreted entirely by the organism. This shell is arched above and has an opening in the center of the lower, flattened surface through which lobose pseudopodia are extended. *Arcella vulgaris* (Fig. 2-33, A) possesses two nuclei. *A. dentata* (Fig. 2-33, B) also has two nuclei, and a shell with dentate projections around the periphery (Fig. 2-33). *A. polypora* may contain as many as 15 nuclei. In prepared specimens a large granular ring of protoplasm appears that stains deeply in haematoxylin; this is known as the chromidial body.

The commonest species of the genus *Difflugia* are *D. oblonga, D. lobostoma,* and *D. corona. D. oblonga* has a bottle-shaped shell with a more-or-less distinct cylindrical neck (Fig. 2-20). *D. lobostoma* has a spherical or ovate shell with a lobed mouth opening. *D. corona* has a spherical shell with a variable number of spines at the end opposite the mouth opening (Fig. 2-33, E). A single nucleus is present. The shells of difflugias are made of cemented sand grains.

The genus *Euglypha* contains several common species with a membranous shell composed of secreted chitinous or siliceous plates, and teeth around the mouth opening. The pseudopodia are filiform. *E.*

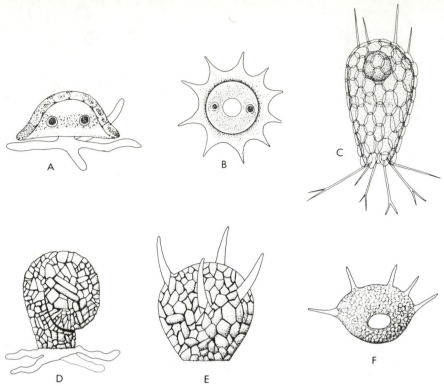

Figure 2-33 Order Arcellinida. Common species of shelled Sarcodina. (A) *Arcella vulgaris*. (B) *Arcella dentata*. (C) *Euglypha acanthophora*. (D) *Lecquereusia modesta*. (E) *Difflugia corona*. (F) *Centropyxis aculeata*. (*A, B, D, E, and F, after Wang; C, after Calkins*.)

acanthophora usually has a few long spines (Fig. 2-33, C). *E. ciliata* has numerous smaller spines and is common in sphagnum moss.

Asexual reproduction in the Arcellinida takes place by binary fission, just as in the Amoebida but is complicated by the presence of a shell. For example, in the case of *Difflugia corona* the organism prior to division accumulates a number of sand grains within its interior. At the time of reproduction the protoplasm absorbs water, swells, and projects through the mouth of the shell. The projecting mass finally attains the size and assumes the shape of another *Difflugia*. The sand grains then rise to the surface, spread over it, and become embedded in a chitinous secretion which hardens and completes the formation of a new shell. By the time the new shell is formed, the nucleus has already divided and one of the daughter nuclei migrates into the newly protruded mass of protoplasm. Finally, the protoplasmic masses separate and two difflugias are formed, one inhabiting the old shell and the other the new shell.

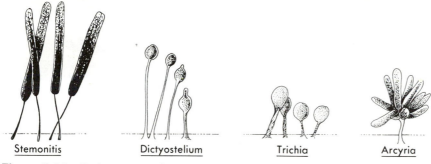

Stemonitis   Dictyostelium   Trichia   Arcyria

Figure 2-34   Order Mycetozoida. Sporangia.

Order Mycetozoida

The order Mycetozoida includes those organisms known as *slime molds*. They have cellulose cyst walls and funguslike fruiting bodies and have been studied as much by botanists as by zoologists. Amoeboid and flagellated stages occur during portions of their life cycles when they are most protozoanlike. The knowledge of groups such as these and the plantlike flagellates has been significant in the increased use of the Kingdom Protista as a classification of forms not differentiated into tissues. The amoeboid stage has lobopodia and many may fuse together and form a *plasmodium* (a multinucleate, amoeboid mass, which is spelled the same as the generic name of malaria). In some species flagellated gametes fuse in pairs and then become multinucleated plasmodia by mitotic division of the nucleus. The plasmodium eventually produces the fruiting body, which is named a *sporangium* (Fig. 2-34). In the sporangium, usually atop a stalk, are produced many haploid, cellulose-walled spores following meiotic division of the nuclei.

The orientation of fruiting bodies appears to be influenced by carbon dioxide concentration. Acrasin, a pheromone responsible for aggregation of individual cells into the plasmodium, has been identified in one species as cyclic AMP (Barkley, 1969) and the release of an enzyme able to hydrolyze AMP has been reported, perhaps functioning as a mechanism to prevent premature aggregation due to exogenous AMP.

AUTOTRACTEA

Order Proteomyxida

The order Proteomyxida contains a number of forms of very doubtful relationships; the only common characteristics are the absence of a shell and the formation of filose or reticulose pseudopodia.

An example of the Proteomyxida is *Pseudospora volvocis* (Fig. 2-35),

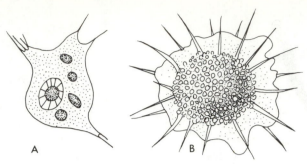

Figure 2-35 Proteomyxida. (A) *Pseudospora volvocis;* amoeboid stage of a parasite of *Volvox.* (B) *Vampyrella lateritia.* (*A, after Robertson; B, after Conn.*)

which often parasitizes colonies of *Volvox.* It occurs in amoeboid, flagellated, and heliozoan stages. In infected colonies of *Volvox* the amoeboid stages can be seen creeping about and devouring the cell individuals of its host. Another genus is *Vampyrella. V. lateritia* (Fig. 2-35, B) has a spherical body and is orange-red in color. It feeds on *Spirogyra* and other algae.

Labrinthula macrocystis are small spindle-shaped protozoans that move with a poorly understood gliding movement. It has been a serious parasite of eel grass along the eastern coast of the United States.

Order Heliozoida

The characteristics of this group may be stated as follows. The body is generally divided into two regions: (1) a cortical layer, which is alveolar in appearance, contains the contractile vacuole and gives rise to the pseudopodia; and (2) a medullary layer, which contains the nuclear apparatus, food vacuoles, and, in some cases, symbiotic algae. Many Heliozoida possess skeletons, which may be either simple or complex in structure and may be composed of various materials. Asexual reproduction is effected by either binary fission or gemmation.

Two of the best known species of Heliozoida are *Actinophrys sol* and *Actinosphaerium eichhorni.* The former occurs among the vegetation in quiet, fresh water. It is spherical and small (about 50 μm in diameter); has a large central nucleus; and possesses many large vacuoles in the ectoplasm. *Actinosphaerium eichhorni* (Fig. 2-36) is large, reaching 1 mm in diameter; has many nuclei in the outer layer of endoplasm and several contractile vacuoles. It is common among the plants in freshwater ponds.

Order Radiolarida

The Radiolarida (Fig. 2-37) are exclusively marine and possess a central capsule, which is a membranous structure that divides the protoplasm of the body into an intra- and extracapsular region. The ex-

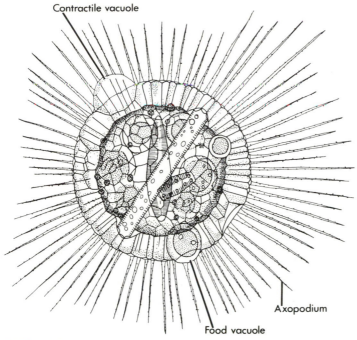

Contractile vacuole

Axopodium

Food vacuole

Figure 2-36 Heliozoida. *Actinosphaerium eichorni*. The endoplasm is crowded with food vacuoles containing diatoms, and nuclei are represented in the figure by the dark areas. (*After Leidy*.)

Figure 2-37 Radiolarida. Siliceous skeletons of several species.

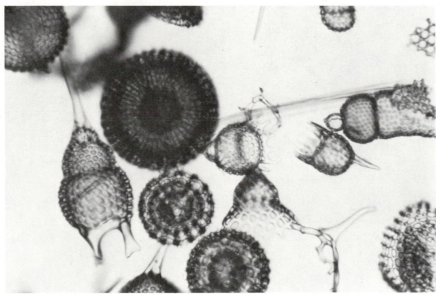

tracapsular region is itself divided into three layers. (1) The assimilative layer contains the food, which is taken in by the axopodia, and various metaplastic granules. (2) The calymma contains a large number of vacuoles and is supposed to have a hydrostatic function. In some species this layer also contains a number of yellow symbiotic dinoflagellates. (3) The external layer surrounds the body and from it the axopodia arise. In the majority of species reproduction is associated with the formation of flagellated swarm spores, which arise by a process of multiple fission within the intracapsular region. Several thousand species are planktonic in the ocean. Most have skeletons or spicules of siliceous material or else strontium sulphate. Because the skeletons are not dissolved by dilute acid, they selectively accumulate in some of the deeper parts of the ocean where acids have removed the calcareous parts of other organisms.

Figure 2-38 Foraminiferida shells. (A) *Globigerina bulloides,* with globular chambers and long spines. (B) *Rotalia beccari,* upper surface showing chambers arranged in a helicoid spiral. (C) *Nummulites laevigatus,* cross section of megalospheric form showing chambers in a spiral plane. (D) *Nodosaria,* with chambers in a linear arrangement. (E) *Nodosaria,* diagram of longitudinal section. (F) *Gumbelina,* with chambers added on alternate sides along the plane of growth. (*A–C, after several authors; D–F, original.*)

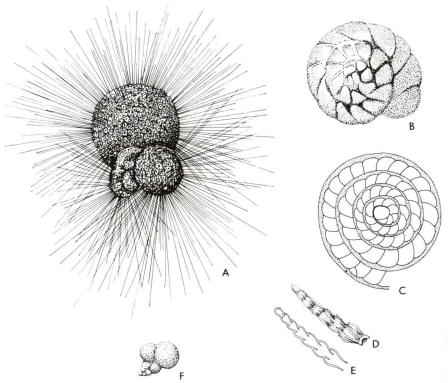

All the Foraminiferida possess shells, which may be chitinous, calcareous, or of foreign materials cemented together. Few have siliceous shells. Shells, or tests, usually have many chambers formed by addition of new and larger chambers to the old following division. Two structural types commonly exist in the same species: (1) a microspheric form, identifiable by a small first chamber, which gives rise asexually to numerous individuals of the next form; and (2) a megalospheric form, identifiable by a large first chamber, which gives rise to numerous flagellated gametes that fuse to produce more of the microspheric form. If the test has only the main opening, or aperture, from which the reticulopodia are protruded it is said to be imperforate. Perforate species have numerous small openings, through which the reticulopodia extend, in addition to the aperture. Many types of shells are formed in different species (Fig. 2-38), with the coiled arrangement of chambers being one of the most common. The majority of the Foraminiferida are marine forms, although a few such as *Allogromia* (Fig. 2-39) occur in fresh water. Most of them live in the mud of the sea bottom and large areas in the Atlantic Ocean consist of globigerina ooze. So abundant are their shells that they form a large part of the white chalk laid down during the Cretaceous period and the nummulitic

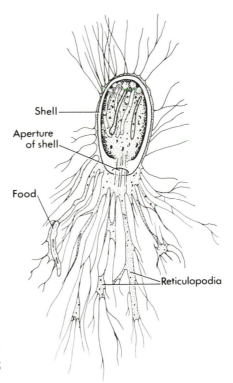

Figure 2-39 Foraminiferida. *Allogromia*, a freshwater species, showing capture of food. (*From Woodruff.*)

limestone of the Eocene. *Globigerina bulloides* (Fig. 2-38) is a cosmopolitan species that is pelagic in habit and also lives in the mud at the bottom of the sea to depths of 3,000 fathoms. Because the pelagic species of the past showered ancient sea beds with their skeletons, which are small enough to be brought up with well drillings, the identification of distinctive species from sedimentary rocks of different ages is now important to geologists for stratigraphic correlation. The Pliocene–Pleistocene boundary in deep-sea sediments is marked by abrupt changes in the Foraminiferida fauna (Ericson, Ewing, and Wollin, 1963). The abundance and variety of species in the geological record, with some in existence for long periods of time and others for shorter periods, makes them one of the more useful groups of fossils for aging rocks.

Estimates of paleoclimate and other variations in environmental conditions have been made from foram diversity (Buzas and Gibson, 1969; Ruddiman, 1969), temperature optima (Lynts and Judd, 1971), oxygen isotopic analysis of shells (Emiliani, 1971), and secondary calcification of the test (Srinivasan and Kennett, 1974).

Sporozoa (Superclasses Apicomplexa, Myxospora, and Microspora)

The sporozoa are all parasitic. Parasites can lose the characteristics of free-living relatives, as adaptation to the host occurs, and thus present great difficulty in determining phylogenetic relationships for classification. Recent confirmation of the coccidian nature of *Toxoplasma* may justify expectation of further clarification of relationships within the sporozoa. For example—do host relationships and mitochondrial morphology indicate that the piroplasms are derived from the Kinetoplastida and should be classified near them because of their lack of spore formation?

Monocystis and *Plasmodium,* the animal most destructive of human life, are used as principal examples of the sporozoa.

MONOCYSTIS

Monocystis is selected as a type of the superclass Sporozoa because it illustrates many of the characteristics of this group and is an easily obtained parasite in the seminal vesicles of the common earthworm. The stages that are usually present are the trophozoite (Fig. 2-40, A), cysts containing two specimens (B), or gametes and spores in various phases of development (C, D, E), and isolated spores (F, G).

The life cycle of *Monocystis* is briefly as follows. *Spores* (Fig. 2-40, G), are probably released from the seminal vesicles of the earthworm by the death and decay of the earthworm. Each spore contains eight

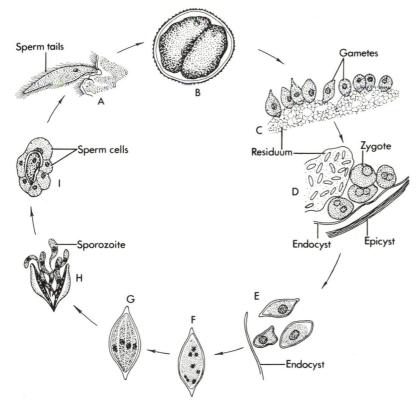

Figure 2-40 Gregarinida. Life cycle of *Monocystis,* parasitic in the seminal vesicle of the earthworm. (A) A mature individual attached to the sperm-funnel of the earthworm. (B) Two mature individuals joined side by side. (C) Gametes formed within the cyst. (D) Conjugation of gametes to form zygotes. (E) Zygotes that have become spores. (F) A single spore containing eight nuclei. (G) A fully developed spore containing eight sporozoites. (H) The eight sporozoites escaping from the sporocyst. (I) Young trophozoite among sperm-mother cells. (*After various authors.*)

sporozoites, which are released from the spore following ingestion by another worm and action of its digestive juices on the spore. Each sporozoite penetrates a bundle of sperm mother cells (I) of the earthworm, and is then termed a *trophozoite.* Here it lives at the expense of the cells among which it lies. The spermatozoa of the earthworm which are deprived of nourishment by the parasite, slowly shrivel up finally becoming tiny filaments on the surface of the trophozoite (A).

When this stage is reached, two trophozoites come together (B) and are surrounded by a common two-layered cyst wall (D). Each then divides, producing a number of small cells called *gametes* (C). The gametes unite in pairs (D) to form *zygotes.* It is probable that the gametes produced by one of the trophozoites do not fuse with each other, but with gametes produced by the other trophozoite enclosed in the cyst. Each zygote becomes lemon-shaped, and secretes a thin hard wall about itself. It is now known as a *sporoblast* (E). The nucleus of the sporoblast divides successively into two, four, and finally eight daughter nuclei (F); each of these, together with a portion of the cytoplasm, becomes a sporozoite (G, H).

PLASMODIUM VIVAX

One of the best known of all the sporozoa is *Plasmodium vivax*, which causes tertian malarial fever. This minute animal was discovered in 1880 in the blood of malaria patients by a French military doctor, Laveran. It is transmitted from person to person by the bite of certain species of diseased mosquitoes belonging to the genus *Anopheles*. The two most common genera of mosquitos are *Culex* and *Anopheles*. One of the easiest methods of distinguishing one from the other is by observing their position when at rest. It will be found that *Culex*, which does not serve as a host for *Plasmodium*, holds its abdomen approximately parallel to the surface on which it alights, whereas the abdomen of *Anopheles* is held at an angle.

There are three well-known types of malaria; these may be recognized by the intervals between succesive chills. (1) Tertian fever, caused by *Plasmodium vivax*, is characterized by an attack every 48 hours; (2) quartan fever, caused by *Plasmodium malariae*, with an attack every 72 hours, and (3) estivoautumnal or subtertian fever, caused by *Plasmodium falciparum*, produces attacks daily, or more or less constant fever. The life histories of these three species of *Plasmodium* differ very slightly one from another.

Malarial fever is transmitted by diseased female mosquitoes only. The mouth parts of these insects are adapted for piercing (Fig. 11-20). When they have been thrust into the skin of the victim, a little saliva is forced into the wound. This saliva contains a weak poison, which is supposed to prevent the coagulation of the blood and, thus, the clogging of the puncture. Blood is sucked up by the mouth parts into the alimentary canal of the mosquito; this process occupies from 2 to 3½ minutes. With the saliva a number of parasites, which were stored in the salivary glands of the insect, find their way into the wound. These are known as *sporozoites* (Fig. 2-41, A). Each sporozoite penetrates an endothelial cell of the liver (B), where it becomes a trophozoite feeding at the expense of the host cell. The fully grown trophozoite divides by multiple fission into many *merozoites*. The merozoites are released and penetrate other endothelial cells to become trophozoites as before to continue the *exoerythrocytic cycle*. The *erythrocytic cycle* starts when merozoites formed in the exoerythrocytic cycle in the liver cells eventually enter red blood cells to develop as trophozoites. In the red blood corpuscle the trophozoite feeds and grows at the expense of the pro-

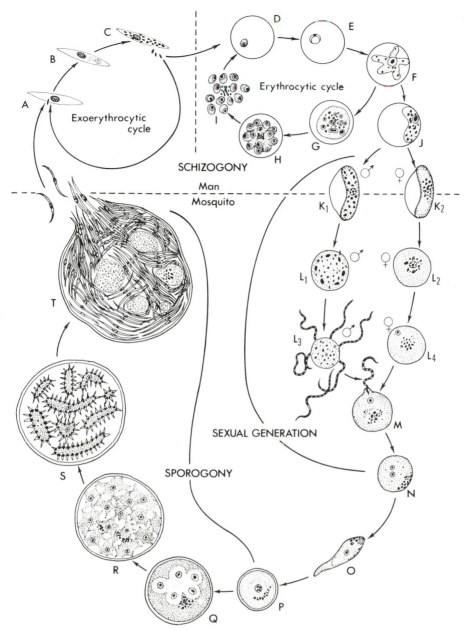

Figure 2-41 *Plasmodium* of humans. Diagram illustrating the life history of the malarial fever parasite. (A–C) Cycle of schizogony in the endothelial cells of the liver. (D–I) Cycle of schizogony in the red blood cells; the duration of this schizogonous cycle is similar to the duration of a cycle of symptoms. (J–N) Sexual generation. (O) Ookinete. (P–T) Sporogony; occurs in the wall of the mosquito alimentary tract. (*From Minchin after various authors, modified.*)

toplasm of the red blood corpuscle. The trophozoite reaches its full size in about 50 hours and is then known as a *schizont;* this undergoes a type of multiple division, thus producing from 15 to 24 daughter cells which are also called *merozoites.* The merozoites are liberated into the blood stream in about 8 hours and immediately attack fresh red blood corpuscles. Some of the merozoites develop into schizonts, but others become sexual cells called *gametocytes.*

When a mosquito bites a malaria patient it sucks up blood containing gametocytes. If the mosquito is of the genus *Culex* or of some genus other than *Anopheles,* the parasites are, in the course of time, destroyed in its stomach; but if the mosquito is a female belonging to the proper species of *Anopheles,* development is completed. The female gametocyte produces a mature *macrogamete,* and within the male gametocyte six to eight flagellalike bodies develop which are *microgametes* comparable to spermatozoa. Fertilization then ensues and the zygote thus formed changes into a motile, wormlike *ookinete* (some sources attribute the wormlike shape to aberrant individuals), which enters the wall of the mosquito's stomach. Here it rounds up into an *oocyst,* which grows at the expense of the surrounding tissue, and after 6 to 7 days undergoes *sporulation* during which hundreds of spindle-shaped sporozoites are produced. These break out into the body cavity of the mosquito and some of them find their way into its salivary glands; here they remain ready to be injected into the next person the mosquito bites.

Development in the blood of humans does not go beyond the macrogamete and microgamete stage of the sexual cycle. The factor limiting further development seems to be the temperature of the person since further development will occur if blood with gametes is removed to lower temperatures. Although the exoerythrocytic and erythrocytic cycles continue in untreated individuals, the infection in the mosquito disappears in a few months.

Until recent years, malaria was the most important of all human diseases, especially in tropical and subtropical regions. Although destruction of the vector with DDT and other means as well as treatment of infected humans with antimalarial drugs has markedly reduced the dangers from malaria, cases still number in the millions each year. Quinine remains one of the most effective therapeutic agents, but numerous synthetic drugs also capable of destroying the parasites in the human body have been discovered. Widespread use of some of the newer antimalarial drugs has been followed with the increase of a few resistant strains of malaria, which must then be treated with one of the other antimalarial drugs. Of interest is the resistance to malaria possessed by some people with modified hemoglobin due to the presence in the heterozygous condition of the gene for sickle cell anemia. Many antimalarial drugs are thought to be effective by blocking amino acid uptake. Transfer of drug resistance between species of *Plasmodium*

(Ferone, O'Shea, and Yoeli, 1970) was successful before the recent successes in gene transfer using restriction enzymes and bacterial plasmids.

OTHER SPOROZOA

The sporozoa are all parasitic Protozoa, which usually pass from one host to another in the spore stage. The spore is generally a seedlike body with a covering called the *sporocyst* (Fig. 2-40, F and G). In species whose spores are subjected to air, water, or other agents in their passage from one host to another, these bodies are very resistant, but in other species which are propagated from one host to another directly, either by being inoculated into the new host by a blood-sucking animal or by being eaten by the new host, the spores do not always have such a resistant covering (Fig. 2-41).

Sporozoa are among the most widely distributed of all parasitic animals; members of almost every large group of animals in the animal kingdom are parasitized by one or more species. Infection is very common among the vertebrates, arthropods, mollusks, and worms, and less common among echinoderms, coelenterates, and Protozoa. Comparatively few species are lethal and only a small number seem to be very harmful to their hosts.

The life cycles of the sporozoa are often very complicated. Frequently, there is an alternation of hosts, certain stages being passed in a vertebrate and other stages in an invertebrate which serves as a transmitting agent from one vertebrate host to another. Two types of reproduction are characteristic of the life cycles of most species: (1) multiplicative reproduction or *schizogony*, during which many organisms are formed in a single host; and (2) *sporogony*, a propagative reproduction involving sexual processes and ending in the formation of spores which are usually transferred to another host.

As compared with the other Protozoa, the sporozoa are greatly modified by their parasitic existence. These modifications are represented by the absence of locomotor organs, a mouth, anal opening, excretory pore, and vacuoles. There is usually a single nucleus present. Nutrition is by absorption or by ingestion of host cytoplasm. Many organs of the host may be parasitized, especially the alimentary tract, kidneys, blood, muscle, and connective tissues. When the parasites live inside of cells they are said to be *cytozoic*; when among the cells, *histozoic*; and when in cavities, *coelozoic*.

Classification of the Sporozoa

The groups that are called sporozoa have certain characteristics in common but are not necessarily closely related. It seems best to separate them into three superclasses—Apicomplexa, Myxospora, and Microspora.

Apicomplexa

The superclass Apicomplexa contains one class, Telosporea, with three orders: (1) Gregarinida, (2) Coccida, and (3) Haemosporida. The members of these orders have an apical complex and produce spores that have neither polar capsule nor polar filament nor vesicular nuclei. The spores are produced at the termination of the life of the trophozoite.

Order Gregarinida

The gregarines are common parasites of insects, especially in the digestive tract and body cavity, and less common in other groups of vertebrates and invertebrates. They are, at first, intracellular but later often become free in cavities of the body. Here they sometimes grow to a comparatively enormous size. *Monocystis* has already been described (Fig. 2-40). Another type easily obtained for study is *Gregarina* (Fig. 2-42, B), which occurs in the intestine of grasshoppers, cockroaches, and mealworms. The sporozoites penetrate the epithelial cells of the intestinal wall and the trophozoites that develop from them are at first intracellular; later the trophozoites break out of the epithelial cell to which they are attached for a time by the head or *epimerite*; this, the *cephalont* stage, consists of two parts: the posterior *deutomerite*, which contains the nucleus, and an anterior *protomerite* and *epimerite* (Fig. 2-42, A). When the cephalont becomes detached from the cell, it loses its epimerite and is then known as a *sporont*. Sporonts are motile by an unknown gliding mechanism. The sporonts within the intestine unite end to end, a condition known as *syzygy* (Fig. 2-42, B). Two sporonts conjugate and surround themselves with a wall, thus forming

Figure 2-42   Telosporea. (A) Gregarinida. *Leidyana erratica* attached to intestinal cells of cricket. (B) Gregarinida. *Gregarina blattarum* of the cockroach; two animals attached end to end in syzygy. (C) Coccida. *Isospora hominis* of humans; oocyst containing two spores, each with four sporozoites. (*After various authors.*)

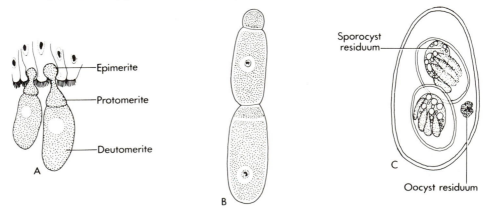

a cyst; they are gametocytes, each of which produces a large number of gametes. The gametes copulate in pairs, thus becoming zygotes, and secrete sporocysts and hence are spores. Within each spore, eight sporozoites are produced. The sporozoites, that are liberated with the aid of digestive enzymes in the intestine of the hosts that have ingested the spores, bring about new infections.

Order Coccida

The Coccida are sporozoa whose life cycle includes both schizogony and sporogony and is passed in a single host. They are parasitic in vertebrates, centipedes, mollusks, insects, annelids, and flatworms. The life cycle of a typical species, *Eimeria stiedae* of the rabbit, is illustrated in Figure 2-43.

The sporozoites that escape from the spores in the intestine of the host (A) penetrate the intestinal wall to reach hepatic portal blood vessels and are swept with the blood to the liver, where they enter the epithelial cells of the bile ducts and become trophozoites (B); each

Figure 2-43 Coccida. *Eimeria stiedae,* life cycle. (A) Sporozoite. (B–D) Cycle of schizogony in the epithelial cells of the bile ducts of the rabbit. (E–J) Gamete production. (K–M) Sporogony.

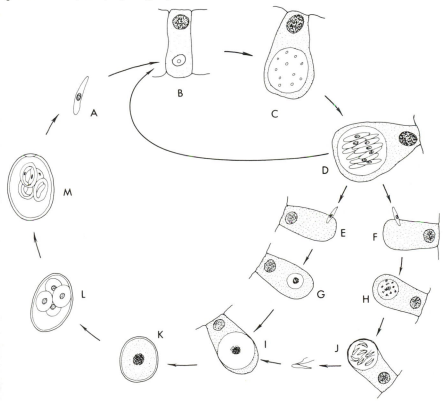

trophozoite is a schizont, which produces a number of daughter merozoites asexually (C, D); this multiplicative process is known as schizogony. The merozoites may enter other epithelial cells, becoming trophozoites, and pass through another period of asexual reproduction (B, C, D), or, after the penetration of epithelial cells, may produce merozoites, which are gametocytes; these penetrate other epithelial cells (E, F), where they develop into macrogametocytes (G, I) or microgametocytes (H, J). Each merozoite produces either one large macrogamete (I) or a large number of microgametes (J). One microgamete copulates with each macrogamete, a process that may be considered fertilization, thus producing a zygote. A wall forms about the zygote, and the body is then known as an oocyst (K). The protoplasm within divides to form four sporoblasts (L), each of which forms a sporocyst about itself, thus becoming a spore (M). Within each spore, two sporozoites develop.

Coccida in Humans. Four species of Coccida have been found in humans. *Isospora hominis* (Fig. 2-42, C) is the most common of the three species of *Isospora,* and all are more common in the tropics. They are apparently parasites of the intestinal mucosa and rarely produce serious disease. Knowledge of them is largely confined to the oocysts, which are present in the feces of infected individuals for a limited time. In most cases they produce a diarrhea of about 2 weeks' duration.

One of the most remarkable recent changes required in classification is the merger of the old order Toxoplasmida into the Coccida. In 1970, Frenkel, Dubey, and Miller, and Sheffield and Melton, both published reports on the coccidian nature of *Toxoplasma,* which produces coccidian oocysts only in a cat host. Infection with this intracellular parasite can be via ingested oocysts from cat litter or merozoites from the flesh of many animals. The most serious human infections are congenital and other early infections that are often fatal to the newborn or cause serious problems in the nervous system. Adult infection is often subclinical. About one third of people tested in the United States have antibodies indicating a prior infection with *T. gondii* (Fig. 2-44).

Sarcocystis apparently has a life cycle also requiring a cat for oocyst formation (Wallace, 1973). However, the trophozoite in other hosts is very large. In this genus the alternation of schizogony and sporogony is obligatory (Mehlhorn and Heydorn, 1978).

Coccida in Lower Animals. The Coccida most easily obtained for study are those in the rabbit, *Eimeria stiedae* (Fig. 2-43). Oocysts of this species may be found in the feces of a large proportion of these animals. They are in an unsporulated stage when passed but sporulation may be observed if the material is placed in a 5 percent aqueous solu-

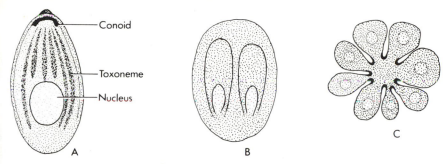

Figure 2-44 *Toxoplasma gondii.* (A) Diagram of *Toxoplasma*. (B) Internal development of two filial cells. (C) Rosette formation of eight filial cells. (*Adapted from electron micrographs by Gavin, Wanko, and Jacobs.*)

tion of potassium bichromate to inhibit the growth of bacteria. Development into sporoblasts and the formation of sporozoites takes place in about 3 days.

Eimeria schubergi is a parasite of centipedes of the genus *Lithobius*. It occurs in the intestine and has a life cycle similar to *E. stiedae*.

Other species of *Eimeria* occur in various species of mammals, birds, reptiles, amphibians, and fish. All of these are pathogenic if their life cycles are similar to that of *E. stiedae* in the rabbit, but most of them do not injure the host very severely. *E. stiedae* frequently brings about the death of rabbits, however, and *E. zurnii* is a cause of diarrhea in cattle. The *Eimeria* of birds, *E. tenella*, is also lethal under certain conditions, especially when epidemics occur among young chickens. Amoeboid host cells seem to aid the sporozoites in reaching the site where development occurs.

Coccidia of the genus *Isospora* also occur in vertebrates of all classes. Cats and dogs are commonly infected with *I. felis* and *I. rivolta*. Other species occur in birds and cold-blooded vertebrates. *Toxoplasma* is closely related to this genus.

The life cycle of the sarcosporidium of rats and mice, *Sarcocystis muris*, is the best known. Spores ingested by mice hatch in the intestine, and liberate amoebulae that penetrate the cells of the intestinal epithelium; here the trophozoites grow and multiply by schizogony. Apparently, the merozoites migrate to the muscles where they become located in the fibers; here growth results in a multinucleate plasmodium which divides by plasmotomy. In the course of time the masses of parasites form long, slender, cylindrical bodies with pointed ends, known as "Miescher's tubes," within which immense numbers of sickle-shaped spores are formed.

A few cases of sarcosporidiosis have been reported from humans; the organisms are found in the muscle usually at autopsy (Fig. 2-47), E). In most cases no serious results are brought about by the infection,

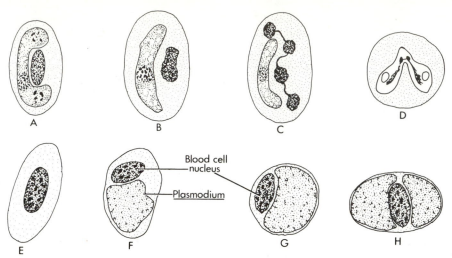

Figure 2-45 Sporozoa of the blood. (A) *Haemoproteus* in the red blood corpuscle of a bird. (B) *Lankesterella* in the red blood corpuscle of a frog. (C) *Karyolysus* in the red blood corpuscle of a frog. (D) *Babesia canis* in the red blood corpuscle of a dog. (E) A normal red blood corpuscle of a bird. (F–H) *Plasmodium relictum* in the red blood cells of a pigeon; (H) shows two parasites. (*A, B, and C, after Hegner; D, after Nuttall; E, F, G, and H, original.*)

although *S. muris,* which spreads throughout the entire body, brings about death in mice, and death sometimes also occurs in sheep as a result of heavy infections.

Haemogregarines are Coccida that occur in the blood of frogs as well as other amphibians, reptiles, birds, and fish. Their life cycles involve a blood-sucking leech or arthropod as a second host. The species *Lankesterella minima* (Fig. 2-45, B) is transmitted to frogs by the leech, *Placobdella. L. minima* sporozoites, when first introduced, penetrate endothelial cells of the frog and produce a generation or more of merozoites followed by a production of macrogametes and microgametes, which unite to form a zygote. The zygote develops into an oocyst within which the sporozoites are formed. When the sporocysts are released, they enter the red blood cells. In the red blood cell the sporozoite develops into the trophozoite, which, as it grows, pushes the host cell nucleus to one side (Fig.2-45, B). *Karyolysus* (Fig. 2-45, C) causes the nucleus of the parasitized red cell to break up into several pieces. The invertebrate host is a mite as is the invertebrate host of *Hepatozoon muris. H. muris* gametocytes develop in white blood cells of the rat and complete the sexual phase of reproduction in a blood-feeding mite. The cycle is completed when a rat eats an infected mite.

Order Haemosporida

The Haemosporida, as the name implies, are sporozoa that live in the blood. They penetrate the blood cells of vertebrates, where they pass through schizogony and go through part of their life cycle in invertebrate hosts, where sporogony occurs. The most important species are the malarial parasites of humans. Representatives of a number of genera may be obtained from common lower animals, for example, *Haemoproteus* from birds; piroplasmas and anaplasmas from cattle suffering from Texas fever; and *Plasmodium* from birds.

Haemoproteus (Fig. 2-45, A) is a genus common in birds, turtles, snakes, and lizards. The best known species is *H. columbae,* a parasite of the common pigeon that has been reported from various parts of the world. The life cycle of this species includes asexual reproduction in the endothelial cells of the blood vessels of the pigeon's lungs and other organs, and the development of male and female gametocytes. The transmitting agent in which sporozoites are formed is a hippoboscid fly, *Lynchia.* Within the red cells the organism grows partly around the red cell nucleus in the form of a halter, hence the term "halteridium" often applied to members of this genus.

Piroplasmas of the genus *Babesia* undergo sporogony after ingestion of the infected blood cells (Fig. 2-45, D) of a mammal by a tick. *B. bigemina* ookinetes penetrate the eggs of ticks where they develop along with the eggs to produce eventually ticks with infected salivary glands containing sporozoites which may infect cattle, the mammalian hosts, when they are bitten by ticks. Piroplasmas are sometimes classified with the Sarcodina or considered of uncertain relationship. Their life cycle is quite comparable to many of the other Haemosporida, although budding or binary fission occurs rather than schizogony. Rudzinska and Trager (1962) have found that the cytoplasm, in electron micrographs, is similar to that of *Plasmodium berghei,* although only a single plasma membrane covers *Babesia* in contrast to two for *Plasmodium,* and ingestion of blood cell cytoplasm seems to be important in the nutrition of both. As suggested earlier, these may be derived from kinetoplastid flagellates.

Malarial parasites of the genus *Plasmodium* occur in lower vertebrates as well as in humans. They have been recorded especially from birds, bats, squirrels, and lizards. The species in birds are of particular interest because they were used by Ross in working out the sexual cycle of malarial parasites in mosquitoes. They are also of great value for purposes of teaching and investigation since they can be transmitted easily from sparrows to captive canaries by blood inoculation, and when the latter are once infected they appear to remain infected throughout their lives. The method of procedure is to prick a vein in the leg and suck up a drop of blood into a syringe containing normal saline solution. This is then injected into the breast muscle or peritoneal cavity of a fresh bird. The average length of the prepatent period is about 5 days. The period of rise in the number of parasites is also about 5 days when from 10 to 5,000 parasites per 10,000 red cells are present. Then a rapid fall in the number of parasites occurs and no more can be found in the blood except after long search. That the birds remain infected is evident from the fact that they suffer relapses, much as humans do, and also by the fact that blood from the birds is infective to fresh birds. Bird malaria is of value for the study of therapeutic agents. *Plasmodium berghei,* which can be maintained in the rat, has been found to ingest the cytoplasm of the host red blood cell in a manner much like food vacuole formation in amoebas.

Myxospora

The Myxospora includes the sporozoa that produce spores in each of which are one to four *polar capsules* and in each polar capsule a coiled *polar filament*. The one order is the Myxosporida. Because of their complex structure, some zoologists think they may represent degenerate metazoans. There are interesting parallels between the polar capsule and coelenterate nematocysts. Both have filaments that seem to be discharged by turning inside out.

Order Myxosporida

The Myxosporida (Fig. 2-47, A and B) are principally parasites of fish. The life cycle is comparatively simple, since there is no intermediate host. The principal organs of the host invaded by the parasite are the gallbladder, gills, kidney, urinary bladder, muscle, integument, connective tissue, spleen, and ovary. The spores are complicated structures (Fig. 2-46). When swallowed by a new host, the polar filaments are thought to turn inside out and fix the spore to the intestinal wall. The two nuclei of the sporoplasm fuse before it hatches to invade the host tissue. Growth to a multinucleate mass occurs and new spores develop within this vegetative stage from portions termed *pansporoblasts* in which several nuclei develop, two of which form the nuclei of the gametes which will form the new sporoplasm. Cells formed by other nuclei of the pansporoblast and their associated cytoplasm are responsible for formation of the polar capsules and valves. The Myxosporida

Figure 2-46 Myxosporida. Diagram of a spore before fusion of the sporoplasm nuclei. (A) Front view. (B) Side view.

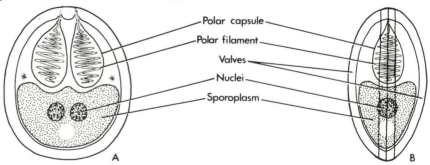

are often parasites with a mild effect on the host, but severe infections may bring about the death of the host.

Microspora

The Microspora are sporozoa lacking an apical complex. Nuclei are vesicular as in the Myxospora, but polar filaments when present are borne by very simple spores. Hosts for both orders are lower vertebrates and invertebrates.

Order Microsporida

The Microsporida (Fig. 2-47, C and D) are characterized by the presence of spores that are extremely small and usually possess only one polar capsule. The hosts of most Microsporida are arthropods; the rest belong to other groups of invertebrates and to the cold-blooded vertebrates. Among the invertebrate hosts are included three species of sporozoa and one ciliate. Certain of the Microsporida are of great economic importance because they bring about the death of animals of value to humans. For example, the species *Nosema bombycis* causes a chronic disease in silkworms known as pebrine. This species is generally distributed throughout the tissues of the host, including the eggs developing in the ovary. These eggs are deposited by the silkworm moth and the larvae that hatch from them are thus infected by so-called "hereditary" transmission. Pasteur was able to control silkworm disease, which once threatened to destroy the silk industry of France, by eliminating infected eggs which he was able to recognize with the aid of a microscope. Another important species economically is *Nosema apis* (Fig. 2-47, C) which causes nosema disease in honey bees. Infection is brought about by the ingestion of spores and is limited to the digestive tract. Other species of Microsporida are to a certain degree

Figure 2-47 Sporozoa. (A) Myxosporida; *Myxidium lieberkuhni,* vegetative stage, from pike. (B) Myxosporida; *Myxidium lieberkuhni,* spore from pike. (C) Microsporida; *Nosema apis,* spores in the cells of the stomach of the honey bee. (D) Microsporida; *Thelohania magna,* young spore from a mosquito larva. (E) Coccida; spores of *Sarcocystis* in human muscle fibers. (*After various authors.*)

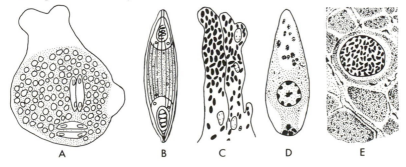

beneficial because they attack harmful insects, for example, the larvae of several species of anopheline mosquitoes that are carriers of human malaria may be infected.

Order Haplosporida

The Haplosporida are parasites of lower vertebrates and invertebrates. The genus *Haplosporidium* occurs in marine and freshwater annelids and mollusks. *Ichthyosporidium* is parasitic in fish. *Bertramia* parasitizes aquatic worms and rotifers.

Superclass Ciliata

TETRAHYMENA PYRIFORMIS

Tetrahymena pyriformis is an inhabitant of fresh water in many parts of the world. It is easily cultured in the laboratory and can be grown in media containing only chemically defined nutrients. Cultures maintained in nutrient media with no other living organisms are said to be maintained *axenically*. Because these ciliates can be grown in pure culture and in chemically defined media, they have been very popular and useful in biological research at the cellular and molecular level. *Tetrahymena* was the first animal to be grown axenically.

General Morphology

T. pyriformis (Fig. 2-48) is an egg-shaped, uniformly ciliated, small, plastic protozoan. The *buccal cavity* is pyriform, much like the shape of the entire cell, and is located near the anterior end. About 20 rows of *cilia* run longitudinally the length of the animal, with two rows originating postorally. The *cytoproct* (cell anus) is located near the posterior in one of the postoral ciliary rows. Two contractile vacuole pores are located several ciliary rows away near the level of the cytoproct. Three membranelles and an undulating membrane are in the buccal cavity. An ovoid *macronucleus* is centrally located. Near the macronucleus is usually a smaller nucleus, the *micronucleus*, which contains the genetic material involved in sexual phenomena. Only ciliates have a macronucleus and a micronucleus, that is, two sizes of nuclei in the same cell. A single *contractile vacuole* is located posterior to the nuclei. Various vacuoles and granules may be visible internally.

Cytology

Ciliates stained with silver stains show the granule at the base of a cilium as a darkly staining object, the *kinetosome*. The kinetosomes are self-duplicating bodies from which the cilium develops. In *Tetrahymena*, the kinetosomes are arranged in longitudinal rows. A fiber arises from each kinetosome and passes to the right and then an-

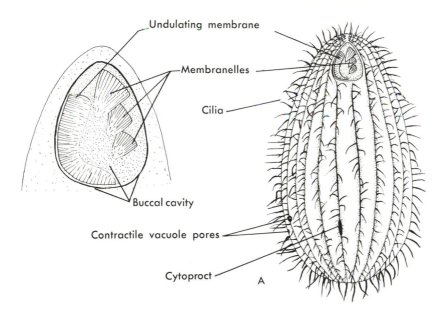

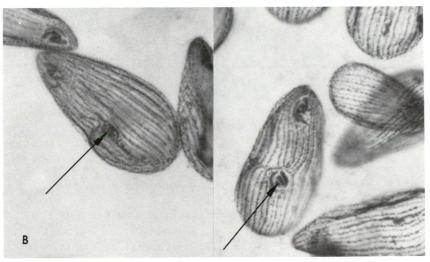

Figure 2-48 *Tetrahymena pyriformis*. (A) Diagram of general anatomy. (B) Organization of kinetosomes in new oral region prior to fission.

teriorly as it closely parallels other fibers from other kinetosomes in the row. This set of parallel fibers is known as the *kinetodesma*. The kinetodesma and the associated row of kinetosomes is a *kinety*. Additional fibers (Fig. 2-49) also embedded in the ectoplasm have been described by Pitelka (1961). The function of the various fibers is not known. They may have a supporting or developmental function. Early speculation that they were important in ciliary control seems discredited (Naitoh and Eckert, 1969) in view of the role played by mem-

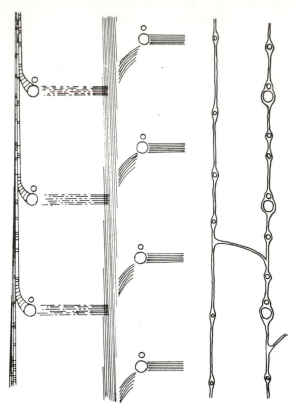

Figure 2-49 Kineties of *Colpidium campylum*. Diagrammatic surface views of three kineties showing features similar to those of *Tetrahymena pyriformis*. Cilia are indicated as empty circles, with parasomal sacs shown as smaller circles anterior to them. Left kinety shows kinetodesmal fibers. Center kinety shows longitudinal, transverse, and postciliary fibers. Right kinety shows primary and secondary meridians of pellicle with small circles indicating protrichocyst attachments. *(From D. R. Pitelka. Fine structure of the silverline and fibrillar systems of three tetrahymenid ciliates. J. Protozool. 8 (1):75–89. Copyright 1961 by the Society of Protozoologists.)*

brane potentials in such control. The cilium (Fig. 2-2) has an ultra-structure similar to the flagellum of the flagellates. Nine peripheral fibers surround two central fibers. These fibers are embedded in an electron transparent material and the surface of the cilium is a membrane continuous with the surface membrane of the cell. The peripheral fibers of the cilium, but not the central fibers, extend into the kinetosome (Fig. 2-50). The buccal cavity of *Tetrahymena* contains three *membranellae* on the left and an *undulating membrane* on the right. These are formed from closely associated sets of kinetosomes from which ciliary fibers that have undergone a degree of fusion arise. In the bottom of the buccal cavity is the *cytostome*. Prior to division a

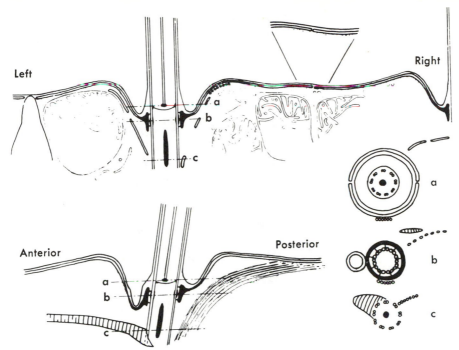

Figure 2-50 Diagrams of cortical structure of *Colpidium campylum*. Top, transverse section; bottom left, longitudinal section; bottom right, three sections through ciliary base at points indicated. Section *a* shows transverse fibrils at bottom, section *b* shows parasomal sac at left, and section *c* shows kinetodesmal fiber to left with postciliary fibrils at right. (*From Dorothy R. Pitelka. Fine structure of the silverline and fibrillar systems of three tetrahymenid ciliates. J. Protozool. 8(1):75–89. Copyright 1961 by the Society of Protozoologists.*)

new mouth arises posterior to, and some distance from, the old mouth but in association with the same kineties.

A *micronucleus,* when present, is usually closely associated with the *macronucleus.* Elliott and others have studied *Tetrahymena* from many parts of the world and found that those lacking a micronucleus are very common. Such *amicronucleate* strains are incapable of sexual reproduction. The macronucleus is centrally located and not visible in unstained specimens. Both nuclei stain with stains that are specific for DNA.

The *contractile vacuole,* located at the posterior, is a single structure, which apparently fills through a system of microtubules that extend into part of the cell (Fig. 2-51). Contractile vacuole pores have a well-defined structure in electron micrographs (Fig. 2-52).

Mitochondria (Figs. 2-51 and 2-53) are of the microtubular type and most abundant near the ectoplasm. *Food vacuoles* circulate throughout the endoplasm in those which are fed on particulate matter, but no food vacuoles are known to be formed in those grown in nutrient solu-

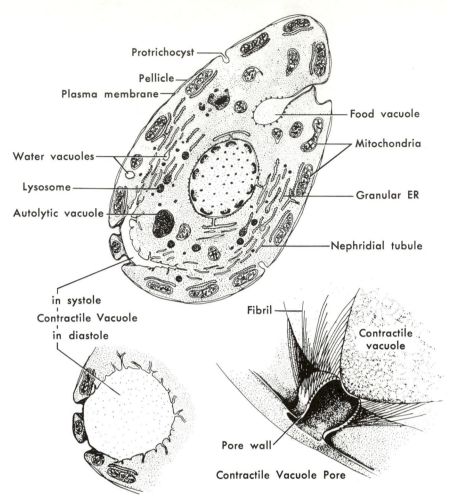

Figure 2-51 Schematic representation of the nephridial system and related structures. The contractile vacuole is shown in systole and diastole. The pore with its associated fibrils and contractile vacuole is shown in three dimensions. Other structures, such as protrichocysts and autolytic vacuoles, are shown. (*From A. M. Elliott and I. J. Bak. The contractile vacuole and related structures in* Tetrahymena pyriformis. *J. Protozool.* **11**(*2*):*250–261. Copyright 1964 by the Society of Protozoologists.*)

tions. Rassmussen and Orias (1975) have developed a mutant strain lacking food vacuoles but able to grow. Lipid droplets (Fig. 2-56, A) are most abundant in the anterior of the cell. Other inclusions are sometimes present, as well as sites of enzyme activity which can be localized by histochemical studies. Various studies such as those of Fennell and Marzke (1954) have shown that such enzymes as the phosphatases vary in abundance and distribution with age of cultures and different culture solutions.

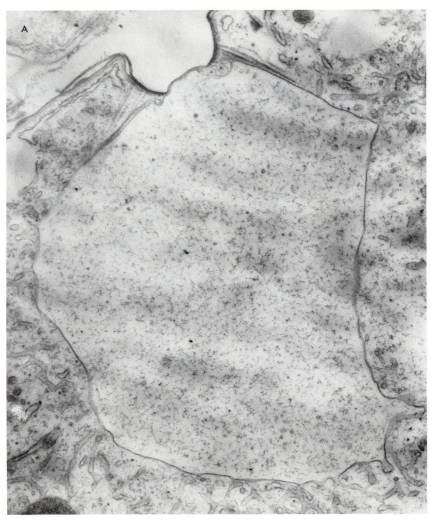

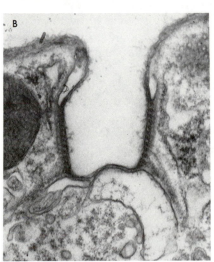

Figure 2-52 Contractile vacuole of *Tetrahymena*. (A) Electron micrograph of the vacuole in diastole, pore to upper end. × 11,250. (B) Vacuole in systole, below pore which is in center. × 18,000. (*From A. M. Elliott and I. J. Bak. The contractile vacuole and related structures in* Tetrahymena pyriformis. J. Protozool. *11(2):250–261. Copyright 1964 by the Society of Protozoologists. Courtesy of Dr. Alfred M. Elliott.*)

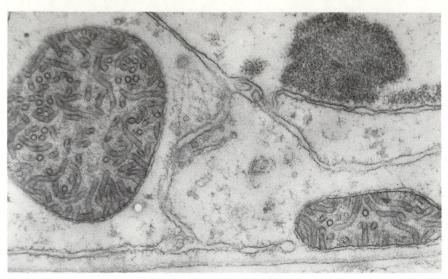

Figure 2-53 Electron micrograph of *Tetrahymena*. Sections of two microtubular type mitochondria are visible. The outer membrane of the macronucleus (upper right) is seen to be continuous with the granular endoplasmic reticulum. × 27,000. (*From A. M. Elliott and I. J. Bak. The contractile vacuole and related structures in* Tetrahymena pyriformis. J. Protozool. *11(2):250–261. Copyright 1964 by the Society of Protozoologists. Courtesy of Dr. Alfred M. Elliott.*)

Physiology

Tetrahymena is of considerable value in physiological research, because it can be grown in pure culture in chemically defined media. The fact that it has a cellular metabolism similar in many ways to the cellular metabolism of man makes it a useful tool. It can utilize nutrients by glycolysis and the tricarboxylic acid cycle. It requires the same amino acids (Fig. 2-54) as those required by humans.

Gibbons and Rowe (1965) have isolated a protein from *Tetrahymena* cilia which has adenosine triphosphatase activity. It may be a contractile element of cilia, but it is different from contractile proteins such as actin and myosin. They have suggested the term *dynein* as a name for the protein.

Viable *Tetrahymena* have been recovered from cells frozen in a protective solution to −20°C, then stored 3 months at the temperature of liquid nitrogen.

Tetrahymena exhibits very sensitive growth responses, either an increase or decrease, to many treatments and substances. Such sensitivity makes them potentially valuable organisms for biological assay of nutrients and other substances.

Growth in numbers of cells following inoculation into fresh media is very rapid (Fig. 2-55). Following a brief lag phase of adjustments to

Amino acids	
Arginine	150
Histidine	110
Isoleucine	100
Leucine	70
Lysine	35
Methionine	35
Phenylalanine	100
Serine	180
Threonine	180
Tryptophane	40
Valine	60

Nucleic Acids	
Adenylic acid	25
Cytidylic acid	25
Guanylic acid	25
Uracil	25

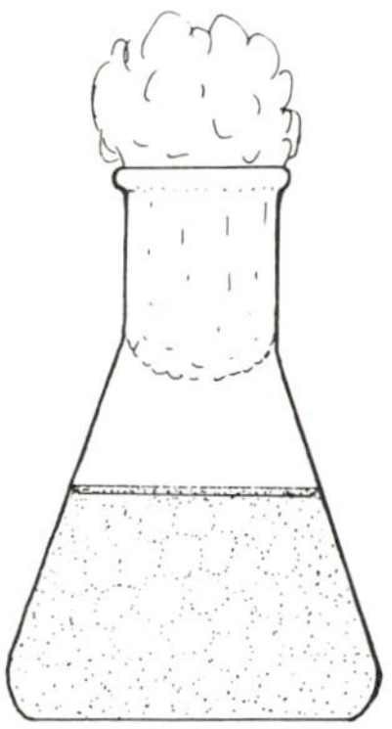

Growth factors	
Ca pantothenate	0.1
Folic acid	0.01
Niacin	0.1
Pyridoxine HCl	2.0
Riboflavin	0.1
Thiamine HCl	1.0
Thioctic acid	0.001

Carbon source	
Glucose	1000.
Sodium acetate	1000.

Inorganic salts	
K_2HPO_4	100.
$MgSO_4 \cdot 7H_2O$	10.
$Zn(NO_3)_2 \cdot 6H_2O$	5.
$FeSO_4 \cdot 7H_2O$	0.5
$CuCl_2 \cdot 2H_2O$	0.5

Figure 2-54 Axenic culture. Growth of *Tetrahymena pyriformis* can be maintained in fortified peptone broth and a number of other bacteria free media. Chemically defined media will also support growth. The ingredients listed above are for Elliott's modification of Kidder and Dewey media. Numerical values are in mg/liter of aqueous solution. The diagrammatic culture indicates the honeycombed pattern frequently assumed by the ciliates as they move in constantly changing formation. Such formations are most noticeable at high population levels.

new culture conditions, a logarithmic growth phase occurs in which the population growth is very rapid until *limiting factors* such as depletion of nutrients, lack of oxygen, or accumulation of waste products slows the rate of increase. Eventually the population dies out after a long period in which the numbers decline because cell deaths exceed cell divisions. Such laboratory growth patterns show interesting parallels to growth phenomena in natural populations. However, the period of decline found in most laboratory situations would be a stable, but fluctuating, level of population size in natural environments due to continuing replenishment of the food supply. Evident in culture conditions is how the unchecked population growth of the logarithmic and following growth phase tends to overshoot the population level, which can be maintained by the culture conditions at that time. Such population growth without checks in the early phases would be rare under natural conditions.

Cells in the logarithmic growth phase can be induced to divide synchronously (Fig. 2-56) with the other cells in the population by a series of temperature shocks (Scherbaum and Zeuthen, 1954). This has simplified the problem of studying the biochemistry of dividing cells.

Taxonomy

Corliss (1954) found that *Tetrahymena* was often referred to as *Glaucoma* or *Colpidium*. Using the infraciliature, especially distribution of

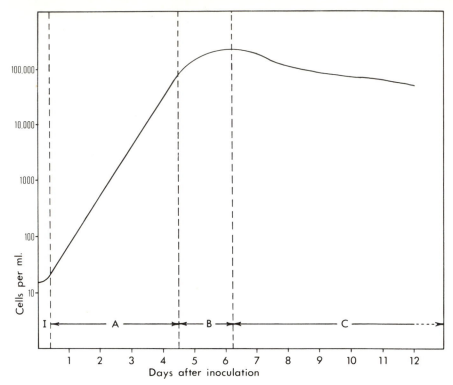

Figure 2-55 Growth curve of *Tetrahymena* populations. *I,* lag phase following inoculation into fresh media. (A) Logarithmic growth phase. (B) Negative acceleration of growth. (C) Period of declining population size.

Figure 2-56 *Tetrahymena pyriformis.* Stained to show lipid droplets. (A) From a culture past the logarithmic phase of growth; (B) and (C) from a culture in the logarithmic growth phase subjected to temperature shocks to synchronize cell divisions. (B) Prior to onset of division. (C) During peak of cell divisions. The darkly stained lipid droplets visible in the anterior of cells in (A) are nearly absent in (B) and (C). Many other cytoplasmic materials not stained in these preparations also have a specific pattern of localization with abundance varying with culture conditions.

100

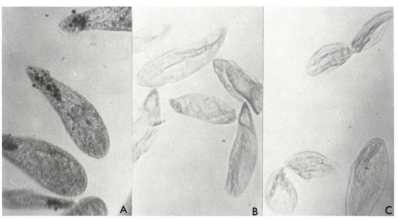

kinetosomes, as well as other characteristics, many strains of *Tetrahymena* are referable to the species *T. pyriformis*. However, reproductive studies of mating types indicate that the species includes a number of reproductively isolated populations. These different populations have been designated as varieties and are comparable to species in other groups. Because they cannot be distinguished by morphological criteria, it is convenient to consider them all *T. pyriformis,* but because they do have physiological differences it is important that experimentalists designate the source of their material when reporting the results of experiments on *Tetrahymena* so that more meaningful comparisons can be made.

Electrophoretic mobility patterns of isozymes have the potential for clarifying the relationships of strains (Borden, Whitt, and Nanney, 1973).

Genetics

Tetrahymena typically has a diploid micronucleus. X rays have been used to induce the development of ones with a haploid micronucleus. *Conjugation* occurs in which micronuclear material is exchanged by two cells of different mating types. Mating types are similar to sexes, although some varieties have numerous mating types. The phenomena of conjugation will be discussed in greater detail for *Paramecium*. Macronuclei are more important in determining phenotype of a cell than micronuclei although the latter determine the characteristics present in the macronucleus following conjugation. That is because the old macronucleus degenerates during conjugation and a new one is formed by the new micronuclei following conjugation. Varieties lacking a micronucleus are common but they divide only by binary fission. In binary fission the macronucleus is thought to divide by *amitosis,* pinching in two with half going to each of the new daughter cells. Because there are many sets of chromosomes in the macronucleus each daughter cell usually gets some of all the needed chromosomes so that more of the same can be formed.

Elliott, Addison, and Carey (1962) found one-third amicronucleate strains in 28 samples containing *Tetrahymena* from 188 samples of European freshwater locations. Several different varieties were found, some of which also occur in America.

PARAMECIUM

Paramecium lives in freshwater ponds and streams and is very easily obtained. Cultures prepared for *Amoeba* will in most cases sooner or later contain a host of paramecia.

General Morphology

If a drop of water containing *Paramecium* (Fig. 2-57) is placed on a slide, the animals may be seen with the naked eye moving rapidly

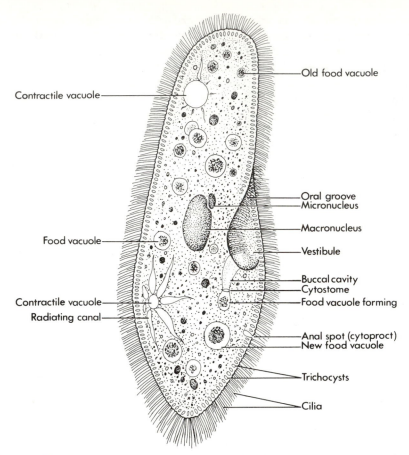

Contractile vacuole

Old food vacuole

Oral groove
Micronucleus

Macronucleus

Vestibule

Food vacuole

Buccal cavity
Cytostome
Food vacuole forming

Contractile vacuole
Radiating canal

Anal spot (cytoproct)
New food vacuole

Trichocysts

Cilia

Figure 2-57 *Paramecium* viewed from the side. (*Modified after several authors.*)

from place to place. Under the microscope they appear cigar-shaped. Close inspection reveals a depression extending from the end directed forward in swimming, obliquely backward and toward the right, ending just posterior to the middle of the animal. This is the *oral groove.* The *vestibule* is a depression in the posterior part of the oral groove which opens into the curving, funnel-like *buccal cavity* (gullet). The oral groove and vestibule have cilia much like those covering the general surface of the cell. The buccal cavity contains organelles of modified cilia somewhat resembling the structures in the buccal cavity of *Tetrahymena.* The buccal cavity of *Paramecium* passes obliquely downward and posteriorly and opens through the *cytostome* into the endoplasm. The oral groove gives the animal an unsymmetrical appearance. Because *Paramecium* swims with the slender but blunt end foremost, we are able to distinguish this as the *anterior end.* The opposite end, which is thicker but more pointed, represents the *posterior end*, whereas the side containing the oral groove may be designated as *oral* or *ventral*, the opposite side *aboral* or *dorsal*. The motile organs are fine threadlike *cilia* regularly arranged over the surface. Two layers of cytoplasm are visible, as in *Amoeba*, an outer comparatively thin clear area, the *ectoplasm*, and a central granular mass, the *endoplasm*. Besides these a distinct *pellicle* is present outside of the ectoplasm. Lying in the ectoplasm are a great number of minute sacs, the *trichocysts*, which discharge long threads to the exterior when properly stimulated. One large *contractile vacuole* is situated near either end of the body, close to the dorsal surface, and a variable number of *food vacuoles* may usually be seen. The *nuclei* are two in number, a large *macronucleus* and a smaller *micronucleus* (some species of *Paramecium* have more than one micronucleus); these are suspended in the cytoplasm near the mouth opening. The *anal spot,* or *cytoproct,* can be observed in living cells only when solid particles are discharged. It is situated just behind the posterior end of the oral groove.

Cytology

The *endoplasm* of *Paramecium* occupies the central part of the body. It appears granular and most of the larger granules are shown by cytochemical reactions to be reserve food particles; they flow from place to place, indicating that the protoplasm is quite fluid. The *ectoplasm* does not contain any of the large granules characteristic of the endoplasm, because its density prevents their entrance. In this respect the two kinds of cytoplasm resemble the ectoplasm and endoplasm of *Amoeba*. If a drop or two of 35 percent alcohol is added to a drop of water containing paramecia, the pellicle will be raised in some of the specimens in the form of a blister. Under the higher powers of the microscope a great number of hexagonal areas can be seen on the surface. The hexagons (Fig. 2-58) are formed by the sides of raised areas around the cilia. The *cilia* (Figs. 2-58 and 2-60) arise from basal granules called *kinetosomes*. The kinetosomes have a regular distribution (Fig. 2-59), except in the buccal cavity, where they are grouped at the bases of three long membranelles and an undulating membrane. From each body kinetosome a fiber passes to the right and anterior to partly parallel similar fibers from other kinetosomes in the same row. These fibers form the *kinetodesma* (Fig. 2-58, A). The kinetodesma and its associated kinetosomes are the *kinety*.

The cilium itself has an ultrastructure similar to the ultrastructure of the flagellum. A "9 + 2" pattern describes the nine double peripheral fibers and two central fibers, which are found in the ciliary cross section (Fig. 2-58, D). The membranelles and undulating membranes of ciliates which have them differ a great deal from the undulating membrane of such flagellates as *Trypanosoma*. The undulating membrane of *Trypanosoma* is powered by the flagellum, which runs through the dis-

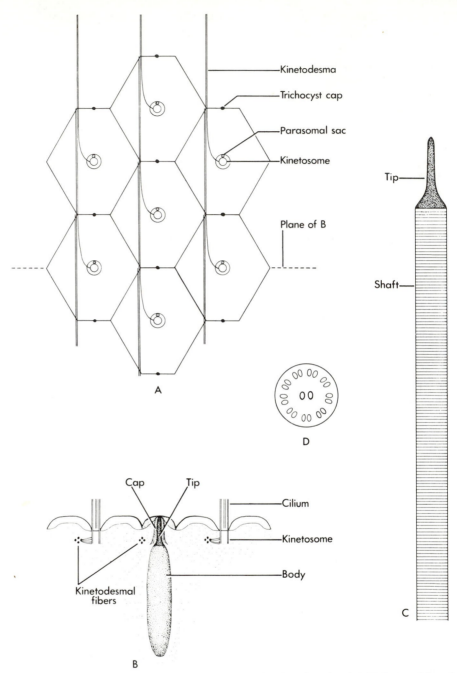

Figure 2-58 *Paramecium.* Diagrams of pellicular details. (A) Relationship of surface hexagons to kineties and trichocysts. (B) Cross section of pellicle showing undischarged trichocyst. (C) Discharged trichocyst. (*Modified after several authors.*)

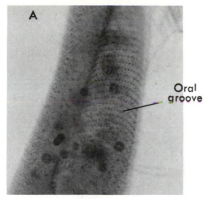

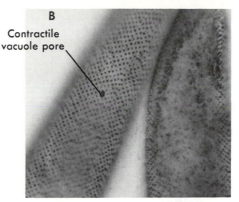

A

Oral
groove

B

Contractile
vacuole pore

Figure 2-59 *Paramecium:* Silver-stained to show the pattern of kinetosomes in the pellicle. Note the transverse rows in the oral groove.

tal end of the membrane, whereas the undulating membrane of ciliates is formed by the close association and fusion of numerous cilia. Cilia develop from the kinetosomes which are self-duplicating bodies.

Interspersed with the kinetosomes of the cilia is a layer of elongate, sacklike bodies with their long axes extending about 4 μm perpendicular to the surface. These are *trichocysts* (Fig. 2-58). During development in a vesicle in the cytoplasm the trichocyst is filled with a layered material (Yusa, 1963). In the mature trichocyst the layering is lost and the structure is capped by an electron-dense part resembling an inverted tack. With proper stimulation trichocysts discharge a filament tipped by the trichocyst cap. The filament is finely striated with the striations oriented across the filament in a direction opposite to that of the striations or layering of the immature trichocysts from which the filaments are formed. The filament extruded from the trichocyst is about eight times the length of the trichocyst. Trichocysts are supposed to function as organs of defense. In some instances they have been reported to provide anchorage to the substrate. Evidence that the trichocysts are weapons of defense is furnished when *Paramecium* encounters another ciliate *Didinium*. If the seizing organ of this protozoan becomes fastened in the *Paramecium*, a great number of trichocysts near the place of the injury are discharged (Fig. 2-61). These produce a substance that becomes jellylike on entering the water; this tends to force the two animals apart, and, if the *Paramecium* is a large one, it frequently succeeds in making its escape.

Two *contractile vacuoles* are present, occupying definite positions,

Figure 2-60 Schematic representation of the beating of a row of cilia. (*After Verworn.*)

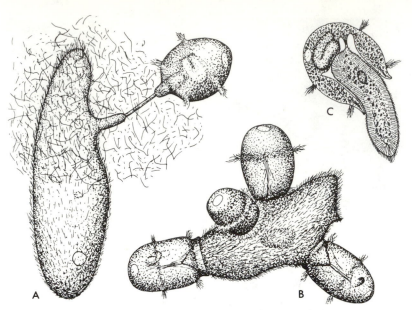

Figure 2-61 *Paramecium* attacked by the ciliate *Didinium nasutum*. (A) *Paramecium* discharges its trichocysts, thus forcing the attacking *Didinium* away. (B) Four small didinia attacking a *Paramecium*. (C) Section of a large *Didinium* swallowing a small *Paramecium*. (*After Mast.*)

one near either end of the body. They lie close to the dorsal surface and communicate with a large portion of the body through a system of *radiating canals,* 6 to 10 in number, which connect with still smaller microtubules that have been demonstrated with the electron microscope. The vacuoles grow in size by the addition of liquid excreted by the protoplasm into the canals and then poured into them. When the full size is reached, the contents are discharged to the exterior through the contractile vacuole pore (Fig. 2-62). The two vacuoles do not contract at the same time, but alternately; the interval between successive contractions is from 10 to 20 seconds.

In *Paramecium trichium* the two contractile vacuoles are permanent structures with vesicles that collect and pour fluids into them and with a long convoluted excretory tubule ending in an excretory pore opening on the aboral surface. The duration of the systole is long compared with that of the disastole due probably to the presence of the excretory tubule. What has been said of the function of the contractile vacuole in *Amoeba* applies as well to that of *Paramecium* (i.e., it acts as an organelle of excretion and respiration, and is probably hydrostatic). Most of the nitrogen excreted by *Paramecium* is in the form of urea, and this substance has been detected in the contractile vacuole. However, the greater part of the excretory matter, including urea, apparently passes by dialysis directly to the exterior through the pellicle. That the primary function of the contractile vacuole is to regulate the

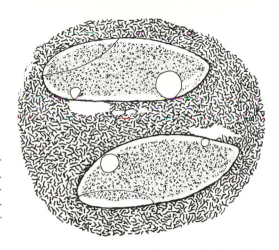

Figure 2-62 *Paramecium:* discharge of contractile vacuoles. The organisms are represented swimming in a solution of india ink. (*After Jennings.*)

water content of the protoplasm is indicated by the correlation between the frequency of pulsation and the rate water is taken in. For example, most long periods between pulsations, up to 6 minutes when actively swimming, occur when little water is ingested, and most short periods occur when the animals are at rest.

Coordination of the cilia as illustrated in Fig. 2-60 appears to be a membrane phenomenon. Calcium and magnesium ion levels are important in regulation with membrane depolarization due to chemical or mechanical stimulation of the anterior end causing beat reversal, whereas stimulation of the posterior and causes increased polarization.

Locomotion

The only movements of *Paramecium* that in any way resemble those of *Amoeba* are seen when the animal passes through a space smaller than its shorter diameter; it will then exhibit an elasticity which allows it to squirm through. In a free field *Paramecium* swims by means of its cilia. They produce a continuous rotation to the left and a simultaneous forward or backward movement. The movement is opposite to the direction that the cilia are inclined. The anterior end moves in an aboral direction due to the powerful beating of the oral cilia. The combination of this aboral swerving and the rotation to the left produces a spiral course of unchanging direction (Fig. 2-63). Rotation is thus effective in enabling an unsymmetrical animal to swim in a straight course through a medium which allows deviations to right or left, and up or down. It is well known that a human being cannot keep a straight course when lost in the woods, although he has a chance to err only to the right or left.

Nutrition

The *food* of *Paramecium* consists principally of bacteria and minute protozoa. The animal does not wait for the food to come within its

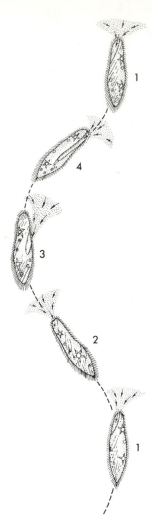

Figure 2-63 *Paramecium:* spiral path; views 1, 2, 3, and 4, successive positions occupied. The dotted areas with small arrows show the currents of water drawn from the front. (*From Jennings.*)

reach, but by continually swimming from place to place is able to enter regions where favorable food conditions prevail. The cilia also aid in bringing in food particles, because a sort of vortex is formed by their arrangement about the oral groove which directs a steady stream of water toward the mouth.

Figure 2-64 illustrates the formation of a *food vacuole*. Food particles that are swept into the mouth are carried down into the buccal cavity by the undulating membrane and membranellae; they are then moved onward by the buccal ciliature and are finally gathered together at the end of the passageway into a vacuole which gradually forms in the endoplasm. When this vacuole has reached a certain size, it is swept away from the extremity of the cytostome by the surrounding protoplasm, and the formation of another vacuole is begun (Fig. 2-65). A food vacuole is a droplet of water with food particles suspended

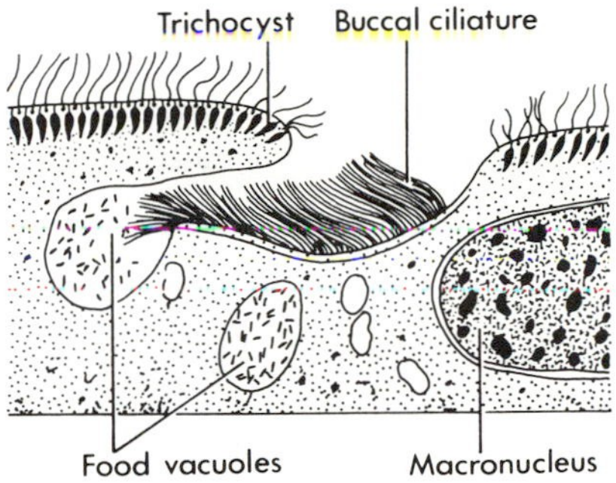

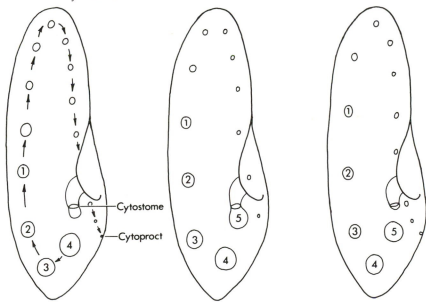

Trichocyst   Buccal ciliature

Figure 2-64 *Paramecium:* formation of food vacuole. Section through cytopharynx, showing the manner in which bacteria are swept into the food vacuole in the process of formation. (*After Maier.*)

Food vacuoles   Macronucleus

within it. As soon as one is separated from the cytostome, it is swept away by the rotary streaming movement of the endoplasm known as *cyclosis*. This carries the food vacuole around a definite course that begins behind the cytostome, passes backward to the posterior end, then forward near the dorsal surface to the anterior end, and finally downward and along the ventral surface toward the mouth. During this journey digestion takes place. At first the vacuole decreases in size as water is removed from it and the acidity in the vacuole increases. Before elimination the vacuole increases in size and the pH approaches the alkaline range. Undigested particles in the vacuole are discharged through the *cytoproct* (Fig 2-57), a structure that is closed at other times in its location on a kinety posterior to the oral groove. The

Figure 2-65 *Paramecium:* successive stages in the formation and movement (cyclosis) of food vacuoles. Numbered vacuoles are numbered in the sequence in which they were formed.

Cytostome

Cytoproct

processes of digestion, excretion, respiration, growth, and other aspects of metabolism are so similar to those described for *Amoeba* that they will not be considered further here. Wichterman (1953) gives a complete account of many aspects of the biology of *Paramecium*.

Reproduction

Paramecium reproduces by simple *binary fission*. This process is interrupted occasionally by a process of nuclear reorganization such as *conjugation, cytogamy,* or *autogamy*. Conjugation is a sexual process, which has been intensively studied.

Binary Fission. In binary fission the animal divides transversely (Fig. 2-66). The first indication of a forthcoming division is seen in the micronucleus, which undergoes mitosis, its substance being equally divided between the two daughter nuclei; these separate and finally come to lie one near either end of the body. Figure 2-66 shows two dividing micronuclei, because there are two of these in *Paramecium aurelia*. The macronucleus elongates and then divides transversely. The buccal cavity produces a bud, which develops into another buccal cavity, these two structures move apart, the old buccal cavity advancing to the ventral middle line of the forepart of the body, and the new one to a similar position in the posterior half. The buccal ciliature

Figure 2-66 *Paramecium aurelia:* binary fission. (*From Newman after Lang.*)

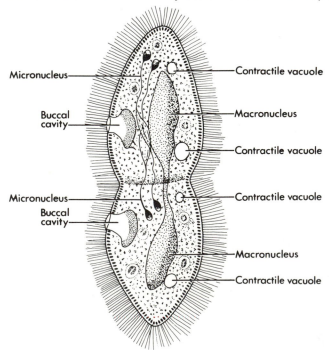

Micronucleus

Buccal cavity

Micronucleus

Buccal cavity

Contractile vacuole

Macronucleus

Contractile vacuole

Contractile vacuole

Macronucleus

Contractile vacuole

remains with the old buccal cavity, whereas new membranellae and an undulating membrane arise in connection with the new buccal cavity. A new contractile vacuole arises near the anterior end of the body, another just back of the middle line. While these events are taking place, a constriction appears near the middle of the longitudinal diameter of the body; this cleavage furrow becomes deeper and deeper until only a slender thread of protoplasm holds the two halves of the body together. This connection is finally severed and the two daughter paramecia are freed from each other. Each contains both macro- and micronuclei, two contractile vacuoles, and a cytostome with cytopharynx. The entire process occupies about 2 hours. The time, however, varies considerably, depending upon the temperature of the water, the quality and quantity of food, and probably other factors. The daughter paramecia increase rapidly in size, and at the end of 24 hours divide again if the temperature remains at 15 to 17°C; if the temperature is raised to 17 to 20°C, two divisions may take place in one day.

Encystment of *Paramecium* has been described, but if it really occurs, it apparently is a rare phenomenon.

Conjugation. At a certain time in the life cycle of *Paramecium* conjugation occurs. Conjugation requires the presence of two paramecia of different mating types. Members of the same *mating type* do not ordinarily enter into conjugation together but can be induced to conjugate by various techniques. One of the most enlightening methods was used in a study where mating and conjugation were induced in those of the same mating type by the introduction of cilia from paramecia of a different mating type. It seems that the surfaces of cilia from similar mating types will not adhere to each other but will adhere to surfaces of cilia of different mating types. Thus, the addition of cilia from those of a different mating type can serve as the glue to hold those of similar mating type together for the process of conjugation to occur. The conjugating pair adhere to each other with their ventral (oral) surfaces in contact and a protoplasmic bridge is constructed between them. As soon as this union is effected, the nuclei pass through a series of stages which have been likened to the maturation processes of metazoan eggs.

Although considerable variation has been observed in different species and strains, some features of conjugation are quite constant in most reports. Reference to Figure 2-67 will help to make clear the following description. Two divisions of the micronuclei occur in each conjugant (1, 2, 3) during which time the old macronucleus starts to disintegrate. During these divisions the haploid number of chromosomes is produced in the micronuclei. One of the haploid micronuclei so formed undergoes another division (4) to form the gametic nuclei, while the remaining micronuclei eventually disintegrate. One of the

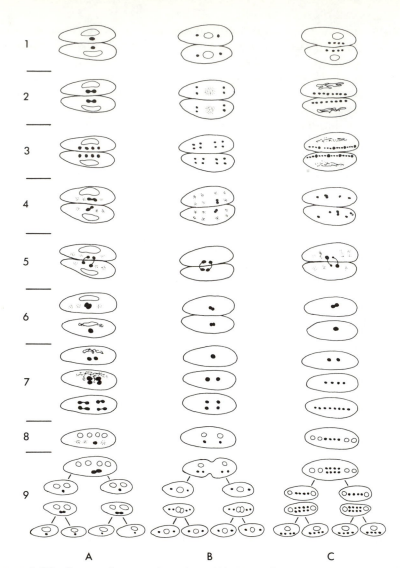

Figure 2-67 *Paramecium:* conjugation. (A) *P. caudatum.* (B) *P. aurelia.* (C) *P. multimicronucleatum.* 1–3, first two prezygotic (meiotic) divisions, degeneration of macronucleus; 4, third prezygotic division (mitotic); 5, fertilization; 6, synkaryon (zygote) formation; 7 and 8, postzygotic divisions and formation of macronuclear anlagen; 8 and 9, last micronuclear divisions accompanied by cell division. (*Modified: A, after Calkins from Wichterman; B, after Woodruff; C, after Barnett.*)

gametic nuclei so formed becomes the migratory male nucleus, which passes to the other conjugant (5), which also has produced a male nucleus for exchange. The male nuclei fuse with the female nuclei (Fig. 2-68) remaining in the conjugants to produce the fusion nucleus of fertilization or *synkaryon* (6, 7). Each ex-conjugant or zygote then undergoes a series of nuclear divisions (7, 9). After two of these divisions, some of the nuclei formed begin to differentiate as macronuclei, while the others remain as micronuclei. Four macronuclei are formed during this process (7, 8, 9), while sufficient micronuclei are formed by division to reconstitute the original number of macronuclei and micronuclei in each of the four cells produced by the two cytoplasmic divisions of each ex-conjugant.

The daughter cells resulting from this process of conjugation are considerably smaller than the original conjugant. They undergo growth and an indefinite number of generations are produced by the transverse division of the four daughter cells resulting from each conjugant.

Cytogamy and *autogamy* resemble conjugation in many respects. In cytogamy the process is nearly identical to conjugation except the synkaryon is formed by the male and female nucleus from the same cell; thus, no cross-fertilization is achieved. Autogamy is almost identical to cytogamy except there is no fusion of cells in pairs; self-fertilization occurs following nuclear events similar to those of conjugation and cytogamy. The events following synkaryon formation are similar in all three processes. Before autogamy was completely described as the meiotic and sexual phenomenon which occurred in *Paramecium aurelia* every 40 or 50 generations, it was known as endomixis and thought to be a simpler method of reorganization than it really is. *P. aurelia* is a complex of 14 sibling species that are virtually indistinguishable except by mating tests (Sonneborn, 1975).

Macronuclear regeneration sometimes occurs after conjugation or autogamy. The fragments of the macronucleus do not disappear but

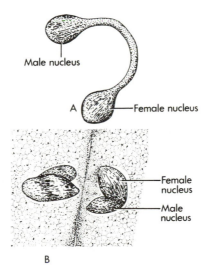

Figure 2-68 *Paramecium:* two views of the micronuclei during conjugation. (A) The spindle formed during the division of the micronucleus, which results in the production of a large female nucleus and a smaller male nucleus. (B) The fusion of the male nucleus of one conjugant with the female nucleus of the other conjugant. (*From Calkins and Cull.*)

grow and are distributed during successive divisions until only one fragment remains per cell as a regenerated and complete macronucleus. Because the fragments of the macronucleus were not formed by mitotic division, the resulting regenerated macronuclei have different genetic characteristics. Because as many as 40 fragments may produce macronuclei by regeneration and they can be functional in the absence of micronuclei, it seems probable that there are at least 40 complete sets of chromosomes in a macronucleus.

Genetics

A number of monohybrid crosses have been performed with paramecia and the results indicate that typical Mendelian inheritance occurs in these protozoans. Artificial induction of autogamy or cytogamy has been a useful tool in these studies by supplying genetically homozygous individuals for use in test crosses and in detecting genes present in the progeny of such crosses. In testing progeny of crosses the heterozygous condition of the micronucleus will be indicated by the presence of all one characteristic after autogamy in about one half of the cases and the contrasting characteristic in the rest of the cases. No change or variability will be present after autogamy of homozygous individuals. Mating type is a characteristic following simple Mendelian inheritance. Expression of most characteristics is dependent on development of the macronucleus.

The most interesting genetic story in paramecia has come from the work of Sonneborn and others. Some paramecia have DNA-containing particles in their cytoplasm which produce a substance capable of killing other paramecia that lack the particles. The particles have been termed *Kappa particles*. The paramecia containing them are referred to as "killers" and the poison produced as *paramecin*. A genetic factor must be present in the macronucleus for kappa particles to be maintained in the cytoplasm (Fig. 2-69). The gene for maintenance of kappa may be present in the macronucleus, but cells will still be sensitive to paramecin if they lack kappa particles. At conjugation some cytoplasm containing kappa may be transferred to sensitive individuals turning them into "killers" if they have the necessary gene for maintenance of kappa. Because the particles must be present to be produced and they can be transferred with some of the cytoplasm during conjugation when cytoplasmic bridges exist, they are an example of cytoplasmic inheritance. Kappa particles are said to have some of the characteristics of viruses, but their dependance for self-duplication upon a nuclear gene makes them an interesting example of nuclear and cytoplasmic interaction.

Behavior

Paramecium is a more active animal than *Amoeba*, swimming across the field of the microscope so rapidly that careful observations are neces-

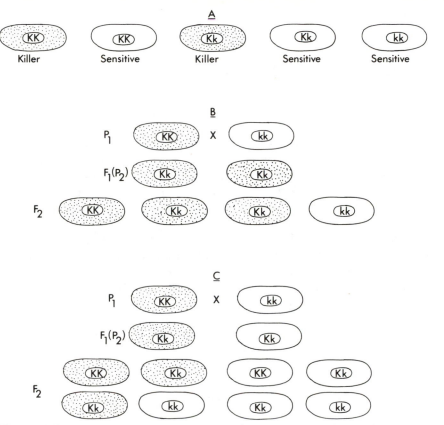

Figure 2-69 *Paramecium:* transmission of Kappa. (A) Killer and sensitive phenotypes for various genotypes and particle combinations. (B) Phenotype ratios when cytoplasmic transfer of Kappa particles occurs. (C) Phenotypes when no cytoplasmic transfer occurs.

sary to discover the details of its movements. As in *Amoeba,* its activities are either *spontaneous,* that is, initiated because of some internal influence, or result from some *external stimulus.* This stimulus is in all cases a *change in the environment.* For example, if a drop of distilled water is added to a drop of ordinary culture water containing a number of paramecia, all of the animals will enter and remain in the distilled water; they are stimulated to a certain kind of activity by the change in the composition of the water. They will soon become acclimated to their new surroundings and will behave themselves within the distilled water in a normal manner until another change in their environment stimulates them to further reactions.

Avoiding Reaction. *Paramecium* responds to stimuli either *negatively* or *positively.* The negative response is known as the *"avoiding reaction";* it takes place in the following manner. When a *Paramecium* re-

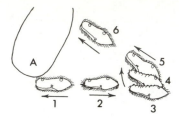

Figure 2-70 *Paramecium:* diagram of the avoiding reaction. (A) Solid object or other source of stimulation. 1–6, successive positions occupied by the animal. (The rotation on the long axis is not shown.) (*From Jennings.*)

ceives an injurious stimulus at its anterior end, it reverses its cilia and swims backward for a short distance out of the region of stimulation; then its rotation decreases in rapidity and it swerves toward the aboral side more strongly than under normal conditions. Its posterior end then becomes a sort of pivot upon which the animal swings about in a circle (Fig. 2-70). During this revolution samples of the surrounding medium are brought into the oral groove. When a sample no longer contains the stimulus, the cilia resume their normal beating and the animal moves forward again. If this once more brings it into the region of the stimulus, the avoiding reaction is repeated; this goes on as long as the animal receives the stimulus. The repetition of the avoiding reaction is very well shown when *Paramecium* enters a drop of $1/50$ percent acetic acid. In attempting to get out of the drop the surrounding water is encountered; to this the avoiding reaction is given and a new direction is taken within the acid, which of course leads to the water and another negative reaction. Figure 2-71, B shows part of the pathway made by a single *Paramecium* under these conditions.

Positive Reaction to Chemicals. If a little acid is placed in the center of a large drop of water containing a number of paramecia, all of the animals in the drop will sooner or later encounter the acid, and having once entered are unable to escape, just as in the case described above. A group is therefore formed in the acid, illustrating what is called a positive reaction (Fig. 2-71, A). This experiment may be repeated

Figure 2-71 *Paramecium:* positive reaction to acids. (A) Collection of paramecia in a drop of $1/50$ percent acetic acid. (B) Path followed by a single paramecium in a drop of acid showing mechanism of the collection shown in (A). Each time the organism comes to the boundary of the acid it gives an avoiding reaction. (*From Jennings.*)

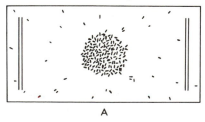
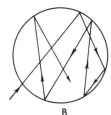

A   B

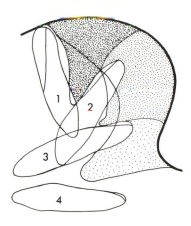

Figure 2-72 *Paramecium:* weak avoiding reaction as a result of contact with acidified ink. 1, 2, 3, 4, successive positions and final escape. (*After Mast and Lashley.*)

using a ½ percent solution of common salt in which are placed a number of specimens. If a drop of ¹/₁₀ percent solution of the same chemical is now added, the paramecia will swim into and directly across it, but on reaching the boundary between the two solutions on the other side of the drop, the avoiding reaction will be given. Soon the weaker solution will contain all of the animals which, having once entered, cannot escape. In many cases, as above, the passage from a strong solution to a weak solution causes no reaction. For certain substances, however, there is a definite strength which seems to suit the *Paramecium* better than any other, and no reaction takes place on entering it. Passage from such a solution to either a weaker or a stronger calls forth the avoiding reaction. The concentration is therefore called the *"optimum."* The optimum concentration depends on the chemical. Movement out of the zone of optimum concentration causes an avoiding reaction (Fig. 2-72). Thus, with the arrival of additional paramecia, there is an accumulation of animals in the optimal region.

Contact. Paramecia may give any of one of three reactions to contact stimuli; the first two are negative, the third positive. (1) If *Paramecium* swims against an obstacle, or if the anterior end, which is more sensitive than the other parts of the body, is touched with a glass rod, the avoiding reaction is given. (2) When any other part of the body is stimulated in a like manner, the animal may simply swim forward. (3) Frequently, a *Paramecium,* upon striking an object when swimming slowly, comes to rest with its cilia in contact with the object (Fig. 2-73). This positive reaction often brings the animals into an environment rich in food.

Light. Paramecia do not respond in any way to ordinary visible *light,* but give the avoiding reaction when ultraviolet rays are thrown upon them; if unable to escape, death ensues in from 10 to 50 seconds.

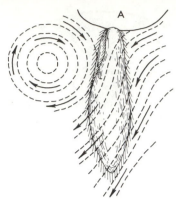

Figure 2-73 *Paramecium* at rest with anterior end against a mass of bacteria A, showing the currents produced by the cilia. (*From Jennings.*)

Temperature. The optimum temperature for *Paramecium* lies, under ordinary conditions, between 24 and 28°C. A number of animals placed on a slide, which is heated at one end, will swim about in all directions, giving the avoiding reaction where stimulated, until they become oriented so as to move toward the cooler end. This is the method of *trial and error;* that is, the animal tries all directions until the one is discovered which allows it to escape from the region of injurious stimulation.

Gravity. Gravity in some unknown way causes paramecia to orient themselves with their anterior ends pointed upward. This brings them near the surface of the water. If a number are equally distributed in a test tube of water, they will find their way to the top. In rich hay infusions paramecia are often so abundant they are visible as a whitish ring in or near the meniscus. Under certain conditions *Paramecium* is positive to gravity.

Currents. If paramecia are placed in running water, they orient themselves with the anterior ends upstream and swim against the current. This is probably caused by the interference of the current with the beating of the cilia, for as soon as an animal reaches a position with anterior end upstream, the water no longer tends to reverse the cilia.

Electric Current. Paramecia may be subjected to an electric current in the laboratory. A weak current causes a movement toward the cathode; a strong current reverses the direction of the beating of the cilia and causes the animals to swim backward toward the anode. Many other interesting phenomena might be cited, but the entire subject is too complex for brief discussion.

Other Remarks. Frequently, *Paramecium* may be stimulated in more than one way at the same time. For example, a specimen that is in contact with a solid is acted upon by gravity and may be acted upon by chemicals, heat, currents of water, and other stimuli. It has been found that gravity always gives way to other stimuli, and that if more than one other factor is at work the one first in the field exerts the greater influence.

Both the spontaneous activities and reactions due to external stimuli are due to changes in the internal condition of the animal. The *physiological condition of Paramecium,* therefore, determines the character of its response. This physiological state is a dynamic condition, changing continually with the processes of metabolism going on within the living substance of the animal. Thus, one physiological state resolves itself into another; this becomes easier and more rapid after it has taken place a number of times, giving us grounds for the belief that stimuli and reactions have a distinct effect upon succeeding responses.

Whether this represents learning may be questioned. Gelber (1958) found that paramecia fed with bacteria introduced on a needle continued to approach an introduced needle containing no bacteria although untrained paramecia did not.

> We may sum up the external factors that produce or determine reactions as follows: (1) The organism may react to a *change,* even though neither beneficial nor injurious. (2) Anything that tends to interfere with the normal current of life activities produces reactions of a certain sort ("negative"). (3) Any change that tends to restore or favor the normal life processes *may* produce reactions of a different sort ("positive"). (4) Changes that in themselves neither interfere with nor assist the normal stream of life processes may produce negative or positive reactions, according as they are usually followed by changes that are injurious or beneficial. (5) Whether a given change shall produce reaction or not often depends on the completeness or incompleteness of the performance of the metabolic processes of the organism under the existing conditions. This makes the behavior fundamentally regulatory.

Behavior is adaptive and of value to the species in most cases, whether due to unknown causes, cell membrane depolarization, simple reflexes, or a complex central nervous system. When this is not true we may expect that the species in question will be replaced by those for which it is true in those circumstances.

Although the preceding generally holds true for changes that the species have been exposed to in the past in their natural environment, it seems likely that their reactions to the many synthetic products (e.g., detergents and insecticides) which humans now introduce into their environment may not yet have been through sufficient selection to have survival value.

OTHER CILIATA

The Ciliata are characterized by the possession of cilia during a part or whole of their life cycle. In most Ciliata the nuclear material is sepa-

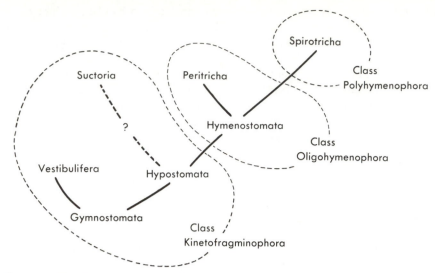

Figure 2-74 Relationship of ciliate subclasses. (*Adapted from Corliss, 1974.*)

rated into a large macronucleus and a smaller micronucleus. Most Ciliata are free-living in fresh water or the sea. Many of them, however, are ectoparasites or endoparasites of vertebrates and invertebrates.

The complex infraciliature with many variations of distinctive type all greatly different from other protozoa indicate that the over 7,000 described species deserve treatment as a phylum as recently proposed by a group of ciliatologists (de Puytorac et al., 1974). Corliss (1974, 1975, 1977) has clarified the relationships involved in the improved classification and a new edition of his ciliate book (Corliss, 1979) is now available with more complete coverage. The seven subclasses following include some of the 23 orders identifiable by the -ida ending, a few suborders identifiable by an -ina ending, and some of the 204 families and 1,123 genera identifiable by the -idae ending or initial capital and italic type, respectively.

Ciliary distribution, infraciliature (fine structure beneath the cilia and plasma membrane), and oral structure are important features in phylogeny. A simplification of ciliate relationships has been adapted from Corliss (1974) in Fig. 2-74.

Gymnostomata

Generally uniformly ciliated. A buccal cavity is absent in gymnostomes. The cytostome opens directly on the surface. In the more primitive gymnostomes, it opens at the anterior pole as in *Coleps hirtus* (Fig. 2-75, A), a barrel-shaped species covered with 20 longitudinal rows of platelets and with three posterior spines. *Lacrymaria*

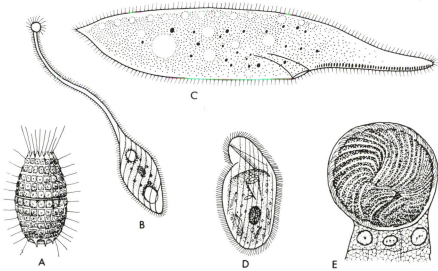

Figure 2-75 Gymnostomata and Hypostomata. (A) *Coleps hirtus*. (B) *Lacrymaria olor*. (C) *Dileptus anser*. (D) *Chilodonella cucullus*. (E) *Amphileptus branchiarum*, rounded up in gill of a frog tadpole. *Dileptus* has many micronuclei (the larger black spherical bodies) and a macronucleus distributed in small granules throughout the endoplasm. (*A, after Conn; B and D, after Wang; C, after Visscher, modified; E, after Wenrich.*)

olor (Fig. 2-75, B) is slender, 400 μm in length, and possesses a long, highly flexible proboscis. *Didinium nasutum* (Fig. 2-61) is barrel-shaped, with cilia restricted to two girdles, and a very expansible cytostome at the end of a proboscislike elevation. It feeds on *Paramecium* and other ciliates, which it engulfs whole. *Dileptus anser* (Fig. 2-75, C) has many micronuclei and the chromatin of the macronucleus distributed in granules throughout the endoplasm.

Parasitic Gymnostomatida species are not as numerous as the widely distributed free-living species. *Amphileptus branchiarum* (Fig. 2-75, E) spends part of its life in the gills of frog tadpoles. Only the marine *Stephanopogon* is in the Primociliatida. *Loxodes* and several others are in the Karyorelictida. *Coleps* and *Prorodon* are among the Prostomatida, whereas most gymnostomes, including *Didinium, Dileptus,* and *Lacrymaria,* are in the Haptorida.

Vestibulifera

Order Colpodida

Colpoda (Fig. 2-76, A), a genus of usually free-living vestibuliferans commonly encountered, is somewhat kidney-shaped. The cytostome is in the bottom of a ciliated depression (vestibulum).

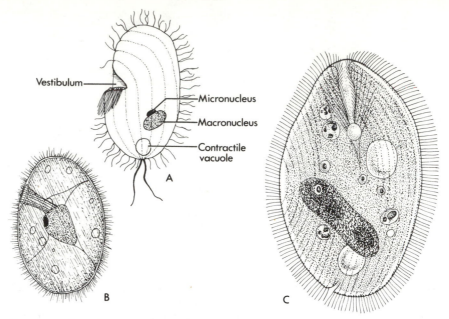

Figure 2-76 Vestibulifera. (A) *Colpoda steini,* free-living. (B) *Isotricha intestinalis,* from the stomach of cattle. (C) *Balantidium coli* from the human intestine. (*A, after several authors; B, after Becker; C, after Wenyon.*)

Order Trichostomatida

Isotricha intestinalis (Fig. 2-76, B) is an inhabitant of the stomach of cattle. *Buxtonella sulcata* lives in the cecum of cattle. *Balantidium* (Fig. 2-76, C) is an interesting genus of parasitic trichostomes recorded from many vertebrates and invertebrates. The species that lives in humans, *B. coli* (Fig. 2-76, C), is of special importance, because it also lives in pigs and monkeys and is responsible for dysenteric conditions in its human host. *B. coli* is present in the majority of pigs in this country but does not cause symptoms in them. It forms cysts, which represent the infective stage, and humans probably become infected by swallowing the cysts from pigs. The occurrence in people in the United States is very rare, and less than 1 percent worldwide.

Order Entodiniomorphida

The order Entodiniomorphida is of special interest, because it contains the most complex of the protozoans. All are endocommensal in herbivores. Simple cilia are lacking, the ciliature being restricted to small zones of membranelles. The structure of *Epidinium ecaudatum* (Fig. 2-77) illustrates the complexity of the group. Of particular interest is the so-called neuromotor apparatus, consisting of a central motorium connected with nerve rings and fibrils that make up a complicated system.

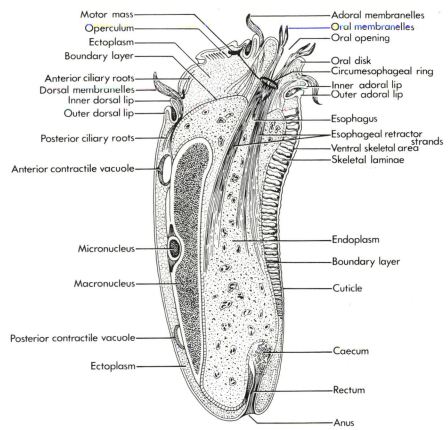

Figure 2-77 *Epidinium ecaudatum.* Diagram to illustrate the complex structure of one of the Entodiniomorphida. (*After Sharp.*)

Ciliates of Cattle. The first and second stomachs of cattle, that is, the rumen and reticulum, are highly infected with Protozoa, mostly ciliates. In this habitat two species of amoebae, five species of flagellates, and about 60 species of ciliates have been reported. Fifty-four species of the ciliates are from the order Entodiniomorphida and most of these are in the genera *Diplodinium, Entodinium,* and *Ostracodinium.* The rumen ciliates are obligate anerobes that feed on bacteria, other Protozoa, and fragments of the host's food. Their exact relations to the host are not known. The host can live without them and does not appear to be harmed by them. They may constitute an important source of protein, because they are readily digested after they pass the upper chambers of the stomach. They do not form cysts, so that transmission to new hosts requires contamination of food with saliva or rumen juice from an infected individual.

Hypostomata

This subclass contains some generalized ciliates, such as *Nassula* and and *Microthorax,* whose ancestors may have given rise to the remaining ciliate groups. The most specialized of the six orders would appear to be the Chonotrichida.

Suctoria

Order Suctorida

The Suctorida are typically sedentary forms that are devoid of any ciliary mechanism in the "adult" stage and have no buccal cavity, but capture and ingest food by means of characteristic tentacles. They reveal their relationship to the Ciliata by the production of free-swimming ciliated larvae, and by the possession of a typical ciliate nuclear apparatus consisting of separate macronuclei and micronuclei. All species possess capitate or suctorial tentacles each of which, as a rule, ends in a suckerlike knob. When a small ciliate, upon which the suctorian feeds, strikes the end of the tentacle, it is held fast by this suckerlike organ, and the general body substance of the prey is sucked down the tentacle into the body. Reproduction takes place typically by the formation of ciliated embryos formed from buds or processes constricted off, either externally or into an internal pouch, from the parent suctorian. After swimming around for a period of time, the ciliated embryos settle down, become attached, and are transformed into typical suctoria.

Free-living species (Fig. 2-78, A-C) belonging to this subclass are frequently encountered in vegetation gathered from ponds and in infusions. *Podophrya collini* is an American species with a stalk. *Sphaerophrya magna* has a spherical body without a stalk.

Spoon et al. (1976) found a large number of microtubules arranged longitudinally in the tentacles of *Heliophrya erhardi.* Food vacuoles of its prey were not ingested with the cytoplasm.

Large numbers of Suctoria may be regarded as parasites in the broad sense of the term. Among these are ectocommensals that attach themselves more or less regularly to some species of aquatic animal; for example, *Discophrya elongata* prefers the shells of gastropods. Endocommensals live in natural cavities in their hosts; for example, *Trichophrya salparum* lives in the pharyngeal cavities of certain tunicates. The true parasites also live both on the surface and in the interior of their hosts. *Ophryodendron abietinum* is an example of a true ectoparasite (Fig. 2-78, E). The genus *Sphaerophrya* includes endoparasites that live within the bodies of ciliates, such as *Stentor, Paramecium,* and *Stylonychia* (Fig. 2-78, D). A sort of brood pouch is formed within the host in which the "embryo" suctorian lives with

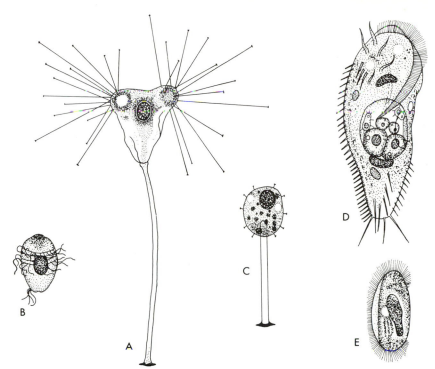

Figure 2-78 Suctoria. (A) *Tokophrya lemnarum:* typical specimen. (B) *T. lemnarum:* larva with four bands of cilia around body and a tuft of cilia at one end. (C) *T. lemnarum:* larva after fixation, with pedicel and small tentacles. (D) *Sphaerophrya:* an endoparasitic species within a brood pouch in the body of a ciliate. (E) *Ophryodendron abietinum:* ciliated embryo. (*A–C, after Noble; D, after Stein; E, after Martin.*)

tentacles retracted. Ciliated larvae are formed in the brood pouch, separate from the host, and develop into free-living individuals.

Hymenostomata

Order Hymenostomatida

Uniform body ciliation, (Fig. 2-79) as well as the possession of an undulating membrane and three membranelles in the buccal cavity, is typical of the species in the order Hymenostomatida. The most familiar genus is *Paramecium*. A number of species of *Paramecium* have been described: *P. aurelia* possess two micronuclei; *P. multimicronucleatum* has four or more micronuclei; and *P. busaria* contains Zoochlorellae as symbionts. The eight species of *Paramecium* may be separated into two groups according to their general shape. Species in the Aurelia group, consisting of *P. caudatum, P. aurelia,* and *P. multimicronucleatum,* are spindle-shaped and the body length is more than three times the width. The Bursaria group, consisting of *P. calkinsi, P.*

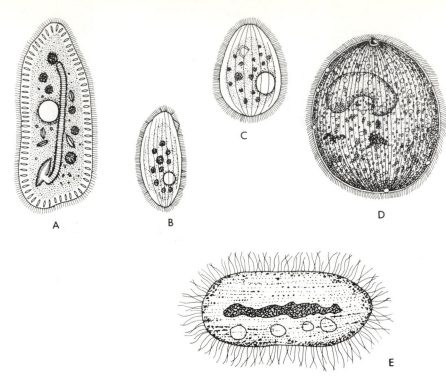

Figure 2-79 Hymenostomata. (A) *Frontonia leucas.* (B) *Colpidium cucullulus.* (C) *Glaucoma scintillans.* (D) *Ichthyophthirius multifiliis.* (E) *Anoplophrya marylandensis.* (A–C) Free-living; (D) parasite from the skin of fish; (E) from the intestine of the earthworm. (*A–C, after Wang; D, after Butschli; E, after Conkling.*)

woodruffi, P. trichium, P. bursaria, and *P. polycaryum,* have blunter ends and the body length is less than three times the width.

Other interesting hymenostomes include the much studied *Tetrahymena pyriformis; Frontonia leucas* (Fig. 2-79, A), a very large species about 300 μm in length; *Colpidium* (Fig. 2-79, B), common in infusions and with a somewhat concave oral surface; and the parasite, *Ichthyophthirius multifiliis* (Figs. 2-79, D), a very large species that attacks the skin and gills of freshwater fish, often causing their death.

Order Scuticociliatida

The order Scuticociliatida includes the Philasterina, Pleuronematina, and Thigmotrichina. *Pleuronema* has a conspicuous projecting membrane along a large, lateral peristome.

Order Astomatida

The order Astomatida contains uniformly ciliated, parasitic, and cytostomeless ciliates (Fig. 2-79, E), which have held a prominent position in phylogenetic speculation. Their similarity to the planula larvae of coelenterates and the acoelous flatworms may be overemphasized, because the absence of a mouth may be a secondary parasitic modification. They are primarily parasites of oligochaetes.

Peritrichea

Order Peritrichida

Members of this order typically lack body cilia in the adult stage but their oral ciliature is conspicuously developed. Most are *sessile* (sedentary). Sessile animals are attached relatively permanently to other objects. Most peritrichs are attached by a stalk that is secreted by a region at the aboral end known as the *scopula*. The feeding apparatus is highly specialized. The adoral spiral zone is wound around the peristomial disk, which can be completely contracted over the cytostome by means of myonemes on the margin. These *myonemes* are strands of cytoplasm specialized as contractile organelles. The adoral cilia are generally modified to form two undulating membranes, which run parallel to each other. Instead of the adoral zone leading directly to the cytostome, it leads into a vestibule supplied with an undulating membrane and into which the cytoproct and contractile vacuole as well as the cytostome open. Stalked species produce a *telotroch* under certain conditions. The telotroch is important in dispersal of the peritrichs, because it is formed from a cell that grows a basal girdle of cilia, loses its mouth structures, and becomes detached from the stalk to swim free until a new place of attachment is found.

Species of the genus *Vorticella* (Fig. 2-80) are common in fresh water and very abundant in activated sludge, where, with other ciliates, they are important in producing a quality effluent (Curds, 1973). *Vorticella campanula* has a body shaped like an inverted bell, has a contractile stalk, is always solitary, and is yellowish or greenish. *Carchesium polypinum* (Fig. 2-81, A) is a colonial species with branching stalks that contract independently of one another. *Epistylis flavicans* (Fig. 2-81, B) is also colonial; its stalks are dichotomous and not contractile. *Zoothamnium* resembles *Carchesium*, but the entire colony contracts simultaneously. Species from these and related genera occur commonly on the surface of crustaceans and the shells of mollusks in either fresh or salt water.

Orphrydium versatile is a sessile colonial species, green from zoochlorellae, embedded in a gelatinous mass that may reach several centimeters in diameter. Often abundant (Fig. 2-82) on shallow vegetation in northern lakes, it is seldom reported—perhaps because it resembles *Nostoc* algae balls to zoologists.

In addition to the sessile peritrichs there are a large number of free-swimming, but parasitic, ones. One of the common species is *Trichodina pediculus* (Fig. 2-81, C). It may be found gliding along the body

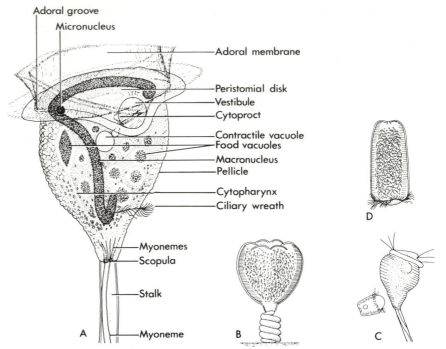

Figure 2-80 *Vorticella.* (A) Diagram showing structure of a specimen. (B) *Vorticella campanula,* contracted. (C) *Vorticella striata* with sexual zooid. (D) *Vorticella campanula,* telotroch or free-swimming swarm stage. (*After Noland and Finley.*)

surface of the freshwater *Hydra,* tadpoles, and other aquatic animals. Other species are ectozoic on the freshwater flatworms of the genus *Planaria* and on the gills of certain fishes and mollusks. *Trichodina* is short and hour-glass-shaped, with a spiral band of adoral cilia leading into a cytopharynx, a ribbon-shaped nucleus, and one contractile vacuole. The aboral surface is modified into a disklike organ of attachment consisting of an inner ring of radially arranged teeth, an outer band of cilia, and a peripheral fold, the *velum.*

Spirotrichea

Order Heterotrichida

The species of Heterotrichida may have cilia covering the entire body or portions of it, but there is an adoral zone of either large cilia or membranelles. Many of the large well-known ciliates living in fresh water and in the sea belong to this order. Both free-living and parasitic species are included. In several of the larger well-known ones such as *Stentor, Blepharisma,* and *Spirostomum* the macronucleus may take on a beaded appearance after division but returns to a more compact state prior to the next division (Fig. 2-83).

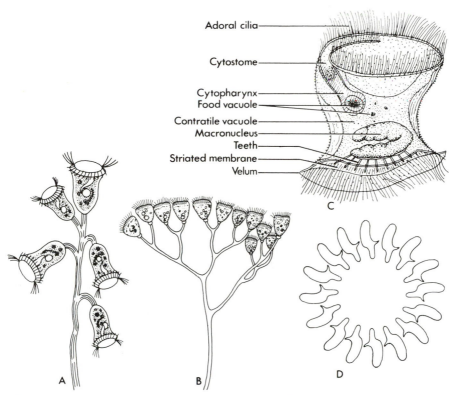

Adoral cilia

Cytostome

Cytopharynx
Food vacuole
Contratile vacuole
Macronucleus
Teeth
Striated membrane
Velum

A   B   C   D

Figure 2-81 Peritrichida. (A) *Carchesium polypinum.* (B) *Epistylis flavicans.* (C) *Trichodina pediculus,* an epizoic parasite that lives on *Hydra* and other aquatic animals. (D) Teeth from aboral disk of *Trichodina.* (*A and B, after Wang; C, after Clark; D, modified after Shtein.*)

Figure 2-82 *Ophrydium versatile,* numbers of colonies and their mean diameter from transects (10 m²) in Campbell Lake, Michigan. Average size is reduced by many new, small colonies in late summer. Over 90 percent of the colonies were on the alga, *Chara.* (*From data by J. Price.*)

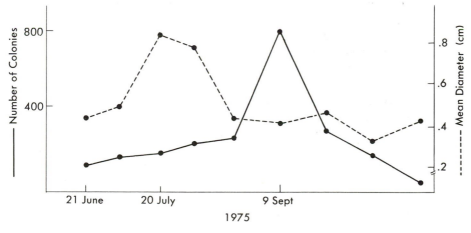

129

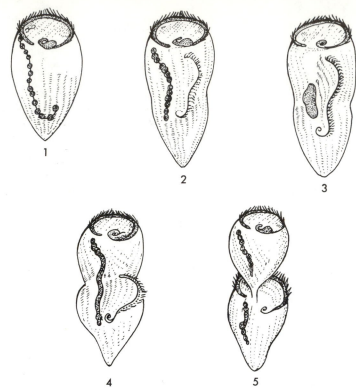

Figure 2-83 Fission of *Stentor coeruleus.* (*Adapted from Tarter, 1961.*)

Stentor coeruleus (Figs. 2-83 and 2-84) is a large bluish ciliate, visible to the naked eye, that has been an especially useful subject for studying the developmental biology of organisms at the cellular level. Much of the research described in *The Biology of Stentor* (Tarter, 1961) is with *S. coeruleus.* The beaded, or nodular, macronucleus makes the removal and transplantation of portions simpler for studies of macronuclear function. The ready fusion of cut portions of cells makes grafting of portions of these large cells possible. Some phenomena demonstrated with microsurgical techniques on these large cells are a polarity in disoriented parts of the ectoderm, an axial gradient in regeneration of parts, induction of a different developmental stage in heteroplastic grafts, and interdependence of nucleus and cytoplasm. *Stentor* is often attached to solid objects, but can swim freely through the water. *Stentor polymorphus* lives either attached or free-swimming. When free-swimming, it is oval or pyriform in shape; when attached, it assumes a trumpet shape. Its body is usually green because of the presence of zoochlorellae. *Stentor* ingests a wide variety of food organisms of the proper size, sometimes including others of the same species.

Blepharisma lateritia (Fig. 2-85, A) is a free-living heterotrich that contains a pigment associated with the kinetosomes, which renders

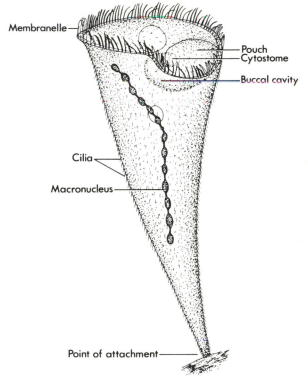

Membranelle

Pouch
Cytostome

Buccal cavity

Cilia

Macronucleus

Point of attachment

Figure 2-84 *Stentor coeruleus* in normal extended position, as when attached to a solid object. (*After Schaeffer, modified.*)

Figure 2-85 Heterotrichida. (A) *Blepharisma lateritia.* (B) *Spirostomum ambiguum.* (C) *Bursaria truncatella.* (D) *Nyctotherus cordiformis,* a parasitic species from the rectum of the frog. (*A–C, after Wang; D, after Wenyon.*)

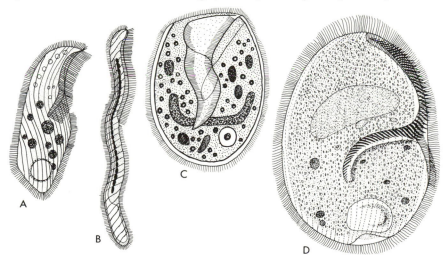

A

B

C

D

131

the body rose-red. *Spirostomum ambiguum* (Fig. 2-85, B) is a long slender species with a moniliform nucleus. *Bursaria truncatella* (Fig. 2-85, C) is a large broad species with a long curved, bandlike macronucleus and many micronuclei.

There are numerous parsitic species in addition to the many free-living Heterotrichida such as those just described. An interesting example is the genus *Nyctotherus,* which may easily be obtained for study from cockroaches and frogs. *N. ovalis* occurs in the intestine of cockroaches. The most common species in frogs is *N. cordiformis* (Fig. 2-85, D). This species has a long peristome extending from the anterior end to the center of the body and leading into a long curved buccal cavity. Other species occur in frogs, myriapods, water beetles, crustaceans, crinoids, and fish.

Order Hypotrichida

The hypotrichs are typically modified for a creeping mode of life and are strikingly dorsoventrally flattened with the ventral surface usually supplied with cirri (groups of cilia fused together), which are used as "legs" for creeping over the substratum. Their movements are jerky and rapid on surfaces but comparable to other ciliates when swimming above the bottom. The peristomial groove is well developed and possesses a very highly differentiated adoral zone of membranelles. The buccal cavity is supplied with an undulating membrane.

There are a large number of free-living species of hypotrichs (Fig. 2-86). Many are common in infusions. Some of the species that may be encountered are as follows. *Stylonychia mytilus* (Fig. 2-86, A and B) occurs in both fresh and salt water; it has two macronuclei and three

Figure 2-86 Free-living Hypotrichida. (A) *Stylonychia mytilus.* (B) *Stylonychia mytilus.* (C) *Oxytricha bifaria.* (D) *Uroleptus musculus.* (E) *Euplotes charon.* (F) *Aspidisca costata.* (B, after Kent; others, after Wang.)

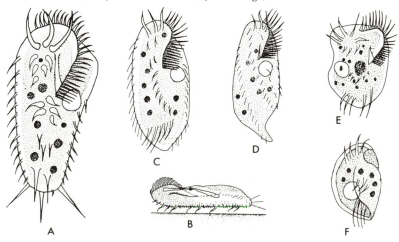

prominent caudal cirri. *Oxytricha bifaria* (Fig. 2-86, C) also occurs in both fresh and salt water; it has marginal cilia but no caudal cirri. *Uroleptus musculus* (Fig. 2-86, D) has two median rows of fine cirri, three anterior cirri, but no anal cirri. *Euplotes charon* (Fig. 2-86, E) is a marine species with an ovoidal body ventrally flattened, nine or more large cirri opposite the peristomial groove, and about nine cirri at the posterior end. *Euplotes* is a genus of ciliates that has been the object of much research. They are large and have a C-shaped macronucleus. The hypotrichs are thought to be the most highly evolved group of ciliates.

An interesting ectoparasitic hypotrich is commonly encountered on *Hydra,* the freshwater polyp. *Kerona pediculus* creeps about on the surface of *Hydra* with untiring activity. It is dependent upon its host for food, but, besides the cells of the host, will ingest other Protozoa, especially euglenoids; the latter, however, are not ingested if the *Kerona* is detached from its host.

Protozoa in General

The knowledge of the morphology, physiology, and classification of various types of Protozoa obtained from the study of the preceding pages makes it possible for us at this point to discuss some of the characteristics common to all Protozoa and to refer to some of the interesting biological facts and principles illustrated by this phylum.

Ecology

Protozoa live in almost every conceivable type of habitat, and their relations to their surroundings provide a fascinating study in ecology. As noted previously, moisture is absolutely necessary for the continued existence of Protozoa; hence, we find them in freshwater streams and bodies of fresh and salt water of all sizes, in the soil to a depth of 20 centimeters or more, within the bodies of other animals, and in moist places almost everywhere.

Each species plays some role in the trophic structure, or energy cycle, of the community of organisms in which they occur. Some of the chlorophyll-containing flagellates are *producers* capable of fixing and storing energy from the sun. Most are not photosynthetic but, like all other animals, are *consumers*. Consumers use the energy of compounds produced by photosynthetic organisms and stored in organic compounds. If the organic compounds are still stored in the plant producing them, the animal using them is a *primary consumer* or *herbivore*. If they have been transferred and stored in another animal, the animal using them from this source is a *secondary consumer;* when use causes the death of the source animal the user is a *carnivore,* when death of the source is not immediate or follows use the user is a *parasite*.

Protozoa that live suspended in the water without access to the shore or bottom constitute a large part of the plankton. Prominent in the plankton are the Phytomastigophorea, some of which occasionally undergo sudden increases in abundance to produce a "bloom" visible to the eye. Sarcodinids of the class Autotractea are in many cases planktonic. Few ciliates are found in the plankton of fresh water, although the marine plankton is nearly the only community occupied by the ciliate order Tintinnida. Many protozoans that are not truly planktonic may be found in the water near the bottom, near vegetation, or near floating debris where they have strayed from their source of food. Many Protozoa are temporarily or permanently attached to fixed or moving objects such as rocks or the bodies of aquatic animals. Shelled species are often unable to swim and must creep about on solid objects in the water. Each species is more or less restricted to a certain type of environment just as are higher animals. Factors limiting them to their environment are often complex interactions of biological and physical features. For more information on this aspect, ecology texts and journals should be consulted. Some have broad tolerances to environmental conditions and are, for example, found in both fresh and salt water. Most are much more specific in their requirements.

Kolkwitz and Marsson stand out with early efforts to use protozoan species as indicators of specific decompositional states for fresh water. Bamforth, Bick, Cairns, Curds, Lee, Muller, Small, and Stout are among many clarifying the ecological role of protozoa with experimental or field studies.

The Protozoa that are best known to students and teachers are those that appear when pond weed is brought into the laboratory and allowed to decay under fresh water. Less well known are the species included in freshwater and marine plankton, those that live in the soil and other moist places, and those that parasitize plants and other animals. The small sizes of protozoans enable them to exist in minute spaces; hence, they are more ubiquitous than any other type of animal.

The number of species of Protozoa that one may expect to find is indicated by the observation that the surface water of a typical pond in the botanical gardens of the University of Pennsylvania during the period of one year contained 27 species of Sarcodina, 31 species of Mastigophora, and 109 species of Ciliata. The seasonal distribution of these organisms appeared to be influenced principally by temperature, the oxygen content and hydrogen ion concentration of the water, and the relative amount of dissolved acids in the water at different times of the year. The greatest number of most of the species occurred during September and October.

Protozoa of the Soil. That Protozoa live in the soil is not generally realized. They appear, however, to be present in all soils, being most abundant in the top 15 cm and few in number at depths below 25 cm. Several hundred species have been recorded from soil; most of these occur in fresh water also. The types most commonly encountered in order of abundance are small flagellates, amoebae, shelled amoebae, and ciliates. Some of the common genera are the flagellates, *Cercomonas, Oikomonas, Heteromita,* and *Spiromonas;* the Sarcodina, *Naegleria, Hartmanella, Amoeba,* and *Difflugia;* and the ciliates, *Enchelys, Colpidium, Colpoda,* and *Gonostomum.* The number of Protozoa in the soil depends chiefly on the size of the soil particles and the amount of organic material present. Cutler reports in 10 samples of soil, 5 from a manured plot and 5 from one that had not been manured for many years, 1,690 amoebae, 7,460 flagellates, and 15 ciliates per gram. There appears to be a definite relation between the number of Protozoa and bacteria present since the latter serve as food for the Protozoa. Soil Protozoa multiply readily in a medium rich in organic matter, such as a hay infusion, and can easily be cultivated in the laboratory by adding a few grams of soil to such a medium in a covered dish. Dragesco (1973) found that tropical regions were poor in numbers and species of Protozoa, although many of the ciliates were new species.

General Morphology

Form and Size. Every species of protozoan has a definite form and the size of the individuals vary within definite, usually narrow, limits. Even *Amoeba proteus,* which is often erroneously described as a shapeless mass of protoplasm, has size and form characteristics that separate it from other species of amoebae. Form is determined largely by the consistency of the cytoplasm, by the production of limiting membranes, shells, and skeletons, and by the character of the activities of the organisms. Thus, the changes in shape of amoebae are possible because of the absence of a cuticle, and the great elasticity of such species as *Euglena* is the result of a thin pleated pellicle. Shells and tests such as those of *Arcella* and *Diffugia* obviously determine the shape of those species that secrete them and are so constant in form as to furnish excellent criteria for taxonomomic purposes. The skeletons of some species are very elaborate structures of calcium carbonate or silica. Form is also correlated with the organism's mode of life. Protozoa that are suspended in the water are often spherical; those that move in a definite direction, such as *Paramecium,* are ellipsoidal and the longer axis lies in the path of progression; attached species, like *Vorticella,* are more or less radial in symmetry; species that creep about on the surface of solid objects, such as *Oxytricha,* may be dorsoventrally flattened; and a few species, as in *Giardia,* are bilaterally symmetrical. Size is a characteristic that is extraordinarily constant and consequently very useful as an aid to identification. However, starvation can cause a great reduction in cell volume. Since linear dimensions are proportional to the cube root of volume measurements, even a reduction to one-eighth volume would only cause a reduction to one-half length if size proportions were constant.

Cytoplasm. The cytoplasm of Protozoa is mostly colorless, but some species are tinted, for example *Stentor coeruleus,* which is blue, and *Blepharisma lateritia,* which is rose-red. Two types of cytoplasm are usually distinguishable, the peripheral ectoplasm which is denser, hyaline, and homogeneous, and the central endoplasm which is more fluid and often granulated or alveolated. That these two types of cytoplasm may change from one to the other has been demonstrated in *Amoeba proteus,* but in many species they appear to be persistent.

Nuclei. Most Protozoa possess a single nucleus, but many of them have two (*Arcella vulgaris*) or many (*Opalina ranarum*) of the same type, or two of different types (Ciliata in general). The characteristic shape of the nucleus is spherical, but many other forms are common; for example, the oval nucleus of *Paramecium,* the kidney-shaped nucleus of *Balantidium coli,* and the moniliform nucleus of *Spirostomum.* There are two principal types of nuclear structure: the vesicular type in which the chromatin is concentrated in a single mass or grain (*Arcella*), and the granular type in which the chromatin is distributed in grains throughout the entire nucleus (*Amoeba*). Although some protozoan nuclei may be described as strictly of the vesicular or granular type the majority are intermediate in structure and partake of the characters of both types. The two types of nuclei possessed by ciliates are known as macronuclei and micronuclei; the former are polyploid and control the vegetative functions, and the latter are diploid and control the reproductive processes of the organism. Another peculiar nuclear condition exhibited by the Ciliata is the distributed nucleus of *Dileptus,* in which the chromatin granules are scattered throughout the cytoplasm. In the *Opalinata* the combination nucleus of *Opalina* contains a group of large chromosomes, trophic in function, and a second group of small chromosomes reproductive in function. As in higher animals the most important nuclear element is the DNA-containing chromatin. In some Protozoa chromatin occurs in the form of granules, called chromidia, located in the cytoplasm either scattered or in a ring, as in *Arcella.* These chromidia are probably of nuclear origin.

NUCLEAR DIVISION. The macronuclei of the Ciliata appear to divide usually by direct or amitotic division as in *Paramecium.* In a few species of Protozoa, for example in *Ephelota* among the Suctorida, the macronucleus branches and new nuclei arise by the pinching off of buds from these branches. But the majority of protozoan nuclei divide by some type of mitosis. Chromosomes arise, divide, and are equally distributed to the daughter cells as in metazoan cells. In some species, mitosis occurs within the nuclear membrane.

Vacuoles. Contractile vacuoles, food vacuoles, and stationary vacuoles containing fluid may be present in the bodies of Protozoa. As a rule, contractile vacuoles are found in freshwater species and are absent in a great many parasitic and marine forms. Their function appears to be to regulate the osmotic pressure of the cell by the elimination of a portion of the water constantly being taken into the cell by imbibition and osmosis. They are also concerned with the elimination of nitrogenous wastes and possibly with the elimination of carbon dioxide (respiration).

Mitochondria. Aerobic respiration is associated with the mitochondria, which are of variable shape with a double membrane covering them and inner membranes shaped as either internal plates or tubules. Most protozoans have tubules in the interior of the mitochondria. Mitochondria seem to be closely associated with evergy-using structures such as locomotor organelles and contractile vacuoles.

Membranes. In addition to the surface membranes of protozoan cells, internal membranes are frequently abundant in their electron micrographs. Some of the membranes are closely packed as in the golgi apparatus, others seem more scattered in arrangement as in the endoplasmic reticulum. Many of the membranes have small granules along their surface. Membranes are thought to be of great significance in carrying enzyme systems in orderly arrangement for efficient cellular metabolism.

Microtubules. Axostyles, axonemes, suctorian tentacles, and rods of the cytopharynx of some ciliates have parallel arrays of hollow tubules about 230 angstroms in diameter as important structural constituents. They seem to form from globular subunits at specific sites such as the kinetosome.

Plastids and Other Inclusions. The principal inclusions that are of more or less constant occurrence in Protozoa are chromatophores, pyrenoids, stigmata, crystals, pigments, and symbiotic organisms. Chromatophores are characteristic of holophytic Protozoa. They vary in shape and number but are constant in these respects in the different species. They contain green chlorophyll but may be colored yellow, brown, or red by pigments. Pyrenoids are the centers around which are formed paramylum bodies or other starchlike substances. They may be free or embedded in chromatophores. Stigmata or eye spots are bodies colored red by pigment and often accompanied by a lens-like paramylum body. Pigments of various colors may be present in

Protozoa. Thus, malarial organisms elaborate a brownish pigment as a product of the digestion of hemoglobin in blood. Crystals may also be constant inclusions in certain Protozoa. For example, several species of freshwater amoebae contain crystals that differ in shape from one another but are constant in shape for each species. Oil droplets are elaborated by some species, such as *Uroglena*.

Organelles of Motion and Locomotion. The ectoplasm is the layer in most intimate contact with the environment and is the seat of the organelles of motion and locomotion. In the Sarcodina movement is effected by means of temporary extrusions of the protoplasm, called pseudopodia, which are always formed from the ectoplasm although the endoplasm may flow in later. They are of two functional types: hydraulic and autotracteous. In the hydraulic type, which may be either lobopodia or filopodia, the endoplasm flows into the region of weakest tensile strength when propelled hydraulically by contraction of other portions of the ectoplasm. In the autotracteous type, which may be axopodia or reticulopodia, the propulsive force is not known although movement may occur in both directions on opposite sides of the pseudopodium, and thus seems to be due to a different mechanism than in the contraction-hydraulic type.

Flagella and cilia, which characterize the Mastigophora, Opalinata, and Ciliata, have a fundamental similarity of ultrastructure. They are vibratile threadlike processes with an internal structure of two central fibers and nine peripheral fibers, which arise from a basal granule, the kinetosome. The mechanism from which they derive their movement is not known, but the way they impart movement to the organism or the surrounding material is known to follow the laws of hydrodynamics. Flagella are comparatively long and usually few in number. Flagella may be divided into two types: tractella and pulsella. The first type is situated at the end that is anterior when the organism is in motion and drags the body along; the latter is generally situated posteriorly and pushes or propels the body forward. In a number of the parasitic flagellates, frequently a thin layer of protoplasm connects a given flagellum, thoughout the greater part of its length, with the body and is known as an undulating membrane. Cilia, which are the organs of locomotion in the class Ciliata, are also slender threadlike processes, which differ from flagella chiefly in that they are very much smaller in comparison to the body of the protozoan, occur in much greater numbers, and differ in their type of movement. Cilia are generally arranged in rows, and there is a definite coordination between the beat of the individual cilia in each row. There are several modifications of the cilia, such as cirri, membranellae, and undulating membranes. These are all produced by the fusion of a number of cilia. Mention should also be made of certain contractile elements known as myonemes, which are musclelike and occur in the body of certain of the Mastigophora, Ciliata, and sporozoa.

Organelles for the Capture and Ingestion of Food. The organelles of locomotion also play the principal rôle in the capture and ingestion of food. Thus, *Amoeba proteus* engulfs solid food bodies and is even able to capture other protozoa and cut in two such organisms as *Paramecium* and *Frontonia* by means of its pseudopodia. Holozoic flagellates that ingest solid particles, in many cases use their flagella to drive them into the cytostome. The ciliates possess the most complicated mechanism for capturing and ingesting food; this consists of cilia modified in length and arrangement or fused together into cirri or undulating membranes which may extend down into the cytopharynx. In the Suctorea, tentacles are provided which paralyze organisms, usually ciliates, that come in contact with them and suck the protoplasm out of the prey and pass it down into the endoplasm leaving very little but the periplast behind. Many other specialized structures are present among the Protozoa that aid in the capture and ingestion of food, such as the trichocysts of *Paramecium* and the myonemes of *Vorticella*. Many free-living Protozoa and practically all of the parasitic Protozoa do not ingest solid particles and have no organelles for that purpose; they manufacture food by photosynthesis or absorb it through the surface of the body.

General Physiology

Nutrition. The principal types of nutrition exhibited by Protozoa are holozoic, holophytic, and saprozoic. Protozoa, such as *Amoeba* and *Paramecium,* that feed on solid particles are said to be holozoic. As a rule, each species is capable of food selection, for example, the holotrichous ciliate, *Actinobolina radians,* pays no attention to many small ciliates and flagellates but immediately captures *Halteria grandinella* whenever an opportunity is afforded; and *Didinium nasutum* has a predilection for *Paramecium*. The ingested food of holozoic Protozoa is confined in food vacuoles and digested and assimilated as described in *Amoeba* and *Paramecium*. Undigested material is cast out at any point on the surface (*Amoeba*) or through a definite cytoproct (*Paramecium*).

Holophytic nutrition is particularly common among the Phytomastigophorea. The exclusively holophytic forms (notably some Chloromonadida) require only the presence of sunlight, certain dissolved minerals, and CO_2 to build up and synthesize their complex food materials. Many of the other chromatophore-bearing species, however, live and reproduce much more readily in a medium containing some organic matter. Consequently, they must, to a certain extent, be saprophytic. Certain species of Protozoa, such as *Paramecium bursaria* and

Stentor polymorphus, may obtain nutriment from the green algae (*zoochlorellae*) that live within them.

Saprozoic nutrition involves the absorption through the general body wall or through some specialized region of the body of dissolved organic material in the medium. Dissolved proteins and carbohydrates resulting from the disintegration of plants and animals, as in an infusion, probably furnish the food for this type of organism. Many parasites, such as *Entamoeba histolytica* and *Balantidium coli,* ingest solid particles and are thus holozoic, but they may also absorb through the body surface food digested by the host or the disintegrated protoplasm of the host. Other parasites, such as *Opalina* and *Giardia*, do not possess any structural modifiations for the ingestion of solid particles and must obtain all their nutriment by absorption.

Secretion. Various types of secretions are produced by Protozoa. Some of them are as follows. The endoplasm of holozoic species secretes digestive enzymes; Suctorea secrete a poison that paralyzes other Ciliata that come in contact with their tentacles; many species secrete shells (*Arcella*) or skeletons (*Radiolaria*); the pseudopodia of certain Sarcodina are mucilaginous (*Difflugia*) and enable the organisms to adhere to the substratum; arcellas secrete gas bubbles which lower their weight and bring them to the surface of the water; the parasite of amoebic dysentery (*Entamoeba histolytica*) secretes a proteolytic enzyme which dissolves the cells of the intestinal wall of its human host.

Excretion. Soluble waste materials are excreted either by diffusion through the general body surface, or, after accumulation in the contractile vacuole, by emptying from it to the exterior. Another very effective method of eliminating waste materials from the body is to throw them out of solution by precipitation. The crystals described in *Amoeba proteus* and the pigment in *Actinosphaerium* may be of this nature. The pigment of certain blood-inhabiting sporozoa is an insoluble waste formed from the digestion of hemoglobin.

Respiration. Respiration, in its broadest sense, includes any process by which an organism liberates the potential energy represented in the complex chemicals of its body. It may be effected either aerobically by oxidation or anaerobically by the splitting up of complex chemical substances into simple compounds. The end result, in either case, is the same. Most free-living Protozoa live in an environment containing more or less dissolved oxygen. Through the general body surface the animal absorbs the oxygen and eliminates the carbon dioxide and water which result from the oxidation. The contractile vacuole is probably also effective in eliminating these wastes. Many parasitic forms live in environments which are partially or totally lacking in oxygen. Here respiration must be of the anaerobic type.

Circulation. The protozoan body is so small that the circulation of nutritive material, secretions, excretions, etc., takes place by streaming movements of the cytoplasm and hence no circulatory system is necessary. Thus, food vacuoles are distributed through the body of *Paramecium* by cyclosis, and in a suctorian, the protoplasm of its prey passes through the tentacles and into the body.

Motion and Locomotion. Movements among the Protozoa are due to pseudopodia, flagella, cilia, and contractile elements such as myonemes. Amoebid movement has been described in *Amoeba proteus.* As noted, pseudopida are of two chief types—lobopodia and axopodia. The former are the common type and occur in the amoebae. They consist of an extrusion of the body with no specific structural differentiation. Axopodia, on the other hand, found in such forms as *Actinophrys,* are capable of swinging or bending movements and possess a secreted axis, which is either rigid or elastic in nature. Mastigophora move by means of flagella. The commonest type of flagellum is the tractellum, a flagellum which drags rather than propels the organism. In some species the entire length of the flagellum is thrown into motion and the organism may progress very rapidly, as in the hemoflagellates; in other species only about one third of the distal end vibrates, often causing the organism to progress in a series of jumps. In certain species, such as *Peranema,* almost the entire flagellum is held stiff and only a very small part of the distal end moves. Mastigophora with flagella may move by means of pseudopodia, and some of the highly developed Euglenida, such as *Euglena,* periodically lose their flagella and move about by crawling along the substratum with a kind of rocking movement. Forms with a semiresistant periplast, especially the Euglenida, often exhibit peristaltic waves of contraction which run down the length of the body. Such movements are termed metabolic, or euglenoid, because they are especially common in *Euglena.*

Movement among the Ciliata is effected by cilia. The action of cilia in locomotion has been described in *Paramecium.* As a rule, cilia are situated in rows, and their movement is concerted. Each cilium contracts a short interval after the one in front of it and before the one behind it. Thus, viewed from the side, a moving row of cilia looks as if successive waves were passing over it. Furthermore, adjacent cilia of different rows beat in unison. When seen from above a ciliated surface has the appearance of a wheat field in the wind; that is, successive waves follow one another across the ciliated surface (Fig. 2-60). Hypotrichous ciliates are capable of swimming, but spend most of their time creeping on the substratum by means of cirri on the ventral surface which represent bunches of fused cilia and act practically as legs. In *Euplotes,* the cirri are aided in creeping by the peristomial membranelles.

The motility systems of Protozoa with actin and myosinlike proteins interacting with high-energy phosphate compounds and divalent cations indicate that the movement of Protozoa has much in common with the movement of higher forms.

Behavior

The reactions of Protozoa to stimuli, which taken together constitute what is known as their behavior, have attracted much study. Some of the results have already been noted in *Amoeba, Euglena,* and *Paramecium.* In general, the reaction of an animal to a stimulus involves three systems: first, a sense receptor; second, a conducting system; and third, an effector. Such structures are not visibly differentiated in forms like *Amoeba,* but the portion of the body that receives the stimulus is also the one that reacts so that conduction is probably little more than a general effect of the stimulus on a localized portion of the body.

The flagellates have progressed markedly in the development of a reflex type of behavior (receptor-conductor-effector), due largely to the presence of definite organs of locomotion and more or less permanent body form. This is evident from a comparison of the behavior of *Amoeba* and *Euglena.* Of special interest is the relation of the stigma to light stimulation.

The reflex type of reaction is developed to a much higher state in the Ciliata. The reactions of *Paramecium* to various stimuli have already been described. In *Stentor* there is fairly clear evidence that one portion of the body receives the stimulus, that this is transmitted to definite cilia, which in turn produce a definite motor response. Fibrillar systems are present in the ciliate ectoplasm, but little evidence is available to clarify their function. Definite neuromotor fibrils and centers are known in the more highly specialized ciliates such as *Euplotes.*

In conclusion, we can do no better than to insert a quotation from Jennings (1906).

> All together, there is no evidence of the existence of differences of fundamental character between the behavior of the Protozoa and that of the lower Metazoa. The study of behavior lends no support to the view that the life activities are of an essentially different character in the Protozoa and the Metazoa. The behavior of the Protozoa appears to be no more and no less machine-like than that of the Metazoa; similar principles govern both.

Reproduction

Asexual Reproduction. The common method of reproduction in the Protozoa is that of simple or binary fission. During this process first the nucleus, then the cytoplasm divides. In some cases the nucleus undergoes a series of divisions after which the body divides into as many parts as there are nuclei. Such a process is known as multiple fission. In other cases the division of the cytoplasm only takes place at special periods in the life cycle so that, as a result, the typical condition of the organism is that of a plasmodium. When division of such a plasmodium does take place, it generally forms two or more multinucleate bodies. Such a process is known as plasmotomy.

Budding is a type of asexual reproduction that occurs regularly in certain species. A bud is a part of the cytoplasm of the parent containing a nucleus of parental origin. Buds cut off from the outside of the parent are exogenous. This occurs, for example, in certain amoebae (*Endamoeba patuxent* of the oyster), in the flagellate *Noctiluca*, in sporozoa, and in *Ephelota* and most other Suctoria. Endogenous or internal budding is not as common as the exogenous type. It consists in the separation of a mass of cytoplasm, containing a nucleus, from the rest of the organism within the body. It occurs commonly in certain sporozoa and a few Suctoria.

Cell Cycle. Molecular and genetic events of dividing Protozoa are often described with a terminology describing portions of the cell division cycle. In *Tetrahymena* (Cameron, 1973) the division period (D) is followed by a macronuclear DNA synthesis period (S). An interval of growth (G_1) occurs after D and before S. The interval after synthesis and before division is the growth period (G_2). The averaged time for a complete cell generation (GT) includes the time for all parts of the cell cycle which may be overlapping or apportioned differently in different species. For laboratory cultures of *Tetrahymena* under favorable conditions GT is usually over 3 hours but can be less than 2 hours.

Sexual Reproduction. Conjugation, or the temporary union of two individuals of the same species involving an exchange of nuclear material, accompanies sexual reproduction in many Protozoa, especially Ciliata. This process has been described fully in *Paramecium*. When the conjugants are of the same size as in *Paramecium*, they are said to be isogamous; when of different sizes, as in *Chilodon cucullulus*, anisogamous. The permanent union of two protozoan individuals is called copulation. The organisms that unite are usually smaller than normal vegetative individuals and are known as gametes. The gametes may be of the same size (isogamous) as in *Monocystis*, or may differ in size (anisogamous) as in *Plasmodium vivax*. In the latter case they correspond to the egg cells and spermatozoa of the Metazoa. The body formed by the fusion of two gametes is a zygote. The life cycles of many of the Protozoa, especially the sporozoa, are sometimes very complicated because of the sexual reproduction involved.

Special Morphology and Physiology

Various topics of interest in connection with Protozoa are referred to briefly here. Further information may be found in books on protozoology or in scientific journals.

Colonial Protozoa and Metazoa. Although Protozoa are defined as unicellular animals, many of them are colonial in habit and others at certain stages in their life cycles consist of many cells. The colonial forms, as we have seen, may consist of a variable number of cells irregularly arranged (*Carchesium*) or of a definite number arranged in a definite way (Volvocidae). The members of the colony may be entirely independent, except for the common gelatinous matrix in which they are embedded (*Pandorina*), or may be connected by protoplasmic strands (*Volvox*). The cells may be of one kind (*Spondylomorum*) or may be differentiated into somatic cells and germ cells (*Volvox*). The somatic cells of Protozoa in general differ from somatic cells in the Metazoa, because there is no histological differentiation among them and they are comparatively independent. Multicellular stages occur regularly in the life cycles of certain Protozoa. For example, in the gregarine, *Ophryocystis,* each member of a pair at the time of copulation divides into two cells, a small uninucleate gamete and a larger binucleate cell which forms a protecting envelope. A theory favored by many zoologists is that the Metazoa evolved from the Protozoa through stages resembling somewhat the colonial Protozoa of today. Perhaps the most fundamental difference between protozoan and metazoan cells is the fact that the Protozoa are able with one cell to carry on all the life processes characteristic of the Metazoa; whereas metazoan cells are limited to special functions.

Conjugation and Autogamy. The significance of conjugation cannot be definitely stated. Some investigators believe that *Paramecium* and other ciliates pass through a life cycle containing three distinct stages. The period of (1) youth is characterized by rapid cell multiplication and growth; (2) maturity, by less frequent cell division, sexual maturity, and the cessation of growth; and (3) old age, by degeneration and natural death. Death is avoided by conjugation, which rejuvenates the senescent animals.

The amitotic division of the macronucleus, which occurs during binary fission, may segregate into divided macronuclei unequal proportions of the various chromosomes. Succeeding divisions could further increase the imbalance to a point at which the macronucleus can no longer function properly, and the reconstitution of the macronucleus following conjugation or autogamy is essential to survival. Dysart has shown that the macronuclei of various tetrahymenids contain from 15 to 900 times the amount of DNA (an important genetic material in the chromosomes) found in the haploid micronuclei. In view of the absence of a spindle for precise division of the chromosomes of the macronucleus, it is not unlikely that after many divisions the balance of chromosomes originally present would be lost in some cells.

A favorable balance may be restored by either conjugation or autogamy. It may be noted at this point that Weismann many years ago suggested that Protozoa do not die a natural death, because the two daughter cells resulting from binary division share the parental protoplasm between them and the parent does not die but simply loses her identity in her offspring.

Mutualism (Symbiosis). Symbiosis, which formerly was used to imply the permanent association of two specifically distinct organisms to their mutual advantage, has taken on a meaning to biologists of an intimate association of more than one kind of organism, which may be (1) disadvantageous to one (parasitism), (2) essential to one but of no advantage or disadvantage to the other (commensalism), or (3) essential or useful to both (mutualism). Mutualism occurs in many protozoans such as the ciliates *Paramecium busaria* and *Stentor polymorphus* and certain Radiolarida, which contain symbiotic algae (zoochlorellae) or dinoflagellates (zooxanthellae). Some Protozoa are obligatory symbionts of higher organisms, e.g., flagellates in the intestine of termites.

Cellular Metabolism. Increased knowledge of culture methods for many protozoans has made possible the rapid advance in our understanding of their physiology at the molecular level. *Tetrahymena pyriformis* has been an outstanding species in revealing details of molecular biology. The presence of many metabolic pathways in common with humans have made the Protozoa of practical value in experimental studies relating to the biology of humans.

Ultrastructure. The electron microscope has contributed a great deal of information in recent years documenting the fundamental unity and diversity of cells and organisms. The prevalence of membranes in the cytoplasm of organisms was a feature not anticipated. The homology of cilia and flagella is of special interest.

The early contributions from transmission electron microscopy were summarized by Pitelka (1963) and are mostly limited to internal anatomy as revealed by sections. Scanning electron microscopy has developed rapidly and can provide clear surface detail with great depth of field. Excellent electron micrographs are a common feature of most journals reporting current protozoological research.

The History of Our Knowledge of Protozoa

Protozoa were unknown before microscopes were invented. They were first seen and described by Anton von Leeuwenhoek (1632–1723).

About 30 species described by him are recognizable today. The first one was a *Vorticella* described in 1675. Later, in 1681, he found in his own stools and described the first protozoan from a human, now known as *Giardia lamblia*. Protozoa soon became favorite material for study and played a prominent role in the development of biology, but only a few names and discoveries can be included here. Joblot (1718) showed that no Protozoa appeared in a hay infusion unless exposed to the atmosphere, thus disproving the theory of the spontaneous generation of Protozoa. He also discovered (1754) the contractile vacuole and the importance of the arrangement of cilia. Part of the life history of the vorticellas was worked out by Trembley (1747). The nature of the Protozoa was being discussed at this time and was not conceded by Linnaeus to be of the animal type until 1767. Order was established for the Protozoa by O. F. Müller (1786), who adopted binomial nomenclature and named nearly 400 species, of which 150 remain in the group today. The term Infusoria was applied to the entire phylum. Ehrenberg's large monograph on *The Infusoria as Complete Organisms* appeared in 1836. He, like many other investigators at this time, believed Protozoa to possess eyes and various internal organs characteristic of higher animals. The Sarcodina were studied particularly by Dujardin (1801–1860), who called their protoplasm "sarcode." The term "protoplasm" was applied to the cell substance by Purkinje in 1840 and by von Mohl in 1846. The unicellular nature of the Protozoa was established by von Siebold in 1848.

From this time on many students of the Protozoa advanced our knowledge of the phylum and published large monographs on the various types. Some of the more important are included in the references at the end of this chapter. Corliss (1978, 1979) provides photographs and statements of the achievements of many deceased protozoologists, including Robert Hegner, the original author of this book.

Parasitic Protozoa did not attract particular attention until it was discovered that certain species were pathogenic to lower animals or to people. The discovery of each new pathogenic species has been followed by a flood of literature representing the results of investigation by many students. Among the disease-producing parasitic Protozoa of humans should be mentioned *Entamoeba histolytica* of amoebic dysentery and amoebic liver abscess, the trypanosomes of African sleeping sickness and of Chagas' disease in South America, the leishmanias of oriental sore and kala-azar, the intestinal flagellates, *Balantidium coli* of ciliate dysentery, and the malarial parasites. Important agents of animal diseases are the trypanosomes of dourine, nagana, surra, mal-de-caderas, and murrina in cattle, horses, and mules, the coccidia of birds and rabbits and the piroplasmas of Texas fever in cattle, and similar diseases.

Protozoa and Metazoa

Zoologists usually assume that Metazoa have evolved from Protozoa, and, granted the truth of evolution, the assumption that the many-celled animals developed from protozoanlike ancestors seems reasonable. How this took place during the remote ages in the history of life upon the earth is a difficult and unsolved problem. Corliss (1972) draws attention to some of the most significant speculation regarding this protozoan role. Among Protozoa, as already described, are species consisting of many cells forming colonies in some of which the cells have become differentiated into somatic cells and germ cells. In certain species the somatic cells are of several types that perform different functions. Such an organism is really a metazoan. It is impossible to state definitely, however, that the Metazoa have evolved by the aggregation of similar cells into colonies and their subsequent differentiation although this seems quite probable. The possibility has also been suggested that the Metazoa have arisen from multinucleate Protozoa in which each nucleus assumed control of that portion of the cytoplasm immediately surrounding it and the so-called energids thus formed developed different functions. The superficial resemblance of the astome ciliates to the planula larvae of coelenterates and the acoelous flatworms has led some to consider the ciliates as the protozoans from which the metazoans arose. The peculiar nuclear condition of the ciliates indicates that if such a hypothetical origin occurred, it would have been from ciliates more primitive than the existing ones.

Protista

Protozoa and algae are so closely related through the Phytomastigophorea that they seem to be the point of divergence of plants and animals. Because the position of other protozoans as animals or plants may be questionable (viz., the fungilike Mycetozoida and some questionable sporozoa), the Protista concept has arisen. Those using the Kingdom Protista include in it all of the unicellular nucleated organisms. Higher plants and animals are left in separate kingdoms as before. The point of the fundamental unity of organisms can be made without disrupting the conventional classification, although many prefer the multi-kingdom approach of Whittaker (1969).

Classification of the Protozoa

The classification presented here is much abbreviated and arranged to be useful to one who is not specializing in protozoology. It includes particularly those groups most frequently encountered by zoologists.

With some changes, the flagellates follow Honigberg et al. (1964); sarcodinids follow Jahn and Bovee (1965); sporozoa follow Levine, as followed by Cheng (1973) and others; and ciliates are after Corliss

(1974, 1975, 1977), interpreting and modifying de Puytorac et al. (1974).

Phylum Protozoa

Unicellular and closely related animals.

Subphylum **Plasmodroma.** With one type of nucleus.

SUPERCLASS I. MASTIGOPHORA. With flagella.

CLASS 1. PHYTOMASTIGOPHOREA. Plantlike flagellates; one to four flagella.

ORDER 1. CHRYSOMONADIDA. One or two yellowish chromatophores; food stored as lipid or leucosin.

ORDER 2. SILICOFLAGELLIDA. Brown chromatophores; siliceous endoskeleton.

ORDER 3. COCCOLITHOPHORIDA. Yellow-brown chromatophores; body covering of calcareous platelets.

ORDER 4. HETEROCHLORIDA. Two unequal flagella; yellow-green chromatophores; food stored as lipid or leucosin.

ORDER 5. CRYPTOMONADIDA. Two flagella originating in depression; food stored as starch or amyloid substances; chromatophores, when present, often brown, green, or red.

ORDER 6. DINOFLAGELLIDA. Many with a thickened cuticle which forms a lorica; two flagella that usually lie in two grooves in the lorica, the longitudinal one in the sulcus, and the transverse one in the girdle; food stored as starch and lipid.

ORDER 7. EBRIIDA. No chromatophores; siliceous endoskeleton.

ORDER 8. EUGLENIDA. Green chromatophores; food stored as paramylum; metaboly.

ORDER 9. CHLOROMONADIDA. Numerous green chromatophores; two flagella, one trailing; food stored as glycogen and lipids.

ORDER 10. VOLVOCIDA. Chromatophore green, often cup-shaped; colonial forms common; food stored as starch.

CLASS 2. ZOOMASTIGOPHOREA. Animal-like flagellates; one to many flagella.

ORDER 1. CHOANOFLAGELLIDA. Collar surrounding base of single anterior flagellum; some colonial.

ORDER 2. BICOSOECIDA. Two flagella, one attaching cell to lorica; some with collar.

ORDER 3. RHIZOMASTIGIDA. One to many flagella and pseudopodia.

ORDER 4. KINETOPLASTIDA. Usually one or two flagella and a nucleic acid containing kinetoplast; most parasitic.

ORDER 5. RETORTAMONADIDA. Two to four flagella; cytostome; all parasitic.

ORDER 6. DIPLOMONADIDA. Eight flagella; bilateral symmetry; two nuclei; most parasitic.

ORDER 7. OXYMONADIDA. Four or more flagella; one or more axostyles; division spindle intranuclear; all parasitic.

ORDER 8. TRICHOMONADIDA. Usually four to six flagella; one or more axostyles; division spindle extranuclear; all parasitic.

ORDER 9. HYPERMASTIGIDA. Many flagella; one nucleus; all entozoic in insects.

SUPERCLASS II. OPALINATA. With many uniformly distributed flagella giving a ciliatelike appearance; one type of nucleus; all entozoic in amphibians. One order, Opalinida.

SUPERCLASS III. SARCODINA. With pseudopodia.

CLASS 1. HYDRAULEA. Pseudopodial movement with protoplasmic flow in only one direction at a time; no axial structure in pseudopodium.

ORDER 1. AMOEBIDA. Without test or shell; usually uninucleate, multinucleate condition not the result of aggregation of individuals.

ORDER 2. HELIOZOIDA. Azopodia or filopodia; spherical which pseudopodia are extended.

ORDER 3. MYCETOZOIDA. Without test; complex life cycle, including large multinucleate stage followed by spore formation.

CLASS 2. AUTOTRACTEA. Movement of protoplasm may occur in opposite directions on the sides of thin pseudopodia, which frequently contain an axial structure.

ORDER 1. PROTEOMYXIDA. Without test; most parasitic.

ORDER 2. HELIOZOIDA. Axopodia or filopodia; sperical symmetry; no central capsule; mostly freshwater forms.

ORDER 3. RADIOLARIDA. Skeleton siliceous or of strontium sulfate; central capsule; all marine.

ORDER 4. FORAMINIFERIDA. Test of chitinlike material, usually greatly thickened by the deposition of calcium carbonate; usually many perforations for extension of pseudopodia; no central capsule; nearly all marine.

SUPERCLASS IV. APICOMPLEXA. No locomotor organelles, with an apical complex, and all are parasites. Formation of spores by adults, ending life of the parent; reproduction both asexual and sexual; spores without polar capsules or polar filaments. One class, Telosporea.

ORDER 1. GREGARINIDA. Coelozoic; reproduction usually by sporogony following association of gamonts in syzygy.

ORDER 2. COCCIDA. Cytozoic; alternation of schizogony

and sporogony; sporocyst and oocyst wall surround sporozoites.

ORDER 3. HAEMOSPORIDA. Cytozoic; alternation of schizogony and sporogony; sporozoites naked; two host life cycle.

SUPERCLASS V. MYXOSPORA. Spores large; usually with two polar capsules and polar filaments; mostly parasites of fish, none free-living. One order, MYXOSPORIDA.

SUPERCLASS VI. MICROSPORA. Spores small; usually only one or no polar capsule; parasites, usually of fish or invertebrates.

ORDER 1. MICROSPORIDA. Usually one polar capsule.

ORDER 2. HAPLOSPORIDA. No polar capsule.

Subphylum (or Phylum) **Ciliophora.** With cilia; usually with two sizes of nuclei and a complex infraciliature.

SUPERCLASS I. CILIATA. With the characteristics of the subphylum.

CLASS 1. KINETOFRAGMINOPHORA. Often with "fragments" of kineties for specialized ciliature.

SUBCLASS 1. GYMNOSTOMATA. Uniform ciliation; cytostome at surface, usually at apex of cell.

ORDER 1. PRIMOCILIATIDA. Homokaryotic.

ORDER 2. KARYORELICTIDA. Diploid, nondividing macronucleus.

ORDER 3. PROSTOMATIDA. Polyploid macronucleus.

ORDER 4. HAPTORIDA. Apical cytopharynx often eversible, with toxicysts.

ORDER 5. PLEUROSTOMATIDA. Cytostome a lateral slit.

SUBCLASS 2. VESTIBULIFERA. Usually with a vestibulum having distinct ciliature.

ORDER 1. TRICHOSTOMATIDA. Uniform ciliation, vestibulum apical.

ORDER 2. COLPODIDA. Vestibular ciliature differentiated from kineties.

ORDER 3. ENTODINIOMORPHIDA. Ciliature in tufts.

SUBCLASS 3. HYPOSTOMATA. Cytostome ventral.

ORDER 1. SYNHYMENIIDA. With circumoral ciliature.

ORDER 2. NASSULIDA. With preoral suture.

ORDER 3. CYRTOPHORIDA. Somatic ciliature all ventral.

ORDER 4. CHONOTRICHIDA. Sessile; vase-shaped; only atrial ciliature.

ORDER 5. RHYNCHODIDA. Rostrate, limited ciliature, invertebrate symbionts.

ORDER 6. APOSTOMATIDA. Spiraled ciliature, cytostome may be lacking.

SUBCLASS 4. SUCTORIA. With sucking tentacles, adult sessile and lacking cilia. One order, SUCTORIDA.

CLASS 2. OLIGOHYMENOPHORA. Usually with a complex oral ciliature.

    SUBCLASS 1. HYMENOSTOMATA. Body ciliature uniform.

      ORDER 1. HYMENOSTOMATIDA. Distinct buccal cavity with membranelles.

      ORDER 2. SCUTICOCILIATIDA. With scuticus during stomatogenesis.

      ORDER 3. ASTOMATIDA. Hymenostomelike ciliation, no mouth, endosymbionts.

    SUBCLASS 2. PERITRICHA. Prominent apical oral ciliature, somatic ciliature a temporary posterior circle, many stalked. One order, PERITRICHIDA.

CLASS 3. POLYHYMENOPHORA. With prominent adoral zone of buccal membranelles.

    SUBCLASS SPIROTRICHA. With characteristics of class.

      ORDER 1. HETEROTRICHIDA. Uniform body ciliation.

      ORDER 2. ODONTOSTOMATIDA. Reduced body ciliation, with only eight or nine buccal membranelles.

      ORDER 3. OLIGOTRICHIDA. Reduced body ciliation, with numerous buccal membranelles, many are loricate.

      ORDER 4. HYPOTRICHIDA. Flattened; cirri as body ciliature.

REFERENCES TO LITERATURE ON THE PROTOZOA

The references listed below are for the most part books and some of the more significant journal articles. In these will be found bibliographies to the literature on Protozoa contained in scientific journals and monographs. Current literature can best be obtained from journals that often include papers on Protozoa and from bibliographies. Among the former are the *Journal of Protozoology, Archiv fur Protistenkunde, Biological Bulletin, Journal of Parasitology,* and *Transactions of the American Microscopical Society.*

A number of books on particular groups are available. Protozoology texts include Kudo (1966), good for generic descriptions; Manwell (1961), excellent for general features; and Grell (1973) and Sleigh (1973), with more ultrastructural and molecular information.

ADLER, P. N., L. S. DAVIDOW, and C. E. HOLT. 1975. Life cycle variants *Physarum polycephalum* that lack the amoeba stage. *Science,* **190**:65–67.

ALLEN, R. D. 1962. Amoeboid movement. *Sci. Am.,* **206**:112–122.

———, D. FRANCIS, and R. ZEH. 1971. Direct test of the positive pressure gradient theory of pseudopod extension and retraction in amoebae. *Science,* **174**:1237–1240.

BARKLEY, D. S. 1969. Adenosine-3′,5′-phosphate: identification as acrasin in a species of cellular slime mold. *Science,* **165**:1133–1134.

BLACKWELDER, R. E. 1963. *Classification of the Animal Kingdom*. Southern Illinois Univ. Press, Carbondale. 94 pp.

BORDEN, D., G. S. WHITT, and D. L. NANNEY. 1973. Isozymic heterogeneity in *Tetrahymena* strains. *Science*, **181**:279–280.

BOVEE, E. C., and T. L. JAHN. 1965. Mechanisms of movement in taxonomy of Sarcodina. II. The organization of subclasses and orders in relationship to the classes Autotractea and Hydraulea. *Am. Midl. Nat.*, **73**:293–298.

———, and T. L. JAHN. 1973. Taxonomy and phylogeny. Pp. 37–82 in K. W. Jeon, Ed. *The Biology of Amoeba*. Academic Press, New York.

BUETOW, D. E. (Ed.), 1968. *The Biology of Euglena*. 2 vols. Academic Press, New York. 361 pp.; 417 pp.

BUZAS, M. A., and T. G. GIBSON. 1969. Species diversity: benthonic Foraminifera in western North Atlantic. *Science,* **163**:72–75.

CAIRNS, J., JR., K. L. DICKSON, and W. H. YONGUE, JR. 1971. The consequences of nonselective periodic removal of portions of fresh-water protozoan communities. *Trans. Am. Micr. Soc.*, **90**:71–80.

CAMERON, I. L. 1973. Growth characteristics of *Tetrahymena*. Pp. 199–226 in A. M. Elliott, Ed. *Biology of Tetrahymena*. Dowden, Hutchinson & Ross, Stroudsburg.

CHALKLEY, A. W., and G. E. DANIEL. 1933. The relation between the form of the living cell and the nuclear phases of division in *Amoeba proteus*. *Physiol. Zool.*, **6**.

CHALLEY, J. R., and W. C. BURNS. 1959. The invasion of the cecal mucosa by *Eimeria tenella* sporozoites and their transport by macrophages. *J. Protozool.*, **6**:238–241.

CHENG T. C. 1973. *General Parasitology*. Academic Press, New York. 965 pp.

CLEVELAND, L. R. 1957. Correlation between the molting period of *Cryptocercus* and sexuality in its Protozoa. *J. Protozool.*, **4**:168–175.

———. 1958. Photographs of fertilization in *Trichonympha grandis*. *J. Protozool.*, **5**:115–122.

CONKLIN, K. A., and S. C. CHOU. 1970. Antimalarials: effects on *in vivo* and *in vitro* protein synthesis. *Science,* **170**:1213–1214.

CORLISS, J. O. 1954. The literature on *Tetrahymena:* its history, growth, and recent trends. *J. Protozool.*, **1**:156–169.

———. 1972. The ciliate Protozoa and other organisms: some unresolved questions of major phylogenetic significance. *Am. Zoologist,* **12**:739–753.

———. 1974. The changing world of ciliate systematics: historical analysis of past efforts and a newly proposed phylogenetic scheme of classification for the protistan phylum Ciliophora. *Syst. Zool.,* **23**:91–138.

———. 1975. Taxonomic characterization of the suprafamilial groups in a revision of recently proposed schemes of classification for the phylum Ciliophora. *Trans. Am. Micr. Soc.*, **94**:224–267.

———. 1977. Annotated assignment of families and genera to the orders and classes currently comprising the Corlissian scheme of higher classification for the phylum Ciliophora. *Trans. Am. Micr. Soc.*, **96**:104–140.

———. 1978. A salute to fifty-four great microscopists of the past: a pictorial footnote to the history of protozoology. Part 1. *Trans. Am. Micr. Soc.,* **97**:419–458.

———. 1979. *The Ciliated Protozoa*. 2nd Ed. Pergamon, Oxford. 455 pp.

————, et al. 1973. The role of protozoa in some ecological problems. *Am. Zoologist,* **13**:144–232.

CURDS, C. R. 1973. The role of protozoa in the activated-sludge process. *Am. Zoologist,* **13**:161–169.

CUSHMAN, J. A. 1955. *Foraminifera.* Harvard Univ. Press, Cambridge. 605 pp.

DIEHN, B., J. R. FONSECA, and T. L. JAHN. 1975. High speed cinemicrography of the direct photophobic response of *Euglena* and the mechanism of negative phototaxis. *J. Protozool.,* **22**:492–494.

DOGIEL, V. A., J. I. POLJANSKIJ, and E. M. CHEJSIN. 1965. *General Protozoology.* Oxford Univ. Press, London. 747 pp.

DRAGESCO, J. 1973. Quelques données écologiques sur les Ciliés libres de l'Afrique. *Am. Zoologist,* **13**:231–232.

ELLIOTT, A. M. (Ed.). 1973. *Biology of Tetrahymena.* Dowden, Hutchinson & Ross, Stroudsburg. 508 pp.

————, M. A. ADDISON, and S. E. CAREY. 1962. Distribution of *Tetrahymena pyriformis* in Europe. *J. Protozool.,* **9**:135–141.

————, and I. J. BAK. 1964. The contractile vacuole and related structures in *Tetrahymena pyriformis. J. Protozool.,* **11**:250–261.

EMILIANI, C. 1971. Depth habitats of growth stages of pelagic Foraminifera. *Science,* **173**:1122–1124.

ERICSON, D. B., M. EWING, and G. WOLLIN. 1963. Pliocene-Pleistocene boundary in deep-sea sediments. *Science,* **139**:727–737.

FARR, M. M., and D. J. DORAN. 1962. Comparative excystation of four species of poultry coccidia. *J. Protozool.,* **9**:403–407.

FENNELL, R. A., and F. O. MARZKE. 1954. The relation between vitamins, inorganic salts and the histochemical characteristics of *Tetrahymena geleii* W. *J. Morphol.,* **94**:587–616.

FERONE, R., M. O'SHEA, and M. YOELI. 1970. Altered dihydrofolate reductase associated with drug-resistance transfer between rodent plasmodia. *Science,* **167**:1263–1264.

FOISSNER, W., and N. WILBERT. 1979. Morphologie, Infraciliatur und Okologie der limnischen Tintinnina: *Tintinnidium fluviatile* Stein, *Tintinnidium pusillum* Entz, *Tintinnopsis cylindrata* Daday and *Codonella cratera* (Leidy) (Ciliophora, Polyhymenophora). *J. Protozool.,* **26**:90–103.

FRANKEL, J., and N. E. WILLIAMS. 1973. Cortical development in *Tetrahymena.* Pp. 375–409 in A. M., Elliott, Ed. *Biology of Tetrahymena.* Dowden, Hutchinson & Ross, Stroudsburg.

FRENKEL, J. K., J. P. DUBEY, and N. L. MILLER. 1970. *Toxoplasma gondii* in cats: fecal stages identified as coccidian oocysts. *Science,* **167**:893–896.

GAVIN, M. A., T. WANKO, and L. JACOBS. 1962. Electron microscope studies of reproducing and interkinetic *Toxoplasma. J. Protozool.,* **9**:222–234.

GELBER, B. 1958. Retention in *Paramecium aurelia. J. Comp. Physiol. Psychol.,* **1958**:110–115.

GIBBONS, I. R., and A. J. ROWE. 1965. Dynein: a protein with adenosine triphosphatase activity from cilia. *Science,* **149**:424–426.

GIESE, S. C. 1973. *Blepharisma.* Stanford Univ. Press, Stanford. 366 pp.

GOLDACRE, R. J., and I. J. LORCH. 1950. Folding and unfolding of protein molecules in relation to protoplasmic streaming, amoeboid movement and osmotic work. *Nature,* **166**:497–500.

GRELL, K. G. 1973. *Protozoology*. Springer-Verlag, Berlin. 554 pp.

GRIFFIN, J. L. 1972. Temperature tolerance of pathogenic and non-pathogenic free-living amoebas. *Science,* **178**:869–870.

GUTTMAN, H. N. 1971. Internal cellular details of *Euglena gracilis* visualized by scanning electron microscopy. *Science,* **171**:290–292.

HONIGBERG, B. M., et al. 1964. A revised classification of the Phylum Protozoa. *J. Protozool.,* **11**:7–20.

HOARE, C. H. 1972. *The Trypanosomes of Mammals*. Blackwell, Oxford. 749 pp.

HYMAN, L. H. 1940. *The Invertebrates: Protozoa Through Ctenophora*. Vol. I. McGraw-Hill, New York. 726 pp.

JAHN, T. L., and E. C. BOVEE. 1965. Mechanisms of movement in taxonomy of Sarcodina. I. As a basis for a new major dichotomy into two classes, Autotractea and Hydraulea. *Am. Midl. Nat.,* **73**:30–40.

———, and F. F. JAHN. 1949. *How to Know the Protozoa*. Brown, Dubuque. 234 pp.

———, M. D. LANDMAN, and J. R. FONSECA. 1964. The mechanism of locomotion of flagellates. II. Function of the mastigonemes of *Ochromonas*. *J. Protozool.,* **11**:291–296.

JARRETT, R. M., and L. N. EDMUNDS, JR. 1970. Persisting circadian rhythm of cell division in a photosynthetic mutant of *Euglena*. *Science,* **167**:1730–1733.

JENNINGS, H. S. 1906. *Behavior of the Lower Organisms*. Columbia Univ. Press, New York. 380 pp.

JEON, K. W. (Ed.). 1973. *The Biology of Amoeba*. Academic Press, New York. 628 pp.

———, I. J. LORCH, and J. F. DANIELLI. 1970. Reassembly of living cells from dissociated components. *Science,* **167**:1626–1627.

JEPPS, M. W. 1956. *The Protozoa, Sarcodina*. Oliver & Boyd, Edinburgh. 183 pp.

KAHL, A. 1932–1935. *Urtiere oder Protozoa. I. Wimpertiere oder Ciliata*. Pts. 18, 21, 25, and 30 of *Die Tierwelt Deutschland*. Gustav Fischer, Jena. 886 pp.

KATES, J. R., and L. GOLDSTEIN. 1964. A comparison of the protein composition of three species of amoebae. *J. Protozool.,* **11**:30–35.

KUDO, R. R. 1966. *Protozoology*. 5th Ed. Thomas, Springfield. 1174 pp.

LEVINE, N. D. 1961. *Protozoan Parasites of Domestic Animals and Man*. Burgess, Minneapolis. 412 pp.

———, et al. 1962. Survival of *Tritrichomonas foetus* stored at −28 and −95°C after freezing in the presence of glycerol. *J. Protozool.,* **9**:347–350.

LUDVIK, J., et al. (Eds.). 1963. *Progress in Protozoology*. Academic Press, New York. 623 + 105 pp.

LUNDIN, F. C., and L. S. WEST. 1962. *The Free-Living Protozoa of the Upper Peninsula of Michigan*. Northern Michigan College Press, Marquette. 192 pp.

LYNTS, G. W., and J. B. JUDD. 1971. Late Pleistocene paleotemperatures at Tongue of the Ocean, Bahamas. *Science,* **171**:1143–1144.

MACKINNON, D. L., and R. S. HAWES. 1961. *An Introduction to the Study of Protozoa*. Clarendon Press, Oxford. 506 pp.

MANWELL, R. D. 1961. *Introduction to Protozoology*. St. Martin's Press, New York. 642 pp.

MARGULIS, L. 1971. The origin of plant and animal cells. *Am. Scientist,* **59**:230–235.

MAST, S. O. 1931. Locomotion in *Amoeba proteus. Protoplasma,* **14**:321–330.

————, and R. A. FENNELL. 1938. The relation between temperature, salts, hydrogen ion concentration and frequency of ingestion of food by *Amoeba. Physiol. Zool.,* **11**:1–18.

MEHLHORN, H., and A. O. HEYDORN. 1978. The Sarcosporidia (Protozoa, Sporozoa): life cycle and fine structure. *Adv. Parasitol.,* **16**:43–72.

MOLYNEUX, D. H. 1977. Vector relationships in the Trypanosomatidae. *Adv. Parasitol.,* **15**:1–82.

MURRAY, A. W., M. SPISZMAN, and D. E. ATKINSON. 1971. Adenosine 3′5′-monophosphate phosphodiesterase in the growth medium of *Physarum polycephalum. Science,* **171**:496–498.

NAITOH, Y., and R. ECKERT. 1969. Ciliary orientation: controlled by cell membrane or by intracellular fibrils? *Science,* **166**:1633–1635.

NICHOLS, K. M., and R. RIKMENSPOEL. 1978. Control of flagellar motion in *Chlamydomonas* and *Euglena* by mechanical microinjection of Mg^{2+} and Ca^{2+} and by electric current injection. *J. Cell Sci.,* **29**:233–247.

NOLAND, L. E., and H. E. FINLEY. 1931. Studies on the taxonomy of the genus *Vorticella. Trans. Am. Micr. Soc.,* **50**:81–123.

PASTER, Z., and B. C. ABBOTT. 1970. Gibberellic acid: a growth factor in the unicellular alga *Gymnodinium breve. Science,* **169**:600–601.

PITELKA, D. R. 1961. Fine structure of the silverline and fibrillar systems of three tetrahymenid ciliates. *J. Protozool.,* **8**:75–89.

————. 1963. *Electron-Microscopic Structure of Protozoa.* Macmillan, New York. 269 pp.

————, and C. N. SCHOOLEY. 1955. Comparative morphology of some protistan flagella. *Univ. Calif. Publ. Zool.,* **61**:79–128.

PUYTORAC, P. de, et al. 1974. Proposition d'une classification du phylum Ciliophora Doflein, 1901. *C.R. Acad. Sci.,* **278**:2799–2802.

RASMUSSEN, L., and E. ORIAS. 1975. *Tetrahymena:* growth without phagocytosis. *Science,* **190**:464–465.

RATCLIFFE, H. L. 1927. Mitosis and cell division in *Euglena spirogyra* Ehrenberg. *Biol. Bull.,* **53**:109–122.

RINALDI, R. A., and T. L. JAHN. 1963. On the mechanism of ameboid movement. *J. Protozool.,* **10**:344–357.

RUDDIMAN, W. F. 1969. Recent planktonic Foraminifera: dominance and diversity in North Atlantic surface sediments. *Science,* **164**:1164–1167.

RUDZINSKA, M. A., and W. TRAGER. 1962. Intracellular phagotrophy in *Babesia rodhaini* as revealed by electron microscopy. *J. Protozool.,* **9**:279–288.

SANDON, H. 1927. *The Composition and Distribution of the Protozoan Fauna of the Soil.* Oliver & Boyd, London. 237 pp.

SCHAEFFER, A. A. 1920. *Amoeboid Movement.* Princeton. 156 pp.

SCHERBAUM, O., and E. ZEUTHEN. 1954. Induction of synchronous cell division in mass cultures of *Tetrahymena piriformis. Exp. Cell Res.,* **6**:221–227.

SCHMIDT-NIELSEN, B., and C. R. SCHRAUGER. 1963. *Amoeba proteus:* studying the contractile vacuole by micropuncture. *Science,* **139**:606–607.

SEAMAN, G. R. 1949. The presence of the tricarboxylic acid cycle in the ciliate *Colpidium campylum. Biol. Bull.,* **96**:257–262.

SHEFFIELD, H. G., and M. J. MELTON. 1970. *Toxoplasma gondii:* the oocyst, sporozoite, and infection of cultured cells. *Science,* **167:**892–893.

SHORTT, H. E., and P. C. GARNHAM. 1948. The pre-erythrocytic development of *Plasmodium cynomolgi* and *Plasmodium vivax. Trans. Roy. Soc. Trop. Med. Hyg.,* **41:**785–794.

SLEIGH, M. A. 1973. *The Biology of Protozoa.* Edward Arnold, London. 315 pp.

SMALL, E. B. 1967. The Scuticociliatida, a new order of the class Ciliatea (phylum Protozoa, subphylum Ciliophora). *Trans. Am. Micr. Soc.,* **86:**345–370.

———, and D. S. MARSZALEK. 1969. Scanning electron microscopy of fixed, frozen, and dried Protozoa. *Science,* **163:**1064–1065.

SONNEBORN, T. M. 1975. The *Parmecium aurelia* complex of fourteen sibling species. *Trans. Am. Micr. Soc.,* **94:**155–178.

SPOON, D. M., G. B. CHAPMAN, R. S. CHENG, and S. F. ZANE. 1976. Observations on the behavior and feeding mechanisms of the suctorian *Heliophrya erhardi* (Rieder) Matthes preying on *Paramecium. Trans. Am. Micr. Soc.,* **95:**443–462.

SRINIVASAN, M. S., and J. P. KENNETT. 1974. Secondary calcification of the planktonic foraminifer *Neogloboquadrina pachyderma* as a climatic index. *Science,* **186:**630–632.

STRICKLER, J. E., and C. L. PATTON. 1975. Adenosine 3′,5′-monophosphate in reproducing and differentiated trypanosomes. *Science,* **190:**1110–1112.

TARTER, V. 1961. *The Biology of Stentor.* Pergamon Press, New York. 413 pp.

WAGTENDONK, W. J. VAN (Ed.). 1974. *Paramecium—A Current Survey.* Elsevier, New York. 499 pp.

WALLACE, G. D. 1973. *Sarcocystis* in mice inoculated with *Toxoplasma*-like oocysts from cat feces. *Science,* **180:**1375–1377.

WARTON, A., and B. M. HONIGBERG. 1979. Structure of trichomonads as revealed by scanning electron microscopy. *J. Protozool.,* **26:**56–62.

WHITTAKER, R. H. 1969. New concepts of kingdoms of organisms. *Science,* **163:**150–160.

WICHTERMAN, R. 1953. *The Biology of Paramecium.* McGraw-Hill, New York. 527 pp.

WILLEY, R. L., W. R. BOWEN, and E. DURBAN. 1970. Symbiosis between *Euglena* and damselfly nymphs is seasonal. *Science,* **170:**80–81.

WOLKEN, J. 1967. *Euglena.* 2nd Ed. Appleton-Century-Crofts, New York. 204 pp.

YUSA, A. 1957. The morphology and morphogenesis of the buccal organelles in *Paramecium* with particular reference to their systematic significance. *J. Protozool.,* **4:**128–142.

———. 1963. An electron microscope study on regeneration of trichocysts in *Paramecium caudatum. J. Protozool.,* **70:**253–262.

3

Phylum Porifera

The sponges have a cellular grade of construction not much more complicated than some of the colonial Protozoa. Most sponges are marine but a few live in fresh water. The bath sponges that are familiar to all of us consist of the skeleton of certain Porifera that live in the sea.

Sponges are usually attached and stationary (sessile) animals in the adult stage, distribution being brought about largely by the actively swimming ciliated larvae or by currents of water, which carry the young from place to place before they become attached. The thousands of different species vary greatly in shape, size, structure, and geographical distribution. They were for centuries considered to be plants, and their animal nature was not accepted until about 1857.

Many zoologists think the sponges are remnants of a group of animals that separated from the other Metazoa at an early stage in the evolutionary series and have remained in their primitive condition ever since. For this reason the Metazoa are sometimes arranged in two groups, one containing the sponges and designated the Parazoa and the other the Metazoa proper.

The view of Tuzet (1963), shown in Figure 3-1, B, that sponges are in the mainstream of metazoan evolution is one that logic may compel us to accept if we agree that existing sponges have diverged and specialized as have the Metazoa. Some reasons and evidence include:

1. If sponges are a lateral offshoot (Fig. 3-1, A), then we must postulate some primitive aggregation of cells for the question mark in Figure 3-1, A. Ockham's razor—the simplest explanation is the best

157

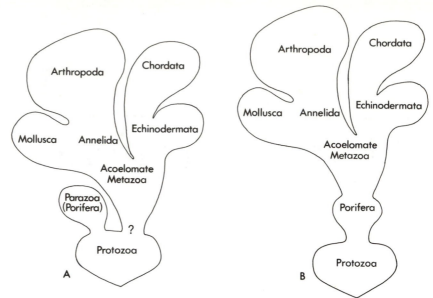

Figure 3-1 Diagram of suggested phylogenetic positions of the Porifera. (A) Conventional view of sponges as an isolated subkingdom, the Parazoa. (B) Tuzet's view of sponges in the mainstream of metazoan evolution.

until proven false—indicates dropping such an unnecessary postulate.

2. Negative evidence such as lack of mouth, locomotion, or comparable embryology is not compelling evidence denying a position in the mainstream for a form expected to be a rudimentary ancestor.

3. Comparable cell types—flagellated cells, amoebocytes, epidermal cells—are found in both sponges and metazoans.

4. Collagen, the important connective tissue fiber of higher forms, is essentially similar to the collagen called spongin in sponges.

5. Immunocompetence has been demonstrated in sponges (Hildemann, Johnson, and Jokiel, 1979).

6. A more sophisticated feeding system and its associated sensory and motility systems would have little basis for selection until higher Metazoa existed. The sponge was adapted to the early premetazoan conditions of unicellular particulate food.

7. Features of sponges (for example, myocytes and carbonate secretion) could be improved through selection for systems of higher organisms.

In view of the preceding it is reasonable to anticipate further evidence of an important phylogenetic position for ancestral sponges with hopes for a clarification or resolution of this issue.

Sponges lack organs but do have some groupings of cells into tissues. Recent interest in sponge regeneration, reorganization, grafting, and development is partially based on similarities to connective tissue of humans and the hope that principles of tissue transplantation may be uncovered.

Sponges are not easy to classify but are arranged in four classes as follows.

CLASS I. CALCAREA. With spicules of calcium carbonate.

CLASS II. HYALOSPONGEA (= HEXACTINELLIDA). With siliceous triaxon spicules.

CLASS III. DEMOSPONGEA. Usually with spicules of silicon, not triaxon, or with spongin, or with both spicules and spongin.

CLASS IV. SCLEROSPONGEA. Skeleton with spongin, siliceous spicules, and a basal carbonate portion.

Class I. Calcarea

LEUCOSOLENIA

Leucosolenia (Fig. 3-2) is a sponge that will serve to illustrate the structure of the most simple members of the phylum. It is found growing on the rocks near the seashore just below low-tide mark, and consists of a number of horizontal tubes from which branches extend up into the water. These branches have an opening, the *osculum,* at the distal end, and *buds* and *branches* projecting from their sides. The buds and branches are hollow, possessing a single spongocoel, which communicates with the horizontal tubes. The entire mass is a colony of animals, and the tissues connected with a single osculum may be considered an individual sponge.

If a branch is examined under a microscope, it will be found to contain a large number of three-pronged (triradiate) *spicules*, which are embedded in the soft tissues of the body wall (Fig. 3-3); these serve to strengthen the body and hold it upright. The application of acid results in the dissolution of these spicules and the production of an ef-

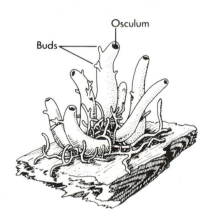

Figure 3-2 *Leucosolenia:* a small colony. *(From Woodruff.)*

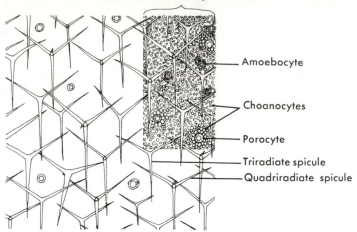

Area with inner cell layer indicated

Amoebocyte

Choanocytes

Porocyte

Triradiate spicule

Quadriradiate spicule

Figure 3-3 Portion of the body wall of *Leucosolenia,* surface view of selected details.

fervescence, thus proving them to be composed of calcium carbonate. The body wall is so flimsy that it is difficult to study even under the best conditions. It is made up of two layers of cells (Fig. 3-4): an outer layer, the *epidermis* (=*pinacoderm*), and an inner *choanocyte layer* (=*choanoderm*). These layers, as will be shown later are not comparable to the epidermis and gastrodermis of the Coelenterata and other Metazoa. Between these two layers is a jellylike substance (*mesohyl*) similar to the mesoglea of *Hydra* in which are many *amoebalike wandering* cells (*amoebocytes*).

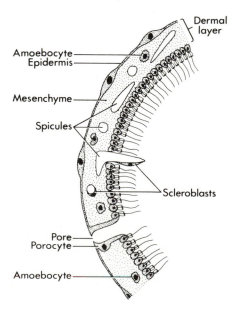

Dermal layer

Amoebocyte
Epidermis

Mesenchyme

Spicules

Scleroblasts

Pore
Porocyte

Amoebocyte

Figure 3-4 Structure of body wall of an ascon sponge. (*From Kerr.*)

The choanocyte layer is peculiar, since it consists of a single layer of *collar cells,* the *choanocytes* (Fig. 3-4), which resemble the similar cells of the choanoflagellate Protozoa (Fig. 2-11, B) and internal cells of animals from several higher phyla. The collar has tentaclelike thickenings supporting a fine mucus mesh capable of trapping bacterial-sized particles. The food vacuoles formed are passed on to amoebocytes in the mesenchyme (= mesohyl).

If a little coloring matter is placed in the water, it will be drawn into the animal through minute intracellular *incurrent pores,* the *ostia* (Fig. 3-4), in the body wall and will pass out through the openings in a sievelike membrane stretched across the osculum. The osculum is therefore the *exhalant opening,* and not the mouth, as a casual examination might lead one to believe. The course of the current of water in such a sponge is shown by arrows in Figure 3-19, A. The presence of the incurrent pores suggested the name Porifera for members of this phylum.

An extensive specialized terminology has been developed for sponges. Some has been indicated parenthetically. It is necessary and meaningful to the specialist in many cases but confusing to the student, especially when structural or functional concepts are not involved. For example, distinctive names of the layers to emphasize the offshoot position of sponges and different method of embryological origin overlooks the fact that it may be possible for the embryological orgin of homologous structures to undergo change. Any portion of the life-cycle stages of an organism may evolve adaptively with some independence from other stages. The appendage that develops on the egg of *Asellus* (Chapter 10) is an example of such an event. Many new cell types have been described for sponges based on valid ultrastructural criteria.

SCYPHA

Scypha (= *Sycon*) (Fig. 3-5) is a simple sponge inhabiting the salt water along both the coasts near the low-tide mark. Here it is found permanently attached by one end to rocks and other solid objects. Its distribution in space is effected during the early embryonic stages, at which time external flagella are present, enabling it to swim about. It is closely related to the European genus *Grantia.*

Morphology

Scypha varies in length from 1 cm to over 2 cm, and resembles in shape a slender vase that bulges slightly near the center. The distal end of the animal opens to the exterior by a large *excurrent pore,* the *osculum.* This opening is surrounded on all sides by a circlet of long straight needles called *spicules.* Smaller spicules protrude from other

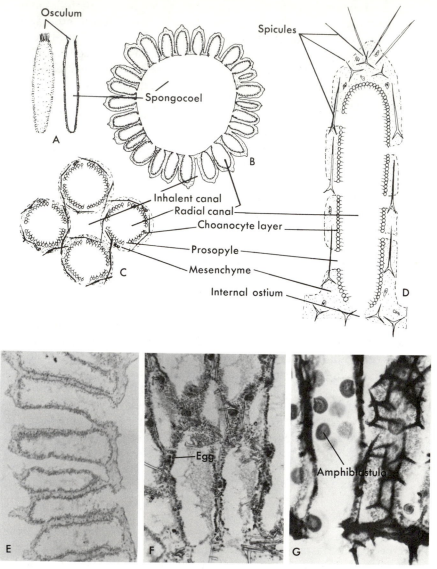

Figure 3-5 *Scypha*. (A) Entire animal and longitudinal section. (B) Cross section. (C) Tangential section through canals. (D) Section along a radial canal. (E) Part of wall in cross section; (F) with eggs; (G) with amphiblastula. (A–D) Diagrammatic, (E–G) photomicrographs.

parts of the body, giving the animal a ciliated appearance. The body wall is perforated by numerous *incurrent pores*.

A specimen of *Scypha* split longitudinally (Fig. 3-5) shows the body to be a hollow sac, one large central cavity, the *spongocoel* being present. The body wall is honeycombed by a great many canals; some of these, the *radial canals,* open to the spongocoel through minute pores, the *apopyles,* and end blindly near the outer surface; others, the *incurrent canals,* open to the outside through incurrent pores or *ostia,* and end blindly near the inner surface of the body wall; still other canals, the *prosopyles,* even smaller than those already noted, connect the radial with the incurrent canal. Figure 3-5, B, shows in cross section the structure of a simple sponge. Discharging into the spongocoel are the radial canals. The body wall is seen to be crowded with both radial and incurrent canals, which have been cut lengthwise. The relations of the various canals to one another are shown in Figure 3-19; here the arrows indicate the direction of the current of water, which enters the incurrent canal, passes through the prosopyles into the radial canal, and thence into the spongocoel, finally escaping from the body by way of the osculum. The surface area of the epithelium covering the body, and lining the internal cavities, is enormously increased by the canal system.

Scypha is an animal possessing an outer dermal layer of cells, an inner choanocyte layer, and a middle region containing cells of several varieties (Fig. 3-6). The *epidermis* (pinacoderm) covers the entire outer surface of the body, and lines the incurrent canals. It is composed externally of a single layer of flat cells. Numerous prosopyles occur con-

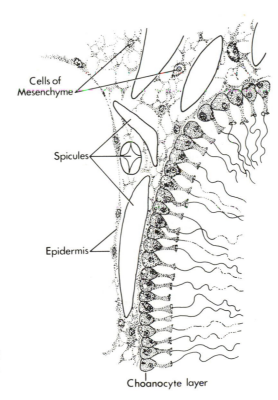

Cells of Mesenchyme

Spicules

Epidermis

Choanocyte layer

Figure 3-6 *Grantia:* Part of body wall. *(From Dahlgren and Kepner.)*

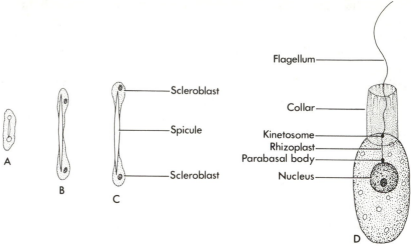

Figure 3-7 Sponge cells, diagrammatic. (A–C) Scleroblasts secreting a monaxon spicule. (D) Choanocyte, showing protoplasmic fingers of the collar.

necting the incurrent with the radial canals. The prosopyle is an intercellular perforation. Cells, called *scleroblasts,* which produce spicules, are also considered constituents of the dermal layer.

The *choanocyte layer* (choanoderm) lines the radial canals. It consists of one layer of collared flagellated cells (Figs. 3-6 and 3-7), the *choanocytes.* The collar of these cells consists of from 20 to 30 protoplasmic fingers fused side by side and capable of contraction. No collar cells are present in the epithelium lining the spongocoel. The flagella of the collar cells create the current of water continually flowing through the body wall into the spongocoel and out of the osculum.

The *middle region* (mesohyl) of the body wall, although not so definite nor firm in structure as the outer and inner epithelia, contains cells and much noncellular material. *Amoebocytes,* which ingest food or act as storage cells, are found here; some serve as the *reproductive cells,* which always arise in the middle layer.

The soft body wall of *Scypha* is supported and protected by a skeleton composed of a great number of *spicules* of carbonate of lime. Four varieties of spicules are always present: (1) long straight monaxon rods guarding the osculum, (2) short straight monaxon rods surrounding the incurrent pores, (3) triradiate spicules always found embedded in the body wall, and (4) T-shaped spicules lining the gastral cavity; four- and five-rayed spicules may also be present. Spicules are built up within cells called *scleroblasts* (Fig. 3-7), which form part of the inner stratum of the dermal layer. A slender organic axial thread is first built up within the cell; around this is deposited the calcareous matter; the whole spicule is then insheathed by an envelope of organic matter like that composing the axial thread. Scleroblasts begin spicule formation as binucleate cells, which form one axis of a spicule. The cell divides to form two cells as the spicule grows, one cell working at each end of the axis. Thus, a pair of cells participate in the formation of each axis, or two cells produce a monaxon spicule and six scleroblasts are needed to produce a triradiate spicule. This view of spicule formation has been questioned by some observers.

Nutrition

Scypha lives upon the minute organisms and small particles of organic matter that are drawn into the incurrent canals by the current of water produced by the beating of the collar-cell flagella. Food particles are engulfed by the collar cells. *Digestion,* as in the Protozoa, is intracellular, food vacuoles being formed. The distribution of the nutriment is accomplished by the passage of digested food, or food vacuoles, from cell to cell, aided by the amoeboid wandering cells of the middle layer.

Excretory matter is discharged through the general body surface and osculum assisted probably by the amoeboid wandering cells, and possibly by the collar cells also. *Respiration* likewise takes place, in the absence of special organs, through the cells of the body wall.

Sponges do not possess differentiated *nervous organs,* but are able to respond to certain stimuli. The pores and oscula are surrounded by contractile cells, called *myocytes,* which are able to close these openings. Apparently, these cells respond to direct stimulation, and, since primitive nervous cells are present, they, therefore, represent what may be considered the very beginning of a neuromuscular mechanism.

Reproduction

Reproduction in *Scypha* takes place by both sexual and asexual methods. In the latter case, a bud arises near the point of attachment, finally becomes free, and takes up a separate existence.

The *sexual reproductive cells* develop in the middle layer of the body wall. Both eggs and sperms occur in a single individual (i.e., *Scypha* is *monoecious* or *hermaphroditic*). In *spermatogenesis* the primordial germ cell, called a spermatogonium, divides, producing a covering cell, the spermatocyst, and a central sperm mother cell, the spermatocyte. The latter forms a number of spermatids by mitosis; these transform into spermatozoa. The *ova* arise by the growth of *oogonia* of the middle layer, which are nourished by neighboring cells. The oogonium divides twice to produce four ova. Spermatozoa are carried to the ova by choanocytes, which become modified into amoeboid cells when the spermatozoans adhere to them.

Embryology

The development of the fertilized egg (Fig. 3-8) has been observed in *Scypha.* The egg segments by three vertical divisions into a pyramidal

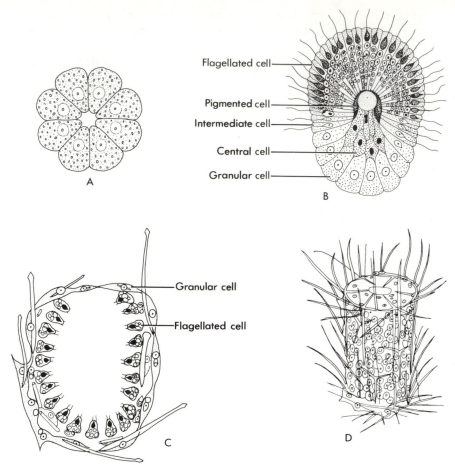

Figure 3-8 Embryology of sponges. (A) An egg of *Sycandra* that has reached the eight-cell stage. (B) An amphiblastula larva of *Leucosolenia*. (C) Young *Leucosolenia*, 4 days old. (D) Older stage in *Sycon raphanus*. (*A, D, after Schulze; B, C, after Minchin.*)

plate of eight cells. A horizontal division now cuts off a small cell from the top of each of the eight, the result being a layer of eight large cells, or macromeres, crowned by a layer of eight small cells, or micromeres. The cells now become arranged about a central cavity, producing a *blastulalike sphere*. The small cells multiply rapidly and develop flagella extending into the interior cavity, while the macromeres become granular. A mouthlike opening into the central cavity develops between the micromeres. The embryo is then known as a *stomatoblastula* and is able to engulf maternal amoebocytes for nutrition. Then follows a process of inversion in which the embryo turns itself inside out through the mouth. The inverted embryo, with flagella on the micromeres now directed outward and one hemisphere still composed of eight macromeres, is called an *amphiblastula*. The mass of cells then becomes disk-shaped by the pushing in of the flagellated cells. Two layers are thus formed, between which the middle layer arises. The invaginated side soon becomes attached, and the embryo lengthens into a cylinder at the distal end of which a new opening, the osculum, appears. In the meantime, spicules and the canal system arise in the body wall.

The larva of some Calcinea becomes a *parenchymula* or stereogastrula in which the central cavity is solidly filled with cells invaginated from one end of the outer completely flagellated layer.

Class III. Demospongea

SPONGILLA

Spongilla (Figs. 3-9 and 3-10) is a common, widely distributed, freshwater sponge. It has a more complex level of canal system development than either *Leucosolenia* or *Scypha*. The choanocytes are restricted to small chambers in the interior. The flagellated chambers propel the water through the canal system of the sponge. The sponges are encrusting in habit and are commonly found on sticks that have been submerged for some time.

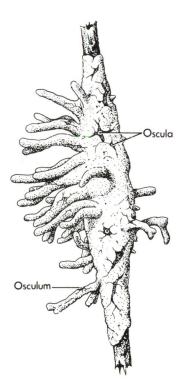

Oscula

Osculum

Figure 3-9 *Spongilla lacustris,* a freshwater sponge. (*After Weltner.*)

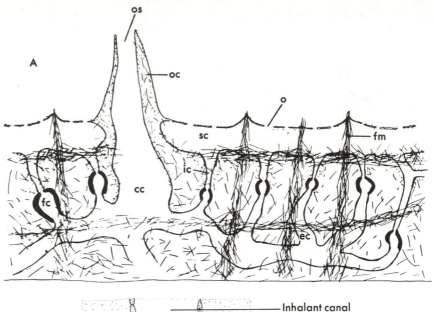

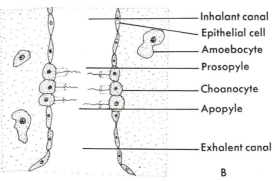

	Inhalant canal
	Epithelial cell
	Amoebocyte
	Prosopyle
	Choanocyte
	Apopyle
	Exhalent canal

B

Figure 3-10 *Spongilla.* (A) Diagrammatic section of peripheral portion of a typical freshwater sponge. Heavy black lines represent layers of choanocytes lining flagellated chambers. Delicate matrix of spongin not shown. cc, central cavity; ec, excurrent canal; fc, flagellated chamber; fm, fascicle of megascleres; ic, incurrent canal; o, ostium; oc, oscular chimney; os, osculum; sc, subdermal cavity. (B) Enlarged diagram of a flagellated chamber. (*A, from Robert W. Pennak.* Fresh-water Invertebrates of the United States, *2nd edition, copyright 1978 by John Wiley & Sons, Inc., New York.*)

Siliceous spicules of *Spongilla* are of the monaxon (one axis, or rod-like) type and of two general sizes. Larger skeletal spicules, called *megascleres,* are held together by spongin fibers and provide support for the colony. Smaller flesh spicules, called *microscleres,* are embedded in cells of the sponge.

Water enters the colony through numerous dermal pores or *ostia.* From subdermal cavities it enters the flagellated chambers through incurrent canals and leaves through excurrent canals to reach the spongocoel. A small amount may leave through contractile vacuoles observed in epidermal cells. The spongocoel opens to the exterior by a large opening, the *osculum*. The living portion of *Spongilla* is never more than a few centimeters thick, but a colony may encrust a many times greater area. An individual of the colony may be considered that portion of the sponge associated with one osculum. Bergquist (1978) notes the long history of concern over what part of a sponge is the individual. She favors the view that a sponge is an individual rather than a colony because its activities are well integrated as a biological unit. That view is certainly valid for nonencrusting marine sponges. There may be some value to consider them colonies and philosophize about the value of cooperation among individuals.

Reproduction

Asexual reproduction can result from regeneration of fragments by vegetative growth or by the development of new sponges from specialized structures termed gemmules. Sexual reproduction in *Spongilla* is similar to that process in *Scypha*. However, sexes are separate but are known to change in some instances following winter dormancy (Gilbert and Simpson, 1976), an obvious advantage to sessile animals, especially if derived asexually from a single parent.

Gemmules (Fig. 3-11) are asexual reproductive bodies resistant to adverse environmental conditions. They are formed by the sponge during the summer and fall and stored in its basal portion. *Amoebocytes* (trophocytes and thesocytes) loaded with nutrients provide a core material capable of developing into a new sponge. Other amoebocytes secrete a *chitinous covering* and reinforcing *spicules*. A *micropyle*, or opening in the gemmule covering, provides an escape route for the

Figure 3-11 Gemmules of freshwater sponges. (A) *Heteromeyenia*. (B and C) *Spongilla*.

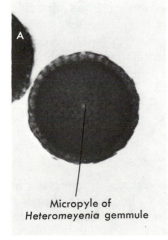
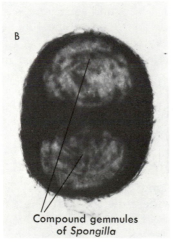

Micropyle of
Heteromeyenia gemmule

Compound gemmules
of Spongilla

thesocytes after the winter quiescent period. Gemmules may be released from their location in the sponge by the decay of the parent during the late fall or winter. They provide an important means of dispersal for freshwater sponges, because they can be carried passively by a wide variety of agents. Because the gemmules are distinctive in structure and spicule types for the different species, they have a great value in species identification. Organization of parts of the sponge, other than gemmules and spicules, shows too great variability to be of as much value in identification.

Embryology

Development of the sponge from the amphiblastula of *Spongilla* differs from that described for *Scypha* in that the flagellated cells from the exterior do not produce the choanocytes. The choanocytes that line the flagellated chambers develop from cells in the interior of the amphiblastula larva.

Other Porifera

CLASS I. CALCAREA

The Calcarea are all marine and live in shallow water. They are comparatively simple in structure. Six orders and about 150 species are known. As their name indicates, their chief characteristic is the presence of calcareous spicules. Subdivision of the class is based on type of canal system, position of the choanocyte nuclei and their relationship to the flagella, and larval development.

The simplest of all sponges occur in two families, the Leucosoleniidae and the Clathrinidae (Fig. 3-12). The characteristics have been presented by *Leucosolenia* as a type. The body wall is thin, the spongocoel is lined with collar cells, and each pore is a perforation in a single cell.

Scypha has served to introduce the more complex Calcarea. The body wall is thickened, the spongocoel is lined with flattened epithelial cells, and the collar cells are located in radial canals or chambers. *Scypha ciliata* occurs from Rhode Island to Greenland; several species occur along the Pacific Coast. *Leucandra taylori* lives off the shore of Vancouver Island, *L. heathi* is in the low intertidal zone of California, and *Amphoriscus thompsoni* is in the Gulf of St. Lawrence.

CLASS II. HYALOSPONGEA (HEXACTINELLEA)

To this class belong the glass sponges. They are mostly deep-sea inhabitants and widely distributed. Their fossil remains are abundant and indicate that they have always been present in large numbers. They are characterized by the possession of siliceous spicules of the

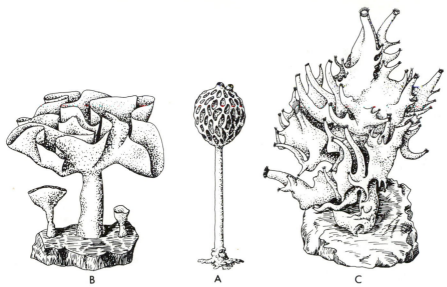

Figure 3-12 Types of Calcarea. (A) Calcinea. *Clathrina lacunosa*. (B) Calcaronea. *Grantia labyrinthica;* three stages of growth. (C) Calcaronea. *Leucandra aspera*. (*A and C, after Minchin; B, after Dendy.*)

triaxon type, that is, with six rays or a multiple of six. The body is often tubular or basket-shaped and the spicules may form a continuous skeleton resembling spun glass. Venus' flower basket, *Euplectella aspergillum* (Fig. 3-13), is especially abundant near the Philippine Islands; its skeleton is often seen in museums. It has a long, curved, cylindrical body held up by a framework of spicules. In nature it is fastened in the mud of the sea bottom by a mass of long siliceous threads at the posterior end. The body wall, as in all Hyalospongea, contains simple thimble-shaped flagellated chambers. The large

Figure 3-13 Hyalospongea. *Euplectella:* the "Venus' flower basket" sponge.

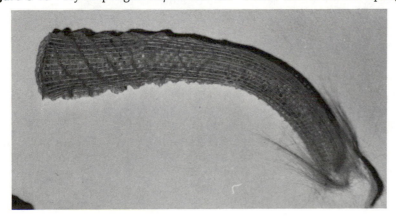

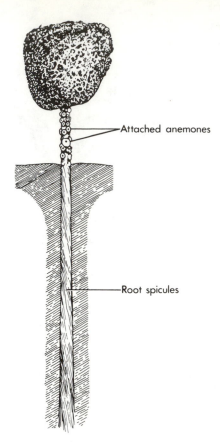

Attached anemones

Root spicules

Figure 3-14 Hyalospongea. *Hyalonema. (From Kerr.)*

spongocoel often serves as a residence for a pair of decapod crustaceans, *Spongicola venusta. Euplectella suberea* occurs in the West Indies. Another American hexactinellid is *Hyalonema longissimum* (Fig. 3-14), which lives off the New England coast at depths of from 60 to 95 fathoms.

Most of the biology of the hexactinellids is poorly known.

CLASS III. DEMOSPONGEA

These are the dominant Porifera of the present time, being widespread in nature and very numerous in species and individuals. Most of the sponges that are generally seen belong to this class. They are often massive and brightly colored, with complicated canal systems connected with small spherical flagellated chambers. The skeleton may consist of siliceous spicules, but these are not triaxon as in the Hyalospongea. Spongin, a substance similar to collagen, may provide all of the skeleton or be combined with siliceous spicules. Some spicules are straight and needle-shaped (monaxon); others have eight rays (te-

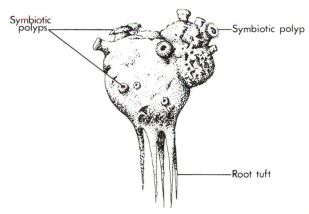

Symbiotic polyps

Symbiotic polyp

Root tuft

Figure 3-15 Demospongea. Order Choristida. *Thenea muricata*. (*After Minchin.*)

traxon). Some have a parenchymula larval stage, some an amphiblastula larva, and some have not been observed to have larval stages.

Order Choristida

Sponges in the order Choristida (Fig. 3-15) live mostly at depths of 50 to 200 fathoms and are usually attached to the bottom by tufts of spicules. They are usually compact and rounded. The outer, dermal layer of the body, the cortex, is thickened and contains skeletal spicules, megascleres, which unite to form the supporting framework. Other spicules, microscleres, lie scattered in the mesoglea. These spicules are modifications of the tetraxon type. Among the members of this order from America are *Thenea echinata*, which occurs on the New England coast north of Cape Cod and is shaped like a mushroom; and *Geodia mülleri*, which is found in the West Indies and is irregularly lobed.

Order Hadromerida

Hadromerid sponges (Fig. 3-17) have monaxon spicules. Their spongin is not in the form of fibers. They are firm but not as flexible as many sponges. *Suberites ficus* is an American species that ranges from Virginia to Maine with a related sponge on the West Coast. It has a bright yellow, elongated body that may reach a length of 15 cm and is often attached by one edge to the shells of hermit crabs. Another species with a similar range is *Polymastia robusta*. This is a yellowish or grayish species of irregular form with fingerlike branches. It reaches a diameter of 30 cm. *Cliona celata*, the sulfur sponge, occurs along our eastern and western coasts. It is bright yellow in color and may form a mass up to 20 cm in diameter. The young clionas are able to bore their way into calcareous rocks or the shells of mollusks forming tunnels in which they live.

Order Poecilosclerida

This is a large order with skeleton of monactine spicules combined with spongin. *Microciona prolifera* is a bright red species that forms large digitate masses on shells and stone along the East Coast, including Long Island Sound. *Mycale adherens* can be found on the shells of *Pecten* along the West Coast. *Mycale fibrexilis* occurs in shallow water on East Coast docks. It is yellowish brown in color and irregular in shape; it may be covered with hydroids and algae.

Order Haplosclerida

These are sponges with spicules and spongin skeleton forming a triangular mesh pattern. *Haliclona oculata* (Fig. 3-16) is the finger sponge, so named because of its being shaped like a hand with many fingers. In addition to large numbers of marine species this order includes the only freshwater sponges, the cosmopolitan family Spongillidae and the African family Potamolepidae.

Freshwater Sponges (Spongillidae). These usually encrust stones, sticks, and plants; but may also occur on the soft mud bottom and in highly colored water and even in waters contaminated by industrial and organic wastes. Freshwater sponges are usually yellow, brown, or green. The green color is due to the presence of Zoochlorellae. These sponges reproduce by the formation of gemmules and the characteristics of these gemmules are the chief means of identification. More

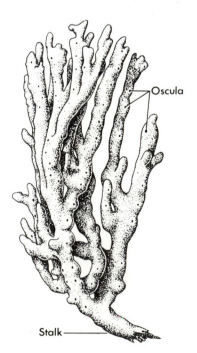

Oscula

Stalk

Figure 3-16 Demospongea. Order Haplosclerida. *Haliclona oculata.* (*After Minchin.*)

174 *Phylum Porifera*

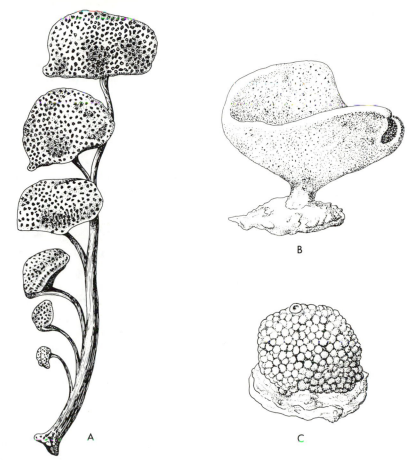

Figure 3-17 Demospongea. Order Axinellida: (A) *Esperiopsis challengeri;* (B) *Phakellia ventilabrum.* Order Hadromerida: (C) *Tethya lyncurium.* (*A, after Ridley, from Minchin; B and C, after Minchin.*)

than 20 species occur in this country. Some of the more common are as follows. *Spongilla lacustris* (Fig. 3-9) is the most abundant species. It prefers running water, is a branching form that lives in sunlight, and is usually green. *Eunapius fragilis* is also a common species, but prefers standing water and avoids the light. All freshwater sponges, except *Spongilla* and *Eunapius,* have birotulate gemmule spicules. Such spicules have a whorl (of spines, usually) at each end of a shaft (Fig. 3-20). *Ephydatia mulleri* is cushion-shaped and rarely branched. It is usually yellow or brown in color and prefers standing water. *Anheteromeyenia ryderi* lives in shallow flowing water, is light green, and is massive in type. *Trochospongilla pennsylvanica* lives in shallow water, is gray or green, and is only about 6 mm in diameter. It is averse to light and lives usually under stones and roots. *Heteromeyenia*

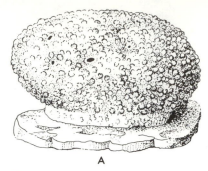
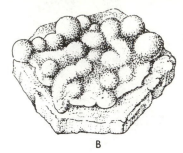

Figure 3-18 Demospongea. (A) Dictyoceratida. *Spongia officinalis*, a bath sponge. (B) Homosclerophorida. *Oscarella lobularis*. (*After Schulze, from Minchin*.)

tentasperma is an irregularly shaped sponge, yellowish green, that encrusts plants, roots, and stones. It occurs in the eastern United States.

Order Dictyoceratida

Everyone is familiar with the skeletons of some members of this order known as the bath sponges or commercial sponges (excluding the synthetic sponges that have largely replaced them in commerce). They occur in tropical and subtropical seas such as the Mediterranean and Gulf of Mexico. The skeleton, exclusively of spongin, left after the tissue is rotted and washed away, forms beautiful anastomosing patterns. About a dozen species are used commercially, and new sponge develops from the tissue left behind at harvest in many instances. The resulting sponge often has a basal cavity formed by overgrowth from the outer ring of healthier sponge tissue. *Spongia, Hippospongia,* and *Coscinoderma* are useful genera. *Spongia officinalis* (Fig. 3-18) has two useful subspecies from the Mediterranean. The glove sponge, *S. tubulifera,* is an American species of little commercial value.

Order Dendroceratida

Sponges of this order also lack spicules but the spongin forms a dendritic pattern or is completely lacking in some encrusting forms such as *Halisarca dujardini,* a slime sponge with elongate, saclike flagellated chambers. It is an American species living on red algae at a depth of about 5 fathoms off the coast of Rhode Island. *H. sacra* occurs off the coast of California.

Sponges in General

External Features

Sponges may be simple, thin-walled tubular structures like *Leucosolenia* and *Scypha* or massive and more or less irregular in shape. Many sponges are indefinite masses of tissue encrusting the stones, shells, sticks, or plants to which they are attached; others are more regular in shape and attached to the sea bottom by means of masses of spicules. The form exhibited by the members of each species may vary somewhat but is nevertheless constant. Some are branched like trees; others are shaped like a glove, or a cup, or a dome. Sponges vary in size from species no larger than a pinhead to species that attain a diameter of 1 meter and a height of 2 meters. Certain sponges appear to be ciliated because of the spicules that protrude from the body. Some sponges are white or gray, but most are colored red, yellow, orange, or green. Carotenoids are the most common pigments causing coloration other than the green due to zoochlorellae in green sponges.

Canal Systems

If it were not for the development of elaborate canal systems sponges would have remained in the simple condition of *Leucosolenia* and would never have been able to become massive in size. The canal system serves much the same purposes as the circulatory system does in higher animals; it furnishes a highway for food through the body and for the transportation of excretory matter out of the body. Three morphological types are usually recognized. The canals of the Hyalospongea are less well defined and seem to form a fourth type.

The ascon type (Fig. 3-19, A) consists of incurrent pores, a spongocoel lined with collar cells, and an osculum.

The sycon type (Fig. 3-19, B) is more complicated; the water flows through the dermal pores (ostia) into the incurrent canals; then through the chamber pores (prosopyles) into radial canals lined with collar cells; from here it is propelled by the flagella of the choanocytes into the spongocoel, finally passing out through the osculum.

In the leucon (rhagon) type (Fig. 3-19, C) there are three distinct parts: (1) the water passes through the dermal ostia and by way of incurrent canals reaches (2) a number of small chambers lined with choanocytes, thence it is carried through (3) an excurrent system to the spongocoel and finally out of the osculum.

The exhalent aperture of the choanocytic chamber often has a *central cell* (Fig. 3-20) that regulates flow by changing position. The central cell has deep surface grooves in which the flagella lie.

Skeletal Systems. All sponges, except those belonging to two families of Demospongea, are provided with a skeleton. This may consist of calcium carbonate or silicon in the form of spicules or of spongin in the form of fibers more or less closely united or a combination of these materials with a massive basal carbonate formation. Siliceous spicules are composed of opal, a form of hydrated silica similar to quartz in its chemical reactions. Spicules are of various types and hence of value in arranging sponges into groups. As already noted, the spicules in the class Calcarea are calcareous; those in the class Hyalospongea are of

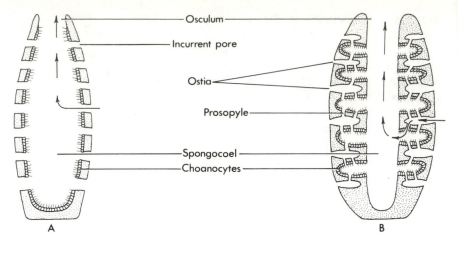

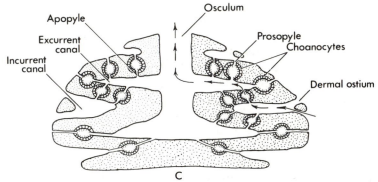

Figure 3-19 Sponge canal system types. (A) *Ascon* type. (B) *Sycon* type. (C) *Rhagon* type.

the triaxon type; and those in the class Demospongea are not triaxon, some being straight and needle-shaped (monaxon) and others eight-rayed (tetraxon). In the Demospongea, spongin may be present with or without spicules in addition. Spongin is a substance chemically allied to collagen. It is secreted by flask-shaped cells called spongoblasts. Spicules are deposited in cells called scleroblasts and more than one cell may take part in the formation of a single spicule. The carbonate of lime or silicon is, of course, extracted from the surrounding water by the cells. Some of the principal types of spicules are illustrated in Figure 3-21. The arrangement of spongin fibers can be observed easily by placing a minute piece of a bath sponge under the microscope. Massive sponges could never have evolved if it were not for spicules and spongin which form a skeleton that holds the body in shape preventing it from collapsing into a mass of jellylike consistency in which canals and flagellated chambers could not possibly exist.

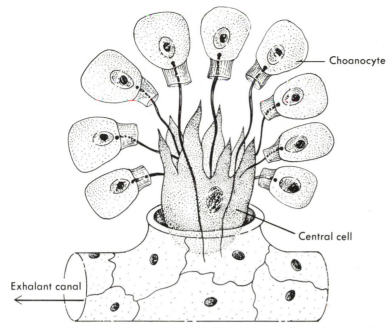

Figure 3-20 Choanocyte chamber of *Suberites massa*. (*Adapted from Connes et al. in Bergquist.*)

Histology

Sponges are many-celled animals in which the somatic cells are differentiated into types for the performance of special functions; that is, division of labor among somatic cells has developed. This is a distinct and important advance over the condition existing even among such complex Protozoa as *Volvox*.

The cells of sponges may be separated into three groups: (1) those of the pinacoderm, or epidermal covering layer; (2) those of the choanoderm, or choanocyte layer; and (3) those of the mesohyl, or mesenchyme, dispersed through the somewhat jellylike middle layer. The principal cells of the outer layer are flattened epithelial cells called *pinacocytes.* They have moderate powers of contraction. Tuzet (1963) has observed them entering the mesenchyme to become nerve cells. Histochemical and structural, but no physiological evidence of nerve cells exists for sponges. *Porocytes,* which are pierced by an opening which passes through them, conduct water through the three layers in the primitive sponges. Porocytes can constrict sufficiently to close the pore. They are thought to be formed from pinacocytes. The choanocyte layer, composed of choanocytes, lacks the basement membrane typical of the epithelial layers of many higher groups.

The mesohyl cells are amoebocytes and reproductive cells. The

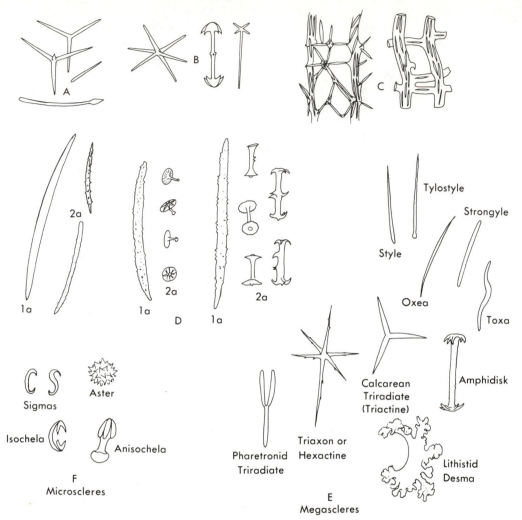

Figure 3-21 Spicules of sponges. (A) Calcareous spicules (*Leucosolenia*). (B) Siliceous spicules (*Hyalonema*). (C) Siliceous spicules with spongin (*Pachychalina* and *Chalina*). (D) Spicules of freshwater sponges: 1, skeleton spicule; 2, gemmule spicules; a, *Eunapius fragilis*; b, *Trochospongilla pennsylvanica*; c, *Anheteromeyenia ryderi*. (E) Megascleres, some structural types. (F) Microscleres, some structural types. (*A–C from Kerr; D, after Old; E and F, after Bergquist, Hyman, and others.*)

amoebocytes are of several types: archaeocytes, undifferentiated cells capable of becoming other cell types; collencytes, lophocytes, or spongocytes, all secreters of spongin or collagen; sclerocytes that secrete spicular material; and, when gemmules form, thesocytes, making up the internal mass of a gemmule and trophocytes to provide nutrients for the thesocytes. Thesocytes are binucleate in some dormant gemmules. Bipolar and multipolar neurons containing neurosecretory granules have been identified in the mesenchyme near the osculum of *Sycon* (Welsch and Storch, 1976).

Nutrition

Sponges obtain their food in the form of minute organic particles, living or lifeless, which enter the incurrent pores suspended in water, and are carried into the flagellated chambers. The current of water flowing through the canal system of the sponge is created by the flagella of the choanocytes which beat continuously. The choanocytes also ingest the food particles either at the side or within the collar; a food vacuole is formed and passed on, digestion takes place, and the undigested residue is cast out from amoebocytes. The nutriment is passed from cell to cell and is circulated to a certain extent by the amoeboid wandering cells of the mesenchyme. It is necessary for the sponge to live in water that is circulating more or less, hence we find these animals in clear, not muddy water. Because the current of water passing through the sponge carries with it the excretions from the body, it is important that the water passing out of the osculum be forced away from the body, since it contains no more food particles but is loaded with wastes. Sponges are therefore so organized and so adapted to their environment that they are able to separate efficiently the water forced out of the osculum from that drawn in through the incurrent pores. This is accomplished by different species in different ways. The force and direction of the jet from the osculum is an important feature as is the nature of the currents at the bottom of the sea. It has been experimentally demonstrated that in *Hymeniacidon* the pressure within the flagellated chambers is equivalent to a column of water 3.5 to 4.0 mm in height. In another sponge, *Spinosella,* the osculum discharged about 0.9 cubic mm of water per second, or 78 liters per day. An average-sized specimen of *Spinosella* with 20 oscula would thus pass about 1½ cubic meters of water through its canal system per day.

Behavior

Sponges are comparatively inactive animals. The larvae are ciliated and swim about but soon become attached to some solid object and remain fixed for life. At the time of fixation the young sponge may move for a short distance as a result of movements of the cells of attachment and thus reach a position where light and other conditions are more satisfactory. Thus, the young sponge that develops from a gemmule of *Spongilla* may move up the side of an aquarium. Certain adult sponges are capable of slight bending movements.

Activity is more pronounced within the body of sponges. This is due to the presence of contractile cells, the myocytes, which are arranged about the osculum and pores and in the wall of the canals in the form of sphincterlike bands. Stimuli of various sorts applied directly to these sphincters result in the opening and closing of the osculum, pores, and canals and in changes in the flow of water. Thus, the osculum of *Hymeniacidon* closes when the animal is transferred from moving into quiet water but opens when a current of water strikes it directly. Transmission to other oscula does not occur. There appears, however, to be present a sluggish type of transmission through the protoplasm. It would seem that, in most sponges, this is transmission without highly differentiated nervous elements such as are present in neural transmission. Studies of the behavior of sponges indicate that no true nervous elements exist.

Reproduction and Growth

Reproduction in sponges is either asexual or sexual. By the asexual method there are produced buds and gemmules. Buds may be set free to take up a separate existence, or, as in *Leucosolenia,* may remain attached to the parent sponge. Often a complex assemblage of individuals is produced that may become very large.

Some marine (for example, *Suberites, Cliona,* and *Haliclona*) and most freshwater sponges have a peculiar method of reproduction by the formation of gemmules. A number of cells in the middle layer of the body wall gather into a ball and become surrounded by a layer of cells which secrete a shell reinforced by spicules. In *Spongilla* the gemmules may possess a peripheral layer containing air cells that float them on the surface, and also rodlike spicules. These gemmules (Fig. 3-11) are formed during the summer and autumn. Gemmule hatching is usually improved by exposure to low temperatures. In the spring they develop into new sponges and are hence of value in carrying the race through a period of adverse conditions, such as the winter season. Young sponges that chance to become attached near each other tend to coalesce. Fusion may be limited to asexually produced colonies of the same parental stock.

In sexual reproduction the eggs and spermatozoa are derived as in *Scypha* from amoeboid wandering cells or modified choanocytes in the mesohyl. Reiswig (1970) has observed massive releases of sperm from a tropical sponge. The sperm lacked a recognizable midpiece between head and tail. A flagellated larva is produced from a holoblastic egg. This larva swims about for a period of time, thus effecting the dispersal of the species, then becomes fixed and passes through many changes, finally developing ostia and an osculum which are necessary for the nutritive processes and growth.

A peculiarity in the embryology of certain sponges is this: the flagellated cells of the larva do not become the outer (dermal) epithelium as do the flagellated cells of the larval coelenterate, but produce the inner layer of choanocytes; and the inner cells do not become the inner epithelium, as do the similarly situated cells in the coelenterate, but produce the dermal layer. It, therefore, seems impossible to homologize the ectoderm and endoderm of coelenterates and other Metazoa with the layers in the sponge larva, because the outer layer (ectoderm?) of the latter becomes the inner layer (endoderm ?) of the adult sponge. The outer layer is consequently termed "dermal epithelium" instead of "ectoderm," and the inner epithelium instead of "endoderm" is the choanocyte layer. With no digestive tract the choanocyte layer cannot properly be called gastrodermis.

The size and rate of growth of sponges depend on the species and largely on the nature of the environment. Annual sponges, that is, those that live one year or less are naturally smaller than perennials. Sponges may increase in weight as much as 40 percent in a single day. Cuttings of bath sponges in Florida increased from 41 cubic cm to 205 cubic cm in 2 months. Certain commercial sponges, such as the wool sponges of the Carribean Sea, grow to be 75 cm in diameter and 25 years of age. David Brigham has photographed barrel-shaped sponges of unknown age about 2 meters in height and 1 meter in diameter at a depth of about 45 meters near Grand Cayman Island.

Ecology

Sponges are sensitive to environmental conditions but are often widely distributed in comparable favorable habitats. *Spongilla lacustris* is nearly cosmopolitan. Gemmules are an important aid to such worldwide dispersal. Poirrier (1977) found that gemmule spicule morphology may vary with ecological conditions. Some species with restricted distribution may be ecomorphs and not valid species.

Sponges are fed upon by some fish, turtles, echinoderms, and arthropods. Far larger numbers do not feed upon them. They often are associated with other species of invertebrates, usually by providing hiding places, sometimes by growing upon them, and one group of crabs even have their posterior legs modified to hold a piece of sponge over their backs for camouflage. Some sponges are very toxic to other animals.

Terpenoids, bromopyrroles, and benzoquinones may be among compounds more important than spicules in preventing grazing by carnivores. Such compounds may be important in preventing overgrowth by other organisms or in reducing site competition, a special need of sessile organisms.

Jackson, Goreau, and Hartman (1971) indicate that utilization of cryptic environments is a mechanism corraline sponges use to reduce competition.

Sponges are mostly beneficial to us. They may destroy oysters and other bivalves by covering their shells and depriving them of food or by boring in and weakening their shells, but, on the other hand, they supply us with the sponges of commerce. Sponge culture, or the growing of sponges from cuttings, has been developed with some success.

The study of sponge regulation, regeneration, grafting, development, ultrastructure, and biochemistry holds promise of revealing basic biological facts of value to medical science, especially in the areas of diseases of connective tissue, tissue transplantation, and biologically active organic compounds.

Aggregates of Sponge Cells

Experiments with some sponges (*Hymeniacidon, Microciona,* etc.) prove that when a specimen is broken up and strained through fine bolting cloth so as to dissociate the cells, the cells will fuse on the bottom of a dish to form plasmodia. These plasmodia in the course of several months acquire canals, flagellated chambers, and a skeleton and develop reproductive bodies. The aggregation of these cells appears to be due to amoebocytes from the middle layer which move about and gather in all the cells they encounter. When cells from two species are intermingled those of each species fuse with one another but not with those of the other species.

Diffusable, surface-active molecules are important in this regulation. Since differences between individuals of the same species can prevent successful grafting of fragments (Hildemann, Johnson, and Jokiel, 1979), it appears that immunologic responses may also be involved in rejection or acceptance at the cellular level in sponges.

Position of Sponges in the Animal Kingdom

As already noted, although sponges are undoubtedly multicellular animals, they are often separated from the Metazoa and placed in an isolated group, the Parazoa, near the foot of the family tree. This is because they were supposed to have branched off from the main evolutionary series at this point and to have given rise to nothing but sponges.

The recent suggestion of Tuzet (1963) that the Porifera gave rise to the other Metazoa (Fig. 3-1, B) via the Coelenterata has some evidence to support it also. The choanocyte, or collared flagellated cells, are not restricted to sponges, as is commonly stated. Similar cells have been reported in some echinoderms. Cells with a single flagellum occur in the coelenterates. Spermatogenesis and oogenesis resemble the same processes in other metazoans. Histological and cytological details show similarities with the metazoans, especially the recently established chemical and physical resemblance between spongin and collagen (an important constituent of vertebrate connective tissue) and demonstration of primitive nerve cells. In either view the development of organ systems and movement characteristic of the remaining phyla justify retention of separate subkingdoms for Protozoa, Parazoa, and Metazoa.

The protozoan origin of the sponges seems clearly indicated when we examine some of the flagellated Protozoa, especially those with collar cells and colonial in habit, like *Proterospongia* (Fig. 3-22). This

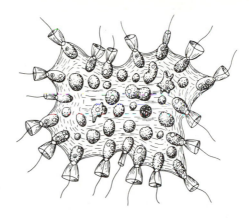

Figure 3-22 *Proterospongia haeck-eli. (After Kent.)*

protozoan not only possesses choanocytes like those of sponges but cells within the colony that resemble those in the middle layer of sponges. It is not difficult to imagine such an organism developing into a sponge.

Knowledge of living Sclerospongea has clarified poriferan relationships of some fossil members and the Stromatoporoidea and Chaetetida.

If the origin of Metazoa from sponges is accepted, it is not hard to imagine a calcarean sponge gaining size with the aid of support from septal tracts of spicules and vertical fusion of chambers, perhaps feeding spongelike and with nematocyst-to-be protozoan toxicysts while supplementing nutrition and carbonate secreting abilities with endosymbiotic zoochlorellae and zooxanthellae until progeny reached anthozoan status.

The large sponges of today have siliceous skeletons. Perhaps calcarean sponges of large size are not viable as an evolutionary strategy today because vertebrate stomach acids would destroy calcareous spicules before damage occurred and vertebrate livers can effectively adapt to detoxify many otherwise toxic compounds.

For those weighting embryology heavily in phylogenetic considerations, the sponges have provided a wide variety of larvae to suit any theory. Perhaps the neatest for a mainstream view of sponges are the very planulalike (Fig. 3-23) larvae in some.

The History of Our Knowledge of Porifera

Although sponges have been recognized since at least the time of Aristotle (384–322 B.C.), they were not proved to be animals until 1857. Ellis in 1765 appears to have been the first to describe the current of water entering and leaving a sponge. This was worked out more fully by Grant (1825), who observed the incurrent pores and oscula and suggested that the currents were due to ciliary action. After the recognition of the animal nature of sponges, their position in the animal kingdom became an important subject. The embryology of sponges as

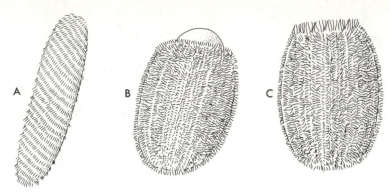

Figure 3-23 Some Demospongea larvae, showing variations in patterns of ciliation. (A) *Halichondria*. (B) *Lissodendoryx*. (C) *Haliclona*. (*Adapted from Bergquist, 1978.*)

worked out by Schulze (1878), Bütschli (1884), Solas (1884), Minchin (1897), Maas (1898), and Delage (1898) indicated fundamental differences between them and other Metazoa which has led to their being considered a separate group and called Parazoa.

Brien, Bergquist, Hartman, Levi, and Simpson are among the leading active sponge specialists. An excellent modern review of sponges is provided by Bergquist (1978).

Classification

Sponges constitute a distinct phylum of animals. They are all multicellular and the somatic cells are arranged in two layers (diploblastic), an outer dermal and an inner choanocyte layer. The cells in these layers are further differentiated for various functions. The body is asymmetrical or radially symmetrical; it is perforated by incurrent pores usually opening into canals or chambers lined with flagellated collar cells (choanocytes), and is, in most species, supported by a skeleton of spicules or spongin or both. The classes and orders are separated for the most part on the basis of composition, structure, and arrangement of skeletal materials; canal systems; type of larvae and reproduction.

The classification of sponges has undergone considerable recent change and more is to be expected. Clustering of orders into subclasses has been done on what seems an artificial basis and hence has been omitted in the Demospongea. Members of the aspiculate Keratosa are now largely in the Dictyoceratida and Dendroceratida. The Class Sclerospongea is new, since the last edition, for recently discovered sponges in cryptic reef habitats with significant fossil affinities. The arrangement below is adapted from Bergquist (1978).

Sessile filter-feeding metazoans with choanocytes and usually a collagenous and mineral spicule supporting system for various types of canal systems.

Class I. CALCAREA. Spicules of calcium carbonate, some triactinal.

SUBCLASS 1. CALCINEA. Blastulae larvae, equiangular triactine spicules.

ORDER 1. CLATHRINIDA. Ascon-type spongocoel.

ORDER 2. LEUCETTIDA. Sycon- or leucon-type canal systems.

SUBCLASS 2. CALCARONEA. Amphiblastulae larvae, triactines not equiangular.

ORDER 1. LEUCOSOLENIIDA. Ascon-type spongocoel.

ORDER 2. SYCETTIDA. Sycon- or leucon-type canal systems.

SUBCLASS 3. PHARETRONIDA. Usually with triactines having two parallel rays (tuning-fork type).

ORDER 1. INOZOIDA. Organized as a single unit.

ORDER 2. SPHINCTOZOIDA. "Segmented" appearance, most are fossils.

Class II. HYALOSPONGEA (HEXACTINELLEA). Siliceous skeleton with hexactine spicules.

ORDER 1. AMPHIDISCOSIDA. With birotulate microscleres.

ORDER 2. LYSSACINIDA. With hexaster-type microscleres.

ORDER 3. DICTYONIDA. With hexaster-type microscleres and major hexactine spicules united in a compact regular framework.

Class III. DEMOSPONGEA. Spicules, when present, all siliceous but not hexactine; no carbonate skeletal material except by inclusion of debris.

ORDER 1. HOMOSCLEROPHORIDA. Spicules very small, amphiblastulae larvae.

ORDER 2. CHORISTIDA. Aster microscleres, tetractine megascleres showing some radial arrangement.

ORDER 3. SPIROPHORIDA. Sigmoid microscleres, radial skeleton arrangement, larval stage omitted in development.

ORDER 4. LITHISTIDA. Desma-type megascleres present.

ORDER 5. HADROMERIDA. Monactine megascleres in a radial arrangement, spongin not in a fiberous arrangement.

ORDER 6. AXINELLIDA. With axial skeletal condensation of spongin and spicules, parenchymellae larvae.

ORDER 7. HALICHONDRIDA. Monaxon megascleres, no microscleres, parenchymellae larvae completely ciliated.

ORDER 8. POECILOSCLERIDA. Monaxon megascleres; microscleres usually modifications of chelaes, sigmas, or toxas; parenchymellae larvae completely ciliated.

ORDER 9. HAPLOSCLERIDA. Megascleres typically oxeas or strongyles, larva incompletely ciliated, includes the freshwater Spongillidae and Potamolepidae besides many marine families.

ORDER 10. DICTYOCERATIDA. Spicules lacking, spongin in an anastomosing pattern.

ORDER 11. DENDROCERATIDA. Spicules lacking, spongin in dendritic pattern when present.

ORDER 12. VERONGIDA Bergquist (1978, p. 178). Spicules lacking, spongin fibers support a collagenous matrix, complex histology with many different cell types.

CLASS IV. SCLEROSPONGEA. Skeleton spongin and siliceous spicules, also a basal mass of calcium carbonate. In cryptic habitats on coral reefs.

ORDER 1. CERATOPORELLIDA. Basal carbonate mass aragonitic.

ORDER 2. TABULOSPONGIDA. Basal carbonate mass calcitic.

REFERENCES TO LITERATURE ON THE PORIFERA

BERGQUIST, P. R. 1978. *Sponges.* Univ. of California Press, Berkeley. 268 pp.

FRY, W. G. (Ed.). 1970. *The Biology of the Porifera.* Academic Press, New York. 512 pp.

GILBERT, J. J. 1975. Field experiments on gemmulation in the fresh-water sponge *Spongilla lacustris. Trans. Am. Micr. Soc.,* **94**:347–356.

———, and T. L. SIMPSON. 1976. Sex reversal in a fresh-water sponge. *J. Exp. Zool.,* **195**:145–151.

HARRISON, F. W., and R. R. COWDEN (Eds.). 1976. *Aspects of Sponge Biology.* Academic Press, New York. 354 pp.

HILDEMANN, W. H., I. S. JOHNSON, and P. L. JOKIEL. 1979. Immunocompetence in the lowest metazoan phylum: transplantation immunity in sponges. *Science,* **204**:420–422.

HYMAN, L. H. 1940. *The Invertebrates: Protozoa Through Ctenophora.* Vol. I. McGraw-Hill, New York. 726 pp.

JACKSON, J., T. F. GOREAU, and W. D. HARTMAN. 1971. Recent brachiopod-coralline sponge communities and their paleoecological significance. *Science,* **173**:623–625.

JEWELL, M. 1959. Porifera. pp. 298–312. In W. T. EDMONDSON. *Fresh-Water Biology.* 2nd Ed. Wiley, New York. 1248 pp.

LEVI, C. 1963. Gastrulation and larval phylogeny in sponges. pp. 375–382. In E. DOUGHERTY *The Lower Metazoa.* Univ. of California Press, Berkeley. 478 pp.

PARIS, J. 1961. Greffes et sérologie chez les Éponges siliceuses. *Vie et Milieu,* Suppl. No. 11. 74 pp.

PENNAK, R. W. 1978. *Fresh-water Invertebrates of the United States.* 2nd Ed. Wiley, New York. 803 pp.

PENNY, J. T., and A. A. RACEK. 1968. Comprehensive revision of a worldwide collection of fresh-water sponges (Porifera: Spongillidae). *U.S. Nat. Mus. Bull.,* **272**:1–184.

POIRRIER, M. A. 1977. Systematic and ecological studies of *Anheteromeyenia ryderi* (Porifera: Spongillidae) in Louisiana. *Trans. Am. Micr. Soc.*, **96**:62–67.

REISWIG, H. M. 1970. Porifera: sudden sperm release by tropical Demospongiae. *Science*, **170**:538–539.

SIMPSON, T. L., and P. E. FELL. 1974. Dormancy among the Porifera: gemmule formation and germination in fresh-water and marine sponges. *Trans. Am. Micr. Soc.*, **93**:544–577.

TUZET, O. 1963. The phylogeny of sponges according to embryological, histological, and serological data, and their affinities with the Protozoa and the Cnidaria. pp. 129–148. In E. DOUGHERTY. *The Lower Metazoa*. Univ. of California Press, Berkeley. 478 pp.

WELSCH, U., and V. STORCH. 1976 *Comparative Animal Cytology and Histology*. Univ. of Washington Press, Seattle. 343 pp.

4

Phylum Coelenterata

The phylum Coelenterata (or Cnidaria) includes the polyps, jellyfishes, sea anemones, and corals. All of these animals have a body wall consisting of two layers of cells, between which is a jellylike substance, the mesoglea. Within the body is a single gastrovascular cavity, or coelenteron. Because of the presence of two cellular layers, coelenterates are said to have a tissue-level organization. They are also acoelomates; that is, they do not possess a second body cavity, the coelom. All coelenterates are provided with nematocysts.

This phylum contains three classes, as follows:

CLASS I. HYDROZOA. This class includes the freshwater polyps, the small jellyfishes, the hydroid zoophytes, and a few stony corals.

CLASS II. SCYPHOZOA. Most of the large jellyfishes are placed in this class.

CLASS III. ANTHOZOA. In this class are included the sea anemones, and most of the stony and horny corals.

In the following pages the freshwater hydroid, *Hydra,* is described in detail because of its simplicity, abundance, and the ease with which it may be collected. *Obelia* represents a marine hydroid and *Gonionemus,* a hydroid jellyfish. The class Scyphozoa is introduced by the common large jellyfish, *Aurelia,* and the Anthozoa by a sea anemone, *Metridium,* and a coral, *Astrangia.*

Class I. Hydrozoa

HYDRA—A FRESHWATER HYDROZOAN

Hydra (Fig. 4-1) is a simple metazoan abundant in freshwater ponds and streams. If a quantity of aquatic vegetation is gathered and placed in glass dishes full of water, these little freshwater polyps may be found clinging to the plants and the sides and bottom of the dish. They are easily seen with the naked eye, being from 2 to 20 mm in length, and may be likened to a short thick thread frazzled at the unattached, distal end. The great variation in length is due to the fact that both body and tentacles are capable of remarkable expansion and contraction because of the presence of specialized muscle fibrils in many of the cells. A number of species of hydras are recognized by zoologists.

External Characters

The *body* of *Hydra* resembles an elastic tube which varies in length and thickness accordingly as the animal is extended or contracted; in the former case it may reach a length of 2 cm. At the distal end is a circlet of usually from five to nine slender, fingerlike projections called *tentacles*. The diameter of the body is frequently increased at certain points by a distention due to the ingestion of large particles of food. Different species of *Hydra* differ in color but the color often depends on the character of the food and hence is not a constant feature. The

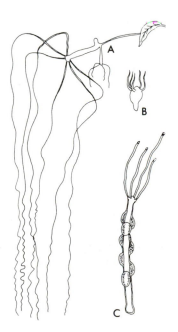

Figure 4-1 *Hydra oligactis.* (A) Hanging from a leaf with tentacles well extended. (B) Position assumed when contracted. (C) Male bearing many testes. (*After Hyman.*)

part of the body which is usually attached to some object is known as the *foot* or *basal disk* and is referred to as the *proximal end.* The foot not only anchors the animal when at rest, but also serves as a locomotor organ. In the common brown species, *Hydra oligactis,* the proximal region is a slender stalk and the distal region constitutes a sort of stomach. A conical elevation, the *hypostome,* occupies the distal end of the body. It is surrounded by the tentacles and has at the top an opening, the *mouth.* This mouth is a circular orifice and the underlying gastrodermis is folded and star-shaped, having clefts running out from the center toward each arm.

The *tentacles* are capable of remarkable expansion, and may stretch out from small blunt projections to very thin threads 7 cm or more in length; in this condition they are so thin as to be barely visible even with a lens. They move independently capturing food and bringing it into the mouth. Their number varies considerably. Six hundred specimens of *Chlorohydra viridissima* possessed from 4 to 12 tentacles each. These occurred in the following proportions: 54 percent had eight; 24 percent, seven; 15 percent, nine; very few animals possessed a greater number than nine, and only occasionally was one found with less than seven. The number of tentacles increases with the size and age of the animal, although unfavorable conditions and extreme age result in a decrease.

Frequently, specimens of *Hydra* are found which possess *buds* in various stages of development (Figs. 4-1 and 4-2). Several buds are often found on a single animal, and these in turn may bear buds before detachment from the parent. In this way a sort of primitive *Hydra* colony is formed, resembling somewhat the asexual colonies of some of the more complex coelenterates to be described later. In *Hydra oligactis* there is a rather definite budding zone where stalk and body meet.

Reproductive organs may be observed on specimens of *Hydra* in the autumn and winter. Both an *ovary* and *testes* are produced on a single individual in some species; the former is knoblike, occupying a position about one third the length of the animal above the basal disk; the testes, usually two or more in number, are conical elevations projecting from the distal third of the body (Figs. 4-1 and 4-2).

Morphology

Hydra (Fig. 4-2) is a diploblastic animal consisting of two cellular layers, an outer thin layer, the *epidermis* (ectoderm), and an inner layer, the *gastrodermis* (endoderm), twice as thick as the outer. Both layers are composed of *epithelial* cells. A thin layer, formed of a jelly-like substance, the *mesoglea,* separates epidermis from gastrodermis. Not only the body wall, but also the tentacles, possess these three definite regions. Both body and tentacles are hollow, the single central space being known as the *gastrovascular cavity.* At the bases of the

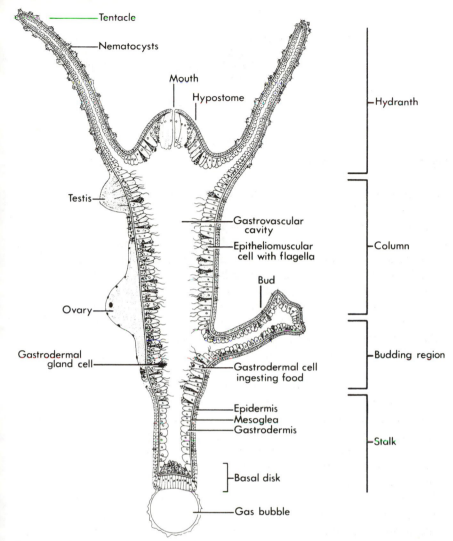

Labels in figure:
Tentacle
Nematocysts
Mouth
Hypostome
Hydranth
Testis
Gastrovascular cavity
Epitheliomuscular cell with flagella
Column
Bud
Ovary
Gastrodermal gland cell
Gastrodermal cell ingesting food
Budding region
Epidermis
Mesoglea
Gastrodermis
Stalk
Basal disk
Gas bubble

Figure 4-2 *Hydra.* Longitudinal section. (*After Kepner and Miller, modified.*)

tentacles are sphincters that are capable of shutting off the connection between the cavity of the body and the cavities of the tentacles. Because of this, injurious material within the enteron may be prevented from reaching the tentacular cavities.

Campbell (1974) found that the cells of the tissues are generally displaced in an oral direction by one of several types of movements in a tissue renewal process. Many interstitial cells seem to maintain their own relative position in the face of the general tissue movement. These movements maintain the integrity of the general anatomy, otherwise endangered by loss of cells from mouth, tentacles, and from budding.

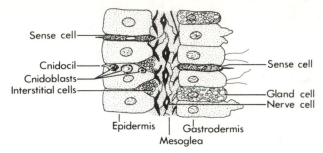

Figure 4-3 *Hydra*. Cross section of the body wall. (*After Kukenthal, modified.*)

Epidermis (Fig. 4-3). The epidermis is primarily *protective* and *sensory*, containing structures characteristic of these functions. Slight differences in structure are observable between the epidermis of the tentacles and that of the body wall, and the latter differs from that of the basal disk. In the epidermis of the body wall are two principal kinds of cells, large epitheliomuscular cells, and small interstitial cells. The *epitheliomuscular* cells (Fig. 4-4) are shaped like inverted cones. At their inner ends are one or more comparatively long (sometimes 0.38 mm) unstriped *contractile fibers* which form a thin longitudinal muscular layer. These muscle fibers explain the remarkable power of contraction exhibited by *Hydra* when stimulated. Near the middle of each cell, embedded in the alveolar cytoplasm, is a nucleus containing one or two nucleoli and a network of chromatin.

The *interstitial cells* are an important cell type for maintenance of the *Hydra*. They are small cells, abundant near the base of the cells forming the epidermis and gastrodermis, and are capable of developing into the other cell types as well as forming nematocysts. The general epithelial cells of ectoderm and endoderm are capable of reproducing their type in the absence of interstitial cells (Marcum and Campbell, 1978). However, irradiated hydras which would otherwise die because the interstitial cells are killed, will recover if they are supplied with a graft containing living interstitial cells. Thus, they can be considered

Figure 4-4 *Hydra oligactis.* Cellular elements enlarged. (A) Epitheliomuscular cell. (B) Secretory cell. (C) Sensory cell. (*After Burck.*)

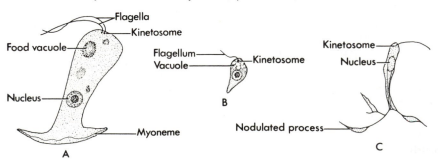

embryonic cells of *Hydra*. In the fall (September and October) some develop into reproductive cells.

Nematocysts or stinging capsules are present on all parts of the body of *Hydra* except the basal disk, being most numerous on the tentacles (Fig. 4-2). Each develops in an interstitial cell known as a *cnidoblast*. These in turn are embedded in little surface tubercles, which give the animal a rough-appearing outline. The tubercles are epidermal cells, each of which usually possesses one or more large nematocysts surrounded by a number of a smaller variety. Four kinds of nematocysts occur in *Hydra*. The largest is known as a *stenotele* (or penetrant) and is 0.013 mm long and 0.007 mm thick; before being discharged it is pear-shaped and occupies almost the entire cell in which it lies (Fig. 4-5). Within it is a *coiled tube* at whose base are three large and a number of small *spines*. A second type is known as a *holotrichous isorhiza* (streptoline glutinant); this type is large and cylindrical and pointed at the end where the thread is discharged. The thread, when discharged, bears a spiral row of minute barbs and tends to coil. *Atrichous isorhizas* (stereoline glutinants) are smaller and oval with a thread that is straight and devoid of barbs when discharged. *Desmonemes* (volvents) are small, pyriform nematocysts containing a thick, smooth thread in a single loop; the thread forms a tight coil when discharged. Projecting from the cnidoblast near the outer end of the nematocyst is a trigger-like spine, the *cnidocil*. Nematocysts may be exploded by adding a little acetic acid, or better, methyl green, to the water. The tube that is coiled within them is then everted. First, the base of the tube with the

Figure 4-5 *Hydra:* nematocysts. (A) Undischarged. (B) Discharged. (C) Discharged but retained within cnidoblasts. (*After Schulze.*)

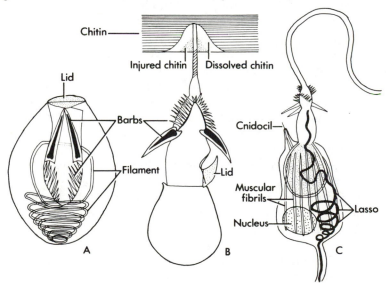

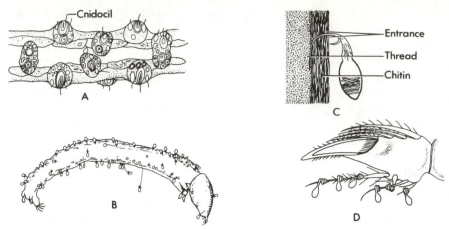

Figure 4-6 *Hydra:* nematocysts and their action. (A) Portion of a tentacle, showing the batteries of nematocysts. (B) Insect larva covered with nematocysts as a result of capture by *Hydra.* (C) A nematocyst piercing the chitinous covering of an insect. (D) Nematocysts holding a small animal by coiling about its spines. (*A and B, from Jennings; C and D, from Toppe.*)

spines appears, and then the rest of the tube rapidly turns inside out (Fig. 4-52). Nematocysts are probably able to penetrate the tissues of other animals only when contact is made before eversion is completed. Even the extremely firm chitinous covering of insects may be punctured by these structures (Fig. 4-6, C). Touching the cnidocil was for a long time supposed to cause the explosion of the nematocysts, and for this reason it is known as a "trigger."

However, mechanical shocks alone have little influence upon some nematocysts. Internal pressure produced either by distortion or by osmosis, is effective. For this reason chemicals which increase the osmotic pressure within the cnidoblast cause the eversion of the threadlike tube. Contraction of a musclelike covering around the nematocyst is said to produce external pressure.

Neurons have been observed with synapses on nematocyst-containing cells of *Hydra.* The same neuron can also have a synapse on an epitheliomuscular cell (Westfall, Yamataka, and Enos, 1971).

Discharge of nematocysts seems more readily accomplished by a proper combination of chemical and mechanical stimuli. Lipid-like substances are thought to be among the chemicals involved in natural discharge. Penetrants and desmonemes discharge readily when stimulated by food organisms. Atrichous isorhizas are easily discharged by mechanical shocks and their adhesive threads enable tentacles to adhere to the substrate during locomotion. An animal when "shot" by nematocysts is immediately paralyzed, and sometimes killed, by a poison called hypnotoxin which is injected into it through the tube.

Nematocysts are developed from interstitial cells, each cell producing one nematocyst. A nematocyst, carried by its mother cell or cnidoblast, moves inside an epitheliomuscular cell and to the surface. The outer part is produced to form a cnidocil which pierces the surface. The cnidocil seems to be derived from a ciliary-type structure. Many workers prefer to call the mature nematocyst-bearing cnidoblast a *nematocyte*. Since the tube of the nematocyst cannot be returned to the capsule, nor another one be developed by the cnidoblast, new capsules must be formed from interstitial cells to replace those already exploded.

The *basal disk* differs somewhat in function from the rest of the body. It is the point by which *Hydra* attaches itself to solid objects, and for this purpose secretes a sticky substance. Epitheliomuscular cells and a few interstitial cells are present, but no nematocysts are to be found here. The columnar epitheliomuscular cells are not only provided with contractile fibers at their bases, but, being secretory, also contain a large number of small refringent granules, as shown in Figure 4-7. Certain ectoderm cells of the basal disk secrete a gas which may be confined within the mucus, thus forming a bubble (Figs. 4-2 and 4-7). Such a gas bubble may lift the *Hydra* to the surface, where it breaks and spreads out like a raft from which the animal hangs suspended in the water.

The *tentacles* (Fig. 4-2) are provided with an epidermis consisting of large flat cells, thin at the edges and thick in the center. The thicker portions give the surface of the tentacle a lumpy appearance. In the center of each thickening is a nucleus around which are embedded sometimes as many as 12 nematocysts each in its own cnidoblast (Fig. 4-6, A). The cnidocils projecting from the cnidoblasts resemble groups

Figure 4-7 *Hydra*. Longitudinal section of the basal region, showing the gastrodermal cells containing food vacuoles and the mucus-secreting epidermal cells containing mucus granules with which the animal attaches itself to the substratum. Formation of a gas bubble is also shown. (*Modified from Kepner and Thomas*.)

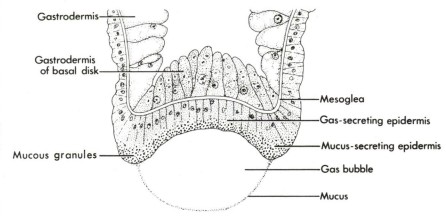

Gastrodermis

Gastrodermis of basal disk

Mucous granules

Mesoglea

Gas-secreting epidermis

Mucus-secreting epidermis

Gas bubble

Mucus

of cilia. Each cnidoblast is drawn out at its base into a contractile fibril which enters the longitudinal muscular sheet at the base of the epidermal cells.

Gastrodermis (Fig. 4-2). The inner layer of cells, the gastrodermis, occupies about two thirds of the body wall. Its functions are *digestive* and *secretory*. The digestive cells are long and club-shaped, with transverse muscular fibrils at their base, forming a circular sheet of contractile substance. Many of them are provided with from one to five flagella, which are nontapering, lashlike processes that arise from a kinetosome (Fig. 4-4). The epitheliomuscular cells possess one or more flagella; the secreting cells, one or two; and the sensory cells, one. Besides flagella *pseudopodia* may also be thrust out from the free end (Fig. 4-3). The flagella create currents in the gastrovascular fluid, and the pseudopodia capture solid food particles. The gastrodermis of the hypostome contains secreting cells that appear to produce a mucous secretion for lubrication during the ingestion of food. The internal structure of these cells differs before and after the animal is fed. In a starving *Hydra* large vacuoles appear, almost completely filling the cell, the protoplasm being reduced to a thin layer near the cell membrane; after a meal, however, the cells are gorged with nutritive spheres, many of which, especially the oil globules, migrate into the epidermis and are stored near the periphery.

The *glandular cells* are smaller than the digestive cells and lack the contractile fibrils at their base (Fig. 4-3). They also differ in appearance according to their metabolic activity: some are filled with large vacuoles containing secretory matter, whereas others, having discharged their secretion, appear crowded with fine granules. Interstitial cells are found lying at the base of the other gastrodermal cells.

The tentacles contain no gland cells. The gastrodermis of the basal disk is provided with only a few glandular cells.

Mesoglea (Figs. 4-2 and 4-3). The mesoglea in *Hydra* is so thin as to be difficult to find, even when highly magnified; in some of the other coelenterates this layer is very thick, constituting by far the largest part of the body. The protein portion of the mesoglea is largely collagen.

Nervous System. *Hydra* possesses a nervous system, but complicated staining methods are necessary to make it visible. In the epidermis there is a sort of plexus of nerve cells connected by nerve fibers (Figs. 4-4 and 4-8, A). Sensory cells in the surface layer of cells (Fig. 4-3) serve as external organs of stimulation and are in direct continuity with fibers from the nerve cells. Some of the nerve cells send processes to the muscle fibers of the epitheliomuscular cells, and are therefore motor in function (Fig. 4-8, B). The lack of a brain and the meshlike

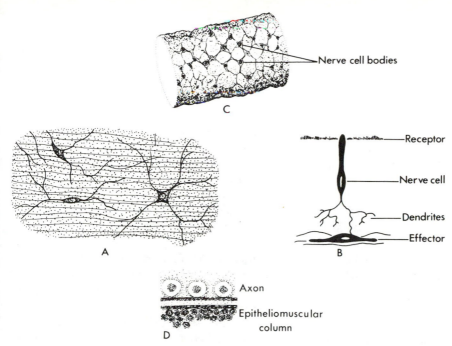

Figure 4-8 Nerve cells of *Hydra*. (A) Plexus of nerve cells in the ectoderm; the parallel lines represent longitudinal muscle fibers. (B) Receptor-effector system of a hydrozoan. (C) Gross view of nerve net. (D) Generalized neuromuscular junction. (*A, after Schneider; B, after Parker; D, adapted from electron micrographs by Westfall, 1973.*)

spacing of the nerve cells with their processes allows us to describe the system as a *nerve net.*

The nerve cell processes, or *neurites,* make such intimate contact with other neurites that the junctions were not known to be synaptic until electron microscopic studies showed that synapses in hydras were of two types. One type is symmetrical and presumably may transmit impulses in either direction. The other type of synapse is asymmetrical and presumably transmits impulses in only one direction. Nerve cell processes are known to have junctions with some nematocytes in *Hydra* (Westfall, Yamataka, and Enos, 1971). The gastrodermis of the body also contains nerve cells, but not so many as are present in the epidermis (Fig. 4-3).

Nutrition

Food. The food of *Hydra* consists principally of small animals that live in the water. Of these may be mentioned small crustaceans such as *Cyclops,* annelids, and insect larvae. *Hydra* normally rests with its basal disk attached to some object and its body and tentacles extended out into the water. In this position it occupies a considerable amount of hunting territory. Any small aquatic animal swimming in touch with a tentacle is at once shot full of nematocysts (Fig. 4-6, B), which not only seem to paralyze it, but also to hold it firmly. The viscid surface of the tentacle aids in making sure that the victim does not escape.

Ingestion. Ingestion takes place as follows: First, the tentacle, which has captured the prey, bends toward the mouth with its load of food. The other tentacles not only assist in this, but may use their nematocysts in quieting the victim. The mouth often begins to open before the food has reached it. The edges of the mouth gradually enclose the organism and force it into the gastrovascular cavity. The body wall contracts behind the food and forces it downs. Frequently, organisms many times the size of the *Hydra* are successfully ingested.

Reactions to Food. It is not uncommon to find hydras that will not react to food when it is presented to them. This is due to the fact that these animals will eat only when a certain interval of time has elapsed since their last meal. The physiological condition of *Hydra,* therefore, determines its reponse to the food stimulus. The collision of an aquatic organism with the tentacle of *Hydra* is not sufficient to cause the food-taking reaction, since it has been found that not only a mechanical stimulus, but also a chemical stimulus must be present. A very hungry *Hydra* will even go through the characteristic movements when it is excited by the chemical stimulus alone. This has been shown by the following experiment. When the tentacles and hypostome of a moderately hungry *Hydra* are brought into contact with a piece of filter paper, which has been soaked for a time in the same culture medium, there is no response. If the filter paper is then soaked in beef juice and offered to the *Hydra,* the usual food reactions are given. Reduced glutathione in low concentrations will also produce feeding reactions.

Beef juice alone calls forth no response in a moderately hungry animal; but does inaugurate the normal reflex, if a very hungry specimen is selected for the experiment. The conclusion reached is that well-fed hydras will not respond to either mechanical or chemical stimuli when acting alone or in combination; that moderately hungry animals will react to a combination of the two, and that hungry animals will exhibit food-taking movements even if a chemical stimulus alone is employed.

These types of responses have been quantified in studies of mouth opening and other feeding responses to reduced glutathione (Fig. 4-9). At low concentrations the response is somewhat proportional to concentration until maximum response is reached. Stimulation beyond that point does not increase the duration of the response.

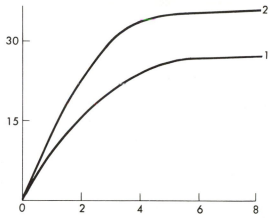

Figure 4-9 Feeding response of *Hydra* to weak solutions of glutathione: 1, starved for 1 day; 2, starved for 2 days. Ordinate, minutes duration of feeding reflex; abscissa, concentration of glutathione in molarity × 10^{-6}. (*From data of Lenhoff, 1961.*)

Digestion. Immediately after the ingestion of the food, the gland cells in the gastrodermis show signs of great activity; their nuclei enlarge and become granular. This is probably the beginning of the process of formation of enzymes, which are discharged into the gastrovascular cavity and begin at once the dissolution of the food. The action of the digestive juices is made more effective by the churning of the food as the animal expands and contracts. The cilia extending out into the central cavity also aid in the dissolution of the food by creating currents. This method of digestion differs from that of *Amoeba* and *Paramecium* in being carried on outside the cell; that is, extracellular. Intracellular digestion also takes place in *Hydra;* the pseudopodia thrust out by the gastrodermal cells seize and engulf particles of food, which are dissolved within the cells. However, most of the food is digested in the gastrovascular cavity. The digested food is absorbed by the gastrodermal cells; part of it, especially the oil globules, is passed over to the epidermis, where it is stored.

Egestion. All insoluble material is egested from the mouth. This is accomplished by a speedy expulsion, which throws the debris to some distance.

Symbiosis

One species of hydra, *Chlorohydra viridissima,* is green, because of the presence within the gastrodermal cells of a unicellular alga, *Chlorella vulgaris.* As in *Paramecium bursaria,* the plant uses some of the waste products of metabolism of the hydra, and the hydra uses some of the oxygen resulting from the process of photosynthesis in the plant. In addition, about 10 percent of the carbon dioxide fixed by the alga becomes incorporated in the protoplasm of the hydra. This symbiotic relationship is one of mutualism.

Among the parasites of *Hydra* are *Hydramoeba hydroxena* (Fig. 2-32) and the ciliates *Kerona pediculus* and *Trichodina pediculus* (Fig. 2-81, C).

Behavior

Chlorohydra viridissima gives a more prompt and decisive response when stimulated than any species of *Hydra*, and for this reason its behavior has been studied more thoroughly than that of the others. The following paragraphs have been compiled largely from experiments upon green hydras, although enough work has been done with other forms to prove that their reactions are practically the same, only more sluggish.

Normal Position of Hydra (Fig. 4-1). Hydras may be found attached to the sides or bottom of an aquarium, to parts of water plants, or hanging from the surface film. Usually, they are near the top, where more oxygen can be obtained from the water than at greater depths. If attached to the bottom, the body is usually held upright; if to the sides, the body is in most cases horizontal, the hypostome generally being lower than the foot; and if to the surface film, the body is allowed to hang directly downward. Suspension from the surface film may be compared with that of a needle placed on the surface of the water. Threads of a gelatinous substance, extending out from the basal disk, help sustain the body, and in some cases a gas bubble attached to the foot keeps the animal afloat. The position of rest in every case gives the *Hydra* the greatest opportunities for capturing food, because in this condition it has control of a large amount of territory.

Spontaneous Movements. All the movements of Hydra are the result of contraction of one or more sets of muscle fibers, and are produced by two kinds of stimuli, internal, or spontaneous, and external. Spontaneous movements may be observed when the animal is attached and

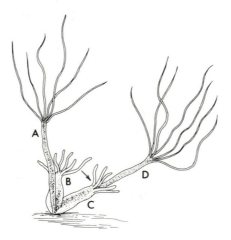

Figure 4-10 Spontaneous changes of position in an undisturbed *Hydra*, side view. The extended animal *A* contracts *B*, bends to a new position *C*, and then extends *D*. (*From Jennings*.)

undisturbed. At intervals of several minutes the body, or tentacles, or both, contract suddenly and rapidly, and then slowly expand in a new direction. Hungry specimens are more active than well-fed individuals. The result is to bring the animal into a new part of its surroundings, where more food may be present (Fig. 4-10). These movements finally cease, and the animal's position is changed by locomotion.

The spontaneous contractions are of two types. One is a short series of contractions with the stimulus originating beneath the area of the hypostome. The other spontaneous contractions are periodic ones with the stimuli originating from a number of locations, including the basal region. Although the two systems are to some extent independent, interaction does occur (McCullough, 1965).

Locomotion. Movement from place to place is effected in several ways. In most cases the animal bends over (Fig. 4-11, A) and attaches itself to the substratum by its tentacles (B), probably with the aid of atrichous isorhiza nematocysts. The basal disk is then released and the animal contracts (C). It then expands (D), bends over in some other direction and attaches its foot (E). The tentacles now loosen their hold and an upright position is regained (F). The whole process has been likened to the looping locomotion of a measuring worm. At other times the animal moves from place to place while inverted by using its tentacles as legs. Locomotion may also result from the gliding of the foot along the substratum, and considerable distances are sometimes covered in this way. In *Hydra oligactis* the tentacles may be attached to some object, the basal disk freed, and the body drawn up to the object by the contraction of the tentacles.

Movements are dependent on the contraction of the longitudinal muscle fibers at the base of the epidermal cells and the circular muscle

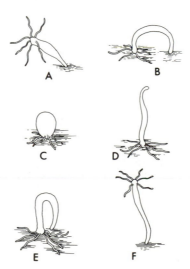

Figure 4-11 Locomotion in *Hydra:* moving like a measuring worm. (*After Wagner.*)

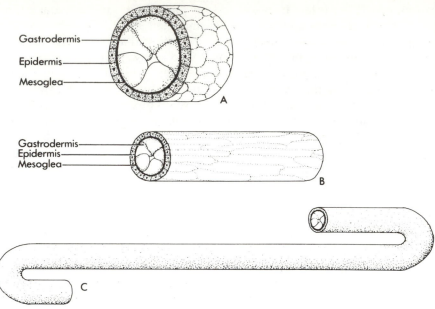

Figure 4-12 Extension of *Hydra* tentacles when constant volume is maintained. (A) Tentacle with longitudinal muscles fully contracted. (B) Same tentacle with circular muscles contracted to one-half maximum length. (C) Size that the same segment of tentacle would be if circular muscles contracted to one-quarter maximum length. (*Based on observations of Engemann, 1966.*)

fibers at the base of the gastrodermal cells. Shortening of the longitudinal muscles is accompanied by a simultaneous decrease in length and increase in diameter of the hydra. Fluid, in the gastrovascular cavity and in the cells themselves, serves to transmit the force hydraulically. Analysis of structure and movement in the tentacles indicates that the gastrodermal cells are important in the independent action of the tentacles. They are large and vacuolated and nearly occlude the cavity of the tentacles. Thus, the fluid in the cells when put under pressure by contraction of the circular muscle of the cells can only act locally and cause extension of that part of the tentacle. Because tentacles with a portion amputated are able to extend immediately, it appears that fluid pressure in the gastrovascular cavity is not significant in the extension of tentacles. The tentacles maintain a constant volume (Fig. 4-12) and their movements are probably independent of fluid pressure in the gastrovascular cavity of the body.

Reactions to External Stimuli

Contact. *Hydra* reacts to various kinds of special stimuli. Reaction to contact accounts for its temporary fixed condition. The attachment while in the resting attitude is a result of this reaction, and not a response to gravity, because hydras have the longitudinal axis of the body directed at every possible angle regardless of the force of gravity. Mechanical shocks, such as the jarring of the watch glass containing a specimen, or the agitation of the surface of the water, cause a rapid contraction of a part or all of the animal. This is followed by a gradual expansion until the original condition is regained.

Mechanical stimuli may be *localized* or *nonlocalized*. That just noted is of the latter type. *Local stimulation* may be accomplished by touching the body or tentacles with the end of a fine glass rod. It has been noted that the stimulation of one tentacle may cause the contraction of all the tentacles, or even the contraction of both tentacles and body. This shows that there must be some sort of transmission of stimuli from one tentacle to another and to the body. The structure of the nervous system would make this possible.

Light. Moderate amounts of light inhibit contractions of hydra. Intense light causes contractions which can be inhibited by reduced glutathione or some foods (Rushforth, Krohn, and Brown, 1964). Normal lighting produces no marked response, although the final result is quite decisive. If a dish containing hydras is placed so that the illumination is unequal on different sides, the animals will collect in the brightest region, unless the light is too strong, in which case they will congregate in a place where the light is less intense. *Hydra* therefore has an optimum with regard to light. The movement into or out of a certain area is accomplished by a method of "trial and error." When put in a dark place *Hydra* becomes restless and moves about in no definite direction; but if white light is encountered, its locomotion becomes less rapid and finally ceases altogether. The value to the organism of such a reaction is considerable, because the small animals that serve as food for it are attracted to well-lighted areas. *Colored lights* have the same effect as darkness; blue, however, is preferred by *Hydra* to white.

Temperature. The reactions of *Hydra* to changes in temperature are also indefinite, although in many cases they enable the animal to escape from a heated region. No locomotory change is produced by temperatures below 31°C; at this temperature, however, the basal disk is released and the animal takes up a new position either away from the heated area or further into it. In the former case the *Hydra* escapes, in the latter it may escape if subsequent movements take it away from the injurious heat, otherwise it perishes. *Hydra* does not move from place to place if the temperature is lowered; it contracts less rapidly, and finally ceases all its movements when the freezing point is approached.

Electric Current. An attached *Hydra,* when subjected to a weak constant electric current, bends toward the anode, its body finally becoming oriented with the basal disk toward the cathode and the anterior end toward the anode side. The entire animal then contracts. In an animal attached by the tentacles a similar bending occurs, but the basal disk in this case is directed toward the anode. These reactions are caused by local contractions on the anode side for which the electric current is directly responsible.

Hydra shows no reactions to currents of water. When placed in a current of water, it neither orients itself in a definite way nor moves either up or down stream.

General Remarks

It is evident from the above outline of the reactions of *Hydra* to stimuli that the only movements involved are produced by contraction and expansion of the body when attached, and by undirected changes of position. Being radially symmetrical, the body may be flexed in any direction.

Local stimuli, such as the application of heat or a chemical to a limited area of the body, causes a contraction of the part affected and a bending in that direction. This results in the movement of the tentacular region toward the stimuli, and the contraction of the entire animal follows, thus carrying it out of the influence of the stimulus. The bending produced by a mild stimulus applied to a tentacle is an efficient reflex for putting more of the tentacle and nematocysts around a small food organism to subdue it.

Nonlocalized stimuli, such as the jarring of the vessel containing the animals, produces, immediately, a contraction of the entire body, which, in most cases, is beneficial, since it removes it from an injurious agent. If, however, this simple contraction is not effective, as in the case of a constant application of heat, the *Hydra* usually resorts to some other reaction (e.g., locomotion), which often enables it to escape from the injurious stimulus.

Finally, it should be remembered that the physiological condition of the animal determines to a large extent the kind of reactions produced, not only spontaneously, but also by external stimuli. "It decides whether *Hydra* shall creep upward to the surface and toward the light, or shall sink to the bottom; how it shall react to chemicals and to solid objects; whether it shall remain quiet in a certain position, or shall reverse this position and undertake a laborious tour of exploration" (Jennings).

Reproduction

Reproduction takes place in *Hydra* both *asexually* and *sexually;* in the former case by budding, in the latter, by the production of a fertilized egg. Longitudinal and transverse division have been described in *Hydra,* but apparently are not methods of reproduction; they are simply processes that enable an abnormal animal to regain its normal shape.

Budding (Figs. 4-1 and 4-2). Asexual reproduction by budding is easily observed in the laboratory. Superficially the bud appears first as a slight bulge in the body wall. This pushes out rapidly into a stalk which soon develops a circlet of blunt tentacles about its distal end. The cavities of both stalk and tentacles are at all times directly connected with that of the parent. When full grown, the bud becomes detached and leads a separate existence. The details of the process are briefly as follows. The interstitial cells in a certain region increase in number and volume, producing a slight outbulging of the epidermis. The growing region is located at the point where the edges of the protrusion meet the body wall. Here the cells are well fed and multiply actively. The epidermal and gastrodermal cells of the parent give rise to the corresponding cells of the bud. When the bud is fully grown, the epidermal cells at its proximal end secrete a sticky substance, which is used later for its attachment. The gastrodermal cells in the same region then unite, separating the cavity of the bud from that of the parent. Finally, the bud becomes detached. The food supply determines the rate of growth of the bud, and a bud may be entirely absorbed by a starving animal.

The tentacles of the bud arise first as two outgrowths opposite each other. The third tentacle develops between these on the side toward the oral end of the parent; the fourth, opposite the third and the fifth and sixth on either side of the third.

Sexual Reproduction. Whether or not there are definite germ cells in the adult *Hydra* is still open to question. So far as is known, both ova and spermatozoa arise from indifferent interstitial cells. Hydras are, with the exception of a few that are hermaphroditic, either male or female. In *H. oligactis* the sex organs are located on the body and not on the stalk. There may be as many as 20 or 30 testes, low and rounded in shape and devoid of nipples. The sexual state can be induced in this species by lowering the temperature, which accounts for the appearance of sex organs in the autumn. During asexual reproduction, sex is inherited, since buds produce the same kind of sex organs as the parent, and all members of a clone are of the same sex.

Spermatogenesis. The male cells of *Hydra* are formed in little conical elevations called *testes*, which project from the surface of the body (Figs. 4-1, C and 4-2). The testis arises within the epidermis from interstitial cells. A single interstitial cell divides mitotically; then adjacent interstitial cells also divide, multiplication continuing until the epidermis becomes distended. An indefinite number of long mul-

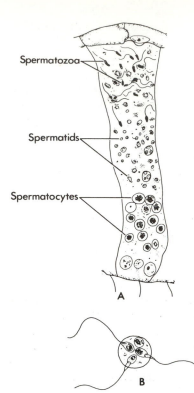

Figure 4-13 Parts of the testis of *Hydra*. (A) A single cyst. (B) Developing spermatozoa. (*From Tannreuther.*)

tinucleated cysts (Fig. 4-13, A) are formed within the testis, each cyst being the product of a single or several interstitial cells. Each interstitial cell is a primordial germ cell; it gives rise by mitosis to a variable number of spermatogonia, which contain 12 chromosomes, the somatic number. Reduction in the number of chromosomes to six occurs just after the spermatogonia have divided to form the primary spermatocytes. The latter give rise to secondary spermatocytes which divide at once, producing spermatids. These two spermatocyte divisions take place without the formation of cell membranes; that is, each primary spermatocyte develops into a four-nucleated cell, which represents the four spermatids. Within this cell the spermatids transform into spermatozoa (Fig. 4-13), B). A single cyst may contain representatives of all of these cell generations—spermatogonia, primary spermatocytes, secondary spermatocytes, spermatids, and spermatozoa. The mature spermatozoa break out of the vesicle in which they are formed, and swim about in the distal end of the cyst (Fig. 4-13, A); they finally reach the outside by way of a minute temporary opening in the end of the cyst. The mature spermatozoa swim about in the water searching for an egg; their activity continues from 1 to 3 days.

Oogenesis. The egg is first distinguished from the interstitial cells of the epidermis by its slightly greater size, its spherical shape, and the

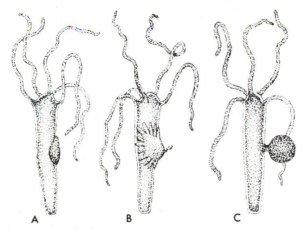

Figure 4-14 *Chlorohydra viridissima.* Stages in the growth of the egg. (A) Young oogonium. A cross section through this oogonium is shown in Fig. 3-14. (B) Oocyte with many protoplasmic extensions. (C) Primary oocyte with complete amount of yolk. (*After Kepner and Looper.*)

comparatively large volume of its nucleus. As the eggs grow in size, the neighboring interstitial cells increase in number by mitosis, and also become larger. The whole structure may at this time be called an ovary (Figs. 4-2, 4-14, and 4-15). The nourishment of the egg is at first similar to that of the other epidermal cells, but later disintegration and resorption of adjacent interstitial cells takes place. Yolk is elaborated from material that enters from the gastrodermis. Usually, only one egg is developed in a single ovary (Fig. 4-15), but sometimes two may arise and complete their development in ovaries side by side. In cases, however, when two or more eggs are contained in one ovary, their ad-

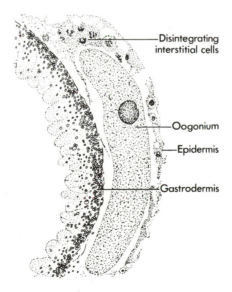

Disintegrating interstitial cells

Oogonium

Epidermis

Gastrodermis

Figure 4-15 *Chlorohydra viridissima.* Cross section through the oogonium shown in Fig. 4-14, A (*After Kepner and Looper.*)

jacent walls dissolve and one of the nuclei survives while the others disintegrate. As the ovum grows it becomes lobed, showing distinct protoplasmic fingers (Fig. 4-14, B); these are drawn in when it has reached its full size. The egg is now nearly spherical (Fig. 4-14, C) and is surrounded by a single layer of epidermal cells. Maturation then takes place. Two polar bodies are formed, the first being larger than the second. During maturation the number of chromosomes is reduced from the somatic number, 12, to six. This occurs at the end of the growth period. Now an opening appears in the epidermis and the egg is forced out, finally becoming free on all sides except where attached to the animal.

Fertilization. Fertilization usually occurs within 2 hours. Several sperm penetrate the egg membrane, but only one enters the egg itself. If not fertilized within 24 hours, the egg becomes sterile. The sperm brings a nucleus containing six chromosomes into the egg. The male and female nuclei unite, forming the fusion nucleus.

Embryology. *Cleavage,* which now begins, is total and regular. A well-defined cleavage cavity is present at the end of the third cleavage (i.e., the eight-celled stage). When the *blastula* is completed, it resembles a hollow sphere with a single layer of epithelial cells composing its wall. These cells may be called the primitive ectoderm. By mitotic division they form endoderm cells, which drop into the cleavage cavity, completely filling it. The *gastrula,* therefore, is a solid sphere of cells differentiated into a single outer layer, the ectoderm, and an irregular central mass, the endoderm. The ectoderm secretes around the gastrula two envelopes. The outer is a thick chitinous shell covered with sharp projections; the inner is a thin gelatinous membrane.

Hatching. The embryo in this condition separates from the parent and falls to the bottom, where it remains unchanged for several weeks. Then interstitial cells make their appearance. A subsequent resting period is followed by the breaking away of the outer chitinous envelope and the elongation of the escaped embryo. Mesoglea is now secreted by the ectoderm and endoderm cells; a circlet of tentacles arises at one end and a mouth appears in their midst. The young *Hydra* thus formed soon grows into the adult condition. The ectoderm of the embryo matures into the epidermis of the adult while the endoderm becomes the gastrodermis. Although the terms ectoderm and endoderm have been used by many to designate the layers of adult coelenterates, they are best reserved for use in the early developmental stages of animals.

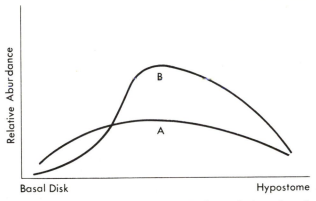

Figure 4-16 Cell growth and division in *Hydra,* relative abundance in different regions of the body. Based on autoradiographic studies of thymidine incorporation into DNA prior to nuclear division. (A) Percentage of labeled nuclei relative to position. (B) Regional distribution of labeled nuclei. (*Adapted from data of Campbell, 1965.*)

Growth

Some doubt may be cast on the role of the interstitial cells in replacing all cell types, because Campbell (1965) has found that most cell types except cnidoblasts undergo division. Autoradiographs of *Hydra* fed labeled thymidine (Fig. 4-16) show uptake of thymidine, and presumably replication of DNA, by nuclei in all regions of *Hydra*. The species investigated was a stalked one and little growth occurred in the stalk. The continual loss of cells from the basal disk and the use of cells in bud formation causes a slow growth of cells from the subhypostomal area toward the stalk.

Depression. Hydras, both in nature and in laboratory cultures, appear to undergo stages of depression during which the tentacles shorten and gradually disappear and the body of the animal becomes shorter due to the disintegration of the tissues at the distal end. Finally only a ball-like mass of cells representing the basal end remains, and this soon disintegrates. Depression appears to be due to a lowered metabolic state, which may be produced by rich feeding, high temperature, senescence, fouling of the culture water, lack of oxygen, or transfer to clean fresh water. Recovery from depression may occur spontaneously or may be induced by transferring the animals to a culture in which other hydras are flourishing. Often, within an hour the body begins to elongate and tentacles begin to appear.

Regeneration. An account of the phenomenon of regeneration is appropriate at this place, since the power of animals to restore lost parts was first discovered in *Hydra* by Trembley in 1740. This investigator

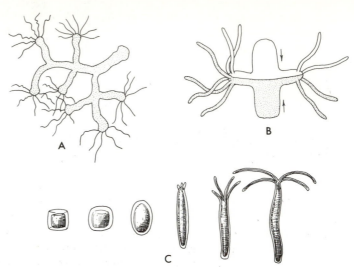

Figure 4-17 *Hydra.* Regeneration and grafting. (A) A seven-headed *Hydra* resulting from regeneration after the distal ends were split lengthwise. (B) Parts of two hydras grafted together. (C) The regeneration of an entire *Hydra* from a small piece of the stalk. (*After several authors.*)

found that if hydras were cut into two, three, or four pieces, each part would grow into an entire animal (Fig. 4-17, C). Other experimental results obtained by Trembley are that the hypostome together with the tentacles, if cut off, produce a new individual; that each piece of a *Hydra,* split longitudinally into two or four parts, becomes a perfect polyp; that when the head end is split in two and the parts separated slightly a two-headed animal results.

The old observation that a hydra turned inside out will have the cells migrate through the mesoglea to regain their original inner or outer orientation has been refuted by the observations of Macklin (1968) that the animal returns itself to the original position by either slow eversion or rapid contractions.

Regeneration may be defined as the replacing of an entire organism by a part of the same. It takes place not only in *Hydra,* but in many other coelenterates, and in some of the representatives of almost every phylum of the animal kingdom. *Hydra* is a species that has been widely used for experimentation. Pieces of *Hydra* that measure ¹/₆ mm or more in diameter are capable of becoming entire animals (Fig. 4-17, C). The tissues in some cases restore the lost parts by a multiplication of their cells; in other cases, they are worked over directly into a new but smaller individual.

Experiments have shown that when pieces of hydras too small to regenerate are allowed to come into contact with one another they fuse. The gastrodermis appears to initiate this fusion and controls the process. The gastrodermis of one piece may fuse with the gastrodermis of an adjoining piece rapidly. Epidermis does not fuse with epidermis nor with gastrodermis, and mesoglea appears to have little or no power of fusion. Fusion, as indicated by tissue culture experiments, appears to be the result of the interlacing of amoeboid processes sent out by the gastrodermal cells of the pieces in contact. As a result of this fusion a platelike body is formed which may, within a period of about 12 hours, develop into a sac. This sac becomes attached at a certain point which is the basal disk; it then elongates; tentacles bud out at the distal end and a normal hydra results (Papenfuss).

Grafting. Parts of one *Hydra* may easily be grafted upon another (Fig. 4-17, B). In this way many bizarre effects have been produced. Parts of two hydras of two species have also been successfully united. Lenhoff (1965) has found that normal and mutant hydras grafted together eventually produce by budding a range of hydras having the characteristics from normal to mutant.

Space will not permit a detailed account of the many interesting questions involved in the phenomena of regeneration, but enough has been given to indicate the nature of the process. The benefit to the animal of the ability to regenerate lost parts is obvious. Such an animal, in many cases, will succeed in the struggle for existence under adverse conditions. Regeneration takes place continually in most animals; for example, new cells are produced in the epidermis of humans to take the place of those that are no longer able to perform their proper functions. Both internal and external factors have an influence upon the rate of regeneration and upon the character of the new part.

OBELIA—A COLONIAL HYDROZOAN

Obelia (Fig. 4-18) is a colonial coelenterate, that lives in the sea, where it is usually attached to rocks, to wharves, or to *Laminaria*, *Rhodymenia*, and other algae. It may be found in shallow water and to a depth of 40 fathoms along the coast of northern Europe, from Long Island Sound to Labrador, and the Pacific Coast.

Anatomy and Physiology

An *Obelia* colony consists of a basal stem, the *hydrorhiza*, which is attached to the substratum; this gives off at intervals upright branches, known as *hydrocauli*. At every bend in the zigzag hydrocaulus a side branch arises. The stem of this side branch is ringed and is expanded at the end into a hydralike structure, the *hydranth*. A single polyp consists of a hydranth and the part of the stalk between the hydranth and the point of origin of the preceding branch. Full-grown colonies usually bear reproductive members (*gonangia*) in the angles where the hydranths arise from the hydrocaulus.

The *Obelia* colony, as just described and as illustrated in Figure 4-18,

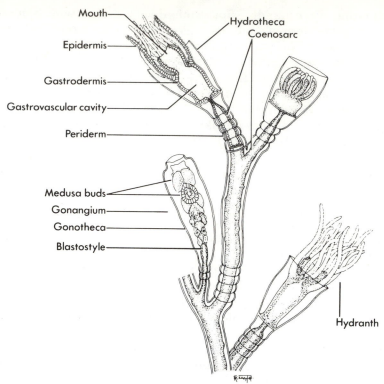

Figure 4-18 *Obelia.* The colonial polyp, or hydroid, stage. Portion showing both reproductive polyps (gonangia) and nutritive polyps (hydranths).

resembles the structure that would be built up by a budding *Hydra* if the buds were to remain attached to the parent and in turn produce fixed buds.

All of the soft parts of the *Obelia* colony are protected by a chitinous covering called the *periderm* (= *perisarc*); this is ringed at various places and is expanded into cup-shaped *hydrothecae* to accommodate the hydranths, and into *gonothecae* to enclose the reproductive members. A shelf that extends across the base of the hydrotheca serves to support the hydranth. The soft parts of the hydrocaulus and of the stalks of the hydranths constitute the *coenosarc* and are attached to the periderm by minute projections. The coenosarcal cavities of the hydrocaulus open into those of the branches and thence into the hydranths, producing in this way a common *gastrovascular cavity.*

A longitudinal section of a hydranth and its stalk (Fig. 4-18) shows the coenosarc to consist of two layers of cells—an outer layer, the *epidermis,* and inner layer, the *gastrodermis.* These layers are continued into the hydranth. The *mouth* is situated in the center of the large knoblike *hypostome,* and the *tentacles,* about 30 in number, are arranged around the base of the hypostome in a single circle. Each ten-

214 *Phylum Coelenterata*

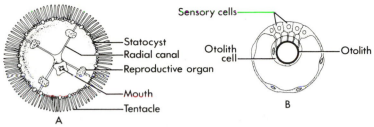

Figure 4-19 *Obelia*. (A) Medusa. (B) Statocyst. (*A, after Parker and Haswell; B, after Kerr.*)

tacle is solid, consisting of an outer layer of epidermal cells and a single axial row of gastrodermal cells; at the extremity are a large number of one type of nematocyst. The hydranth captures, ingests, and digests food as in *Hydra.*

The reproductive polyps arise, as do the hydranths, as buds from the hydrocaulus, and represent modified hydranths. The central axis of each is called a *blastostyle,* and together with the gonothecal covering is known as the *gonangium.* The blastostyle gives rise to *medusa-buds,* which soon become detached and pass out of the gonotheca through the opening in the distal end.

Some of the *medusae* of *Obelia* (Fig. 4-19, A) produce eggs, and others produce spermatozoa. The fertilized eggs develop into colonies like that which gave rise to the medusae. The medusae provide for the dispersal of the species, because they swim about in the water and establish colonies in new habitats. The medusa of *Obelia* is shaped like an umbrella with a fringe of tentacles and a number of organs of equilibrium on the edge (Fig. 4-19, B). Hanging down from the center is the *manubrium* with the mouth at the end. The gastrovascular cavity extends out from the cavity of the manubrium into four *radial canals* on which are situated the reproductive organs.

The *germ cells* of the medusae of *Obelia* arise in the epidermis of the manubrium, and then migrate along the radial canals to the reproductive organs. When mature, they break out into the water. The eggs are *fertilized* by spermatozoa which have escaped from other medusae. *Cleavage* is similar to that of *Hydra,* and a hollow *blastula* and solid *gastrulalike* structure are formed. The gastrulalike structure soon becomes ciliated and elongates into a free swimming larva called a *planula.* This soon acquires a central cavity, becomes fixed to some object, and proceeds to found a new colony. The physiological processes of *Obelia* are similar to those described in *Hydra.*

GONIONEMUS—A HYDROZOAN MEDUSA

The structure of a hydrozoan jellyfish or medusa may be illustrated by *Gonionemus* (Fig. 4-20). This jellyfish is common along the eastern and

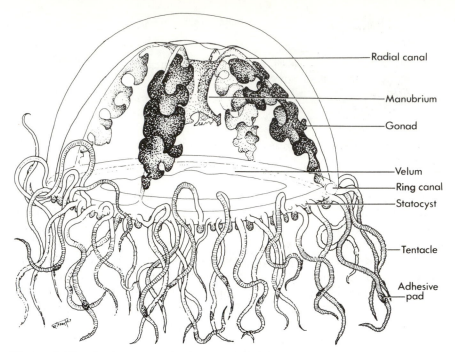

Radial canal

Manubrium

Gonad

Velum

Ring canal

Statocyst

Tentacle

Adhesive pad

Figure 4-20 *Gonionemus:* a hydrozoan jellyfish.

western coasts of the United States. It measures about 1½ cm in diameter, without including the fringe of tentacles around the margin. In general form it is similar to the medusa of *Obelia* (Fig. 4-19, A). The convex or aboral surface is called the *exumbrella;* the concave, or oral surface, the *subumbrella.* The subumbrella is partly closed by a membraneous ledge called the *velum.* Water is taken into the subumbrellar cavity and is then forced out through the central opening in the velum by the contraction of the body; this propels the animal in the opposite direction, thus enabling it to swim about.

The *tentacles,* which vary in number from 16 to more than 80, are capable of considerable contraction. Near their tips are *adhesive pads* at a point where the tentacle bends at a sharp angle. Hanging down into the subumbrellar cavity is the *manubrium* with the *mouth* at the end surrounded by four frilled *oral lobes.* The mouth opens into a *gastrovascular cavity,* which consists of a central *stomach* and four *radial canals.* The radial canals enter a *circumferential canal,* which lies near the margin of the umbrella.

The cellular layers in *Gonionemus* are similar to those in *Hydra,* but the *mesoglea* is extremely thick and gives the animal a jellylike consistency. Scattered about beneath the epidermis are many *nerve cells,* and about the velum is a *nerve ring. Sensory cells* with a tactile function are abundant on the tentacles. The margin of the umbrella is supplied with two kinds of sense organs: (1) at the base of the tentacles are

Figure 4-21 *Gonionemus:* hydralike stage in development. One of the tentacles is carrying a worm (W) to the mouth. The tentacles are in a contracted state. (*After Perkins.*)

round bodies, which contain pigmented endoderm cells and communicate with the circumferential canal; (2) between the bases of the tentacles are small outgrowths, which are organs of equilibrium, the *statocysts. Muscle fibers,* both exumbrellar and subumbrellar, are present, giving the animal the power of *locomotion.*

Suspended beneath the radial canals are the sinuously folded *reproductive organs* or *gonads. Gonionemus* is dioecious, one individual producing either eggs or spermatozoa. These reproductive cells break out directly into the water, where *fertilization* takes place. A ciliated *planula* develops from the egg as in *Obelia.* This soon becomes fixed to some object, and a mouth appears at the unattached end. Then four tentacles grow out around the mouth and the hydralike larva is able to feed (Fig. 4-21). Other similar hydralike polyps bud from its walls. The medusae are thought to arise from these polyps as lateral buds.

The physiological processes of *Gonionemus* are very similar to those of *Hydra.* Food is captured with the aid of nematocysts while the animal is swimming and conveyed to the mouth by the tentacles, or the *Gonionemus* lies with its aboral surface up, and any food that drops on its tentacles or manubrium are similarly engulfed. Only one type of nematocyst occurs in *Gonionemus.*

OTHER HYDROZOA

Most members of this class are colonial, and typical species exist as polyps, which give rise by budding to free or sessile medusae. The medusae (with a few exceptions, e.g., *Obelia*) possess a velum and a nerve ring. There are no mesenteries in the polyp; the tentacles are usually solid; and the reproductive cells are ectodermal and discharged directly to the exterior. Four orders and a number of suborders are included here.

Order 1. Hydroida

Hydrozoa with an attached hydroid stage.

Suborder 1. Anthomedusae (Gymnoblastea). Perisarc usually covering coenosarc but not the polyps and reproductive individuals; medusae with gonads on the manubrium and with ocelli.

Hydra lives in fresh water; is solitary; has from 4 to 12 hollow tentacles; perisarc absent; no medusa. Most of the commonly occurring hydras of North America are dioecious, although a few species are hermaphroditic. *Hydra americana* is easily distinguished by the tentacles, of the extended individual, which are shorter than the column. *H. oligactis* and *H. pseudoligactis* are large hydras with a distinct stalk at the basal end. The male of the latter species has nipples on the testes and the threads within the holotrichous isorhizas are coiled transversely. *H. oligactis* lacks nipples on the testes, and the threads are coiled lengthwise in the isorhizas. *H. carnea* is small, reddish brown, and the theca of the embryo has short spines. *H. littoralis*, found along shores and in moving water, has long spines on the theca of the embryo.

Chlorohydra viridissima, the green hydra, is small and hermaphroditic. Albino strains have been produced in the laboratory. It seems likely that the green hydra must acquire its symbiotic algae after hatching, because the reproductive structures develop in the epidermis, which lacks the algae found in the gastrodermis. Many of those hatched from eggs are albino.

Clava (Fig. 4-22, A) forms a nonbranching colony; the polyps rise from a filiform hydrorhiza and the tentacles are scattered irregularly over the hydranth; sporosacs are present at the base of the tentacles. *C. leptostyla* has reddish hydranths, about 20 tentacles, and pink (male) or purple (female) sporosacs; on *Fucus,* piles, etc.; Long Island northward.

Figure 4-22   Anthomedusae. (A) *Clava:* a hydroid bearing sporosacs. (B) *Hydractinia:* four types of individuals. (*After Allman, from Fowler.*)

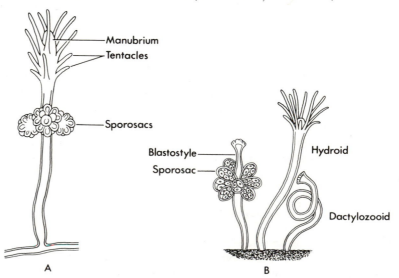

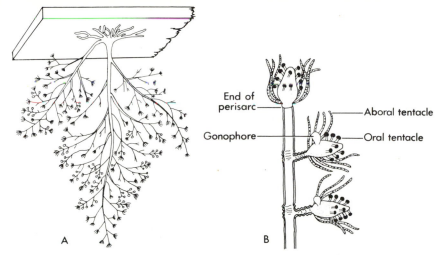

Figure 4-23 Anthomedusae. (A) *Bougainvillia;* colony, natural size, hanging from floating object. (B) *Pennaria;* tip of branch. (*A, after Allman; B, after Borradaile.*)

Bougainvillia (Fig. 4-23, A) forms an arborescent colony; the hydranths possess a single whorl of filiform tentacles; the free-swimming medusa has four clusters of tentacles. *B. carolinensis* colonies are about 30 cm high, each hydranth with prominent hypostome and 12 tentacles; the medusa is about 4 mm in diameter; on *Fucus,* piles, etc.; Cape Cod southward.

Eudendrium forms a branching colony; the polyps rise from a reticulated hydrorhiza; the hypostome is trumpet-shaped and has a single whorl of filiform tentacles; no medusae but sporosacs present. *E. ramosum* forms large colonies about 12 cm high; about 20 tentacles; sporosacs red (male) or orange (female); on rocks, piles, etc.; North Carolina northward.

Hydractinia (Fig. 4-22, B) forms an encrusting colony; the polyps are nutritive, reproductive, or defensive in function; no medusae. *H. echinata* forms a colony about 10 mm high; no tentacles are present on the reproductive polyps; on *Fucus,* rocks, piles, hermit crab shells, etc.; Atlantic Coast; similar to *H. milleri* of the Pacific Coast.

Podocoryne resembles *Hydractinia* but has a medusa with 8 or more long tentacles. *P. fulgurus* has a medusa about 1 mm high and 4 oral and 8 marginal tentacles; may be phosphorescent; Massachusetts to North Carolina.

Pennaria (Fig. 4-23, B) forms a regularly branching colony; the hydranth possesses 10 to 12 filiform basal tentacles and knobbed tentacles on the hypostome; medusa either sessile or free. *P. tiarella* colonies are about 12 cm high; bright pink in color; medusae buds on side of hydranth, sessile or free; on rocks, piles, etc.; Maine southward.

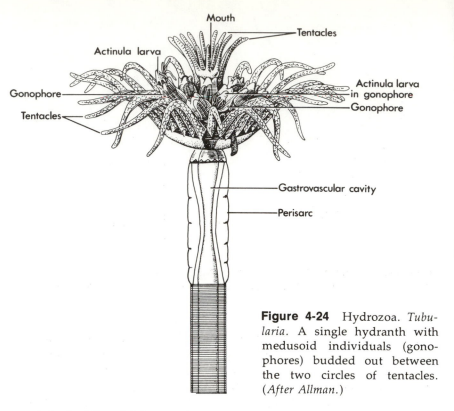

Figure 4-24 Hydrozoa. *Tubularia*. A single hydranth with medusoid individuals (gonophores) budded out between the two circles of tentacles. (*After Allman.*)

Tubularia (Fig. 4-24) is solitary or forms large, pink colonies; the hydranth has one basal and one distal whorl of filiform tentacles; medusae attached to polyp; free-swimming, hydroidlike bodies are formed by medusoids. *T. crocea* forms dense tuftlike colonies about 9 cm high; with 20 to 24 basal tentacles; on piles, etc.; Massachusetts southward; California.

Suborder 2. Leptomedusae (Calyptoblastea). Periderm covering coenosarc and becoming hydrothecae over nutritive polyps and gonothecae over reproductive polyps; medusae with gonads on the radial canals; usually statocysts.

Obelia has already been described in detail.

Sertularia has opposing pairs of hydrothecae along the stem; not joined to the stem by a stalk. *S. pumila* forms a branched colony about 3 cm high; gonangia oval; on *Fucus,* etc.; New Jersey northward. *S. furcata* is common on eelgrass, Pacific Coast.

Campanularia (Fig. 4-25) forms a simple or branched colony; hydrothecae bell-shaped and without operculum. *C. flexuosa* forms an irregularly branched colony about 25 mm high; stem annulated near base of branches; stalks of hydrothecae annulated; on piles, etc.; Long Island northward.

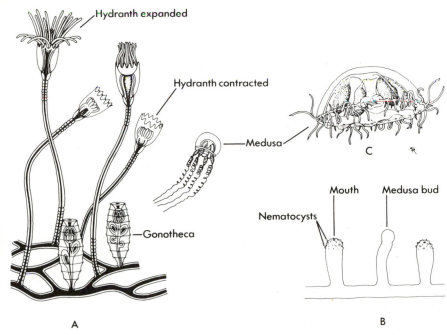

Figure 4-25 Hydroida. (A) Leptomedusae. *Campanularia*. A colony with three types of individuals. (B) Limnomedusae. *Craspedacusta sowerbyi*. Polyp stage of the freshwater jellyfish. (C) *C. sowerbyi*, medusa stage. (*A, after All-man.*)

Suborder 3. Limnomedusae. Includes the freshwater medusae and forms such as *Gonionemus*, already described in detail.

Craspedacusta sowerbyi (Fig. 4-25) lives in fresh water; has a small, colonial hydroid stage which lacks tentacles; medusae bud from the colony and develop up to 300 tentacles in four size groups; lithocysts up to 200 or more; manubrium long. The medusa has considerable resemblance to the medusa of *Gonionemus*. *Craspedacusta* occurs widely but sporadically in lakes and ponds in the United States. Populations of only one sex seem to occur more frequently than those with both sexes. Before the relationship of the hydroid and medusoid stage was known, the hydroid stage of *Craspedacusta* was named *Microhydra ryderi*.

The smallest of hydrozoans appear to be the freshwater *Calpasoma*, a tentacle-bearing form, otherwise resembling a *Craspedacusta* polyp, and the marine *Microhydrula*. *Microhydrula* is only known from the sand layer of marine aquaria.

Suborder 4. Hydrocorallinae. Hydrozoa with branched hydrorhiza; calcareous exoskeleton; polyps of two types, nutritive (gastrozooids) and for capturing prey (dactylozooids), arising from skeletal pits; me-

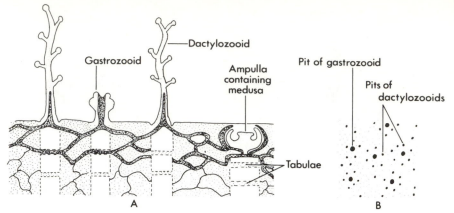

Figure 4-26 Hydrocorallina. *Millepora*. (A) Diagrammatic section. Living canals shown in black; degenerating canals, by lines; skeleton, by stippling. (B) Surface view of skeleton. (*A, from Borradaile and Potts, after Hickson.*)

dusoids from buds of the coenosarc, mostly confined to skeletal pits (ampullae). *Millepora* has separate pores for gastrozooids and dactylozooids. A common member of the coral reef fauna, *Millepora* (Fig. 4-26) forms a massive base from which irregular branches arise; polyps very contractile; gastrozooids with four or five knobbed tentacles; dactylozooids; medusa simple, with four or five rudimentary tentacles. *M. alcicornis* lives on the Florida coast; it is sometimes called the stinging coral because of its powerful nematocysts.

Millepora is in the Milleporina. The remaining hydrocorals are in the Stylasterina. These close relatives of the Milleporina have branching or encrusting calcareous exoskeletons. Tabulae are lacking in the polyp tubes, instead, a calcareous style extends up from the bottom of the tube into the base of the polyp. The polyp tubes contain grooves around the periphery, which are occupied by the dactylozooids surrounding the gastrozooid.

Order 2. Actinulida

Small hydrozoans with external ciliation that resemble an actinula larva (Fig. 4-53, E). The few species found since their discovery in 1927 are all marine interstitial inhabitants bearing statocysts. Two known genera are European. *Halammohydra* has an aboral adhesive organ; *Otohydra* lacks an adhesive organ. Development is direct.

Order 3. Trachylinida

Hydrozoa without alternation of generations; medusae usually develop directly from eggs; tentaculocysts present; gonads on radial canals or on floor of gastric cavity.

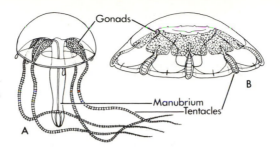

Figure 4-27 (A) Trachymedusae. *Liriope scutigera*. (B)
Narcomedusae. *Cunina oc-
tonaria*. (*A, after Fewkes; B,
after Brooks.*)

Suborder 1. Trachymedusae. Marginal tentacles on edge of umbrella;
gonads on radial canals; sense tentacles in pits or vesicles.

Liriope (Fig. 4-27, A) is hemispherical and has a long manubrium
and six or eight closed lithocysts. *L. exigua* is about 2 cm in diameter;
Gulf Stream, New England.

Aglantha has eight radial canals, eight gonads on the radial canals,
and free lithocysts. *A. digitale* is about 3 cm high and 1.5 cm in diameter; many tentacles; gonads long; North Atlantic.

Suborder 2. Narcomedusae. Marginal tentacles aboral from edge of
umbrella; gonads on oral wall of stomach; sense tentacles not enclosed. No manubrium.

Cunina is rather flat and transparent; the larvae live as parasites
within the umbrella of the mother or other medusae. *C. lativentris* is
about 1.5 cm in diameter; 10 to 12 marginal lobes, tentacles and gastric
pouches; four lithocysts on each lobe; Atlantic and Mediterranean.

C. octonaria (Fig. 4-27, B) has eight marginal lobes, tentacles, and
gastric pouches; larvae produce larvae by budding; about 7 mm in diameter; cosmopolitan.

Suborder 3. Pteromedusae. Contains the genus, *Tetraplatia*, which is
elongated along an oral–aboral axis. It possesses four equatorial
swimming flaps, each with two statocysts. The nematocysts are all
atrichous isorhizas, a condition found in one ctenophore. Other resemblances of trachyline medusae to the Ctenophora can be found in
the genus *Hydroctena*.

Order 4. Siphonophorida

Pelagic, colonial Hydrozoa highly polymorphic; medusoid and hydroid members of the colony bud off from a coenosarc that arises from
a planula larva. A *pneumatophore*, a sac filled with gas secreted by part
of the epithelium, serves as a float for the colony, which may contain
the following types of individuals.

1. *Hydrophyllium:* leaf-shaped; shields other members of the colony.

2. *Nectophore* (or *nectocalyx*): medusoid member modified for swimming; provides propulsion for the colony.

3. *Gastrozooid:* a hydroid member of the colony; mouth for ingestion of food; usually a single basally attached tentacle.

4. *Dactylozooid:* a hydroid member that lacks a mouth; long and tentacular; armed with nematocysts for defense and capture of prey.

5. *Gonozooid:* a medusoid member that produces the eggs or sperm.

Nanomia bijuga is a small siphonophore, which is so abundant at depths of about 300 meters over much of the ocean that depth-recording echo-sounders can detect their depth. Gas secreted within the pneumatophore of *Nanomia* is principally carbon monoxide. The bubble produced is less than 2 mm in diameter but effective in producing the sonic reverberations of the deep scattering layer of the ocean.

Physalia (Fig. 4-28) has a bladderlike pneumatophore with a dorsal crest up to 12 cm in length that may contract, forcing out the gas through a dorsal pore and thus bringing about submergence; a fresh

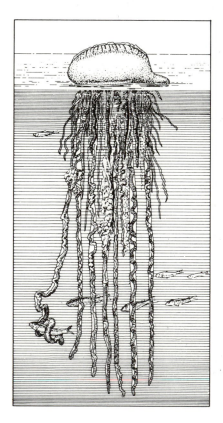

Figure 4-28 Siphonophorida. *Physalia*, The Portuguese man-of-war, capturing a fish. (*After Newman.*)

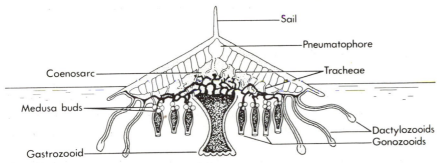

Figure 4-29 Siphonophorida. *Velella*. Vertical section. The cavity of the pneumatophore is in white; the network of the gastrodermal tubes are shown very dark. (*After Haeckel, modified.*)

supply of gas may be secreted bringing the colony to the surface again. There are no nectophores, movement from place to place being due to currents in the water or winds against the pneumatophore. *P. pelagica*, the Portuguese man-of-war, has stinging dactylozooids up to 16 m long; it occurs in the Gulf Stream from Florida northward; specimens are often cast up on the shores. The dactylozooids are able to catch large fish, which, by contraction, they draw up to the gastrozooids; these enclose the prey in a digestive sac by spreading their lips over it. The dactylozooids are able to contract to one seventieth of their maximum length. They exhibit rhythmic contraction waves, neuromuscular in nature, that travel at an average rate of 121 mm per second.

Velella (Fig. 4-29) forms a colony that resembles a single medusa in appearance. The pneumatophore is a chambered disk from the center of the ventral surface of which hangs a single large gastrozooid surrounded by gonozooids, with a circlet of dactylozooids around the outer edge. Free medusae arise from the gonozooids. *V. mutica* has a pneumatophore about 4 cm long with an elevated ridge on the dorsal surface that serves as a sail; it occurs on the Atlantic Coast, especially southward. *V. lata* is often stranded in large numbers on California beaches.

A new order, Chondrophorida, is sometimes used for *Velella* and similar species because of their considerable differences from other siphonophores and evidence of some affinity with the Anthomedusae.

Siphonophores are especially dangerous to man when contact is made with their nematocysts.

Class II. Scyphozoa

AURELIA—A SCYPHOZOAN MEDUSA

Aurelia (Fig. 4-30) is one of the commonest of the scyphozoan jelly-fishes of both coasts. The species *A. flavidula* ranges from the coast of Maine to Florida. Members of genus may be recognized by the eight shallow lobes of the umbrella margin, and the fringe of many small tentacles.

In structure *Aurelia* differs from *Gonionemus* and other hydrozoan medusae in the absence of a velum, the characteristics of the canal system, the position of the gonads, and the arrangement and morphology of the sense organs.

The *oral lobes* or *lips* of *Aurelia,* which hang down from the square mouth, are long and narrow, with folded margins. The *mouth* opens into a short *gullet,* which leads to the somewhat rectangular *"stomach."* A *gastric pouch* extends laterally from each side of the stomach. Within each gastric pouch is a *gonad* and a row of small *gastric filaments* bearing *nematocysts.* Numerous *radial canals* some of which branch several times, lead from the stomach to a *circumferential canal* at the margin.

The circulation of fluid within the canals is due to the beating of cilia attached to the gastrodermal cells (Fig. 4-31, A); is definite in direction; and occupies a period of about 20 minutes. Water flows

Figure 4-30 Scyphozoa. *Aurelia aurita.* Ventral view. (*After Shipley and Mac-Bride, modified.*)

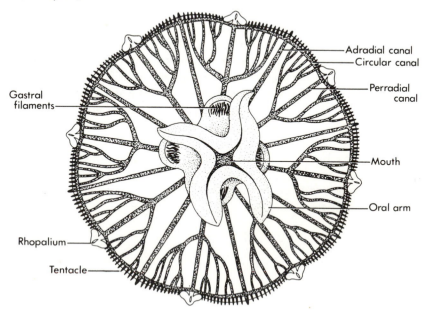

Adradial canal
Circular canal
Perradial canal
Gastral filaments
Mouth
Oral arm
Rhopalium
Tentacle

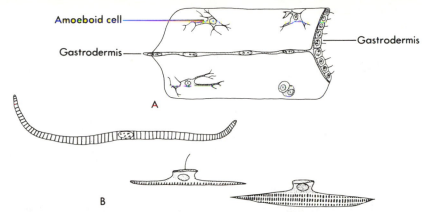

Figure 4-31 Histology of medusae. (A) *Aurelia;* section showing wandering cells, gastrodermal cells and mesoglea. (B) *Lizzia;* above, muscular cell from subumbrella; below, two epitheliomuscular cells from base of tentacle. (*After Hertwig, from Lankester.*)

through the mouth into the gastric cavity; thence into the gastric pouches through the adradial canals to the circular canal; into the interradial and perradial canals and out of exhalent grooves on the oral arms. Food particles carried by the circulating fluid are engulfed by gastrodermal cells.

The epidermal cells are more highly specialized for contraction than those of *Hydra* or *Gonionemus;* their bases are elongated into a cross-striated fiber capable of rapid rhythmic contraction (Fig. 4-31, B). An epidermal nerve net more complex than that of *Hydra* is also present (Fig. 4-32, A). A simple receptor-effector system exists in certain medusae similar to that illustrated in 4-32, B. It consists of a receptor at

Figure 4-32 *Aurelia aurita;* nerve cells. (A) Scattered nerve-ganglion cells from subumbrella. (B) Diagram of receptor-effector system. (*A, after Schafer; B, after Parker.*)

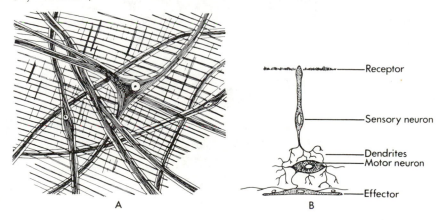

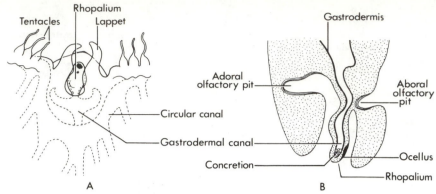

Figure 4-33 *Aurelia aurita;* rhopalium. (A) Oral aspect. (B) Longitudinal section. (*A, after Fowler; B, after Eimer, from Fowler.*)

the body surface ending in a nerve net which is associated with a nerve cell and thence with a muscle cell (effector).

There is a double nerve net in *Aurelia.* A main subumbrellar nerve net stimulates the muscles to cause pulsations of the medusa for locomotion. This net communicates at the rhopalial ganglia with a more diffuse net of the oral and aboral epidermal layer. The latter net controls feeding and other localized activities.

The eight *sense organs* of *Aurelia* lie between the marginal lappets (Figs. 4-30 and 4-33) and are known as *rhopalia.* They contain organs of *equilibrium.* As shown in Figure 4-33, each rhopalium is a hollow projection connected with the gastrodermal canal. It contains a number of calcareous ($CaSO_4.2H_2O$) concretions formed by the gastrodermis, and bears an epidermal pigment spot, the ocellus (Fig.

Figure 4-34 *Aurelia aurita;* section of double eye. (*After Schewiakoff, from Dahlgren and Kepner.*)

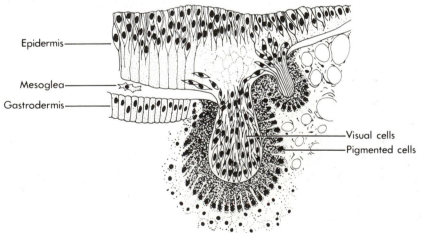

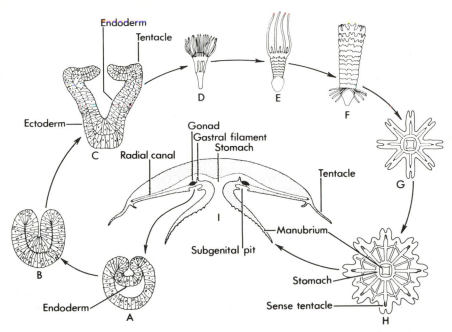

Figure 4-35 *Aurelia:* life history. (A–C) Longitudinal sections through gastrula stages. (D) Scyphistoma. (E and F) Strobila. (G and H) Ephyra. (I) Vertical section through adult. (*A–C, more highly magnified than the other figures; from Kerr.*)

4-34), which is sensitive to light. The rhopalium is protected by an aboral hood and by lateral lappets. *Olfactory pits* are situated near by.

The gonads are frill-like organs lying in the floor of the gastric pouches. They have a pinkish hue in the living animal. The eggs or spermatozoa pass through the stomach and out of the mouth. An *alternation of generations* occurs in *Aurelia,* but the hydroid stage is subordinate. The eggs develop into free-swimming planulae which become attached to some object and produce hydralike structures, each of which is called a *scyphistoma* (Fig. 4-35). This buds like *Hydra* during most of the year, but finally a peculiar process called *strobilization* takes place. The scyphistoma divides into disks which cause it to resemble a pile of saucers. Each disk develops tentacles, and, separating from those below it, swims away as a minute medusa called an *ephyra.* The ephyra gradually develops into an adult jellyfish.

Hamner and Jenssen (1974) found a population of *Aurelia aurita* in Tomales Bay, California, that went from a 2-cm bell diameter in March to a 12-cm bell diameter 3 months later, in June. They maintained laboratory populations and found numerous facts about growth and degrowth (Fig. 4-36). Starvation and feeding produced size-dependent effects. Normal growth resumed with feeding of starved animals if size did not go below 2-cm bell diameter. Sexual maturity and breed-

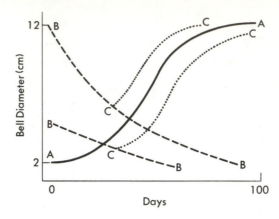

Figure 4-36 Growth and degrowth of *Aurelia aurita*. (A) Growth of a natural population. (B) Degrowth of a starved population. (C) Growth of a previously starved group when fed. Ordinate, bell diameter in centimeters; abscissa, time in days. (*Adapted from data of Hamner and Jensen, 1974.*)

ing were dependent on size, not age. However, spermatogenesis continued in cells, once it started, until sperm were formed, although the rest of the animal and gonad were undergoing degrowth to an immature condition due to starvation. Sexually mature individuals could be maintained in a sexually mature or breeding condition continuously by appropriate feeding. They found that the natural population included some that continued breeding into their second year of life. No determination has yet been made of the possibility that growth and degrowth, perhaps in annual cycles coordinated with circulation with oceanic surface gyres, could alternate to greatly extend the life of an individual.

OTHER SCYPHOZOA

Most of the larger jellyfishes belong to the Scyphozoa. They can be distinguished easily from the hydrozoan medusae by the presence of notches, usually eight in number, in the margin of the umbrella. They are called acraspedote (without velum or craspedon) medusae in contrast to the craspedote (with velum or craspedon) medusae of the Hydrozoa. The Scyphozoa usually range from a few centimeters to 1 meter in diameter, but some have been reported over 2 meters in diameter with tentacles over 28 meters long. They are all marine and are usually found floating near the surface of the sea, though some of them are attached to rocks and weeds. There is an alternation of generations in their life history, but the asexual stage (the scyphistoma, Fig. 4-35, D) is subordinate. The class contains five orders.

Order 1. Stauromedusida

Scyphozoa (Fig. 4-37, A) without rhopalia; tentacles perradial and interradial; umbrella goblet-shaped, temporarily attached by a narrow stalk at aboral pole; a stomodaeum is present, suspended by four mesenteries; no alternation of generations.

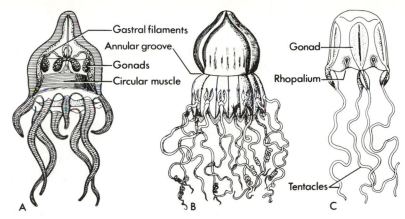

Figure 4-37 Scyphozoa. (A) Order Stauromedusida: *Tercera princeps*. (B) Order Coronatida: *Periphylla hyacinthina*. (C) Order Cubomedusida: *Charybdea marsupialis*. (*From Sedgwick, after Haeckel.*)

Lucernaria has no rhopalioids (adhesive pads) between the eight lobes; stalk cylindrical. *L. quadricornis* is about 7 cm high and 5 cm in diameter; red, gray, or green; 100 or more tentacles on each lobe; Massachusetts northward, Europe.

Haliclystus has a margin with eight lobes bearing many knobbed tentacles and eight rhopalioids between the lobes; stalk quadrate. *H. auricular* is about 3 cm high and 3 cm in diameter; 100 or more tentacles on each lobe; Massachusetts northward, central California to Alaska, Europe.

Order 2. Cubomedusida

Scyphozoa with four perradial rhopalia; tentacles interradial; umbrella four-sided, cup-shaped; no alternation of generations. The potency of their nematocysts has earned them the name sea-wasps. Found in warmer oceans.

Chironex fleckeri is thought to have been responsible for over 50 deaths on northern Australian coasts; if collapse does not occur within 20 minutes, survival is likely (Barnes, 1966).

Charybdea (Fig. 4-37, C) has a bell from 2 to 23 cm high and four interradial tentacles. *C. xaymacana* is transparent; lives in shallow water; is an active swimmer and a voracious feeder, ingesting comparatively large fish.

Order 3. Coronatida

Scyphozoa with four interradial rhopalia; tentacles perradial and adradial; umbrella conical, with transverse constriction; a stomodaeum is present suspended by four mesenteries; no alternation of generations.

Periphylla has 16 marginal lobes, 12 tentacles, and horseshoe-shaped gonads. *P. hyacinthina* (Fig. 4-37, B) is about 8 cm high and 4 cm in diameter; reddish in color; cosmopolitan.

Order 4. Semaeostomida

Scyphozoa with four or more perradial and four or more interradial rhopalia; umbrella disk-shaped; alternation of generations. This order contains most of the Scyphozoa.

Aurelia has already been described in detail.

Dactylometra has a quadrate mouth with four long, oral lobes, 40 hollow marginal tentacles, 48 marginal lobes, and eight rhopalia. *D. quinquecirrha,* the common sea nettle, reaches a diameter of 25 cm; Long Island southward.

Cyanea is a large disk-shaped medusa with no marginal tentacles but eight groups of very long tentacles on the subumbrella; long, large oral lobes. *C. capillata* is our largest jellyfish, ranging from 10 cm to 2 meters in diameter and possessing tentacles over 12 meters long; northern Pacific and North Atlantic south to North Carolina. Agassiz dried a large specimen, washed out the salts, and found the water and salt constituted 99.8 percent of the wet weight.

Order 5. Rhizostomida

Scyphozoa with no tentacles; typically eight rhopalia; mouth obliterated by fusion of the oral arms, secondary mouths remain at the ends of the oral arms where the grooves have not completely fused. *Cassiopeia* has eight mouth arms bearing numerous small appendages; Florida and the West Indies.

Class III. Anthozoa

METRIDIUM—A SEA ANEMONE

Metridium senile (Fig. 4-38) is a sea anemone, which fastens itself to the piles of wharves and to solid objects in tidepools along the North Atlantic Coast and the Pacific Coast. It is a cylindrical animal with an *oral disk* consisting of hollow *tentacles* arranged in a number of circlets about the slitlike *mouth.* The tentacles as well as the body, or *column,* can be expanded and contracted, and the animal's position may be changed by a sort of creeping movement of its *basal disk.* The *skin* is soft but tough and contains no skeletal structures. The tentacles capture small organisms by means of *nematocysts,* and carry the food thus obtained into the mouth. The beating of the *cilia* which cover the tentacles and part of the mouth and *gullet* is necessary to force the food into the *gastrovascular cavity.* At either side of the pharynx, or *stomodaeum,* is a ciliated groove called the *siphonoglyph.* Usually only one or two siphonogylphs are present, but sometimes three occur in a single specimen. A continual stream of water is carried into the body cavity through these siphonoglyphs, thus maintaining a constant supply of oxygenated water.

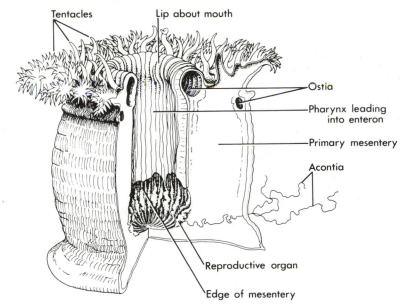

Tentacles

Lip about mouth

Ostia

Pharynx leading into enteron

Primary mesentery

Acontia

Reproductive organ

Edge of mesentery

Figure 4-38 *Metridium*. View of a specimen with one quadrant removed. (*From Woodruff*.)

If a sea anemone is dissected, the central or *gastrovascular cavity* will be found to consist of six radial chambers; these lie between the pharynx or stomodaeum and the body wall, and open into a common basal cavity. The six pairs of thin, double partitions between these chambers are called *primary septa* or *mesenteries* (Fig. 4-39). Water passes from one chamber to another through pores (*ostia*) in these mesenteries. Smaller mesenteries project out from the body wall into the chambers, but do not reach the stomodaeum; these are *secondary septa*. *Tertiary septa* and *quaternary septa* lie between the primaries and secondaries. There is considerable variation in the number, position, and size of the septa.

Each mesentery possesses a longitudinal retractor *muscle band*. The bands of the pairs of mesenteries face each other except those of the primaries opposite the siphonoglyphs. These primaries, which are called *directives*, have the muscle bands on their outer surfaces. The edges of the mesenteries below the stomodaeum are provided with *septal filaments* having a secretory function. Near the base these filaments bear long, delicate threads called *acontia*. The acontia are armed with *gland cells* and *nematocysts*, and can be protruded from the mouth or through minute pores (cinclides) in the body wall. They probably serve as organs of digestion, offense, and defense.

Near the edge of the mesenteries lying parallel to the mesenteric filaments are the *gonads*. The animals are dioecious, and the eggs or spermatozoa are shed into the gastrovascular cavity and pass out

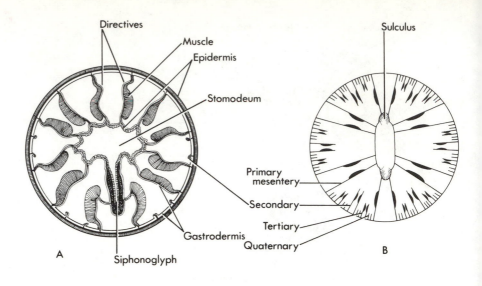

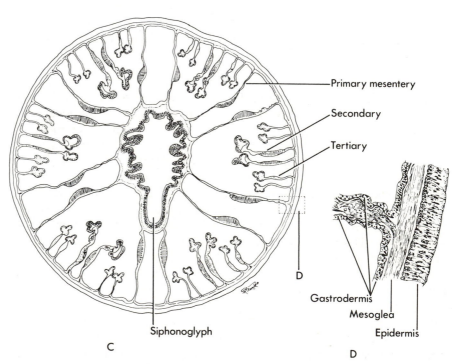

Figure 4-39 Mesenteries (septa) of sea anemones as seen in transverse sections. (A) Section through the stomodeum of *Peachia*. (B) Section through a typical actinian. (C) Transverse section of *Metridium*. (D) Junction of mesentery and body wall of Metridium. (*A, from Kerr; B, from Borradaile and Potts; C and D, original.*)

234

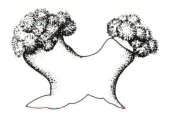

Figure 4-40 *Metridium.* A stage in binary fission. (*After Agassiz and Parker.*)

through the mouth. The fertilized egg develops as in other sea anemones, forming first a free-swimming planula and then, after attaching itself to some object, assuming the shape and structure of the adult.

Asexual reproduction is of common occurrence, new anemones being formed by *budding* or *fragmentation* at the edge of the basal disk. *Longitudinal fission* has also been reported (Fig. 4-40). Fragmentation, or *pedal laceration,* which results when the animal creeps away on the basal disk while a portion remains attached, is a successful method of reproduction because of the great regenerative powers of the basal disk. The presence or absence of directives in the regenerating fragments has considerable influence on the organization that develops (e.g., those without directives may lack a siphonoglyph). Thus, anemones produced by this method of reproduction show great variation.

ASTRANGIA—A CORAL POLYP

Astrangia danae (Fig. 4-41) is a coral polyp inhabiting the waters of our North Atlantic Coast. A number of individuals live together in colonies attached to rocks near the shore. Each polyp looks like a small sea anemone, being cylindrical in shape and possessing a crown of tentacles. The most noticeable difference is the presence of a basal cup of calcium carbonate termed the *corallite* (Fig. 4-42). This structure of calcium carbonate is what we commonly call coral. It is produced by the epidermis of the coral polyp and increases gradually during the life of the animal.

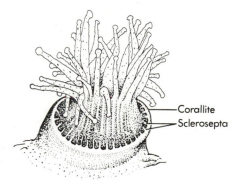

Corallite
Sclerosepta

Figure 4-41 *Astrangia.* A specimen with tentacles extended from the theca. (*From Johnson and Snook.*)

Class III. Anthozoa **235**

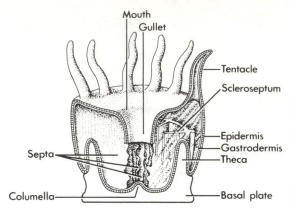

Figure 4-42 Semidiagrammatic view of one half of a simple coral polyp, showing relationship of soft parts to theca (corallite). (*From Shipley and MacBride.*)

The corallite is divided into chambers by a number of radial *sclerosepta*, which are built up between the pairs of mesenteries of the polyp. The center of the cup is occupied by a *columella* formed in part by the fusion of the inner ends of septa, and in part by projections from the base of the polyp. Although *Astrangia* builds a cup less than 1 cm in height, it produces enormous masses of coral in the course of centuries.

The food of *Astrangia* consists of small organisms, such as algae, Protozoa, hydroids, worms, crustaceans, and mollusks. These organisms are ingested as a result of the muscular action of the tentacles and central part of the oral disk. The mucus secreted by the oral disk and cilia of the stomodaeum also play a part in ingestion. A protease of digestive value is secreted into the gastrovascular cavity, but most digestion is intracellular. Food particles are engulfed by the mesenterial filaments in which food vacuoles are formed. The reaction of these vacuoles is at first acid but changes to alkaline. Digestion probably occurs during the period of alkalinity. Some specimens contain zooxanthellae, which utilize some of the waste products of the polyp and release some oxygen that the polyp can use. Some may also be digested by the polyp. The colorful flesh of most corals of shallow water is due to the presence of these flagellates. The symbionts are particularly important to the *hermatypic* (reef-building) corals, in which they provide much of the nutrition and speed carbonate deposition nearly tenfold at times.

OTHER ANTHOZOA

The Anthozoa or Actinozoa are solitary or colonial coelenterates with a polyp stage but no medusae. The polyps have a coelenteron divided into chambers by radially arranged membranes known as mesenteries. If mesenteries are absent, eight pinnately branched tentacles are present. A stomodaeum is present. Most of the Anthozoa secrete a calcareous skeleton, which we know as coral. Two subclasses with several orders in each are presented in the following discussion.

Subclass 1. Zoantharia

Anthozoa with usually many simple hollow tentacles, arranged generally in multiples of five or six; two siphonoglyphs as a rule; mesenteries vary in number; skeleton absent or present; simple or colonial; dimorphism rare.

Order 1. Actiniarida

Zoantharia usually solitary; many complete mesenteries; no skeleton. These are the sea anemones many of which are beautifully colored; in the large *Stoichactis* of the Great Barrier Reef of Australia, gray, white, lilac, green, and yellow may be found on the same individual.

Metridium has already been described in detail.

Edwardsia is a slender, solitary type of sea anemone with 16 tentacles in two circlets of eight each. *E. leidyi* is about 3 cm long and 1.5 mm in diameter; it is parasitic on a ctenophore, *Mnemiopsis leidyi*.

Halcampa has a long, slender body consisting of three parts, a foot region, a central region usually covered with sand, and an oral retractile region; habitat, sand or mud. *Halcampa* species occur on both coasts from the intertidal to depths of over 100 meters. The 2 to 6 cm length is about 10 times width, tentacles number 10 or 12.

Anthoplura xanthogrammica is a large green anemone of California tide pools. *A. elegantissima* is more variable in color. Hart and Crowe (1977) have shown that attached gravel reduces water loss when they are exposed under conditions resembling low tide.

Sagartia possesses a sphincter, acontia, and cinclides; three or four circlets of retractile tentacles; surface smooth; oral disk not lobed. *S. luciae* is about 8 mm long and 6 mm in diameter; four rows of tentacles, 84 in all; olive green with orange longitudinal stripes; Long Island northward.

Order 2. Madreporarida or Scleractinida

Zoantharia usually colonial (Figs. 4-43 and 4-44); many complete mesenteries; calcareous skeleton formed by ectoderm cells. Most of the stony corals belong to this order. Many of the coral polyps are tinted with pink, lilac, yellow, green, violet, red, etc., and give the coral reefs the wonderful color effects for which they are famous.

Astrangia has already been described.

Porites forms a more or less branching porous colony with zooids close together; cup with 12 short septa. *P. porites* occurs in Florida and the West Indies.

Figure 4-43   Corals on the Great Barrier Reef of Australia. (A) A shallow pool showing several types of coral colonies. (B) The underside of a fragment of coral rock covered with many invertebrates. (*Photos by author.*)

Figure 4-44 Coral skeletons showing variations produced by different sizes and arrangements of polyps.

Oculina forms a dendritic compact colony with zooids spirally arranged and widely separated. *O. diffusa* is much branched; cup 3 mm in diameter; North Carolina to Florida.

Meandrina has confluent zooids with septa in rows. *M. sinuosa* forms massive, encrusting colonies up to 25 cm or more in diameter and is known as brain coral; the sinuous ridges on the surface are septa and grooves; Florida and West Indies.

Fungia has no siphonoglyph; solitary; large; concave below and convex above. *F. elegans* is about 6 cm in diameter; Gulf of California.

Acropora is an important genus of reef-building corals. Most species of *Acropora,* such as the stag-horn coral, are branched colonies.

Order 3. Corallimorphida

Large nematocysts; coral-like soft parts, but no skeleton. *Corynactis californica;* California.

Order 4. Antipathida

Colonial Zoantharia (Fig. 4-45, B) with a horny, usually branching axial skeleton, but no calcareous spicules; in all the large seas, usually at a depth of from 50 to 500 fathoms; black corals.

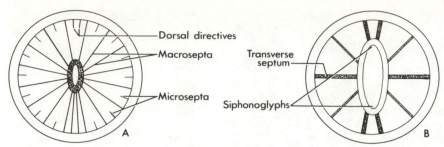

Figure 4-45 Septal arrangement. (A) Zoanthida, with macrosepta and microsepta. The gonads develop on the macrosepta. (B) Antipathida, gonads develop on the transverse septa.

Antipathes forms branching colonies of long slender stalks; polyps with six tentacles. *A. larix* has an axis with many long spines; up to 1 meter long; six longitudinal rows of parallel branches from 3 to 10 cm long along main stalk; West Indies, Mediterranean.

Cirripathes is unbranched. *C. spiralis* has a flexible spiral stalk over 1 meter long; West Indies, Mediterranean, Indian Ocean.

Order 5. Zoanthida

Zoantharia without skeletons; macrosepta alternate with microsepta (Fig. 4-45, A) except in the region of the directives; gonads occur on the macrosepta. Solitary and colonial forms occur. Usually small polyps.

Order 6. Ceriantharida

Large, solitary Zoantharia that live embedded in the sand. *Cerianthus* (Fig. 4-46) is the best known representative. It possesses many tentacles in two rows and one siphonoglyph. The epidermis secretes a long tube of mucus in which the animal lives. Several species are found in California. *C. americanus* occurs from Cape Cod to Florida. It reaches a length of 60 cm and is about 25 mm in diameter. As many as 130 tentacles may be present. Members of this order may have a variety of foreign objects embedded in their secreted tubes.

Subclass 2. Alcyonaria

Anthozoa with eight hollow, pinnate tentacles, and eight complete septa; usually with one siphonoglyph, ventral in position; retractor muscles of the septa all on the side toward the siphonoglyph; nematocysts are atrichous isorhizas; all are colonial and embedded in a *coenenchyme* of mesoglea; polyps are connected in the coenenchyme by a network of gastrodermal tubes, the *solenia,* and at the surface by the epidermis (Fig. 4-47); new polyps arise from solenial buds; many have a horny or calcareous skeleton secreted by the coenenchyme.

Figure 4-46 Ceriantharida. *Cerianthus*, a solitary sea anemone with many tentacles. (*After Andres, from Hickson.*)

Figure 4-47 Alcyonaria. *Alcyonium*. Diagram of section through colony showing extended polyps with pinnate tentacles and coenenchyme. The direction of water circulation is shown by arrows. The mesoglea is indicated by dots and the spicules it contains by small crosses. (*From Borradaile and Potts.*)

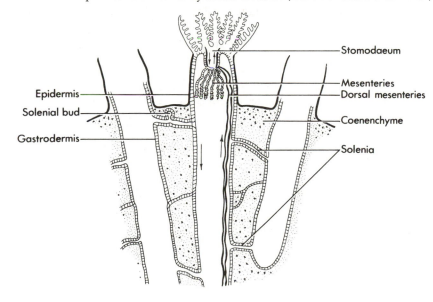

241

Order 1. Alcyonacida

Massive or creeping colonies. May be separated into four suborders.

Suborder 1. Alcyonaceae. The soft corals. Alcyonarians with the polyps embedded in a fleshy coenenchyme. *Alcyonium* (Fig. 4-47) forms a colony of thick, soft, leathery lobes; polyps long but all except distal end buried in the mesoglea. *A. digitatum* is from 4 to 10 cm high; yellow or red; Long Island northward.

Suborder 2. Coenothecaleae. The blue coral, *Heliopora,* of the Indo-Pacific is the only genus. Polyps brown, skeleton blue from iron salts.

Suborder 3. Stoloniferae. Typically with the polyps arising singly from a creeping base. However, in *Tubipora musica* (Fig. 4-48), the organ pipe coral, the polyps are joined together by horizontal bars. The polyps are bright green and when disturbed will retract into their tubes and thus reveal the dull red skeleton. Found on coral reefs in Old and New World.

Suborder 4. Telestaceae. With lateral polyps arising from elongated polyps which are connected by the creeping base.

Order 2. Gorgonacida

Colonial Alcyonaria (Fig. 4-49, C); skeletal axis branched and not perforated by gastrovascular cavities of the zooids; the horny corals such as the sea whips.

Corallium forms the precious red coral of commerce; polyps retractile, white; axis hard, red. *C. nobile* (*rubrum*) is up to 30 cm high; Mediterranean.

Gorgonia produces a fan-shaped colony in which the branches unite to form a network in one plane. *G. flabellum* is the sea fan; up to 50 cm high and wide; yellow or red; West Indies, South Atlantic.

Figure 4-48 Alcyonacida. *Tubipora musica,* skeleton of the organ-pipe coral. (A) Surface view. Note lack of sclerosepta. (B) Perspective view.

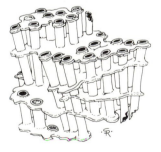

A

B

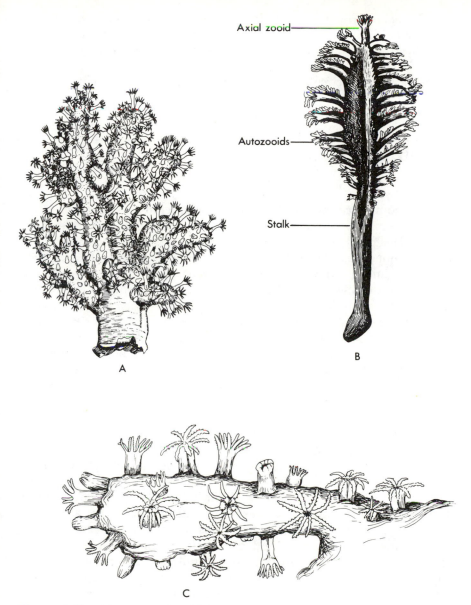

Figure 4-49 Alcyonaria. (A) Alcyonacida. *Alcyonium palmatum,* with polyps extended. (B) Pennatulacida. *Pennatula,* a sea feather or sea pen. (C) Gorgonacida. *Euplexaura marki,* with polyps extended. (*A, from Kerr; B, after Jungerson; C, from Johnson and Snook.*)

Order 3. Pennatulacida

Alcyonaria forming bilaterally symmetrical colonies; zooids usually borne on branches of an axial stem, which is supported by a calcareous or horny skeleton; zooids dimorphic; sea pens and sea feathers. *Pennatula* (Fig. 4-49, B) is called a sea feather; stalk embedded in sand or mud; distal end of stalk (rachis) with paired lateral branches (pinnulae). *P. aculeata* has from 20 to 50 long pinnulae on each side; about 10 cm long; red; South Carolina northward.

Renilla forms a circular or reniform (kidney-shaped) rachis; polyps all dorsal. *R. reniformis* has white polyps; dorsal surface of rachis pink or violet; about 7 cm long; North and South Carolina, West Indies.

Coelenterata in General

Morphology

The foregoing account has shown that coelenterates all possess a body wall composed of three layers, the outer epidermis and inner gastrodermis separated by mesoglea. In the *Hydra* the mesoglea is an essentially noncellular layer. Most have thicker mesoglea containing cells but no organs and thus show a tissue-level organization. Mesoglea is especially extensive in the jellyfish composing most of the jellylike material of the organism. In the Anthozoa the mesoglea is most highly developed as a tissue and contains a variety of cells and fibers. The body wall encloses a single cavity, the *coelenteron* or *gastrovascular cavity*, in which both digestion and circulation take place. In some of the coelenterates, like *Hydra* (Fig. 4-2), this cavity is simple, but in others, like *Aurelia* (Fig. 4-30), it is modified so as to include numerous pouches and branching canals.

All coelenterates possess stinging cells called *nematocysts;* these are organelles of offense, defense, locomotion and, probably, digestion. *Muscle fibrils* (Fig. 4-31) are present in a more-or-less concentrated condition. *Nerve fibers* and *sensory organs* are characteristic structures; they may be few in number and scattered as in *Hydra* (Fig. 4-8), or numerous and concentrated as in *Aurelia* (Fig. 4-32). Never fibers form a network that in some species seems to include parts that are polarized, or capable of transmitting impulses in only one direction across the asymmetrical synapses, and parts which have symmetrical synapses and are nonpolarized (i.e., the impulses may be conducted in any direction over the network). Two networks of nerve fibers may occur in coelenterates, one in the epidermis and one in the mesoglea, with some interconnections. Some coelenterate epithelial tissues lack nerve fibers but are able to conduct impulses.

Hydroid and Medusa Compared. The two principal types of coelenterates are the *polyp* or *hydroid,* and the *jellyfish* or *medusa*. These are fundamentally similar in structure (Fig. 4-50), but are variously modified (Fig. 4-51). Both polyps and medusae are *radially symmetrical*. Although the medusae upon superficial examination appear to be very

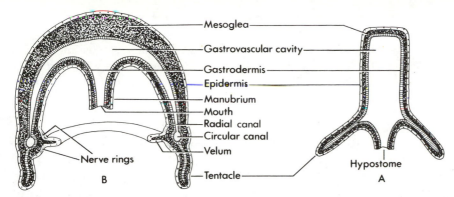

Figure 4-50 Diagrams showing the similarities of an inverted polyp (A) and a medusa (B). (*From Parker and Haswell.*)

different from the polyps or hydroids, they are constructed on the same general plan as the latter. Figure 4-50 illustrates in a diagrammatic fashion the resemblance between the hydrozoan polyp (A) and medusa (B) by means of longitudinal sections. If the medusa were grasped at the center of the aboral surface and elongated, a hydra-like form would result. Both have similar parts, the most noticeable difference being the enormous quantity of mesoglea present in the medusa.

Metagenesis. Metagenesis is the alternation of a generation which reproduces only asexually by division or budding with a generation which reproduces only sexually by means of eggs and spermatozoa. This phenomenon occurs in other groups of the animal kingdom, but finds its best examples among the Hydrozoa. *Obelia* is an excellent illustration of a metagenetic animal. The asexual generation, the colony of polyps (Fig. 4-18), forms buds of two kinds, the hydranths and the gonangia. The medusae (Fig. 4-19, A), or sexual generation, reproduce the colony by means of eggs and spermatozoa.

The polyp and medusa stages are not equally important in all Hydrozoa; for example, *Hydra* has no medusa stage and *Geryonia* no polyp or hydroid stage. Various conditions may be illustrated by different Hydrozoa. In the following list, O represents the fertilized ovum, H, a polyp, M, a medusa, m, an inconspicuous or degenerate medusa, and h, an inconspicuous or degenerate polyp.

1. O—H—O—H—O— (*Hydra*).
2. H—m—O—H—m—O (*Sertularia*).
3. O—H—M—O—H—M—O (*Obelia*).
4. O—h—M—O—h—M—O (*Liriope*).
5. O—M—O—M—O (*Geryonia*).

The past tendency to emphasize the importance of the concept of metagenesis is dwindling in significance. Some zoologists think that the generations represent developmental (larval?) stages in the life cycle leading to the sexually mature adult (the medusae when present). Thus, the complete cycle from egg to egg is one generation. Probably the most significant fact of metagenesis, or different stages in the life cycle of invertebrates, is the ecological advantages it confers on the species.

In those with a sessile stage, the alternation with a planktonic stage provides a means of dispersal as well as a location where a different food source may be involved. When each has a large reproductive capacity, one sexual, the other asexual, the species is less likely to be eliminated due to predation, disease, or competition if a stage occupying another habitat can restock the original habitat.

Genetic Advantages of Metagenesis. The genetic or ecological advantages of retaining favorable genetic combinations by asexual reproduction, or adapting by segregation and recombination of characteristics through sexual reproduction, can enable a species to best meet the stability and adapt to the variability alternately encountered in the environment. Heterozygosity might be bred out by stringent selection of a sexually reproducing population, whereas a favorable genotype could be maintained in an asexually reproducing population. The variability is important for adaptation to new conditions when sexual reproduction resumes. Heterozygosity may produce beneficial phenotypes, as well as tend to be lost in very small sexually reproducing populations. Asexual reproduction may be more important, for the foregoing reasons, as part of the life cycle of short-lived organisms in maintaining their evolutionary fitness.

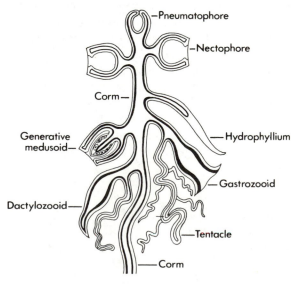

Figure 4-51 Siphonophorida. Diagram showing possible modifications of medusoids and hydroids. The thick black line represents gastrodermis, the thinner line epidermis. (*After Allman.*)

Polymorphism. The division of labor among the cells of a metazoan has already been noted. When division of labor occurs among the members of a colony, the form of the individual is suited to the function it performs. A colony containing two kinds of members is said to be *dimorphic;* one containing more than two kinds, *polymorphic.* Some of the most remarkable cases of polymorphism occur among the Hydrozoa (Fig. 4-51). The "Portuguese man-of-war" (Fig. 4-28), for example, consists of a float with a sail-like crest from which a number of polyps hang down into the water. Some of these polyps are nutritive, others are tactile; some contain batteries of nematocysts, others are male reproductive zooids, and still others give rise to egg-producing medusae. The following outline indicates some of the modifications that may occur among the members of colonial Hydrozoa.

A. Polymorphic modifications of hydrozoan medusae.
   1. Sexual Medusoid.
      a. Structure.   Like typical medusa of Anthomedusae or modified because of arrested development. (The hydrophyllium (bract), pneumatophore, and aurophore are considered by some to be modifications of a part of a medusa.)
      b. Function.   Production of ova or spermatozoa. (Hydrophyllia are protective, pneumatophores are hydrostatic structures, and aurophores of unknown function.
   2. Nectophore.
      a. Structure.   Without tentacles, manubrium, or mouth.
      b. Function.   Locomotion.
B. Polymorphic modifications of hydrozoan hydroids.
   1. Gastrozooid.
      a. Structure.   With large mouth, nematocysts, and tentacle with nematocysts.
      b. Function.   Ingestion of food.
   2. Dactylozooid.
      a. Structure.   Without mouth; with many nematocysts and tentacles.
      b. Function.   Offense and defense.
   3. Blastostyle.
      a. Structure.   Without mouth or tentacles.
      b. Function.   Produces sexual medusoids by budding.

Nematocysts. The Coelenterata is the only phylum producing nematocysts. The most nearly homologous organelles are found in some Protozoa. The greatest varieties of nematocysts are found in some Hydrozoa, as described for *Hydra.* Holotrichous isorhiza nematocysts have been demonstrated to have their tube discharged by eversion of a somewhat helically pleated internal tube containing inner spines (Fig. 4-52), which then take their position on the outer surface. A

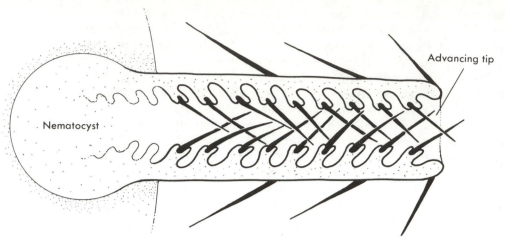

Figure 4-52 Nematocyst eversion, diagrammatic.

stenotele may use a similar set of barbs at the base of the thread to anchor to the prey tissue as the tube unfolds into the prey. Some nematocysts have cholinesterases that may destroy neurotransmitters in host synapses that are struck; this may be a mechanism by which soft-bodied coelenterates subdue stronger organisms. Phosphatases are also present in nematocysts.

Discharge of nematocysts is normally caused by a combination of chemical and mechanical stimuli and probably some neural activity.

Physiology

The late Thomas Goreau, given an informative tribute by Sir M. Yonge (pp. xxi–xxxv in Stoddart and Yonge, 1971), demonstrated the important role of zooxanthellae in nutrition of hermatypic corals. Similar roles seem to be played by such symbionts in other invertebrates.

The *food* of coelenterates consists principally of small, free-swimming animals, which are usually captured by means of nematocysts and carried into the mouth by tentacles and cilia or flagella. *Digestion* is mainly *extracellular,* enzymes being discharged into the gastrovascular cavities for this purpose. Starch digestion has not been observed in coelenterates. Nematocysts are thought to be an important source of proteases in the early stages of digestion. The digested food is transported to various parts of the body by currents in the gastrovascular cavity and is then taken up by the gastrodermal cells; some is passed over to the epidermal cells. Both *respiration* and *excretion* are performed by the general surface of the epidermis and gastrodermis. *Motion* is made possible by muscle fibrils, and many species have also the power of *locomotion.* Planktonic forms may swim by pulsations that expel currents of water, bottom dwelling forms may use various methods of creeping. There is no true skeleton, although fluid in the gastrovascular cavity or in the cells may serve as a hydraulic skeleton (Fig. 4-12), the periderm, and the stony masses built up by coral polyps support the soft tissues to a certain extent. The nervous tissue and sensory organs provide for the perception of various kinds of stimuli and the conduction of impulses from one part of the body to another. Coelenterates are generally sensitive to light intensities, to changes in the temperature, to mechanical stimuli, to chemical stimuli, and to gravity. *Reproduction* is both asexual, by budding and fission, and sexual, by means of eggs and spermatozoa.

Asexual reproduction is characteristic of some coelenterates and rare or absent in others. The most common method is by *budding.* The wall of the hydroid sends out a hollow protrusion which may become either a new hydroid or a medusa. Certain medusae also produce medusae by budding. *Fission* is rare in hydroids and very rare in medusae.

Sexual Reproduction. Both male and female germ cells are rarely developed by a single individual as in certain hydras. Usually a colony produces either ova or spermatozoa, or these originate in different individuals of a single colony. Sometimes one blastostyle may give rise to both kinds of germ cells.

Embryology

Tubularia may be selected to illustrate the embryology of a hydrozoan (Fig. 4-53). Segmentation of the egg results in a hollow *blastula* of cells of different sizes. Cells are divided off from the wall into the cavity of the blastula until this cavity is completely filled. Spaces now appear among the inner mass of cells due to their absorption, and eventually a single central cavity results which will become the *coelenteron* or gastric cavity. The outer layer of cells of this *gastrula* is the *ectoderm* and the inner layer, the *endoderm.* The embryo now becomes disk-shaped and from the edge *aboral tentacles* grow out. Then the embryo elongates into a cylindrical form; a *mouth* opening appears in the center of the end opposite the aboral tentacles; and *oral tentacles* are pushed out around the mouth. This *actinula larva* (Figs. 4-24 and 4-53, E) escapes from the gonophore and sinks to the bottom where it becomes attached by the aboral end. It increases in length; *buds* appear at the sides; and creeping *stolons* grow out from the base. The latter give rise to upright shoots each of which produces daughters by lateral budding.

The embryology of a scyphozoan such as *Aurelia* (Fig. 4-35), differs considerably from that of *Tubularia*. The egg of *Aurelia* undergoes *cleavage* that results in the formation of a hollow, spherical *blastula*. This becomes invaginated into a *gastrula* with two layers of cells, an outer *ectoderm* and an inner *endoderm*, and with a central cavity, the *coelenteron*, which communicates with the outside through a *blastopore.* The ectoderm becomes ciliated and the gastrula changes from

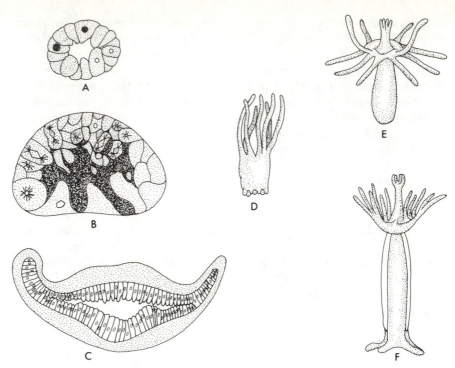

Figure 4-53 *Tubularia.* Stages in the embryology of a hydrozoan. (A) Blastula. (B) Endoderm cells budding off from wall of blastula. (C) Formation of aboral tentacles. (D) Aboral tentacles well developed; oral tentacles appearing. (E) Larva in creeping stage. (F) Larva at time of attachment to substratum. (*A–C, after Brauer; D–F, after Allman.*)

spherical to oval in shape. After a free-swimming existence of 4 or 5 days the *planula* larva attaches at the aboral end and assumes a cup shape. A cone now arises with the blastopore in the center and four *primary tentacles* grow out around the blastopore which may now be called the *mouth.* These tentacles are solid, with a central core of endoderm covered by a layer of ectoderm. The larva changes gradually into the *scyphistoma* stage. Four *secondary tentacles* now grow out alternating with the primary tentacles; next eight more tentacles develop and soon another set of eight all of which form a single circlet of 24 tentacles around the mouth. Meanwhile *nematocysts* are formed in the epidermis and groups of them appear like warts on the tentacles.

Phylogeny

Speculation regarding the origin of the Coelenterata and their differentiation into classes is a fascinating pastime. Three views having some adherents are the (1) protozoan origin of coelenterates by way of the ciliates or flagellates and a planulalike intermediate form; (2) origin from the sponges, as discussed in the previous chapter (Fig. 3-1, B);

and (3) origin from the flatworms. Elaborate arguments for the latter view have been presented by Hadzi (1963) and rejected by Hand (1963). Variations of the first view are most popular and Hyman (1940) describes an actinulalike ancestral type for the coelenterates that is as plausible as any. Interrelationships of the coelenterate classes is also speculative. The Hydrozoa with well-developed medusae and hydroid stages seem nearest the ancestral type with the Trachylinida showing closest affinities to the Scyphozoa and Anthozoa.

Only a few deny that the Coelenterata gave rise to the Ctenophora. They also seem to be near the ancestral type of the Platyhelminthes. The danger of such assumptions is pointed out by Ax (1963),

> In numerous phylogenetic speculations is to be found the erroneous suggestion that one recent great systematic group is derived directly from another. But the undeniable fact is that contemporaneous orders, classes, and phyla have a long evolutionary history extending far back in geological time. Thus it is quite impossible to derive the Turbellaria directly from such recent groups as the Ciliata, Cnidaria, Ctenophora, or Annelida.

As with many invertebrates, the fossil record of the coelenterates, other than corals, is rather meager and inadequate to prove their phylogeny. Even the question of whether their primitive symmetry was radial, as in most adult coelenterates or bilateral, as in the planula larva (and perhaps many corals) is not solved.

The demonstrated antiquity of corals, with relatives in the early Paleozoic, including the extinct Tabulata and Rugosa of the Anthozoa, does not prove that they are older than groups that do not readily fossilize. If some ancestral sponges were also ancestral to coelenterates, it is easy to imagine the calcium-secreting, chambered or septate anthozoans as in the most direct line of descent. The prevalence of utilization of zooxanthellae would also be consistent with early coelenterates' needs prior to the advent of complex prey. A similar interpretation would apply to the simple condition of many anthozoan nematocysts.

The History of Our Knowledge of the Coelenterata

The coelenterates were originally called Zoophyta or plant-animals. They were included by Cuvier, together with the ctenophores and echinoderms, in a division known as Radiata, principally because of their radial symmetry. The study of echinoderm embryology proved these animals to be fundamentally bilaterally symmetrical, and a comparison of the anatomy of adult coelenterates and echinoderms revealed the absence of a coelom in the former, as well as many other profound differences. Leuckart in 1847 coined the term Coelenterata for the group and elevated them to a distinct phylum. The term *polypus* we owe to Trembley, who introduced the name in 1744 because

Hydra resembles somewhat in appearance the octopus, which was known to the ancients as the polyp. Linnaeus and his predecessors called the jellyfish *medusae* since their tentacles reminded them of the curls of *Medusa*. At first the coelenterates and ctenophores were both included in the phylum Coelenterata but the ctenophores are so distinct they are now usually considered as a separate phylum.

Libbie Hyman introduced the use of the terms "gastrodermis" and "epidermis" for the endoderm and ectoderm of coelenterates. She later reversed her opinion. The preference here is for the use of her substitutes, even though the original terms, "endoderm" and "ectoderm," were used for coelenterate tissues prior to their use by embryologists.

Relations of Coelenterates to Society

Coelenterates as a whole are of very little economic importance. They are probably very seldom used as food by people but are eagerly devoured by some fishes and invertebrates. Some corals are used as ornaments and for the manufacture of jewelry. Coral polyps build *fringing reefs, barrier reefs,* and *atolls* (Fig. 4-43). These occur where conditions are favorable, principally in tropical seas, the best known being Bikini Atoll, Eniwetok Atoll, and the Fiji Islands of the Pacific Ocean, the Great Barrier Reef of Australia, and in the Bahama Islands region.

A *fringing* or *shore reef* is a ridge of coral built up from the sea bottom so near the land that no navigable channel exists between it and the shore. Frequently, breaks occur in the reef, and irregular channels and pools are created which are often inhabited by many different kinds of animals, some of them brilliantly colored (Engemann, 1960).

A *barrier reef* is separated from the shore by a wide, deep channel. The Great Barrier Reef of Australia is over 1,700 km long and encloses a channel from 20 to 50 meters deep and in some places 48 km wide. Often a barrier reef entirely surrounds an island.

An *atoll* is a more or less circular reef enclosing a lagoon. Several theories have been advanced to account for the production of atolls. Charles Darwin, who made extensive studies of coral reefs and islands, is responsible for the subsidence theory. According to Darwin, the reef was originally built up around an oceanic island which slowly sank beneath the ocean, leaving the coral reef enclosing a lagoon. This view has been substantiated by borings made on atolls in recent years because coral sediments extend down to great depths, and coral reef development occurs very slowly, if at all, below 100 meters. Reef development also requires warm water, at least 20°C, and a salt content near that of the open ocean. Thus living reefs are restricted to the tropics by temperature. Their need for a high salt content prevents their development near the mouths of large rivers, such as the Amazon, and in the lagoon area of atolls. If once established in shallow water their growth may keep pace with rising sea levels of postglacial ages, or subsiding ocean bottoms or sea mounts.

Tropical storms can greatly modify reefs. Quarrying can also damage reefs, as in India, where Pillai (1971) has estimated that as much as 250 cubic meters of reef material are removed per day for use in production of cement, lime, calcium carbide, and other products. More recent concern for the physical security of some oceanic islands and their reef food chains has centered on the destruction of some reefs by population explosions of the crown-of-thorns starfish, which feeds upon living coral and has seemingly caused reef destruction in some parts of its range (Fig. 12-30). The starfish will be discussed more in Chapter 12.

Coral skeletons from ancient reefs are useful for interpretation of past environments because of growth rings or marks, incorporation of some elements and their isotopes, and distribution, abundance, and related studies.

The study of coral reefs has significance to the petroleum geologist. Ancient reefs incorporated in continental land masses are now important reservoirs of petroleum. The ancient reefs presumably formed under similar ecological conditions to those today. Thus, knowledge of modern ones will aid in predictions of location and extent of ancient ones.

Coral reefs have the most varied and abundant collection of animal life of any natural habitat. They are a scuba divers paradise, but not without dangers.

Only a few coelenterates, such as *Chironex* and *Physalia* have nematocysts potent enough to constitute a severe danger. A small Hawaiian zoantharian was found (Moore and Scheuer, 1971) to have an unusual toxin of very high toxicity that may have prompted the legends that made the tidepool it was in taboo.

Coelenterates are convenient organisms for some lines of research in basic biology. Some which may have application in human medicine deal with development, regeneration, and connective tissue.

REFERENCES TO LITERATURE ON THE COELENTERATA

Ax, P. 1963. Relationships and phylogeny of the Turbellaria. Pp. 191–224 in E. Dougherty. *The Lower Metazoa.* Univ. of California Press, Berkeley. 478 pp.

Barnes, J. H. 1966. Studies on three venomous cubomedusae. Pp. 307–332 in W. J. Rees. *The Cnidaria and Their Evolution.* Academic Press, London. 449 pp.

Brien, P. 1960. The fresh-water hydra. *Am. Scientist,* **48:**461–475.

Burnett, A. L. (Ed.). 1973. *Biology of Hydra.* Academic Press, New York. 466 pp.

Campbell, R. D. 1965. Cell proliferation in *Hydra:* an autoradiographic approach. *Science,* **148:**1231–1232.

————. 1974. Cell movements in *Hydra. Am. Zoologist,* **14**:523–535.

CROWELL, S. (Ed.). 1965. Behavioral Physiology of Coelenterates. *Am. Zoologist,* **5**(3):335–589.

DAKIN, W. J. 1950. *Great Barrier Reef.* Aust. Nat. Travel Assn., Melbourne. 135 pp.

DAVIS, L. E. 1973. Ultrastructure of ganglionic cell development. Pp. 299–317 in Burnett (1973).

ENGEMANN, J. G. 1960. The effect of coloration on Great Barrier Reef animals. *Pap. Mich. Acad. Sci., Arts Letters,* **45**:9–15.

FRASER, L. A. 1962. The histology of the musculature of *Gonionemus. Trans. Am. Micr. Soc.,* **81**:257–262.

HADZI, J. 1963. *The Evolution of the Metazoa.* Macmillan, New York, 499 pp.

HAMNER, W. M., and R. M. JENSEN. 1974. Growth, degrowth, and irreversible cell differentiation in *Aurelia aurita. Am. Zoologist,* **14**:833–849.

HAND, C. 1963. The early worm: a planula. pp. 33–39. In E. DOUGHERTY. *The Lower Metazoa.* Univ. of California Press, Berkeley. 478 pp.

HART, C. E., and J. H. CROWE. 1977. The effect of attached gravel on survival of intertidal anemones. *Trans. Am. Micr. Soc.,* **96**:28–41.

HYMAN, L. H. 1940. *The Invertebrates: Protozoa Through Ctenophora.* Vol. I. McGraw-Hill, New York. 726 pp.

KRAMP, P. L. 1961. Synopsis of the medusae of the world. *J. Mar. Biol. Assoc. U.K.,* **40**:7–469.

LANE, C. E. 1960. The Portuguese Man-of-War. *Sci. Am.* **202**:158–168.

LENHOFF, H. M. 1961. Activation of the feeding reflex in *Hydra littoralis. J. Gen. Phsyiol.,* **45**:331–342.

————. 1965. Cellular segregation and heterocytic dominance in hydra. *Science,* **148**:1105–1107.

————, and W. F. LOOMIS (Eds.). 1961. *The Biology of Hydra and of Some Other Coelenterates.* Univ. of Miami, Coral Gables. 467 pp.

————, L. MUSCATINE, and L. V. DAVIS (Eds.). 1971. *Experimental Coelenterate Biology.* Univ. of Hawaii Press, Honolulu. 281 pp.

LENTZ, T. L. 1966. *The Cell Biology of Hydra.* North-Holland, Amsterdam. 199 pp.

LYTLE, C. F. 1962. *Craspedacusta* in the Southeastern United States. *Tulane Stud. Zool.,* **9**:309–314.

MACKLIN, M. 1968. Reversal of cell layers in hydra: a critical re-appraisal. *Biol. Bull.,* **134**:465–472.

MARCUM, B. A., and R. D. CAMPBELL. 1978. Development of *Hydra* lacking nerve and interstitial cells. *J. Cell Sci.,* **29**:17–33.

McCULLOUGH, C. B. 1965. Pacemaker interaction in *Hydra. Am. Zoologist,* **5**:499–504.

MOORE, R. E., and P. J. SCHEUER. 1971. Palytoxin: a new marine toxin from a coelenterate. *Science,* **172**:495–498.

PARK, H. D., N. E. SHARPLESS, and A. B. ORTMEYER. 1965. Growth and differentiation in *Hydra.* I. The effect of temperature on sexual differentiation in *Hydra littoralis. J. Exp. Zool.,* **160**:247–254.

PICKWELL, G. V., E. G. BARHAM, and J. W. WILTON. 1964. Carbon monoxide production by a bathypelagic siphonophore. *Science,* **144**:860–862.

PILLAI, C. S. Gopinadha. 1971. Composition of the coral fauna of the southeastern coast of India and the Laccadives. Pp. 301–327 in Stoddart and Yonge (1971).

RAHAT, M., and R. D. CAMPBELL. 1974. Three forms of the tentacled and non-tentacled fresh-water coelenterate polyp genera *Craspedacusta* and *Calpasoma*. *Trans. Am. Micr. Soc.*, **93**:235–241.

ROSS, D. M. 1965. Preferential settling of the sea anemone *Stomphia coccinea* on the mussel *Modiolus modiolus*. *Science*, **148**:527–528.

RUSHFORTH, N. B., I. T. KROHN, and L. K. BROWN. 1964. Behavior in *Hydra*: inhibition of the contraction responses of *Hydra pirardi*. *Science*, **145**:602–604.

RUSSELL, F. S. 1970. *The Medusae of the British Isles.* Cambridge Univ. Press, Cambridge. 284 pp.

SMITH, S. V., R. W. BUDDEMEIER, R. C. REDALJE, and J. E. HOUCK. 1979. Strontium-calcium thermometry in coral skeletons. *Science*, **204**:404–407.

SPOON, D. M., and R. S. BLANQUET. 1978. Life cycle and ecology of the minute hydrozoan *Microhydrula*. *Trans. Am. Micr. Soc.*, **97**:208–216.

STODDART, D. R., and SIR M. YONGE (Eds.). 1971. *Regional Variation in Indian Ocean Coral Reefs.* Symp. Zool. Soc. Lond., No. 28. Academic Press, London. 584 pp.

SWEDMARK, B., and G. TEISSIER. 1966. The Actinulida and their evolutionary significance. Pp. 119–132 in W. J. Rees. *The Cnidaria and Their Evolution.* Academic Press, London. 449 pp.

TOTTON, A. K., and H. E. BARGMANN. 1965. *A Synopsis of the Siphonophora.* British Museum, London. 230 pp. + 40 Pl.

WESTFALL, J. A., S. YAMATAKA, and P. D. ENOS. 1971. Ultrastructural evidence of polarized synapses in the nerve net of *Hydra*. *J. Cell Biol.*, **51**:318–323.

———. 1973. Ultrastructural evidence for neuromuscular systems in coelenterates. *Am. Zoologist*, **13**:237–246.

5

Phylum Platyhelminthes

The phylum Platyhelminthes contains a group of animals which at first were included with other wormlike animals in a phylum called Vermes. They are known as flatworms because they are much flattened dorsoventrally. Among them are both free-living and parasitic species; the former live principally in fresh or salt water; the latter are mostly endoparasitic. The parasitic flatworms are known as trematodes or flukes and cestodes or tapeworms. They are widely distributed among humans and other vertebrates and are often pathogenic and sometimes bring about the death of the host. Free-living flatworms may be obtained from ponds and streams or in bodies of salt water. Each species has its own particular habitat, which should be determined before collecting is attempted.

The Platyhelminthes are the most primitive major phylum to show (1) well-developed organ systems in the mesodermal layer between epidermal and digestive epithelium, (2) consistent bilateral symmetry, and (3) head development.

Three classes of Platyhelminthes are recognized: (1) Turbellaria, (2) Trematoda, and (3) Cestoda. Most Turbellaria are free-living and inhabit either fresh or salt water; a few live in moist soil and a few are parasitic. The trematodes and cestodes are all parasitic. Characteristics of the three classes follow.

PHYLUM PLATYHELMINTHES. Unsegmented, triploblastic, bilaterally symmetrical animals; body flattened dorsoventrally; without circulatory system, anus, or coelom; usually hermaphroditic.

256

CLASS I. TURBELLARIA. Mostly free-living; enteron present; body usually covered with cilia; rhabdites usually present; suckers usually absent; life cycles simple.

CLASS II. TREMATODA. Parasitic; enteron present; no cilia in adult, but cuticle present; suckers on ventral surface; typically with complex life cycles.

CLASS III. CESTODA. Endoparasites; enteron absent; no cilia in adult, but cuticle present; proglottids usually formed; typically with more than one host in life cycle.

Class I. Turbellaria

DUGESIA—A FRESHWATER FLATWORM

External Features

Dugesia tigrina (Fig. 5-1) is a common planarian of the class Turbellaria and serves to illustrate the typical organization of a platyhelminth better than any of the Trematoda or Cestoda, because the latter are considerably modified for a parasitic existence. It is a flatworm found only in fresh water, usually clinging to the underside of logs or stones. Like most of the members of the phylum Platyhelminthes, its body is extremely flattened dorsoventrally and exhibits bilateral symmetry. *Dugesia* is broad and blunt at the anterior end and pointed at the posterior end. The length of an adult specimen may exceed 2 cm. The dorsal surface contains many black pigment cells, or *melanophores,* which make it difficult to determine the location of the internal structures in the living animal. The pigment cells may be bleached prior to staining preserved specimens for study of internal structures. After observation at very low magnification of the living animal in a small amount of water, the soft, contractile body is usually placed on a slide, and then pressed out slightly with a cover glass.

Figure 5-1 *Dugesia,* with pharynx extended to feed on a small freshwater crustacean.

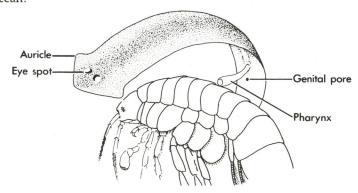

Cephalization, or the concentration of nervous structures in a head region, occurs in Turbellaria such as *Dugesia.* The lateral projections of the triangular head are tactile lobes or *auricles.* Sensory pits may be present on the dorsal posterior region of the auricles. A pair of eye spots are present on the dorsal surface near the anterior end. The *mouth* is in a peculiar position near the middle of the ventral surface. From it the muscular *pharynx* may extend. Posterior to the mouth is a smaller opening, the *genital pore.* The ventral surface of the body is covered with *cilia,* which propel the animal through the water. This is not the only method of locomotion, because muscular contraction is also effective.

Morphology

A study of the structure of the adult and of the early embryonic stages shows *Dugesia* to be a triploblastic animal (i.e., possessing the three germ layers, ectoderm, mesoderm, and endoderm, from which several systems of organs have been derived). The mesoderm lying between the body wall and the intestine forms large cells, the parenchyma or mesenchyme, with long irregular processes among which are large intercellular spaces. There are well-developed muscular, nervous, digestive, excretory, and reproductive systems (Fig. 5-2); these are constructed in such a way as to function without the coordination of a circulatory system, respiratory system, coelom, and anus.

A feature of special interest in the turbellarian epidermis is the presence of *rhabdites.* Rhabdites are thought to transform into secretions capable of supplementing the action of mucus glands in providing secretions to protect the surface. Rhabdites occur only in the Turbellaria and a few Nemertinea. They are intracellular rods that are more numerous in the dorsal epidermis of *Dugesia.* Evidence is lacking for the suggestion that they are structures with a phylogenetic relationship to nematocysts.

Digestive System. The digestive system (Fig. 5-2) consists of a *mouth,* a *pharynx* lying in a muscular sheath, and an *intestine* of three main trunks with a large number of small lateral extensions. The muscular pharynx can be extended as a proboscis (Fig. 5-1) to penetrate and suck up prey contents with the aid of secreted endopeptidase. Digestion is both *extracellular* and *intracellular;* that is, part of the food is digested in the intestinal trunks by secretions from cells in their walls, whereas other food particles are engulfed by pseudopodia thrust out by cells lining the intestine, and are digested inside of the cells in vacuoles. The digested food is absorbed by the walls of the intestinal trunks, and, because branches from these penetrate all parts of the body, no circulatory system is necessary to carry nutriment from one place to another. As in *Hydra,* no *anus* is present, the *feces* being ejected through the mouth.

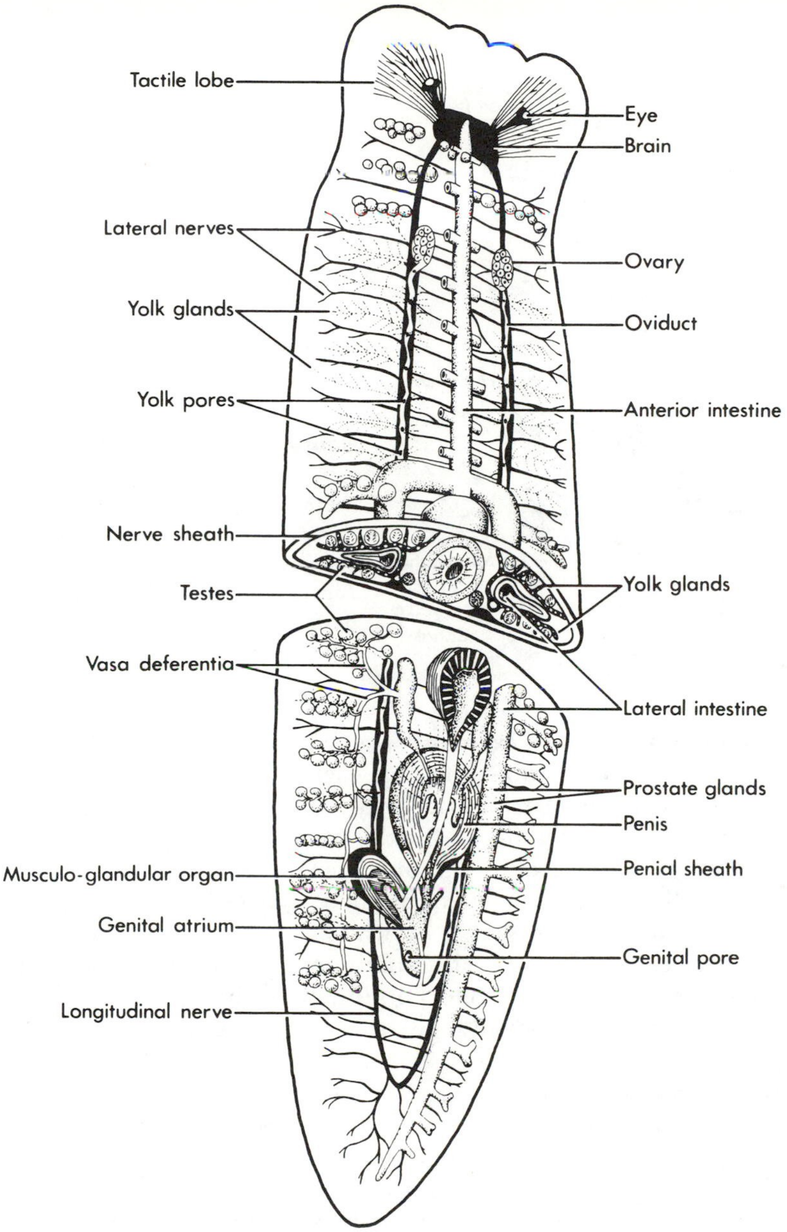
Tactile lobe
Eye
Brain
Lateral nerves
Ovary
Yolk glands
Oviduct
Yolk pores
Anterior intestine
Nerve sheath
Testes
Yolk glands
Vasa deferentia
Lateral intestine
Prostate glands
Penis
Penial sheath
Musculo-glandular organ
Genital atrium
Genital pore
Longitudinal nerve

Figure 5-2 *Procotyla fluviatilis*. Diagram showing the structure of a freshwater turbellarian. (*From Gamble.*)

The food of *Dugesia* consists largely of small living crustaceans, worms, and dead animals. These are captured by the proboscis and ingested after being covered with slime secreted by the slime glands; or the pharynx is attached to the prey, digestive fluid is poured out, and the dissolved particles are pumped into the intestine.

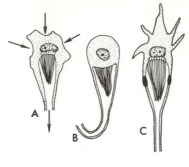

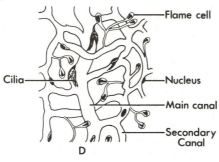

Cilia

Flame cell

Nucleus

Main canal

Secondary Canal

D

Figure 5-3 Flame cells and excretory ducts of Platyhelminthes. (A) From a planarian. (B) From the miracidium stage of the liver fluke, *Fasciola*. (C) From the tapeworm, *Taenia*. (D) Excretory canals ending in flame cells. (*A–C, after Kerr; D, after Benham.*)

Excretory System. The excretory system comprises a pair of longitudinal, much-coiled tubes, one on each side of the body; these are connected near the anterior end by a transverse tube and open to the exterior by two small pores on the dorsal surface. The longitudinal and transverse trunks give off numerous finer tubes, which ramify through all parts of the body, usually ending in a *flame cell* (Fig. 5-3). The flame cell is large and hollow, with a bunch of flickering cilia extending into the central cavity. Because it communicates only with the excretory tubules it is considered excretory in function, though it may also carry on respiratory activities and regulate water content.

Muscular System. The power of changing the shape of its body, which may be observed when *Dugesia* moves from place to place, lies principally in three sets of muscles, a circular layer just beneath the epidermis (Fig. 5-4), external and internal layers of longitudinal muscle fibers, and a set of oblique fibers lying in the mesenchyme.

Nervous System and Sense Organs. *Dugesia* possesses a well-developed nervous system, consisting of a bilobed mass of tissue just beneath the eye spots, called the *brain* (Fig. 5-2) and two lateral *longitudinal nerve cords* connected by transverse nerves. From the brain, nerves pass to various parts of the anterior end of the body, imparting to this region a highly sensitive nature.

Each eye (Fig. 5-5) consists of a highly pigmented retina of a single

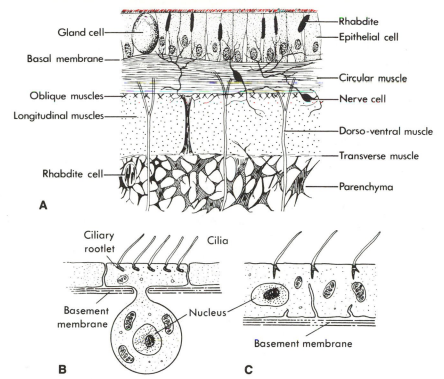

Figure 5-4 Turbellarian epithelium. (A) *Procotyla fluviatilis*, section through the skin as seen by conventional microscopy. (B) *Otoplana*, epithelial cell with subepithelial nuclear portion as determined by electron microscopy. (C) *Gyratrix*, syncytial-type epithelium. (*A, after Hallez, from Petrunkevitch; B and C, after Bedini and Papi, 1974.*)

cup-shaped cell, inside of which are from 2 to 30 nerve cells, the fibrillae of which are in contact with the retina and the opposite ends united into an optic nerve that passes to the brain. There is no lens but the epidermis over the eye is not pigmented. The ciliated pits on

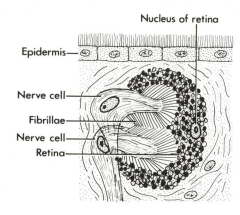

Figure 5-5 Turbellaria. *Planaria.* Histology of the eye. (*After Hesse-Doflein, modified.*)

either side of the head contain special sensory cells with long cilia and are connected with the underlying nervous network. In some Turbellaria tentacles occur near the anterior end with very long cilia; these are probably organs of chemical sense that aid in finding food. Cells, probably of chemical sense, are scattered over the surface of certain rhabdocoeles. An otocyst located above the brain is present in some Turbellaria.

Reproductive System. Reproduction is by *fission* or by the sexual method. Each individual possesses both male and female organs (i.e., it is *hermaphroditic*). The *male organs* (Fig. 5-2) consist of numerous spherical *testes* connected by small tubes called *vasa deferentia;* the *vas deferens* from each side of the body joins the *cirrus* or *penis*, a muscular organ that enters the *genital cloaca*. A *seminal vesicle* lies at the base of the penis, also a number of unicellular *prostate glands*. Spermatozoa originate in the testes, and pass, by way of the *vasa deferentia*, into the seminal vesicle, where they remain until needed for fertilization. *Dugesia* sperm, like most Platyhelminthes spermatozoa, are unlike those found in other phyla in that the flagellum has a "9 + 1" pattern rather than the "9 + 2" pattern typical of most other flagella (Silveira, 1969). The female *reproductive organs* comprise two *ovaries;* two long *oviducts* with many yolk glands entering them; a *vagina*, which opens into the genital cloaca; and the *uterus*, which is also connected with this cavity. The *eggs* originate in the ovary, pass down the oviduct, collecting yolk from the yolk glands on the way, and finally reach the uterus. Here fertilization occurs, and *cocoons* are formed, each containing from four to more than twenty eggs, surrounded by several hundred yolk cells. During copulation spermatozoa from one animal are transferred by the penis to the other member of the pair. The genital organs are so constructed that self-fertilization is prevented and cross-fertilization assured.

The closely related *Dugesia dorotocephala* has diploid chromosome numbers of 8, 16, and 24 (Benazzi, 1975), indicating that 4 may be the minimal haploid complement in that species and perhaps in the genus.

Regeneration and Starvation

Planarians show remarkable powers of regeneration. If an individual is cut in two (Fig. 5-6), the anterior end will regenerate a new tail (*B*, *B¹*), while the posterior part develops a new head (*C*, *C¹*). A crosspiece (*D*) will regenerate both a head at the anterior end, and a new tail at the posterior end (*D¹–D⁴*). The head alone of a planarian will grow into an entire animal (*E–E³*). Pieces cut from various parts of the body will also regenerate completely. No difficulty is experienced in grafting pieces from one animal upon another, and many curious monsters have been produced in this way. New tissue appears to be formed by

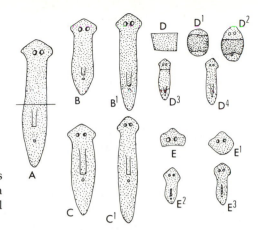

Figure 5-6 Tricladida. Diagrams illustrating the regeneration of entire planarians from small pieces. (*From Morgan.*)

the differentiation of formative cells, most of which migrate through the parenchyma to the region of differentiation.

Montgomery and Coward (1974) found that the minimal-sized piece of planaria capable of regenerating a new planarian contains about 10,000 cells.

If *Dugesia* is starved, it absorbs its internal organs as food, in regular order; first the mature eggs are absorbed, then the yolk glands, the rest of the reproductive organs, parenchyma, intestine, and muscles. The body size decreases during this process; for example, in certain experiments the animals decreased from 13 mm long and 2 mm broad to 3.5 mm long and 0.5 mm broad in 9 months. When starved specimens are given food, they regenerate the lost organs and become normal in size. The structural changes involved in regeneration of planarians involves replacement of missing tissue by new growth and reorganization of old tissue.

Axial Gradients

Dugesia fragments exhibit a *polarity* when regenerating; that is, a tail normally regenerates in the direction of the old tail and the head in the direction that the old head was in when the fragment was in the donor organism. The theory of axial gradients seems to partially explain the control of such proper regenerative responses. The primary axis or axis of polarity is an imaginary line extending from the anterior to the posterior end of the body. In *Planaria* the head has a relatively high rate of metabolism and dominates the rest of the body. Experiments have shown that a gradient of metabolic activity proceeds from the anterior to the posterior end. For example, if planarians are cut into four pieces, the anterior piece will be found to use up more oxygen and give off more carbon dioxide than any of the others; the second piece comes next in its rate of metabolism; the third piece next; and the tail piece gives the lowest rate of all. Thus is demonstrated an axial gradient in the metabolism of the animal from the anterior to the posterior end. This is particularly well brought out in older planarians especially at the time of transverse division.

When a planarian is young, it is relatively short and its whole body, especially the head, has a relatively high rate of metabolism. As it grows older it becomes longer and its whole metabolic rate slows down. When young, the high metabolic rate of the head was able to exercise a dominance, through the transmission of stimuli down the gradient, over the entire length of the animal. With a slowing down of the metabolic rate of the apical end and an increase in the length of the path over which the impulse travels, there comes a time when the apical end can no longer maintain a physiological dominance over the entire axis. At the point where dominance fades out, an independent part of the body arises through what is known as physiological isolation. The isolated piece, the second zooid, has its own gradient, the metabolic rate of the anterior end being the highest. This region now becomes a new apical end or, morphologically speaking, the head of a new zooid. No structural indications of a new individual are visible, however, at this time. The only tests of the presence of a second or third individual are physiological tests. The isolated posterior zooid now forms a new head, with eyes, brain, and other parts. The new head then reorganizes the rest of the piece into a complete new individual.

The physiological dominance that prevents development of a second head seems to be due to a diffusible substance formed in the brain which inhibits development of other heads. When action of the inhibitor can be prevented a bipolar, or two-headed individual may result. Such animals can be produced by chemical treatment and have even been observed in very long individuals in a stock culture (Jenkins, 1963).

Memory Transfer

One of the most interesting aspects of behavior in *Dugesia* is the reported ability of RNA from a trained planarian to elicit responses comparable to trained ones from untrained ones given the RNA. Most of the studies relating to this phenomenon have been done by psychologists. The early studies showed that planarians could be conditioned to respond to stimuli to which they do not ordinarily respond, or the type of response could be modified. The conditioning was done by using electrical shock as an unconditioned stimulus, which causes contraction of the planarian. With the electrical shock was given a light shock which, by itself, was insufficient to cause a contraction. After many exposures to coincident stimuli the planarians would develop a high rate of contractions to the light alone. The learning, or memory, was not in the brain alone, because anterior and posterior halves regenerated from conditioned planarians both had high levels of retention of the conditioning. The fact that planarians fed upon trained planarians showed a higher rate of response to the light shocks alone than untrained planarians indicated a chemical factor was involved. Ribonuclease destroyed the effectiveness of minces of the trained food source. RNA from trained animals was effective in transferring the memory, or, more accurately stated, inducing a higher rate of response in experimental organisms than the rate found in control organisms. Although all studies have not supported this concept, the majority of later studies indicate the effects of conditioning can be transferred by RNA in *Dugesia.*

OTHER TURBELLARIA

Eight orders of Turbellaria are recognized here. The old order Rhabdocoelida has been split into five orders, in all of which the intestine is a single straight tube, which may have short diverticula in one order. The Acoelida have no intestine; the Tricladida have an intestine with three main branches; and the Polycladida have a central cavity with many branches serving as the intestine. The condition of the pharynx and the protonephridia (flame cells) are useful in separating most of the other orders. A simple pharynx occurs in the Catenulida and Macrostomida; a bulbose, muscular pharynx is in most Neorhabdocoelida, Temnocephalida, and Alloeocoelida.

The principal features of the Turbellaria have been illustrated in *Dugesia.* The body is usually soft and less than 2 cm in length, more-or-less completely covered with cilia and with mucuslike secretions of glands from the epidermis and parenchyma. Rod-shaped bodies, called rhabdites, of uncertain function are formed in special cells, among or beneath the epidermal cells, and distributed to the epidermal cells (Fig. 5-4). The parenchyma completely fills the space between the body wall and intestine, hence no coelom is present. The muscular pharynx of many is a conspicuous feature of the digestive system. The excretory system is of the protonephridial type with flame cells or flame bulbs at the inner terminations. From two to numerous longitudinal nerve trunks arise from an anterior brain. Numerous interconnections occur between the trunks. Most species are hermaphroditic but seem to require cross-fertilization. The eggs are laid by many species in capsules attached to solid objects in the water, such as stones or vegetation. Freshwater, marine, and a few terrestrial species occur. Most of them feed on other small animals but a few are parasitic.

Order 1. Acoelida

These (Fig. 5-7) have no muscular pharynx and are without an intestine, the latter being represented by a group of endoderm cells. The Acoelida are small and marine in habitat, and may be found among

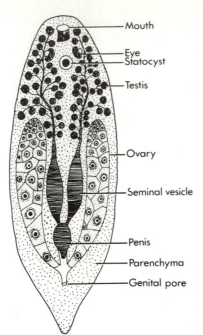

Figure 5-7 Acoela: plan of structure. (*From von Graff.*)

rocks and seaweed. One species, *Convoluta henseni*, is a free-swimming plankton organism; another, *C. roscoffiensis*, has symbiotic algae living in its parenchyma. *Polychoerus carmelensis* is found on *Ulva* in California tidepools. The Acoelida lack excretory structures. A temporary digestive space is formed among the parenchymal cells when food is ingested.

Order 2. Catenulida

Pharynx simple; intestine saclike; one protonephridium, median in position. *Stenostomum* is a cosmopolitan, common genus. It reproduces by transverse divisions, which are completed slowly so that chains of individuals (zooids) (Fig. 5-8) result. Ciliated pits are found on *Stenostomum*. The very similar *Catenula* has a statocyst. Sexual reproduction occurs rarely. Although most are found only in fresh water, forms such as *Retronectes thalia* Sterrer and Rieger (1974) are found in shallow marine sediments around the world.

Order 3. Macrostomida

Pharynx simple; intestine saclike; two protonephridia. Reproduction is as in the Catenulida in *Microstomum lineare* (Fig. 5-8), a species that sometimes possesses nematocysts (Fig. 5-22) derived from ingested hydras. No asexual reproduction occurs in *Macrostomum*, a common genus. Both freshwater and marine species exist in this order.

Figure 5-8 Macrostomida. *Microstomum lineare*, undergoing division. The individual has first divided into two, near its middle, and each of these sections has again divided. Each of the four zooids has again divided into two, and so on, until 16 individuals are here marked out. (*From von Graff.*)

Order 4. Neorhabdocoelida

Pharynx bulbose; intestine saclike; two protonephridia; sexual reproduction only; both freshwater and marine species. *Dalyellia viridis* is large (5 mm), green with zoochlorellae, and common in fresh water in the United States in the springtime. Many different colors, or no pigmentation, occurs in different species of this order. *Kytorhynchus oculatus* Rieger (1974) has a pharynx formed from an anterior invagination of the body wall and is found in sediments off Beaufort, North Carolina.

Order 5. Temnocephalida

Ectocommensal, freshwater turbellarians with internal structures indicating a close relationship to the Neorhabdocoela. Peculiar external features include tentacles and an adhesive disk once thought to indicate a relationship to the Trematoda. External cilia usually lacking. Principally tropical and southern hemisphere in distribution. Usually on crustaceans. *Temnocephala caeca* (Fig. 5-9) occurs on the large, subterranean isopod *Phreatoicopsis terricola* in Australia.

Order 6. Alloeocoelida

Simple, bulbose, or plicate pharynx; intestine with short diverticula; small size; mostly marine. *Plagiostomum* is a common littoral marine genus with some freshwater members.

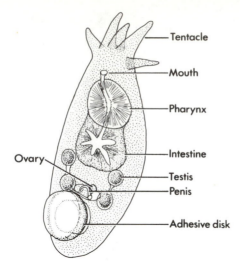

Figure 5-9 Temnocephalida. *Temnocephala caeca*, ectocommensal on *Phreatoicopsis terricola*, a very large subterranean isopod living in moist tunnels in the Ottway Mountains, Victoria, Australia. (*From a specimen collected and prepared by the author.*)

Order 7. Tricladida

Plicate pharynx; intestine with three main branches, each with many diverticula; mouth near the center of the ventral surface; ventral surface ciliated. Three suborders are recognized.

Suborder 1. Paludicola. Freshwater or brackish-water species, including *Dugesia tigrina* and *D. dorotocephala* the common large, black, two-eyed planarians with prominent auricles. The latter species is larger and has pointed auricles. *Procotyla fluviatilis* is a translucent species and its organs are easier to observe than those of *Dugesia;* it has an adhesive disk on the anterior end. *Polycelis coronata* has many eyes arranged somewhat like a coronet, and *Phagocata gracilis* possesses many pharyngeal tubes lying in a common chamber and opening separately into the intestinal tract.

Suborder 2. Maricola. Marine species, a few parasitic, with the mouth located in the posterior part of the body. This group includes *Procerodes wheatlandi,* and *Bdelloura candida,* which attaches itself to the living horseshoe crab, *Limulus. Bdelloura* is of particular value for study because of the absence of pigment and, hence, its internal organs are clearly visible.

Suborder 3. Terricola. These land planarians live in very moist places and resemble small slugs. *Bipalium kewense* is a cosmopolitan species that is an inhabitant of the tropics but may appear in greenhouses in temperate regions on tropical plants; it may reach a length of 25 centimeters.

The order Polycladida (Fig. 5-10) contains marine species only; they have a central digestive cavity, from which branches extend to various parts of the body, and a posteriorly located mouth. Many polyclads are very broad and thin and may reach a length of 10 cm or more. Many tropical species are very colorful. Often, numerous eyes are clustered in patterns useful for species identification. *Hoploplana inquilina* lives on the gastropod *Busycon; Stylochus ellipticus* is a common, large species on the New England coast and possesses many frontal and marginal eye spots; *Notoplana atomata* has two groups of cerebral ocelli of about 30 each, and two groups of dorsal ocelli of about 15 each. *N. acticola* is found under intertidal rocks in California.

Figure 5-10 Polycladida: plan of structure. Ventral view, testis shown only to the left, ovaries only to the right. (*From von Graff.*)

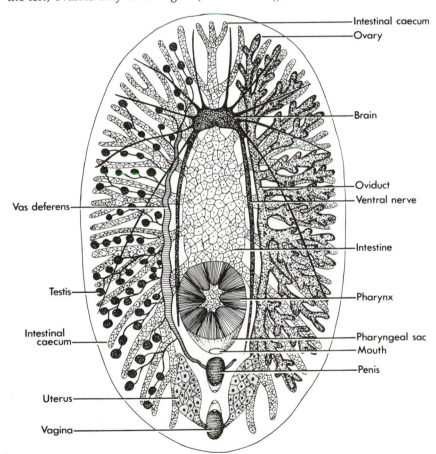

Class II. Trematoda

OPISTHORCHIS SINENSIS (CLONORCHIS SINENSIS)

Opisthorchis sinensis is a liver fluke of humans, and it also occurs in cats and dogs, which serve as reservoir hosts. Although common in the Orient, it does not occur in the western hemisphere. It is selected as the type example for the trematodes because it is less specialized structurally than *Fasciola hepatica*, the sheep liver fluke. *O. sinensis* lives as an adult in the bile ducts of humans and in severe infections can cause death.

Morphology

Figure 5-11 shows the shape and most of the anatomical features of a mature worm. The *mouth* is situated at the anterior end and lies in the middle of a muscular disk, the anterior sucker or *oral sucker*. The *ventral sucker* (*acetabulum*) is a short distance back of the mouth; it serves as an organ of attachment. The *genital pore* opens just anterior to the ventral sucker. The *excretory pore* lies at the posterior end of the body, and another pore, the opening of Laurer's canal, is situated in the middorsal line about one-third the length of the body from the posterior end.

The digestive system is simple. The mouth opens into a short muscular *pharynx*, which leads into another short tube, the *esophagus*. The *intestine* (Fig. 5-11) consists of two branches, one extending from near the anterior to the posterior end of each side of the body. No circulatory system is present but food materials, consisting of blood cells and materials from the lining of the bile ducts, are digested in the intestine, which is close to all regions of the body.

The excretory system does not require the transportative functions of a circulatory system either, because *flame bulbs* are widely distributed in the parenchyma of the body. From the flame bulbs run small excretory vessels that unite eventually to form two main excretory vessels that empty into the *bladder,* located in the midline of the posterior third of the fluke. The bladder opens through the excretory pore at the posterior end of the body.

The nervous system consists principally of a bilobed brain dorsal to the esophagus, three pairs of longitudinal nerves, and a number of transverse nerves.

The muscular system resembles that of *Dugesia*. The suckers have many radial muscles as well as circular and longitudinal muscles on both surfaces. The outer circular muscles and inner longitudinal and diagonal muscles of the body wall are covered by a noncellular *cuticle,* which is thought to be secreted by cells sunken into the parenchyma with processes extending through the muscle layers to the cuticle. The *parenchyma* is a loose tissue lying between the body wall and the

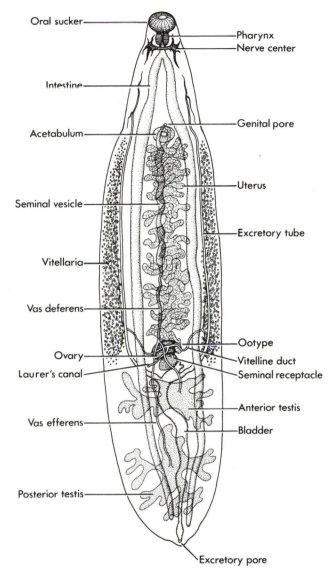

Oral sucker
Intestine
Acetabulum
Seminal vesicle
Vitellaria
Vas deferens
Ovary
Laurer's canal
Vas efferens
Posterior testis

Pharynx
Nerve center
Genital pore
Uterus
Excretory tube
Ootype
Vitelline duct
Seminal receptacle
Anterior testis
Bladder
Excretory pore

Figure 5-11 *Opisthorchis sinensis,* a liver fluke that lives in humans. (*After Faust.*)

alimentary canal; within it are embedded the various internal organs described above, as well as the reproductive system.

Both male and female *reproductive organs* are present in every adult (Fig. 5-11); they are extremely well developed, and, as in *Dugesia,* quite complex. The *male* organs are as follows: (1) a pair of branched *testes* (one anterior to the other) in which the spermatozoa arise; (2) two ducts, the *vasa efferentia,* which unite to form (3) a *vas deferens* that anteriorly enlarges into (4) a thicker, elongate *seminal vesicle,* where spermatozoa are stored until discharged through (5) a narrow *ejaculatory duct,* (6) *genital atrium,* and (7) the *genital pore.* Copulatory organs such as a *penis* or *cirrus* are common in trematodes but lacking in *Opisthorchis.*

The *female organs* are (1) a single *ovary* in which the eggs are produced; (2) a short *oviduct* which receives a duct from (3) *Laurer's canal* and the (4) *seminal receptacle* before entering (5) the *shell gland* or *ootype,* at which place (6) the *vitelline* or *yolk duct* brings in and surrounds the eggs with yolk globules derived from (7) the *vitelline* or *yolk glands,* the shell forms and the eggs pass on into (8) a sinuous, looping *uterus,* which leads to the (9) *genital atrium* and (10) *genital pore.*

Although self-fertilization may occur, we infer from the arrangement of the reproductive system that spermatozoa are introduced through the dorsal pore of *Laurer's canal* and stored in the *seminal receptacle* until leaving to join the *ova* from the *ovary* to produce the fertilized *egg.*

Life Cycle

Opisthorchis sinesis has a complex life cycle with the production of the following stages: the *adult fluke* described above, the *egg,* the ciliated *miracidium,* the *sporocyst,* the *redia,* the *cercaria,* and the *metacercaria,* which becomes the adult in the definitive host. The life cycles of flukes typically involve snails as an *intermediate host* for the sporocyst and redia stages with a vertebrate as the *definitive host* where the sexually mature adults occur. Many variations occur in life cycles of different species, especially in the absence or modification of various pre-adult stages. The life cycle of *Fasciola hepatica* (Fig. 5-12), the sheep liver fluke, was the first trematode life cycle to be determined.

The Egg. The egg of *O. sinensis* (Fig. 5-13, C) is provided with a zygote embedded in a mass of yolk cells. The capsule covering is formed by the fusion of material carried by the yolk cells. The shell is operculated; that is, a lid, or *operculum,* covers one end. A ciliated miracidium develops within the egg. The eggs are passed with the feces of the host.

The Miracidia. Miracidia escape from the eggs in the intestine or rectum of a suitable fresh-water snail host after they are ingested by the snail. The miracidia penetrate the intestinal wall and develop into sporocysts in the surrounding blood spaces.

Sporocysts. As the miracidium enlarges to become the sporocyst, the germ cells at the posterior grow and produce germ balls, which differentiate into the next larval type, the rediae. The sporocysts from which they rupture are thin-walled, saclike structures.

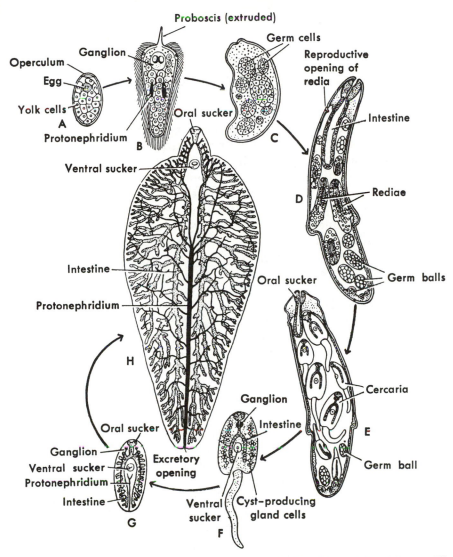

Figure 5-12 *Fasciola hepatica;* life cycle. (A) "Egg." (B) Miracidium. (C) Sporocyst. (D and E) Rediae. (F) Cercaria. (G) Tail-less encysted stage (metacercaria). (H) Adult (neither reproductive organs nor nervous system are shown). (*From Kerr.*)

Rediae. The rediae migrate to the hepatopancreas, or liver, of the snail where they grow and produce cercariae by an asexual process similar to the formation of rediae within the sporocysts. A number of germ balls differentiate simultaneously into cercariae. Rediae can be distinguished from sporocysts by the presence of a pharynx and short saclike intestine.

Cercariae. A cercaria (Fig. 5-14) has an oral sucker and ventral sucker as well as rudiments of other systems. The long, flattened tail used for locomotion after the cercaria emerges from the first intermediate host enables it to swim about in search of a suitable fish, usually a cyprinid, where it bores into the flesh to encyst as a metacercaria.

Metacercariae. Encysted metacercariae undergo loss of cercarial features such as eyespots and tail and further development of adult features such as development of the two intestinal branches and elongation of the excretory bladder. Ingestion of the second intermediate host by the definitive host (a human being or a fish-eating carnivore such as the dog or cat) is followed by release of the juvenile fluke from the metacercarial cyst. The young fluke migrates through the digestive tract and up the bile duct to one of the bile ducts of the liver where it matures within a month.

The *adult* has already been described. A large number of adults could potentially be produced from each egg. The great number of eggs produced by a single fluke is necessary, because the majority of the eggs may not be ingested by the particular kind of snail required, and only a few of the resulting cercariae may find the proper fish host with even fewer reaching the definitive host. The generations within the snail increase the number of larvae that may develop from a single egg. This complicated *life history* may be looked upon as (1) providing a reproductive mechanism with both sexual and asexual features, (2) enabling the fluke to gain access to new hosts, and (3) reducing competition between different stages in the life cycle. The sexual reproductive method provides great numbers of eggs. It also provides a mechanism for variability or stability depending upon selective pressures over long periods of time. The asexual method provides a method of producing many genetically identical progeny, thus rapidly capitalizing on valuable changes which may occur. Because heavy infections with *O. sinensis* can be fatal, it would be of little value to the species to produce heavy reinfections of the same host by direct development of eggs within the host. An intermediate host provides a means of access to other definitive hosts of the same species, that may be especially valuable when strong territoriality is exhibited by the definitive host. Because developmental stages may have nutritional requirements that differ from the adult, they may find it more readily in a different host. More important, it reduces the stress on the definitive host as compared to one that might harbor all stages in various organs. The evolutionary advantage of a compatible relationship with the host is unknown because there are no statistics on the number of parasites that become extinct because they cause the extinction of their host.

The trematodes resemble Turbellaria in many respects but differ from them in others, largely because of specialization for a parasitic existence. Organs of attachment and or reproduction are highly developed but organs of locomotion, sense, and digestion are reduced or lost entirely. Most trematodes are small, ranging from 1 mm to 2 or 3 cm in length. Suckers are present on the ventral surface varying in size and location according to the species. The ectoderm cells of trematodes are sunk into the parenchyma and outwardly are fused as a syncytial cuticula (Fig. 5-24) often armed with spines. Beneath the cuticula is a dermomuscular sac consisting of an outer layer of circular muscles, a thinner middle layer of diagonal muscles, and an inner layer of longitudinal muscles. There are also dorsoventral muscles and muscles connected with the suckers. No coelom is present, the body between the intestine and dermomuscular sac being filled with parenchyma.

The digestive system consists of a mouth located in the oral sucker, a muscular pharynx, and a thin esophagus into which salivary glands pour their secretions; then comes an intestine of two branches more or less elongated and ending blindly. The food of trematodes consists of the intestinal contents of the host, mucus, and other secretions, epithelial cells, etc., according to the species. Undigested particles are egested through the mouth. Excretory matter is collected by the flame cells (Fig. 5-3) from the surrounding parenchyma, passes through capillaries into collecting tubes, and is carried to a bladder, usually near the posterior end, and then expelled through an excretory pore. There are no circulatory nor respiratory systems. The nervous system resembles that of the Turbellaria. There are two cerebral ganglia above the pharynx, connected by a commissure; from these ganglia three pairs of nerve cords arise and pass backward, being located near the dorsal, ventral, and lateral surfaces, respectively. Adult trematodes have no sense organs.

The reproductive organs of trematodes are complex and life cycles usually involve several different hosts, which results in enormously increased powers of reproduction. The production of large numbers of offspring is necessary in parasitic animals because the chances that any one individual will reach a new host are rather slight. Most trematodes are hermaphroditic. The description and illustration of the reproductive organs of the liver fluke indicate their character. The eggs (Fig. 5-13) of one worm may be fertilized by spermatozoa from the same worm, or cross-fertilization may occur. The larvae that hatch from the eggs of ectoparasitic trematodes are ciliated and swim about until they attach themselves to a new host. Endoparasitic trematodes usually pass through a complicated life cycle as in the liver fluke (Fig. 5-12).

The simplest life cycle of a trematode is found in the Monogenea,

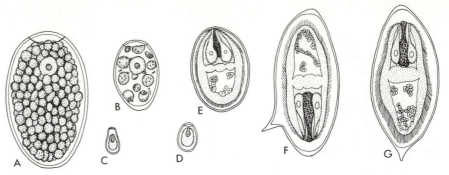

Figure 5-13 Trematoda. Eggs of certain species that live in humans. (A) *Fasciolopsis buski,* an intestinal fluke; China. (B) *Paragonimus westermani,* a lung fluke; Far East, America. (C) *Opisthorchis sinensis,* a liver fluke; Far East. (D) *Metagonimus yokogawai,* an intestinal fluke; China, Japan. (E) *Schistosoma japonicum,* a blood fluke; China, Japan, Philippines. (F) *Schistosoma mansoni,* a blood fluke; Africa, America. (G) *Schistosoma haematobium,* a blood fluke; Near East. (*After Cort.*)

especially in *Gyrodactylus* (Fig. 5-15, F), whose egg develops within the parent to a new worm almost ready to have its own egg hatch before it emerges from the parent to take up its life, generally on the same host.

Variations from the digenetic type of life cycle illustrated by *Opisthorchis* include: a miracidium hatched in the water which actively penetrates a snail after locating it as in *Fasciola hepatica;* daughter sporocysts may be produced by the sporocysts, or the sporocyst stage may be omitted with miracidia developing directly into rediae; daughter rediae may be produced by the redia until a generation of rediae is produced in which the cercariae develop (perhaps the host tissue invaded influences whether germ balls will develop into redia or cercariae as one snail was known to release cercariae over a 7-year period when reinfection was prevented); cercariae (Fig. 5-14) show variation

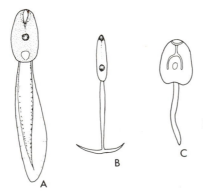

Figure 5-14 Trematoda. Cercarial types. (A) Pleurolophocercous type of *Opisthorchis.* (B) Furcocercous type of *Schistosoma.* (C) Cercaria of *Fasciola hepatica.* (*Modified after several authors.*)

in structure and development, some penetrate the definitive host and develop into mature flukes as do the schistosomes, others become metacercariae encysted on vegetation as does *Fasciola hepatica* or in a second intermediate host, or a tail-less juvenile may occur in a molluscan intermediate host; except in the schistosomes, or blood flukes, which actively penetrate their definitive host, digenetic trematodes reach their definitive hosts when they are ingested by them. The adults of various species localize in particular organs such as the intestine, bile ducts, bladder, lung, mouth, or blood vessels. A sample of the types of adult trematodes is given in Figure 5-15.

An example of one of the more complex life cycles is that of *Alaria mustelae*, which involves four different species of hosts. The sporocysts in (1) the snail, *Planorbula armigera*, produce cercariae which penetrate into (2) tadpoles or frogs in which they become mesocercariae; these when eaten by (3) a mammal, such as a mink, raccoon, or mouse, become metacercariae in the muscles or lungs; the metacercariae grow to the adult state when eaten by (4) another mammal, such as a mink, weasel, cat, dog, or ferret, in the intestines of which they deposit their eggs. From the latter, miracidia hatch which are capable of infecting the snail.

Many larval trematodes, the cercariae, have been described, the adult stages of which are unknown. These cercariae possess certain adult characteristics such as suckers, digestive and excretory systems and temporary larval characteristics, such as tails, stylets, and cystogenous and cephalic glands. Cercariae may be obtained by placing freshwater snails in a glass of water and watching for them with a hand lens as they emerge from these hosts. They have been given names and classified in various ways but their true relations can only be determined when the adults into which they develop become known.

Classification of Trematodes

Three subclasses of Trematoda are accepted: the Aspidobothrea, the Monogenea, and the Digenea.

Subclass 1. Aspidobothrea

Aspidobothrea lack an oral sucker but have a very large, compartmented ventral sucker; usually with only one host, a mollusk or a cold-blooded vertebrate. A common species is *Aspidogaster conchicola* (Fig. 5-15, E) found in the pericardial cavity of freshwater clams.

Subclass 2. Monogenea

The Monogenea are principally ectoparasites of fish, turtles, and amphibians. Each species is usually restricted to a single species of host.

They attach themselves to the gills of fish, but some species live in the mouth, cloaca, bladder, or on the fins. The oral sucker may be reduced or divided into a double structure. The ventral sucker is commonly at the posterior of the body where it may consist of multiple suckers or be supplied with a number of hooks. The variable anterior adhesive structure is termed a *prohaptor,* whereas the posterior structure is termed an *opisthaptor.* Two orders are included in the Monogenea.

Order 1. Monopisthocotylida

The members of this order have double excretory pores that are anterior and dorsal. The opisthaptor is single; the vagina is unpaired; the uterus contains a single egg; and there is no genitointestinal canal. Common genera with species occurring on the gills of many freshwater fish include *Ancyrocephalus, Dactylogyrus,* and *Gyrodactylus* (Fig. 5-15, F), which differ in the number of hooks, anchors, and bars on the opisthaptor.

Order 2. Polyopisthocotylida

Polyopisthocotylida have an opisthaptor with multiple adhesive structures, a double vagina, and a genitointestinal canal. A genus that contains a number of species from North America is *Polystoma* (Fig. 5-15, A). It has six suckers in a circle or in two rows at the posterior end. Species have been described from the mouth cavity and urinary bladder of turtles and frogs. *Polystoma integerrimum* is influenced by the hormonal cycle of its frog host to deposit eggs at the time the frog breeds. Thus, the parasite's eggs hatch in a suitable location for infecting young tadpoles. *Diplozoon* (Fig. 5-15, G) is a curious gill parasite of minnows that during the larval stage fuses with another specimen permanently so that the sperm duct of one enters the yolk duct of the other and they remain in permanent copulation.

Subclass 3. Digenea

The subclass Digenea contains endoparasitic trematodes including all of the species parasitic in people. Their life cycle involves two or more species of hosts. Terms descriptive of the general arrangement of the suckers include *amphistome* when the suckers are at the extreme anterior and posterior ends, *monostome* when only an oral sucker is present, *holostome* when the ventral sucker (acetabulum) is in a median position with an adhesive organ behind it, and *distome* when the ventral sucker is closer to the oral sucker. Over 100 families have been described. Considerable splitting of taxa has taken place to accommodate the growing number of known species. Consequently, superfamilies are often convenient for grouping trematodes, while the higher classification is rather unsettled. Higher classification is gener-

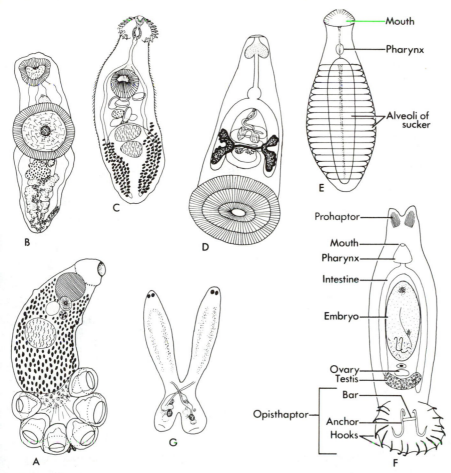

Figure 5-15 Types of adult trematodes. (A) *Polystoma megacotyle*, from the mouth cavity of the turtle. (B) *Gorgodera minima*, from the bladder of the frog. (C) *Stephanoprora gilberti*, from the intestine of aquatic birds. (D) *Megalodiscus temperatus*, from the rectum of the frog. (E) *Aspidogaster*. (F) *Gyrodactylus*, epizoic on fish; viviparous. (G) *Diplozoon*, from gills of fish, two individuals fused in permanent copulation. (*After several authors.*)

ally based on characteristics of developmental stages of the trematodes. A few of the superfamilies and families are included below.

DIVISION 1. ANEPITHELIOCYSTIDIA. Digenea retaining the thin-walled, primitive excretory bladder.

Order 1. Strigeatida

Cercariae with forked tails.

STRIGEIDAE. The Strigeidae are the holostome flukes. Adults are parasitic in the intestine of fish-eating birds; body frequently with an-

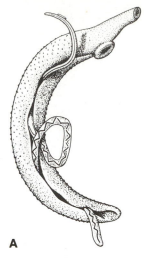

Figure 5-16 (A) *Schistosoma haematobium,* a human blood fluke. The larger male is carrying the more slender female in the gynaecophoric canal. (B) *S. mansoni,* scanning electron micrograph of male anterior. (*A, from Augustine, after Manson-Bahr; B, courtesy of Dr. Harvey Blankespoor.*)

A

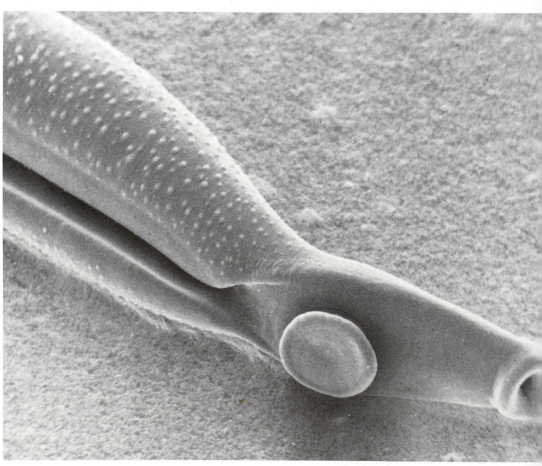

B

terior and posterior divisions; adhesive organ behind the acetabulum; a second intermediate host is required for development of the metacercariae. *Lymnaea stagnalis,* a common pond snail, can serve as either first or second intermediate host for the strigeid of ducks, *Cotylurus,* but presence of sporocysts confers immunity from penetration by cercariae, although cercariae will penetrate uninfected snails to become metacercariae.

SCHISTOSOMATIDAE. Schistosomes are flukes with sexual dimorphism, sexes being separate; blood flukes of birds and mammals; entrance to definitive host gained by the furcocercous cercaria (Fig. 5-14, B) boring into the host. Three species of *Schistosoma* (Fig. 5-16) produce the most serious trematode infections of people. Pulmonate snails serve as the intermediate host and the resulting cercariae penetrate the skin of the definitive host when it enters the water. Cercariae of bird schistosomes will attempt to penetrate the human skin but can enter only far enough to produce the allergic response known as "swimmer's itch."

BUCEPHALIDAE (GASTEROSTOMATIDAE). In the Bucephalidae the anterior sucker is not the oral sucker, the mouth is in the center of the ventral sucker centered on the ventral surface; intestine saclike; genital pore at the posterior end. These forms live in the intestine of both freshwater and marine fish as adults and in bivalve mollusks as larvae. *Bucephalus gracilescens* lives in marine fish and as a larva in the internal organs of the oyster.

Order 2. Echinostomida

Cercariae with unbranched tails.

ECHINOSTOMIDAE. *Echinostoma revolutum* has a ring of spines around the anterior end, adult in intestine of semiaquatic mammals and birds, cosmopolitan.

FASCIOLIDAE. The family Fasciolidae, large flattened flukes with highly branched intestines, includes *Fasciola hepatica* (Figs. 5-12 and 5-17), the sheep liver fluke, and *Fasciolopsis buski,* an intestinal fluke of humans.

PARAMPHISTOMOIDEA. The superfamily Paramphistomoidea contains amphistome flukes that are parasites in the digestive tract of hosts from all classes of air-breathing vertebrates; examples are *Megalodiscus temperatus* (Fig. 5-15, D) in the rectum of frogs and *Gastrodiscoides hominis* in humans.

Order 3. Renicolida

Contains only one genus, adults are bladder and kidney parasites of birds, immature stages are in marine gastropods, excretory bladder Y-shaped.

DIVISION 2. EPITHELIOCYSTIDIA. Digenea with the primitive bladder wall replaced by epithelium from the mesoderm.

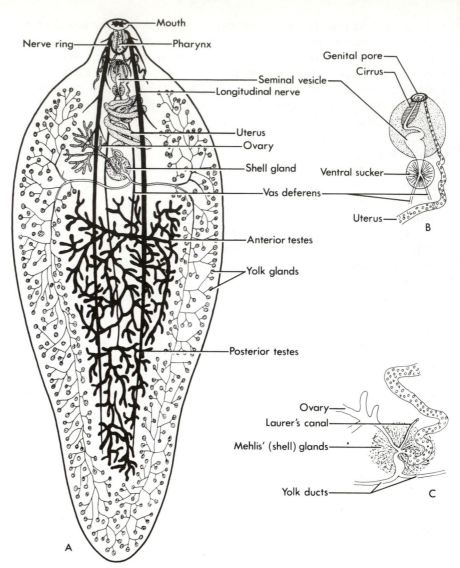

Figure 5-17 *Fasciola hepatica.* (A) Diagram of the reproductive and nervous systems. (B) Male structures in the region of the genital pore. (C) Female ducts. (*Modified after Leuckart and others.*)

Order 1. Plagiorchiida

Cercariae lack caudal excretory vessels. Many of the typical distomes are in this order. Two superfamilies.

PLAGIORCHIOIDEA. The superfamily Plagiorchioidea contains parasites of fish, amphibians, reptiles, birds, and mammals, including two occasionally found in people. The site of infection is usually the digestive tract, but some occur in respiratory, excretory, or reproductive organs. They are distomes with a forked urinary bladder and a spiny cuticle. Flukes of the genus *Haematoloechus* may be found in the lungs of most adult leopard frogs supplied by biological supply houses or caught in their natural habitat.

Haematoloechus medioplexus (Fig. 5-18) has a life history that has been experimentally determined in the laboratory using *Rana pipiens* for the definitive host. It lives as an adult parasite in the lungs of the leopard frog and has larval stages in a snail, *Planorbula armigera*, and certain dragonflies of the genus *Sympetrum*. The parasites produce in great numbers eggs which pass from the lungs through the glottis into the mouth cavity. They are swallowed, pass through the digestive tract, and are voided with the feces. The eggs, containing mature miracidia, are deposited in this way in the shallow water along the margins of ponds and lakes where snails are usually abundant. The miracidia remain alive within the eggs for weeks without hatching, and it is necessary for the snails to eat the eggs in order to become infected. The miracidia are small (25 μm in length) covered with a dense

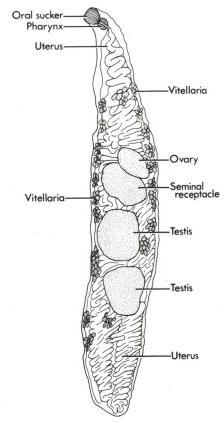

Figure 5-18 *Haematoloechus medioplexus,* a lung fluke from the frog. (*After Cort.*)

coating of long cilia, and move relatively slowly. The elongated, sausage-shaped daughter sporocysts are located in the digestive gland of the snail where they give rise to comparatively few cercariae which make their escape from the snail. Cercariae swimming in the vicinity of a dragonfly nymph are caught by the respiratory currents produced by the gills in the rectum and are swept into the chamber of the branchial basket through the anal opening. They penetrate through the chitin into the tissues of the lamellae, and the metacercariae encyst shortly within thin, structureless, transparent, pliable cysts where they ordinarily attain maximum size in from 14 to 20 days. The encysted metacercariae remain in the lamellae until the nymph transforms into an adult dragonfly. Hundreds may be found in a single insect. They live here in a more or less inactive condition until the host dies in the fall of the year or until such a parasitized dragonfly is eaten by a leopard frog or a common toad. In the stomach of either animal the insect is sufficiently digested to free the metacercariae. These are stimulated to activity by the digestive fluid and creep up the esophagus in measuring-worm fashion. From the esophagus they pass through the glottis into the lungs, where they reach maturity, under optimum conditions, in 28 or 30 days from the time they enter the frog. The flukes grow considerably after attaining maturity.

ALLOCREADIOIDEA. The superfamily Allocreadioidea includes a number of families of habits similar to other plagiorchids. The family TROGLOTREMATIDAE includes the distome lung fluke of mammals, *Paragonimus,* and a monostome, *Collyriclum faba* in the English sparrow. Most members of the family have a scaly cuticle and live in cysts in various organs of the host, two or more per cyst.

GORGODERIDAE. Members of the Gorgoderidae are distome flukes that are parasites in the urinary bladder of amphibians. *Gorgodera minima* (Fig. 5-15, B) is common in the bladder of the frog; species of *Phyllodistomum* have the rear of the body expanded, disklike, posterior to the acetabulum. Sphaeriid clams serve as the molluscan host.

Order 2. Opisthorchida

Cercaria have excretory vessels in the tail; about 25 families.

OPISTHORCHIDAE. The Opisthorchidae are distome flukes such as *Opisthorchis sinensis* (Fig. 5-11) from people; adults in the bile ducts of fisheating mammals, birds, and reptiles; molluscan host commonly *Bythinia;* second intermediate host, usually a minnow.

Reservoir Hosts

The requirement of a particular species of host by a particular parasite, or host–parasite specificity, is not of much significance in the trematodes. Of particular interest from a public health standpoint is that many trematode parasites of humans can complete their life cycle in a mammal other than the human and those thus serve as reservoir hosts.

Some Common Trematodes of People

	Scientific Name	Developmental Stages	Geographical Distribution	Hosts
Intestinal Flukes	*Fasciolopsis buski*	in water-grown vegetables	China, Indo-China, Formosa, Sumatra, India	human, pigs
	Heterophyes heterophyes	in fresh-water fish	Egypt, China, Japan	human, cat, dog, and fox
	Metagonimus yokogawai	in trout	Spain, China, Russia, Japan	human, cat, and dog
Liver Flukes	*Opisthorchis sinensis* (Fig. 5-11)	in freshwater fish	China, Japan, Korea, Indo-China	human, cat, dog, and fish-eating mammals
	Fasciola hepatica (Fig. 5-12)	on vegetation	The Americas, Asia, Europe, Africa	sheep, cattle, pigs, human
	Opisthorchis filineus	in freshwater fish	Baltic countries, Annam, Philippines	human, cat, dog
Lung Flukes	*Paragonimus westermani*	in freshwater crabs or crayfish	Japan, Korea, Formosa, Philippines, The Americas	human, tiger, dog, cat, pig, mink
Blood Flukes	*Schistosoma haematobium* (Fig. 5-16)	in freshwater	Africa, Asia	human, monkey; experimentally in rat, mouse
	Schistosoma mansoni	in freshwater	Africa, West Indies, South America	human, experimentally in other mammals
	Schistosoma japonicum	in freshwater	Eastern Asia	human, cat, dog, pig, etc.

Examples of this can be seen in the table, Some Common Trematodes of People.

Class III. Cestoda

TAENIA—A TAPEWORM

The tapeworm, *Taenia solium,* is a common parasite, which lives as an adult in the alimentary canal of a person. A nearly related species, *Taeniarhynchus* (= *Taenia*) *saginata,* is also a parasite of humans. Each adult tapeworm (Fig. 5-19) is a long flatworm consisting of a knoblike

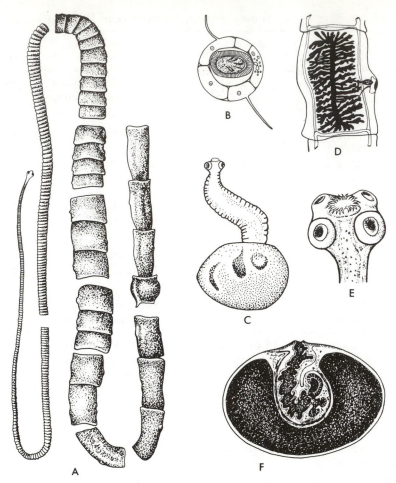

Figure 5-19 Cestoda, tapeworms living in humans. (A) *Taeniarhynchus sagi-nata*, adult; the beef tapeworm. (B) *T. saginata;* egg surrounded by its membrane; note hooks within the egg. (C) Cysticercus or bladderworm with head evaginated and capable of attachment to the intestinal wall. (D) *T. saginata,* ripe proglottid. (E) *Taenia solium,* scolex showing hooks and suckers. (F) Cysticercus in median section showing scolex at bottom of invagination. (*After various authors.*)

head or holdfast organ, the *scolex,* and a great number of similar parts, the *proglottids,* arranged in a linear series. The animal clings to the wall of the alimentary canal by means of *hooks* (lacking in *Taeniarhynchus*) and *suckers* on the scolex. Behind the scolex is a short *neck* followed by a string of proglottids which gradually increase in size from the anterior to the posterior end. The worm may reach a length of 3 meters and contain 800 or 900 proglottids. Since the proglottids are budded off from the neck, those at the posterior end are the oldest. The production of proglottids may be compared to the formation of ephyrae by the scyphistoma of *Aurelia* (Fig. 3-34, E and F), and is called *strobilization*.

The anatomy of the tapeworm is adapted to its parasitic habits (Fig. 5-20). There is *no alimentary canal*, the digested food of the host being absorbed through the body wall. The *nervous system* is similar to that of *Planaria* and the liver fluke, but not so well developed. Longitudinal *excretory tubes*, with branches ending in *flame cells*, open at the posterior end and carry waste matter out of the body.

A *mature proglottid* is almost completely filled with *reproductive organs*; these are shown in Figure 5-20. *Spermatozoa* originate in the spherical *testes*, which are scattered about through the proglottid; they are collected by fine tubes and carried to the *genital pore* by way of the *vas deferens*. *Eggs* arise in the bilobed *ovary* and pass into a tube, the *oviduct*. Yolk from the *yolk gland* enters the oviduct and surrounds the eggs. A shell is then formed in the *shell gland* and the eggs pass into the *uterus*. The eggs have in the meantime been *fertilized* by spermatozoa, which may come from the same proglottid, and move down the vagina. As the proglottids grow older, the uterus becomes distended with eggs and sends off branches, and the rest of the reproductive organs are absorbed. The ripe proglottids break off and pass out of the host with the feces.

The eggs of *Taenia* develop into six-hooked embryos (Fig. 5-19, B) while still within the proglottid. If they are then eaten by a pig, they escape from their envelopes (Fig. 5-19, D) and bore their way through the walls of the alimentary canal into the voluntary muscles, where they form *cysts* (Fig. 5-19, F). A scolex (head?) is developed from the

Figure 5-20 *Taeniarhynchus saginata*. Mature proglottid of the "beef tapeworm" of the human. (*After Leuckart.*)

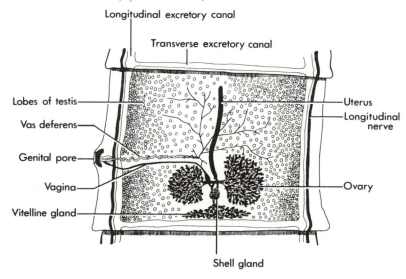

Longitudinal excretory canal

Transverse excretory canal

Lobes of testis

Vas deferens

Genital pore

Vagina

Vitelline gland

Uterus

Longitudinal nerve

Ovary

Shell gland

cyst wall and then becomes everted. The larva is known as a *bladder-worm* or *cysticercus* (Fig. 5-19, C) at this stage. If insufficiently cooked pork containing cysticerci is eaten by a person, the scolex becomes fastened to the wall of the intestine, and a series of proglottids is developed.

OTHER CESTODA

The Cestoda or tapeworms are highly modified for a parasitic existence. They are all endoparasitic and almost all of them live as adults in the digestive tract of vertebrates and as larvae in the tissues of vertebrates and invertebrates. No digestive system is present even in the simple trematodelike cestodes, and nutriment is absorbed through the surface of the body. Most cestodes are shaped like a band or ribbon (Fig. 5-19) and consist of many segments called proglottids (Fig. 5-20); these, however, are not true segments like those of higher invertebrates, such as the annelids. The adults are often several meters in length and consist of a small head or scolex, a short neck, and a strobila or long chain of proglottids. The scolex usually bears suckers, or acetabula, and is sometimes armed with hooklets. The neck is the growing region, from the posterior end of which proglottids are budded off. The proglottids increase in size as they are pushed back and various systems of organs develop in them.

Each proglottid usually possesses a genital pore on one of the lateral borders or on the surface, but some species have a separate uterine pore. The body is covered with a cuticula as in the trematodes and the internal organs lie in a mass of parenchyma cells in which are also embedded calcareous corpuscles. Circular, longitudinal, transverse, and dorsoventral muscle fibers are present as in the trematodes and also nerve bundles in the scolex from which arise longitudinal nerve fibers. The excretory system likewise resembles that of the trematodes.

Tapeworms are hermaphroditic. The reproductive organs vary in different groups; those of *Taenia*, already described, indicate their character. Each proglottid contains a complete set of both male and female organs. Apparently the eggs formed in a proglottid may be fertilized by spermatozoa from the same proglottid, although copulation is known to occur between proglottids of the same worm and of different worms. In some species the eggs are continually being discharged through a uterine pore, but in most species they are stored up in the proglottids which become "gravid," separate from the chain and pass out in the feces of the host. The eggs in these proglottids contain embryos that, when fully developed, are called onchospheres; these are able to continue their development only when ingested by a proper host. The onchospheres escape from the egg and burrow through the intestinal wall into the body cavity or vascular spaces or into certain tissues. The onchospheres of the lower cestodes become spindle-shaped, hooked procercoids, which develop in a second intermediate host into wormlike, hookless plerocercoids. The larvae of certain higher cestodes are called cysticercoids; they have a rudimentary bladder and may possess a tail. The true bladder-worms are (1) the small cysticercus, which gives rise to one scolex; (2) the large coenurus, from which many scolices arise; and (3) the echinococcus or hydatid, which gives rise to daughter and granddaughter cysts in which many scolices are developed from brood pouches (Fig. 5-21, C and D). The bladder-worms are the stage infective to the definitive host and each scolex may give rise to a tapeworm. Tapeworms, when alive, successfully resist the digestive juices of the host, but soon disintegrate when dead. They may live for many years; several cases are on record of tapeworms that lived for 20 years or more.

Classification of the Cestoda

The cestodes are usually divided into two subclasses. One subclass, the Cestodaria lacks division into proglottids and has larvae with 10 instead of the usual six hooks, but is a cestode in that it lacks epidermis and digestive system, has only anterior organs of attachment, and are all endoparasites. Cestodarians are rarely encountered and most are parasites of primitive marine fish.

The other subclass, the Eucestoda, contains most of the cestodes. Most have chains of proglottids in the adults. Of the 11 orders of Eucestoda, only two contain parasites of mammals and humans, the Pseudophyllidea and Cyclophyllidea. Holdfast organs on the scolex may include *hooks*, *suckers*, folds termed *bothria*, or extended and folded *bothridia*.

Order 1. Proteocephalida

Small tapeworms; scolex with four suckers; vitellaria as lateral bands; parasites of fishes, amphibians, and reptiles.

Order 2. Tetraphyllida

Medium-sized tapeworms; scolex with four bothridia; vitellaria in lateral bands; parasites of elasmobranch fishes. *Calliobothrium verticillatum* occurs in the spiral valve of the smooth dogfish.

Order 3. Disculiceptida

Only one genus known from elasmobranch fishes. Scolex with only a single flattened anterior expansion. Life cycle unknown.

Order 4. Lecanicephalida

Scolex with a variable anterior portion and a posterior portion bearing four suckers; parasites of elasmobranch fishes.

Order 5. Pseudophyllida

Small to large tapeworms; scolex with two bothria; vitellaria as scattered follicles; uterine pore opens on surface; parasites of fishes, birds, and mammals. The most important human species is the broad or fish tapeworm, *Dibothriocephalus latus* (Fig. 5-21, A and B), which has two intermediate hosts, freshwater copepods and fish. It may reach a length of 6 meters, live more than 20 years, and may cause various symptoms such as anemia in the human.

Figure 5-21 Cestodes that live in the human. (A) *Dibothriocephalus latus*, the fish tapeworm of humans. (B) *D. latus*, ciliated embryo (coracidium) escaping from egg. (C) *Echinococcus granulosus*, adult stage that lives in the dog. (D) *E. granulosus*, formation of scolices from the germinative membrane of a hydatid cyst. (E) *Hymenolepis nana*. (F) *H. nana*, egg. (*Mostly after Leuckart.*)

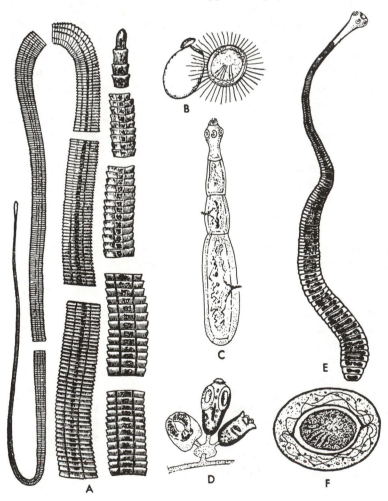

Order 6. Trypanorhynchida

The scolex in this group is very long and consists of a head with two or four bothria and four retractile, spinose proboscides, and a long stalk. The genital pores are marginal. The larvae encyst in teleost fish and the adults parasitize elasmobranchs.

Order 7. Cyclophyllida

Tapeworms with a scolex having four suckers; may have a hooked rostellum; no uterine pore; vitellaria posterior to ovary; genital pore opens laterally; the proglottids break from the strobila when gravid; the eggs are not operculated, and the onchospheres are not ciliated. These are taenioid cestodes. Many of the more important cestode parasites of human and other mammalian species belong to the family Taeniidae from which our type was selected. Among these are *Taenia solium,* the pork tapeworm of humans, *Taeniarhynchus saginata,* the beef tapeworm of humans, *Taenia pisiformis* of cats and dogs, *Echinococcus granulosus* (Fig. 5-21, C) of dogs and other carnivores which produce larval stages known as hydatids in humans and other mammals. Common species that belong to other families are *Raillietina tetragona* in domestic fowls, *Moniezia expansa* in sheep, *Hymenolepis nana* (Fig. 5-21, E) or the dwarf tapeworm of humans, *H. diminuta* of rats and mice, and *Dipylidium caninum* of cats and dogs.

Order 8. Aporida

Scolex variable, usually large with four suckers; no external evidence of proglottid formation; small parasites of swans and ducks.

Order 9. Nippotaeniida

Scolex with one anterior sucker; few proglottids; parasites of fish in Japan and Russia.

Order 10. Caryophyllida

No proglottids formed; parasites of fishes and oligochaetes; thought to represent tapeworms that have developed sexual maturity in a larval procercoid stage; few species.

Order 11. Spathebothrida

Scolex variable; no external proglottid formation; parasites of sturgeons and marine fish.

A few important tapeworms of humans are listed in the table, Some Common Cestodes of People.

Some Common Cestodes of People

Scientific Name	Developmental Stages	Geographical Distribution	Hosts
Dibothriocephalus latus	in copepods and freshwater fish	worldwide, especially Europe, Africa, North America, Japan	human, dog, cat, fox (intestine)
Echinococcus granulosus	in liver, brain, lungs, etc., of human, pig, sheep, etc.	worldwide, especially Iceland, Australia, Argentina	dog and other carnivores (intestine)
Hymenolepis nana	human, rat, mouse	worldwide	human, rat, mouse
Taeniarhynchus saginata	in muscles of cattle	worldwide	human
Taenia solium	in muscles of pig	worldwide	human

Platyhelminthes in General

The flatworms differ greatly among themselves due largely to the fact that the Turbellaria are mostly free-living, whereas the Trematoda and Cestoda are all parasitic in habit. The turbellarians, therefore, exhibit the typical organization of the group, the other two groups being modified considerably for a parasitic existence.

External Features

The Platyhelminthes are flat and unsegmented, with definite anterior and posterior ends and bilateral symmetry. The epidermis is ciliated in the Turbellaria, but in the Trematoda and Cestoda the surface is covered by a thick cuticle (Figs. 5-24 and 5-25) secreted by cells in the parenchyma. Minute projections and pores are visible on the cuticle with the electron microscope. In some species sense organs are present.

Body Wall

Beneath the epidermis or cuticle are two layers of muscle, an outer layer of circular muscles and an inner layer of longitudinal muscles. In the Cestoda there are usually additional well-developed zones of longitudinal, dorsoventral, and transverse muscles embedded in the inner regions of the parenchyma. Filling the area between the body wall and the organs is a mass of parenchymatous tissue (Fig. 5-4), which is very primitive in nature; it is of value in the reparation of lost parts and probably serves for the transportation of nutritive material. There is no coelom.

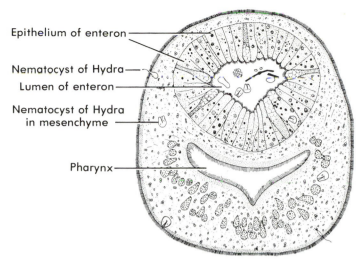

Epithelium of enteron

Nematocyst of Hydra

Lumen of enteron

Nematocyst of Hydra in mesenchyme

Pharynx

Figure 5-22 Turbellaria. *Microstomum caudatum.* Cross section through body near anterior end, showing especially the location of nematocysts appropriated from hydras ingested as food. (*After Kepner.*)

The nematocysts of *Hydra* are sometimes appropriated by a flatworm, *Microstomum caudatum* (Fig. 5-22). The worm eats and digests the *Hydra,* leaving the nematocysts free in the enteron. The endoderm cells of the worm engulf these nematocysts and pass them on to wandering mesenchyme cells which then become cnidophages. The nematocysts are uniformly distributed over the surface of the *Microstomum* within 12 hours after the ingestion of a *Hydra.* The worm appears to devour the *Hydra* not because of its food value but in order to secure the nematocysts which it employs, as does *Hydra,* as weapons of defense and offense. This is indicated by the fact that a worm already supplied with nematocysts does not attack *Hydra* readily although one with few nematocysts does; that a worm with many nematocysts may retain the body of an ingested *Hydra* but egest the nematocysts, whereas a worm with few nematocysts may egest the body and retain the nematocysts; and that *Microstomum* has actually been seen to sting and paralyze other animals with the nematocysts obtained from *Hydra.*

Digestive System

In the Turbellaria and Trematoda there is a saclike intestine with a single opening, which serves both as mouth and anus. In the simplest forms the intestine is unbranched but in others, branches occur that may penetrate to all parts of the body thus rendering a circulatory system unnecessary. The Cestoda have lost the intestine and absorb nutriment through the general surface of the body.

Excretory System

A protonephridial excretory system occurs in almost all of the flatworms and in some is very complicated. The most characteristic feature of this system is the flame cell.

Nervous System

This consists of a network with a concentration of nervous tissue at the anterior end, forming the cerebral ganglia, and several longitudinal nerve cords. Sense organs occur in free-living Turbellaria and may be present in the free stages of Trematoda and Cestoda. They exist as eyes, otocysts, taste cells, tentacles, and ciliated ectodermal pits.

Reproductive System

The flatworms are characterized by a complex reproductive system. The structure and differences have already been described for representatives of the classes. The trematodes and cestodes lay eggs more or less continuously but the turbellarians recognize the seasons in their reproductive activity. As noted above, the latter commonly multiply by asexual reproduction also.

The sperm of most flatworms have two flagella on their tailpiece. The flagella have an unique ultrastructure.

Sexual individuals are all hermaphroditic, with the exception of the schistosomes. In some Turbellaria, for example *Procotyla fluviatilis*, egg-laying proceeds during the entire year; in others, for example *Crenobia alpina*, eggs are produced only during the winter months. Two different kinds of eggs are produced by *Mesostomum*: (1) "winter eggs," which have a thick shell, a large amount of yolk, escape only when the mother dies, and remain dormant for a long period; and (2) "summer eggs," which have a thin shell, a small amount of yolk, escape from the living parent, and develop quickly.

Embryology

The development of the egg of the polyclad, *Hoploplana inquilina*, has been worked out in detail (Surface, 1907) and an account of the embryology of this species is appropriate at this point. This flatworm lives in the mantle cavity of certain gastropods. The eggs are spherical and about 1/10 millimeter in diameter. They are laid in spiral gelatinous capsules containing from 100 to 2,000 eggs each. The early cleavage stages occur at intervals of about 1 hour. On the sixth day the larvae break through the egg membranes and emerge as free-swimming organisms known as Müller's larvae. These swim about for a time and fall to the bottom, where they change into the polyclad form.

Cell Lineage (Fig. 5-23). It is possible to follow the cleavage cells and their descendants from stage to stage until the larval condition is

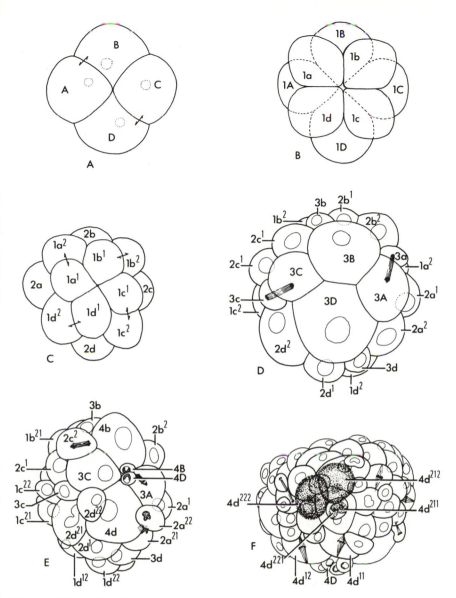

Figure 5-23 Cell lineage of a flatworm, *Hoploplana inquilina*. (A) Four-cell stage. (B) Eight-cell stage with 4 macromeres (1*A*, 1*B*, 1*C*, 1*D*) and 4 micromeres (1*a*, 1*b*, 1*c*, 1*d*). (C) Sixteen-cell stage viewed from animal pole, showing three quartets of micromeres. (D) Thirty-two-cell stage viewed from vegetative pole, showing dexiotropic formation of 3*a* and 3*c*. (E) Thirty-six-cell stage viewed from near animal pole, showing especially cell 4*d*. (F) Stage with more than 100 cells, showing endoderm cells ($4d^{11}$, $4d^{12}$, $4d^{211}$, $4d^{221}$) and mesoderm cells ($4d^{212}$, $4d^{222}$). The number and letter tell which division and from which macromere the cell is derived; the exponents indicate subsequent cell lineage, so that [1] is the daughter cell nearer to and [2] is farther from the animal pole. (*After Surface.*)

reached and in this way to ascertain the fate of each cell. Observations of this kind have been recorded for various types of animals and constitute what is called *cell lineage*. The egg of *Hoploplana* undergoes *total cleavage* into two and these into four cells (A). One of these four, designated by the letter *D*, is larger than the other three, which makes it possible to distinguish the different cells as cleavage advances. When the four cells, labeled *A, B, C, D*, divide to form eight (B), the four near the animal pole are smaller than the other four; they are known as *micromeres* and labeled 1*a*, 1*b*, 1*c*, and 1*d*. At this stage, therefore, a quartet of micromeres are present and a quartet of larger, sister cells, called *macromeres* to which the labels 1*A*, 1*B*, 1*C*, and 1*D*, are applied. All the cells that are descended from any cell of the four-cell stage comprise a *quadrant*. The cleavage spindles during the division from the four-cell to the eight-cell stage are oblique and the micromeres come to lie in the furrows between the macromeres, that is, alternate with them (B). This is known as *spiral cleavage*. In most cases when the egg is viewed from the animal pole, the micromeres lie in the furrow clockwise of the corresponding macromere; that is, cleavage is *dexiotropic* (B). Sometimes the micromeres lie in the furrow counterclockwise of the corresponding macromere, in which case cleavage is said to be *laevotropic*.

The eight-cell stage develops into the 16-cell stage by the laevotropic division of the quartet of micromeres (1*a*, 1*b*, 1*c*, 1*d*, to form $1a^1$, $1a^2$, $1b^1$, $1b^2$, $1c^1$, $1c^2$, $1d^1$, $1d^2$) and the production of a second quartet of micromeres (2*a*, 2*b*, 2*c*, 2*d*) by the macromeres (C). The three quartets of micromeres present at this stage are destined to form the entire ectoderm of the animal, and the macromeres the endoderm and mesoderm. The 32-cell stage (D) is attained by the production of a third quartet of micromeres (3*a*, 3*b*, 3*c*, 3*d*) by a dexiotropic division of the macromeres, and the division of each of the three quartets of micromeres. After the 32-cell stage division is not synchronous, and the various groups of cells must be considered separated.

Cell 4*d*, which results from the division of macromere *D* at about the 40-cell stage, is of particular interest (E) since it gives rise to all of the endoderm ($4d^{11}$, $4d^{12}$, $4d^{211}$, $4d^{221}$) and part of the mesoderm ($4d^{212}$, $4d^{222}$). At about the 100-cell stage, $2a^{112}$, $2b^{112}$, $2c^{112}$, and $2d^{112}$ are set aside as *mesoderm cells*. In many protostomes the 4*d* cell gives rise to only the mesoderm. Primitive ganglion cells are recognizable at about this time ($1a^{112212}$, $1b^{112212}$, $1c^{112212}$, $1d^{112212}$).

Ultrastructure

The transmission electron microscope has made visible several interesting features of the Platyhelminthes. A unique "9 + 1" pattern is present in sperm tails. The central filament appears to have two flattened, helically wound strands forming its cortex. Acoelida sperm flagella have either a "9 + 0" pattern or the traditional "9 + 2" pattern.

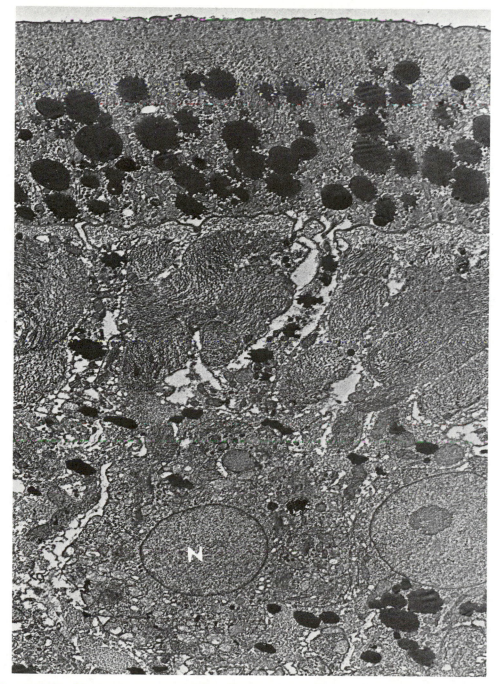

Figure 5-24 Electron micrograph of adult *Acanthoparyphium spinulosum* integument. Integument contains dark globules mingled with many mitochondria; cytoplasmic connection can be seen with some globular material extending down to the integumentary cell having the nucleus (N) visible near the bottom. ×3,500. (*From R. F. Bils and W. E. Martin. 1966. Fine structure and development of the trematode integument.* Trans. Am. Microsc. Soc., *85 (1): 78–88. Courtesy of Dr. Walter E. Martin and Dr. R. F. Bils.*)

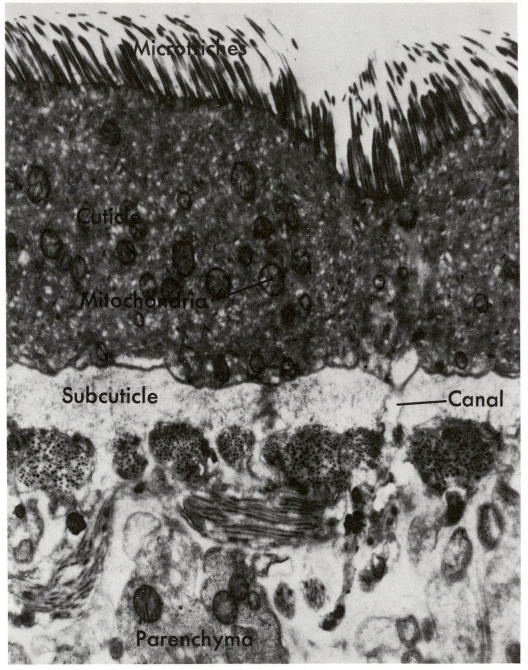

Figure 5-25 Cestode cuticle. Electron micrograph of *Hymenolepis citelli*. ×5,000. (*Courtesy of Dr. Alvin H. Rothman.*)

Ciliary rootlets are often long and striated below the basal granule. A basement membrane lies beneath the epidermis of only the advanced Platyhelminthes; it is absent in Acoelida, Macrostomida, and Catenulida.

In a few Turbellaria and typically in Trematoda (Fig. 5-24) and Cestoda (Fig. 5-25), the nucleus-bearing portion of the epidermal cells is in an enlargement of the cell beneath the basement membrane and among the general body tissues. Epidermal layers of both syncytial and cellular types occur. The trematodes and cestodes have variable surface features. Some are relatively smooth, some have minute projections, and some have rounded or spinous projections (Fig. 5-26).

Blankespoor and van der Schalie (1976) used scanning electron microscopy of sequentially prepared *Gigantobilharzia huronensis* miracidia penetrating *Physa gyrina* feet to show that about 4 minutes are needed to complete penetration of the tegument. Miracidium penetration is illustrated in Figure 5-27.

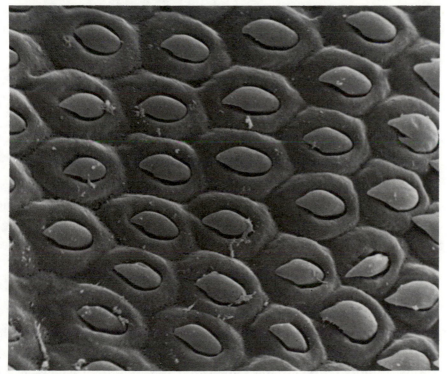

Figure 5-26 Spinous tegument of a digenetic trematode, *Loxogenes. (Scanning electron micrograph courtesy of Dr. Harvey Blankespoor.)*

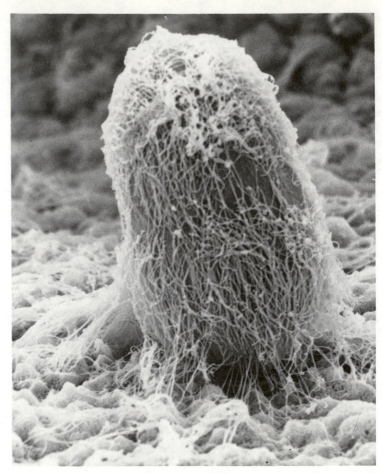

Figure 5-27 *Trichobilharzia occelata* miracidium penetrating epidermis of *Lymnaea stagnalis. (Scanning electron micrograph courtesy of Dr. Harvey Blankespoor.)*

Regeneration and Behavior

The regenerative ability of the turbellarians has contributed much interesting information on development. Of great recent interest is the transmission of some types of learned behavior by RNA in regenerating planarians, which is still questioned by many zoologists.

Phylogeny and Interrelations

The flatworms, especially the Turbellaria, resemble the coelenterates in certain respects, which indicate that they were derived from coelenteratelike ancestors. There is a single opening for the ingestion of food and egestion of waste material; a nerve net reminiscent of the coelenterates; and sense organs, such as otocysts, characteristic of coelenterates. The relation of flatworms to ctenophores is even more striking; polyclads being much like ctenophores that have adopted the creeping habit. In both groups the large macromeres produce ectoderm-forming micromeres; the nervous center is at the upper pole; an ectodermal stomodaeum is at the lower pole; and the primary locomotor organs are eight ciliated bands forming combs of cilia in the adult ctenophore and in Müller's larva. The latter is what we may imagine the ctenophorelike ancestor of the polyclads to resemble. The strong preference for vertebrate definitive hosts by the parasitic flatworms indicates a development contemporary with or subsequent to the development of the vertebrates. Revision of classification is to be expected, especially in the tapeworms, when less reliance is placed on highly variable and adaptive features such as the scolex, and when more reliance is placed on fundamental systems and their arrangement.

The absence of fossils of the Platyhelminthes makes discussion of phylogeny highly speculative. The previously mentioned resemblance of polyclads and ctenophores is not considered of phylogenetic significance by most zoologists. The view of Hadzi, that the Turbellaria gave rise to most of the unsegmented invertebrates, including the coelenterates, does not seem to have been accepted by many. A popular view is that the flatworms arose from a planulalike organism by way of the Acoela. The planula larva of coelenterates then becomes the basis for a close phylogenetic relationship with the flatworms. The general view seems to be that the flatworms have a higher level of development than the coelenterates and thus may be derived from them or from their ancestors.

The Gnathostomula may be near relatives as well as the clearly related Nemertinea. Acanthocephala and Mesozoa also have indications of a closer relationship to the Platyhelminthes to a greater extent than do most other phyla.

Modifications for Parasitism

Within the phylum are a broad range of parasitic modifications. The cestodes may be considered the most highly modified for a parasitic existence and seem to do little damage to their normal hosts. Some features associated with the parasitic modifications encountered in the flatworms include (1) loss of the epidermis and development of a protective cuticle (Figs. 5-24 and 5-25), (2) reduction and loss of a digestive system, (3) development of organs of attachment, (4) development of enlarged reproductive systems or egg-producing capacity, (5) tendency to complex life cycles, and (6) physiological accommodations with the host. Generalized examples of the complex life cycles are presented in the table, Some Parasite Life Cycles.

Stage	Host	Location	Reproduction
Trematode (digenetic)			
Egg	released from vertebrate	vertebrate to water	none, hatches
Miracidium	mollusk	penetrates epidermis	none, grows and transforms
Sporocyst	mollusk	internal organs	asexual, internal budding
Redia	mollusk	hepatopancreas	asexual, internal budding
Cercaria	mollusk, 2nd intermediate	mollusk, water, next host surface	none
Metacercaria	2nd intermediate host	encysts on or under surface	none
Fluke (adult)	definitive (vertebrate)	intestine or cavity of an organ	sexual, egg production
Cestode (taenioid)			
Egg	vertebrate	feces	none, hatches
Onchosphere	intermediate (herbivore)	penetrates intestine	none
Cysticercus	intermediate (herbivore)	general soft tissue	usually none
Tapeworm (adult)	definitive (predatory vertebrate)	intestine	asexual—strobilization; sexual, egg production

Immunological phenomena; physiological interaction with, and dependence on, the host; and ultrastructural features of significance in parasitic existence are among areas of greater activity in current research.

The question of whether parasitism leads to mutualism or commensalism, rather than vice versa, has not been resolved. Probably either adaptation toward increased pathogenicity or toward mutualism occurs under appropriate circumstances. The most frequent result appears to be a very low, almost undetectable, effect on the health of a healthy host in an optimal environment.

Immune responses sometimes terminate infestations and produce resistance to further infestations in vertebrate hosts and/or invertebrate hosts. A few parasitized hosts have improved growth as a result of certain parasitic infestations.

Ecological studies on free-living flatworms have only recently progressed beyond the distributional and general environmental characteristics descriptive stage. Turbellarians have variations in distribution based on rather specific preferences; for example, Thum (1974) found *Notoplana acticola* commonly in the intertidal zone under rocks where poor sorting of particles was evident. They showed seasonal variation in size, and size-related effects in several ways.

Campbell (1973) and Chubb (1977) discuss the seasonal aspect of some parasites. Ecology of the hosts is obviously of significance in the ecology of the parasite. The diversity of the digenetic trematodes is remarkable in view of their almost universal requirement of a gastropod as an intermediate host. As remarkable is the restriction of adult tapeworms to vertebrate intestines and the restriction of the adults of almost half the orders of tapeworms to elasmobranchs or other primitive fishes. Such facts may give hints at the time and place of origin of such parasitic groups. But do adult or larval stages point to the beginnings better?

At one time many parasitologists thought host–parasite specificity was very great so that a parasite in a new host would be given a new name for that reason only. In one study (Campbell, 1973), one fluke species was found in 33 host species of birds of widely varying relationships. On the other hand some, schistosomes for example, do have high host specificity for both intermediate and definitive hosts. High specificity may indicate a long history of parasitism of the species involved.

The Importance of Flatworms

The importance of flatworms is chiefly because of the numerous parasitic species infesting people, their animals, and important wildlife species. The free-living species are of some importance in "the balance of nature" and of more value as basic biological research tools because of their simple level.

The History of Our Knowledge of the Platyhelminthes

Parasitic flatworms, such as tapeworms, were well known to the ancients as inhabitants of humans and domesticated animals. Trematodes were first described clearly in the sixteenth century, but it was not until the close of the seventeenth century that the hydatids of tapeworms were recognized as animals, having been considered previously as abscesses or abnormal growths. Linnaeus placed all of the invertebrates except the arthropods into one class, the Vermes, and allocated the worms to an order, the Intestina, of this class. The fundamental difference between segmented and unsegmented worms was recognized by Lamarck and by Cuvier in the late 1700s. In 1808 Rudolphi proposed the name Entozoa for the unsegmented worms, because at that time most of the species known were internal parasites. Rudolphi recognized five orders: (1) Nematodes, (2) Acanthocephala, (3) Trematodes, (4) Cestodes, and (5) Cystici. The term Turbellaria was coined by Ehrenberg in 1831 for the free-living species of flatworms because the beating of their cilia caused disturbances (turbellae) in the water. In 1744 Trembley published the first figure of a turbellarian in his memoir on *Hydra*. O. F. Müller and Ehrenberg founded our knowledge of the Turbellaria, and Lang and von Graff have contributed more recently. In 1851 Vogt united four orders of flatworms in the class Platelmia, which we now know as the phylum Platyhelminthes. Leuckart, who contributed so much to our knowledge of the parasitic worms, presents the history of the subject up to the year 1879 in his book *The Parasites of Man*. Major contributions to the knowledge of the Turbellaria have been made in recent decades by Libbie Hyman. Many workers have been active contributors to knowledge of the parasitic worms, recent important books having been contributed by Dawes on the Trematoda, by Wardle and McLeod on the tapeworms, and by Yamaguti on both groups.

REFERENCES TO LITERATURE ON THE PLATYHELMINTHES

BEDINI, C., and F. PAPI. 1974. Fine structure of the turbellarian epidermis. Pp. 108–147 in N. Riser and M. Morse. *Biology of the Turbellaria*. McGraw-Hill, New York.

BENAZZI, M. 1975. New karyotype in the American fresh-water planarian *Dugesia dorotocephala*. *Syst. Zool.*, **23**:490–492.

BEST, J. B., A. B. GOODMAN, and A. PIGON. 1969. Fissioning in planarians: control by brain. *Science*, **164**:565–566.

BILS, R. F., and W. E. MARTIN. 1966. Fine structure and development of the trematode integument. *Trans. Am. Micr. Soc.*, **85**:78–88.

BLANKESPOOR, H. D., and H. VAN DER SCHALIE. 1976. Attachment and penetration of miracidia observed by scanning electron microscopy. *Science*, **191**:291–293.

BROOKS, T. J. 1963. *Essentials of Medical Parasitology*. Macmillan, New York. 358 pp.

BURT, D. R. R. 1970. *Platyhelminthes and Parasitism*. American Elsevier, New York. 150 pp.

CAMPBELL, R. A. 1973. Studies on the biology of the life cycle of *Cotylurus flabelliformis* (Trematoda: Strigeidae). *Trans. Am. Micr. Soc.*, **92**:629–640.

CHENG, T. C. 1973. *General Parasitology*. Academic Press, New York. 965 pp.

———, C. N. SHUSTER, and A. H. ANDERSON. 1966. A comparative study of the susceptibility and response of eight species of marine pelecypods to the trematode *Himasthla quissetensis*. *Trans. Am. Micr. Soc.*, **85**:284–295.

CHUBB, J. C. 1977. Seasonal occurrence of helminths in freshwater fishes. Part I. Monogenea. *Adv. Parasitol.*, **15**:133–199.

Costello, D. P., C. Henley, and C. R. Ault. 1969. Microtubules in spermatozoa of *Childia* (Turbellaria, Acoela) revealed by negative staining. *Science,* **163:**678–679.

Dawes, B. 1946. *The Trematoda.* Cambridge Univ. Press, Cambridge. 644 pp.

Erasmus, D. A. 1972. *The Biology of Trematodes.* Crane, Russak, New York. 312 pp.

Hay, E. D., and S. J. Coward. 1975. Fine structure studies on the planarian, *Dugesia. J. Ultrastruct. Res.,* **50:**1–21.

Hendelberg, J. 1974. Spermiogenesis, sperm morphology, and biology of fertilization in the Turbellaria. Pp. 148–164 in N. Riser and M. Morse. *Biology of the Turbellaria.* McGraw-Hill, New York.

Hyman, L. H. 1951. *The Invertebrates.* Vol. II. *Platyhelminthes and Rhynchocoela.* McGraw-Hill, New York. 550 pp.

———, and E. R. Jones. 1959. Turbellaria. Pp. 323–365 in W. T. Edmondson. *Fresh-Water Biology.* 2nd ed. Wiley, New York. 1248 pp.

Jacobson, A. L., C. Fried, and S. D. Horowitz. 1966. Planarians and memory. *Nature,* **209:**599–601.

Jenkins, M. M. 1963. Bipolar planarians in a stock culture. *Science,* **142:**1187.

———, and H. P. Brown. 1964. Copulatory activity and behavior in the planarian *Dugesia dorotocephala* (Woodworth) 1897. *Trans. Am. Micr. Soc.,* **83:**32–40.

Leuckart, R. 1886. *The Parasites of Man.* 771 pp.

Lim, H. K., and D. Heyneman. 1972. Intramolluscan intertrematode antagonism: a review of factors influencing the host-parasite system and its possible role in biological control. *Adv. Parasitol.,* **10:**192–268.

Malek, E. 1977. Natural infection of the snail *Biomphalaria obstructa* in Louisiana with *Ribeiroia ondatrae* and *Echinoparyphium flexum,* with notes on the genus *Psilostomum. Tulane Stud. Zool. Bot.,* **19:**131–136.

Montgomery, J. R., and S. J. Coward. 1974. On the minimal size of a planarian capable of regeneration. *Trans. Am. Micr. Soc.,* **93:**386–391.

Noble, E. R., and G. C. Noble. 1976. *Parasitology.* 4th Ed. Lea & Febiger, Philadelphia. 566 pp.

Oschman, J. L., and P. Gray. 1965. A study of the fine structure of *Convoluta roscoffensis* and its endosymbiotic algae. *Trans. Am. Micr. Soc.,* **84:**368–375.

Pantelouris, E. M. 1965. *The Common Liver Fluke. Fasciola hepatica* L. Pergamon Press, Oxford. 259 pp.

Pennak, R. W. 1978. *Fresh-Water Invertebrates of the United States.* 2nd Ed. Wiley, New York. 803 pp.

Riser, N. W., and M. P. Morse (Eds.). 1974. *Biology of the Turbellaria.* McGraw-Hill, New York. 530 pp.

Rothman, A. H. 1963. Electron microscopic studies of tapeworms. *Trans. Am. Micr. Soc.,* **82:**22–30.

Schell, S. C. 1970. *The Trematodes.* Brown, Dubuque. 355 pp.

Schmidt, G. D. 1970. *The Tapeworms.* Brown, Dubuque. 266 pp.

———, and L. S. Roberts. 1977. *Foundations of Parasitology.* Mosby, St. Louis. 604 pp.

Silveira, M.   1969.   Ultrastructural studies on a "nine plus one" flagellum I. *J. Ultrastruct. Res.*, **26**:274–288. Reprinted in *Biology of Turbellaria: Experimental Advances*, 1973. MSS Information Corp., New York. 175 pp.

Smithers, S. R., and R. J. Terry.   1976.   The immunology of schistosomiasis. *Adv. Parasitol.*, **14**:399–422.

Smyth, J. D.   1969.   *The Physiology of Cestodes.* Freeman, San Francisco. 279 pp.

Surface, F. M.   1907.   The early development of polyclad *Planocera inquilina*. *Proc. Acad. Nat. Sci. Philadelphia*, **59**.

Thum, A. B.   1974.   Reproductive ecology of the polyclad turbellarian *Notoplana acticola* (Boone, 1929) on the central California coast. Pp. 431–445 in N. Riser and M. Morse. *Biology of the Turbellaria*. McGraw-Hill, New York.

Wardle, R. A., and J. A. McLeod.   1952.   *The Zoology of Tapeworms.* Univ. of Minnesota Press, Minneapolis. 780 pp.

———, J. A. McLeod, and S. Radinovsky.   1974.   *Advances in the Zoology of Tapeworms, 1950–1970.* Univ. of Minnesota Press, Minneapolis. 274 pp. (Nine new orders are erected for mostly cyclophyllidean families. They are not used here.)

Wolff, E.   1962.   Recent researches on the regeneration of planaria. In D. Rudnick. *Regeneration.* Ronald Press, New York. 250 pp.

Wright, C. A.   1971.   *Flukes and Snails.* Allen & Unwin, London. 168 pp.

Yamaguti, S.   1958.   *Systema Helminthum.* Vol. I, *The Digenetic Trematodes of Vertebrates.* 1959. Vol. II, *The Cestodes of Vertebrates.* 1963. Vol. IV, *Monogenea and Aspidocotylea.* Wiley, New York.

6

Phylum Aschelminthes

The Aschelminthes constitute a phylum of six diverse classes that have in common a pseudocoelom, bilateral symmetry, a cuticle, a lack of complete segmentation, a digestive tract with an anterior oral and posterior anal opening, no circulatory or respiratory systems, usually separate sexes, and usually a protonephridial system of flame cells or flame bulbs. An anterior nerve mass, which may encircle the digestive tract, gives off two or more longitudinal tracts. Nuclear numbers within species are constant.

The Acanthocephala are sometimes included in this phylum, but they also show relationships to the Platyhelminthes. Because of their distinctive characteristics, they are treated as a separate phylum in the next chapter. The Priapulida have been removed for the same reason.

The body cavity and complete digestive tract represent a considerable advance over previous phyla discussed. The body cavity may be lined with a membrane, but it is never nucleated as is the lining membrane of a true coelom. The digestive tract does not have a muscular wall. The Gnathostomula are probably most like the ancestral type. The Rotifera have long fascinated microscopists with their variety and abundance in fresh water. The Nematoda are of great economic importance because of the many parasitic species of humans, animals, and plants.

The following classification into two subphyla seems to accommodate the classes conveniently in this broadly limited phylum. Many zoologists have preferred to treat each class as a separate phylum because of the great differences in general morphology. The group does

307

have many general common characteristics with some exceptions, and, in the case of the Gnathostomula, an apparent position nearer the presumed ancestral stem, probably paralleling the turbellarian stem, and retention of some primitive characteristics that stand as exceptions to the phylum diagnosis.

Phylum Aschelminthes

Unsegmented, triploblastic, bilaterally symmetrical, pseudocoelomate animals with complete digestive tract and no circulatory or respiratory system.

SUBPHYLUM I. TROCHELMINTHES. Microscopic Aschelminthes; typically with external cilia and toes of a bifurcated foot provided with adhesive glands; muscles as discrete fibers or tracts; reproduction parthenogenetic in many species, one or few eggs produced at a time; protonephridia or cyrtocytes. Marine or fresh water.

CLASS 1. GNATHOSTOMULA. Mouth subapical; pharyngeal region provided with jaws and complex musculature; each epidermal cell with a single cilium, no cuticle; pseudocoel lacking, parenchyma poorly developed; no anus; hermaphroditic; all marine, interstitial in fine sand. One order, Gnathostomulida.

CLASS 2. GASTROTRICHA. Pharynx simple; no segmentation of cuticle, which often has scales or plates which may be spined; hermaphroditic or parthenogenetic females.

CLASS 3. ROTIFERA. Pharynx modified into a special grinding structure, the mastax; cilia typically forming an anterior ring, the corona; sexes separate or parthenogenetic females.

CLASS 4. KINORHYNCHA. With an anterior region that may be everted or inverted; no external cilia; bristly cuticle differentiated into about 13 superficial segments; sexes separate; muscles mostly longitudinal with some segmental arrangement; all marine.

SUBPHYLUM II. NEMATHELMINTHES. Aschelminthes devoid of motor cilia; round elongate bodies; only longitudinal muscles in body wall, contractile fibers in peripheral part of muscle cell, nucleus in expanded inner part; no flame cells. Many are parasitic. The unsegmented roundworms.

CLASS 1. NEMATODA. Lateral epidermal chords present and containing the excretory ducts; pharynx present; female reproductive system opens to exterior. Thousands of species.

CLASS 2. NEMATOMORPHA. Lateral epidermal chords and excretory ducts absent; pharynx undifferentiated; female reproductive system opens into intestine; intestine often degenerate; juveniles parasitic in arthropods.

Class I. Gnathostomula

This newest of major animal groups was first seen in 1928 by Remane, and first described by Ax in 1956. The explosion of interest in them has resulted in additional important contributions by Graebner, Kirsteuer, Riedl, Sterrer, and others. This account is based largely on information in Riedl (1969) and Sterrer (1972).

Gnathostomulids often exceed 6,000 per liter of fine, littoral sand. They are exceeded in numbers only by nematodes and gastrotrichs in this preferred environment. They remain in the sand so tenaciously, adhering to the grains, that many samples were undoubtedly thrown out before they emerged from their home, which had turned anaerobic and smelly from stagnation.

Awareness of their existence and development of extraction techniques has resulted in the rapid description of new species, currently totaling about 100, with hundreds more anticipated.

GNATHOSTOMULA

Morphology

Gnathostomulids are small and wormlike, generally less than 1 mm long and about 50 μm in diameter. The anterior is rounded and bears a *sensorium* consisting of two (more in most other genera) stiff cilia apically and eight larger ciliary structures bilaterally paired around them (Fig. 6-1, A). The tail is tapered. A *single cilium* is found per epidermal cell; these are arranged in a regular pattern.

The *mouth* is ventral, near the anterior end. A *male genital opening* is near the posterior end.

The mouth and *pharynx* contain a basal plate and a pair of lateral *jaws* with complex musculature. The *gut* has a layer of large cells, but no anal opening; when filled, the gut can cause considerable distention of the body.

A thin circular muscle layer is present beneath the epidermis. Internal to the circular muscle layer are several prominent longitudinal muscle fibers. A poorly developed parenchyma fills up the body spaces. The *nervous system* includes an anterior concentration of nervous tissue and a peripheral portion with fibers near the base of the epidermis.

The *ovary* is anterior to a *bursa* and the *paired testes*. A *stylet* is the most conspicuous part of the copulatory apparatus. Hermaphroditism is the rule, but males and females occur, sometimes after temporary resorption of the opposite sex organs.

Ultrastructure

Inclusions in the epidermal cells are of uncertain homology to flatworm rhabdites. Microvilli are present, lining part of the mouth.

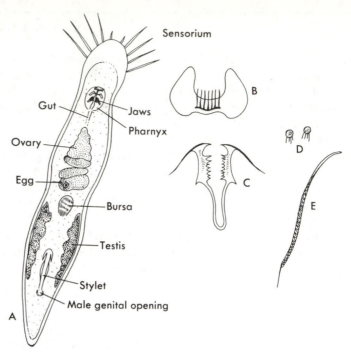

Figure 6-1 Gnathostomulida. (A–D) *Gnathostomula jenneri*. (B) Basal plate. (C) Jaws. (D) Sperm. (E) *Pterognathia swedmarki* sperm. (*After Riedl and Sterrer.*)

Sperm lack flagella but several filaments are present on *Gnathostomula jenneri* sperm. Other sperm types occur in other gnathostomulids. Some have long slender sperm (Fig. 6-1, E) with a single flagellum of the "9 + 2" type. A few peculiar excretory type cells (*cyrtocytes*) are in paired groups near the testis and among the parenchymal cells.

Reproduction and Development

Sperm are injected into the bursal region. The most posterior egg is fertilized and extruded through a rupture in the dorsal body wall posterior to the bursa. Cycles of exhaustion and regeneration appear to occur for the sexual apparatus. The extruded egg rounds up and undergoes a *spiral cleavage*. The first free-swimming stage is completely ciliated, with a tail, intestine, and rudiments of the jaws evident. By the time the length doubles, adult structures are evident.

Ecology

Fungal hyphae appear to be a preferred food. Predators may include small turbellarians and nematodes.

The physical habitat appears to be the dominant factor in their distribution. They are widely distributed geographically in fine, well-sorted sands. They occupy the shallow littoral marine locations to a depth where no oxygen is found. Their tolerance for low oxygen is so great that they do not show peak emergences from stagnating samples of their environment for from 32 to 66 days. Their ability to adher to sand particles that are only two to four times their own diameter is particularly evident at the posterior end. A high proportion of particles under $1/10$ mm in diameter seems to exclude gnathostomulids from the sediment. Particles larger than $1/4$ mm in diameter, when present in quantity, are also associated with reduced abundance. It seems that gnathostomulids have specialized in a location with spaces as small as they can go and spaces too small for their predators. We may expect that this provides ideal conditions for the development of gnathostomulids that will prey upon other gnathostomulids, although they will be difficult to demonstrate.

Phylogeny

Reidl (1969) implies that these peculiar organisms may be considered a class of the Platyhelminthes or Aschelminthes. He and Sterrer preferred the phylum designation until relationships were better understood. The general organizational level is near that of the flatworms. The "9 + 1" pattern in flatworm sperm flagella is so distinctive that it constitutes strong evidence the gnathostomulids developed from their common stock before the differentiation of the Turbellaria. The variability of gnathostomulid sperm, some with strong reseblances to gastrotrich and nematode sperm, indicates the possibility of origin from a Turbellarian ancestor must not be totally discounted. The sperm tail could have been lost. Later gamete competition could have reestablished the flagellum on the sperm from the genetic capacity to produce epidermal cilia. Gamete competition in the minute creatures might tend to large sperm or spermatophores that are capable of blocking passage of other sperm, or toward better motility where size is insufficient to obstruct other sperm.

Posterior adhesion in *Gnathostomula* correlates with development of pedal glands in Trochelminthes and some nematode papillae.

The basal plate, jaws, and pharynx bear a strong homology to the rotifer mastax. The beginnings of strong development of longitudinal muscles suggest affinity with the Nemathelminthes, as do the microvilli in a portion of the digestive tract. The sensorium is suggestive of marine nematode cephalic features, as are the basal plate and jaws possibly related to precursors of the three-lobed nematode mouth.

Ecological considerations are strongly supportive of the Gnathostomula being nearest the stem group of the Aschelminthes. The interstitial fauna is dominated by aschelminths and turbellarians. The rotifers seem to be the freshwater ecological equivalent of the gnathostomulids in the interstitial habitat. Pennak (1978, pp. 3–5) reviews his earlier views of the interstitial zone as an area suitable for marine organisms to evolve to freshwater organisms. The two groups would seem to confirm his hypothesis.

Negative evidence regarding the cuticle, pseudocoel, and anus justifies a major separation. In view of the cluster of parallel features with various aschelminths, the negative evidence may be considered a blend of neotenic, interstitial adaptations, small size adaptations, and features unique to their own selective history since they departed common ancestry. Thus, the view expressed here is that the Gnathostomula are the cement that holds the diverse classes of the Aschelminthes together in a way that must exceed the expectations of Hyman (1951) when she erected the phylum, prior to the first publication on the gnathostomulids.

Many thanks are due Reidl, Sterrer, and others for their fine work detailing this seemingly insignificant group that may be an important phylogenetic find as the closest relative of the ancestral aschelminth that provides possible hints of near flatworm relationships.

Class II. Gastrotricha

CHAETONOTUS

Chaetonotus (Fig. 6-2) is one of the more common gastrotrichs encountered on aquatic vegetation or detritus and in infusions. The body, usually from 100 to 200 μm long, is divided into a *head, neck,* and *trunk* bearing two tail forks. Locomotion is by means of cilia on two ventral areas. Much of the cuticle is divided into scales, many of which bear a spine. Four groups of sensory bristles occur on the head. The tail forks (*caudal furcae*) serve as tubes for paired *adhesive glands.*

The digestive system consists of a straight tube differentiated into an anterior *pharynx* and a posterior *intestine.* The *mouth* is at the extreme anterior end, the *anus,* on the dorsal surface of the posterior end. In some species a rectum may be posterior to a constriction of the intestine.

The *protonephridia* are paired and each consists of a flame cell with a long coiled tubule which opens to the exterior of the midregion of the body. Six pairs of longitudinal strands form the muscular system. Fibers lead from a *brain,* looped over the pharynx, to the sides of the head and the posterior end of the body. Some gastrotrichs have pigmented eyespots.

Only parthenogenetic females and hermaphrodites are known. The *eggs* are very large, sometimes about half the length of the body; they are attached to various aquatic objects and give rise directly to the fully developed organism. The female pore is near the posterior end. The eggs may be *tachyblastic,* which proceed with development imme-

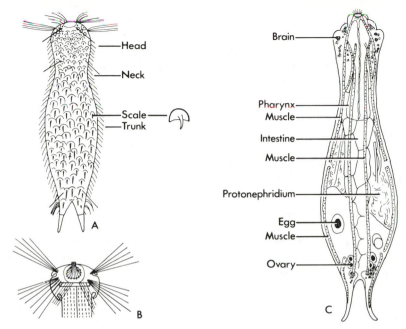

Figure 6-2 Gastrotricha. (A) *Chaetonotus brevispinosus*, dorsal view. (B) *Polymerurus rhomboides*, ventral view of head. (C) *Chaetonotus*, showing internal organs. (*A and C, after Zelinka; B, after Stokes.*)

diately, or *opsiblastic*, which undergo a period of dormancy during which they can survive a variety of unfavorable conditions.

OTHER GASTROTRICHA

The approximately 2,000 species of gastrotrichs may be divided into the following two orders.

Order 1. Macrodasyida

Hermaphroditic; several pairs of adhesive tubes in addition to those of the tail fork; no protonephridia; all marine, in interstitial habitat and surface of sediments, generally overlapping distribution of gnathostomulids and kinorhynchs.

Gagne (1977) describes *Dolichodasys* from Massachusetts as very elongate and with testis anterior to ovaries. Its aberrant sperm have a single filamentous tail. Copulation and sperm transfer occur in at least some from this order.

Order 2. Chaetonotida

Only parthenogenetic females; two or four adhesive glands and tubes at the posterior; one pair of protonephridia; mostly freshwater species. *Chaetonotus* (Fig. 6-2, A and B) is a common genus, which contains about a third of the nearly 40 species of freshwater gastrotrichs in the United States. A study by Sacks (1964) found that *Lepidodermella squammata* laid no more than four eggs during its entire life cycle. The eggs are laid about a day apart during the first 5 days of life. The life span averaged 10 days and those hatched from first eggs and last eggs of individuals seemed to live as long as those from second and third eggs. Bennett (1979) has shown that *L. squammata* is a bacterivore that can get some additional nutrients from algae, if they are available.

Class III. Rotifera

EPIPHANES SENTA

Epiphanes senta (Fig. 6-3) is a commonly studied rotifer which serves to illustrate the class. The Rotifera (Rotatoria), once known as wheel animalcules, are extremely small Metazoa. They were at one time considered Infusoria. Most of them are inhabitants of fresh water, but some are marine and a few parasitic. The anatomy of *Epiphanes* is shown in Figure 6-3. The *head* is provided with *cilia,* which aid in locomotion and draw food into the mouth. The *tail* or *foot* is bifurcated and adheres to objects by means of a secretion from a *cement gland.* The body is usually cylindrical and is covered by a shell-like *cuticle.*

The Protozoa and other minute organisms used as *food* are swept by the cilia through the *mouth* into the *pharynx,* also called the *mastax* or chewing stomach. Here chitinous *jaws* (the *trophi*), which are constantly at work, break up the food. The movements of these jaws easily distinguish a living rotifer from other organisms. The food is *digested* in the glandular *stomach.* Undigested particles pass through the *intestines* into the *cloaca* and out the cloacal opening.

Two coiled tubes, which give off a number of ciliated lobules and enter a bladder, constitute the *excretory system.* The bladder contracts at intervals, forcing the contents out of the cloacal opening. Because the amount of fluid expelled by the bladder is very large, water regulation and respiration are probably also a function of this organ, the oxygen being taken into the animal with the water, which diffuses through the body wall, and the carbonic acid being cast out with the excretory fluid. The body cavity is not a true coelom.

The *sexes* of rotifers are separate. The *female* possesses an *ovary* in which the eggs arise, a *yolk gland,* which supplies the eggs with yolk, and an *oviduct,* which carries the eggs into the cloaca. From here the eggs reach the exterior through the cloacal opening. The *males* are usually smaller than the females, and often degenerate. They possess a *testis* in which the spermatozoa arise, and a *penis* for transferring the spermatozoa to the female.

Two kinds of eggs are produced by rotifers: (1) summer eggs, and

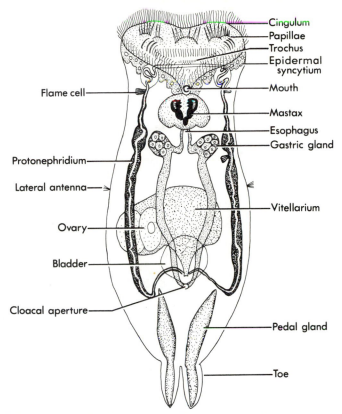

Figure 6-3 Rotifera. *Epiphanes senta.* Female, ventral view. (*After Borradaile and Potts, modified.*)

(2) winter eggs. The *summer eggs*, which develop parthenogenetically, are thin-shelled, and of two sizes; the larger produce females and the smaller males. The *winter eggs*, which are fertilized, have thick shells and develop females. The winter eggs and the summer eggs that develop into males are produced by females whose eggs undergo meiosis. If the egg is unfertilized a male is produced, if the egg is fertilized it develops into a female. Thus, males have only half the chromosomes in their nuclei that female rotifers have. This is shown in Figure 6-4. The fertilized winter egg is the stage most resistant to unfavorable environmental conditions (Gilbert, 1974).

Constancy of Nuclear Numbers

Many rotifers exhibit *eutely*, the phenomenon of each somatic organ having a characteristic number of nuclei that is constant for the species. In most rotifers the number of nuclei is from 900 to 1,000. The following numbers of nuclei have been reported from some groups of nuclei in *Epiphanes senta*. Epidermis, 280; pedal glands, 19 each;

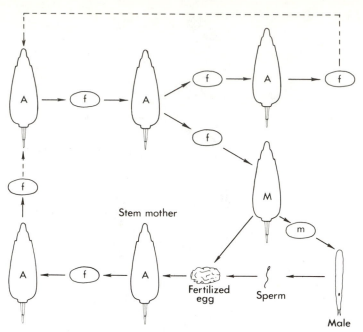

Figure 6-4 Diagram showing the alternation of generations in a rotifer, such as *Lecane inermis*. The most common type of individual, the "amictic" female (A), reproduces exclusively by diploid parthenogenesis. Its eggs (*f*) develop invariably into females, which are usually amictic, so that multiplication by diploid parthenogenesis may continue for many generations (as indicated by the long, broken-lined arrows). But the daughters of the amictic female may be "mictic" (*M*). Their eggs are haploid, and develop parthenogenetically (*m*) into males, or, if fertilized, produce amictic females, the "stem mothers," with which the cycle begins anew. (*After Miller.*)

muscular system, 104; brain, 183; other nervous tissue, 97; digestive tract lining, 159; gastric glands, 6 each; yolk gland, 8; oviduct, 3; protonephridium, 14 each. Six nuclei in a gastric gland and eight nuclei in a yolk gland have proven to be common to most species of rotifers examined. Birky and Field (1966) have found that species of *Asplanchna,* a common planktonic genus, are exceptions to the rule and that variation is subject to environmental control.

Cells of which various organs are constituted do not vary greatly in size within the individuals of a species. Thus, variations in the size of the body or of an organ are generally considered to be due to differences in the number of cells. In adult rotifers the cells are generally fused into a syncytium and thus the number of nuclei represents the approximate number of cells an equivalent cellular tissue would have. In many animals, as in *Epiphanes,* it has been shown that certain organs are almost invariably made up of a definite and constant number of cells or nuclei and in some animals the cell or nuclear number of the entire somatic tissue is constant.

In the differentiated somatic cells of such an animal the power of mitotic cell division appears to be lost; for example, no regeneration occurs in the rotifer, *Stephanoceros* (Fig. 6-6, I), if its arms are injured.

Cell or nuclear constancy occurs in species in which there is determinate cleavage of the egg. Apparently, partial constancy is more frequent than complete constancy. In four genera of Acanthocephala, however, all of the somatic tissues have a constant number of cells although the number may differ in different species. The nematode worm, *Oxyuris curvula*, has a constant number of cells (412) in the excretory and nervous systems and in the connective tissue, but a variable number in other organs. The number of muscle cells in the body wall of seven different species belonging to the genus *Oxyuris* is the same for all, namely 65. Similarly, the cells in the ommatidia of the compound eye of arthropods are not only constant in number but also constant in their location and physiological interrelations. Cell or nuclear constancy may occur also among the germ cells. For example, in the fly, *Miastor,* a definite number of primordial germ cells, and later, of ova, develop in each individual. Many other instances of cell or nuclear constancy are known among the tunicates, insects, annelids, nematodes, turbellarians, trematodes, etc.

OTHER ROTIFERA

The *corona,* or anterior crown of cilia found on most rotifers has many variations in different species. The primitive type, much like *Epiphanes,* was thought to have an anterior circle of cilia, the *trochus,* and another circle just behind the trochus, the *cingulum.* Between the two rings is a ciliated buccal field enclosing the mouth. One line of evolutionary development probably gave rise to a lobed trochus (Fig. 6-5, D) found in the suborder Flosculariacea. Further separation of the trochus into two disks (Fig. 6-5, E) is found in the Bdelloida. Some think the trochal disks developed from papillae in the buccal field. In the Collothecacea the anterior has lost the typical ciliation and only a few bristles may remain.

Rotifers are abundant in the plankton of fresh water. They are even more abundant on the surfaces of aquatic plants and debris. They reach their greatest abundance in numbers in the wet interstices of sandy beaches (Pennak, 1978). They are important primary consumers in aquatic ecosystems.

Williams (1966) found that five genera of ploimate rotifers (Fig. 6-7), *Keratella, Polyarthra, Brachionus, Synchaeta,* and *Trichocerca,* were commonly abundant in collections of planktonic rotifers in some major waterways of the United States. They outnumbered other metazoans

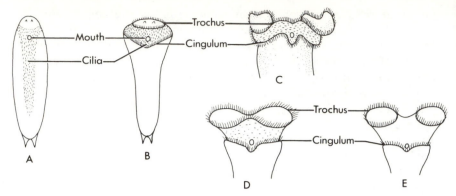

Figure 6-5 Presumed derivation of the rotifer corona. (A) Hypothetical, ventrally ciliated ancestor. (B) Basic ciliation of a rotifer. (C) *Hexarthra*, with reduced buccal field. (D) Lobed floscularian corona. (E) Trochal disks of the bdelloid type. (*Modified from Hyman and Edmondson, after de Beauchamp.*)

30 to 1 but were nearly absent in the winter. More than one genus commonly shared the dominance at one time, although only one species from a genus tended to be among the dominants.

Classification of Rotifers

Rotifers are microscopic in size, frequently with a superficial segmentation. They possess a mastax in the pharynx, the trophi (Fig. 6-6, M, N, and O) of which are often important in species identification. A crown of cilia (the corona) is usually at the anterior end, and the tail is frequently jointed with toes at the posterior end. Three orders are as follows.

Order 1. Seisonida

These rotifers are parasitic on marine crustacea. The corona is greatly reduced; the body is long, narrow, and ringed; the neck is much elongated; and the elongated foot has a terminal perforated disk with which it attaches itself to its host. All species belong to the genus *Seison* (Fig. 6-6, K).

Order 2. Bdelloida

Freshwater species that swim with the corona or creep like a leech or both. Two ovaries occur in these rotifers. Behind the corona is a dorsal proboscis. The body is usually cylindrical and without a lorica; it is composed of rings that may be drawn together like the sections of a telescope. *Rotaria* (Fig. 6-6, G) and *Philodina* (Fig. 6-6, H) are very common bottom-dwelling rotifers that frequently occur in infusions. The corona bears two wheels of cilia on separate retractile lobes (Fig. 6-5, E). Two posterior spurs make it difficult to distinguish the three toes of *Rotaria* or the four toes of *Philodina*. *Philodina roseola* has two eyes behind the proboscis and a slender, rose-colored body that may cause the bottom of shallow ponds to appear red when they are abundant. Many species of this order are able to withstand desiccation and come to life when placed in water, after existing for months in a dried condition.

Figure 6-6 Types of rotifers and their jaws. (A) *Proales werneckii*. (B) *Monommata longiseta*. (C) *Mytilina spinigera*. (D) *Brachionus pala*. (E) *Floscularia ringens*. (F) *Trochosphaera solstitialis*. (G) *Rotaria citrinus*. (H) *Philodina roseola*. (I) *Stephanoceros fimbriatus*. (J) *Conochilus unicornis*. (K) *Seison annulatus*. (L) *Collotheca proboscidea*. (M) Jaws of *Asplanchnopus myrmeleo*. (N) Jaws of *Asplanchna priodonta*. (O) Jaws of *Harringia eupoda*. (K) Seisonida. (G, H) Bdelloida. (A–F, I, J, L) Monogononata. (*From Jennings after various authors.*)

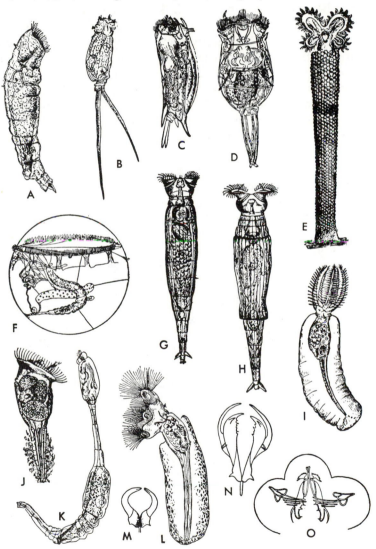

Order 3. Monogonontida

Rotifers with one gonad. Three suborders are quite distinct.

Suborder 1. Ploima. Planktonic or swimming rotifers with normal corona (Figs. 6-3 and 6-5, C). *Epiphanes senta* has already been described. *Keratella* has a sculptured lorica with one or two posterior spines and three pair of anterior spines, there is no foot. *Brachionus* (Fig. 6-6, D) and *Platyias* also are loricate but have a foot, the foot is retractable in *Brachionus*. *Polyarthra* (Fig. 6-7, B) has 12 cuticular paddles along the sides and attached at the anterior. *Trichocerca* is asymmetrical due to a twisted lorica. *Monostyla* has one long toe, which projects from the fused dorsal and ventral plates of the lorica. *Lecane* is similar but has two toes. *Synchaeta* lacks a lorica but has four anterior spines. *Asplanchna herricki* is a large species, saclike in form, and without foot or anus, waste products being cast out through the mouth. It thrusts its forcepslike jaws out of the mouth and seizes other animals with them. The plankton of the Great Lakes sometimes contains enormous numbers of this and related species. The simplest of all rotifers belong to the genus *Proales* Fig. 6-6, A). They are small and cylindrical, with a short foot and toes, and without a tail; the corona is a uniformly ciliated area and does not possess portions bearing longer cilia.

Figure 6-7 Rotifers common in the plankton of temperate lakes and ponds. (A) *Keratella* lorica, (B) *Polyarthra*, (C) *Conochilus*, (D) *Trichocerca*, (E) *Asplanchna*, (F) *Kellicottia* lorica. (*After several sources.*)

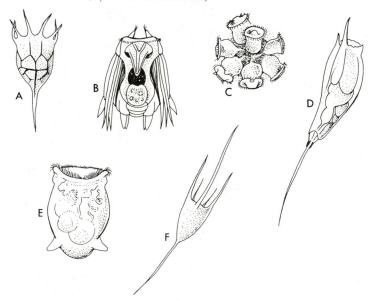

Suborder 2. Flosculariacea. Rotifers that have a large corona with two parallel rows of cilia around the outer edge, the inner row being larger than the outer. Between these two rows is usually a groove, often lined with fine cilia, along which food particles are carried to the mouth. The jaws are of the peculiar malleoramate type. The species may be solitary or colonial, free-swimming or attached. *Floscularia ringens* (Fig. 6-6, E) has a large corona with four large lobes, lives in a tube formed of spherical pellets and is common on water plants. *Conochilus unicornis* (Fig. 6-6, J) is a free-swimming, pelagic species, living in small colonies in which each individual occupies a separate, transparent tube. *Sinantherina* has a heart-shaped corona and forms visible colonies, either floating or attached; *Lacinularia* is similar, but embedded in a gelatinous mass.

Suborder 3. Collothecacea. The adult males and young rotifers placed in this suborder are free-swimming, but the adult females are mostly solitary and attached to water plants by a stalk, which is the modified foot. They live in a transparent tube and usually possess a lobed corona that is much expanded and has the mouth in the center. The cilia at the edge of the corona are often long and do not beat actively but are moved about so as to entangle prey. In these forms the digestive tract is modified for dealing with the large animals captured by the corona. *Collotheca ornata* (Fig. 6-6, L) has a corona with five knobbed lobes, a foot about twice as long as the body but no eyes. *Stephanoceros fimbriatus* (Fig. 6-6, I) also has a corona with five lobes; these are very long and slender and their cilia are long and non-vibratile. *Cupelopagis vorax* (Fig. 6-8) is a large rotifer, lacking vibratile cilia, that feeds on small organisms that blunder into its retractible, funnel-like anterior.

Figure 6-8 Rotifera. *Cupelopagis vorax,* a highly modified collothecacean that uses a funnel to trap food organisms rather than using ciliary currents. (A) Funnel retracted. (B) Funnel extended. (C) Lateral view.

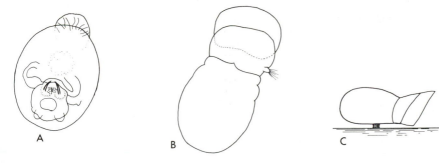

A   B   C

Class IV. Kinorhyncha

This group of about 100 species of microscopic, marine, spiny, and superficially segmented animals have their mouth situated on a protrusible mouth cone. The segmentation is superficial and is not thought to be truly metameric. It is likely that they arose from the gastrotrich stem close to the rotifer line. The Priapulida are thought to have arisen from the Kinorhyncha by some zoologists. A close relationship with nematodes is also claimed.

Thirteen segments (or *zonites*) form the slightly elongate, tapering body. Some internal structures reflect the external segmentation, namely the muscles and the ventral nerve, which has ganglia in the segments of the trunk. The digestive tract (Fig. 6-9, B) is a straight tube with the retractable mouth cone opening through a narrowed esophagus into the stomach–intestine. Several small glands are associated with these portions. A short rectum opens through a posterior anus. Paired protonephridia of the flame cell type open on the eleventh segment. Sexes are separate and the paired gonads open on the last segment. The cuticle is molted as juvenile kinorhynchs grow, and segments are added at the posterior until the adult stage is reached. Kinorhynchs live primarily in marine bottom muds and are of no special economic importance. Higgins (1965) is a leading investigator of the Kinorhyncha. Further study of the Kinorhyncha may clarify their relationship to the nematodes and priapulids.

Figure 6-9 Kinorhyncha. Generalized diagram of anatomy. (A) External, with mouth cone extended. (B) Internal. (C) Anterior with mouth cone retracted.

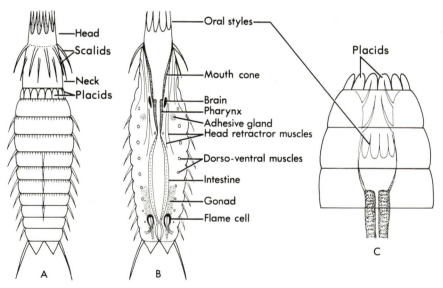

Class V. Nematoda

The Nematoda are called unsegmented roundworms (along with the Nematomorpha and Acanthocephala), to distinguish them from the flatworms and segmented annelids. They are long, slender animals usually with a smooth and glistening surface and tapering somewhat toward one or both ends. Roundworms are very widely distributed and occur in large numbers, both as regards species and individuals, in soil, fresh water and salt water, and as parasites in plants and animals. It has been estimated that the top 6 inches of an acre of arable soil contains thousands of millions of nemas. In size, they range from about 0.5 mm to over 1 meter in length. One can recognize most of them easily because of their peculiar movements. They whip about by contortions of their entire body in a dorsoventral plane without making any forward progress unless solid particles are present for the body to lash against. As a rule, the free-living nemas are slighted in courses in zoology because they are mostly small and not of medical interest. However, they present the morphological and physiological characteristics of the phylum in certain respects better than do the parasitic species, which are modified more or less for their mode of life.

TURBATRIX (ANGUILLULA) ACETI—THE VINEGAR EEL

The vinegar eel, *Turbatrix aceti,* is a favorable free-living nematode for study because it is easily procured at any time of the year in cider vinegar, can be examined in the living condition, and can be fixed and stained easily. It is visible to the naked eye when held before a bright light and exhibits characteristic nematode movements. Heating for a minute kills and straightens vinegar eels, and staining with Delafield's haematoxylin in 70 percent alcohol reveals the internal organs.

The worms are cylindrical, the anterior end being slightly narrowed and the posterior end tapering to a fine point. The female is about 2 mm in length and 0.05 mm in breadth and the male about 1.4 mm in length and 0.03 mm in breadth. The *cuticle* is transparent and transversely striated. The *mouth* is a pore at the anterior end; it opens into a cone-shaped *buccal capsule* about 0.01 mm long. The posterior portion of the buccal capsule is provided with one dorsal and two ventrolateral *teeth.* The *pharynx,* in which the buccal capsule is partly embedded, consists of an anterior thick portion 0.1 mm in length, followed by a narrow neck 0.05 mm in length which leads to a muscular bulb 0.02 mm in diameter in which is a valvular apparatus. The *excretory pore* is difficult to find; it is on a level with the anterior end of the bulb. Just in front of the bulb, the *nerve ring* crosses the pharynx. The *intestine* is a straight tube, about 0.02 mm in diameter in the female and smaller in the male; it extends from the pharynx to the *rectum,* which is short, narrow, and straight and opens into the *cloaca.*

The *male reproductive organs* consist of a testicular tube, a seminal vesicle, a genital pore, a pair of spicules, and an accessory piece. The *testicular tube* forms a loop arising about 0.3 mm from the anogenital pore, extending forward about 0.1 mm and then bending back upon itself to the *genital pore* which opens into the cloaca. It contains reproductive cells in two or three rows near its origin but these form a single row from near the bend to the posterior end. A portion of the tube near the posterior end is modified into a thin-walled *seminal vesicle* in which spermatozoa may be seen. This joins the rectum to form a short cloaca. Near the opening of the cloaca to the outside is a pair of *curved spicules* about 0.37 mm long and an accessory piece, which is a curved plate shaped somewhat like the keel of a boat. The mucles that extend the spicules are conspicuous. Five pairs of *papillae* are present near the cloacal opening; one postanal, dorsal pair; two preanal, central; one adanal, ventral; and one postanal, ventral.

The *female reproductive organs* comprise an ovarian tubule, a seminal receptacle, a uterus, a vagina, and vulva. The *ovarian tubule* arises posterior to the middle of the body, runs forward dorsally about 0.5 mm, where it joins the *uterus,* which runs backward ventrally to the vagina, which opens to the outside through the slitlike *vulva* near the middle of the body. An oval sac extending posteriorly from the vulva is probably a *seminal receptacle.* The ovarian tubule contains a single row of germ cells; these are fertilized near the bend where they become oval in shape and soon begin to segment. A varied number of *eggs* may be present in the uterus but usually about two ovoid and three vermiform embryos may be observed. The thin egg membrane ruptures in the uterus and the young are born in an active condition, that is, the vinegar eel is viviparous. The fertilized eggs hatch in about 8 days, and the larvae, which are 0.2 mm long when hatched, become mature in about 4 weeks. Males and females are equal in number; they may live for 10 months or more and one female may produce as many as 45 larvae. The male may be recognized by the blunter posterior and (as in most aschelminths) smaller size than in the female.

ASCARIS LUMBRICOIDES

External Features

Ascaris lumbricoides is the common roundworm parasitic in the intestine of humans. Roundworms morphologically indistinguishable from this species occur in pigs, but these have been shown to be physiologically different and not infective to humans. The sexes are separate. The female, being the larger, measures from 7 to 25 cm in length and about 6 mm in diameter. The *body,* when alive, may be milk-white or somewhat reddish-yellow and has a characteristic sheen; it has a dorsal and a ventral narrow white stripe running its entire length and a broader *lateral line* is present on either side. The *cuticle* is smooth and marked with fine striations. The *mouth* opening is in the anterior end and is surrounded by one dorsal and two ventral *lips;* these are finely toothed and bear *papillae,* the dorsal lip two and each ventral lip one. Near the posterior end is the *anal opening,* from which, in the male, extends two spicules or *penial setae* of use during copulation. Many ventral preanal and postanal *papillae* are also present in the male. The male can be distinguished from the female by the presence of a bend in the posterior part of the body. In the female, the genital opening, the *vulva,* is located about one third the length of the body from the anterior end.

Internal Anatomy

If an animal is cut open along the dorsal line (Fig. 6-10), it will be found to contain a straight *alimentary canal,* and certain other organs, lying in a central cavity, the pseudocoel. The alimentary canal is very simple, since the food is taken from material already digested by the host whose intestine the worm inhabits. It opens at the posterior end through the *anus,* which is not present in members of the phyla already discussed. A small *buccal vestibule* opens into the muscular *pharynx,* from 10 to 15 mm long, which draws the fluids into the long nonmuscular intestine through the walls of which the nutriment is absorbed. The walls are lined with microvilli (in the pig ascaris, Kessel et al., 1961), which have an ultrastructural resemblance to cilia. They form a tightly packed "brush-border" (Fig. 6-12) with no known motor function. They are now thought to be absorptive or secretory structures. Just before the anal opening is reached, the intestine gradually becomes smaller; this portion is known as the *rectum.* The rectum opens through the *anus* in the female and into the *cloaca* in the male.

Figure 6-10 Nematoda. *Ascaris lumbricoides.* (A) Ventral surface of anterior end of male. (B) Lateral view of posterior end of male. (C) Dissection of female to show internal organs. (*A and B, original; C, from Woodruff.*)

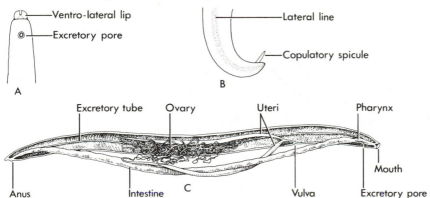

The *excretory system* consists of two longitudinal canals one in each lateral line; these open to the outside by a single pore situated near the anterior end in the ventral body wall.

A ring of *nervous tissue* surrounds the pharynx and gives off two large nerve cords, one dorsal, the other ventral, and a number of other smaller strands and connections.

The *male reproductive organs* are a single coiled threadlike *testis,* from which a *vas deferens* leads to a wider tube, the *seminal vesicle;* this is followed by the short muscular *ejaculatory duct* which opens into the cloaca. In the *female* lies a Y-shaped reproductive system. Each branch of the Y consists of a coiled threadlike *ovary* which is continuous via a short *oviduct* with a larger canal, the *uterus.* The uteri of the two branches unite into a short muscular tube, the *vagina,* which opens to the outside through the genital aperture or *vulva. Fertilization* takes place in the uterus. The egg is then surrounded by a shell of chitin, and passes out through the genital pore. The genital tubules of a female worm may contain as many as 27 million eggs at one time and each mature female lays about 200,000 eggs per day. Because the fertilized eggs in the uterus undergo the early cleavage divisions and have only a few large chromosomes, sections of *Ascaris* uterus are frequently used for the beginning study of mitosis.

Figure 6-11 *Ascaris.* (A) Cross section through region of esophagus. (B) Cross section through midregion of a female. (*A, after Borradaile and Potts, modified; B, original.*)

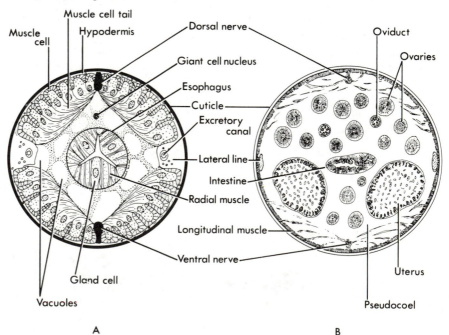

Muscle cell tail
Hypodermis
Muscle cell
Dorsal nerve
Oviduct
Ovaries
Giant cell nucleus
Esophagus
Cuticle
Excretory canal
Lateral line
Intestine
Radial muscle
Longitudinal muscle
Ventral nerve
Gland cell
Vacuoles
Uterus
Pseudocoel

A

B

The relations of the various organs to one another, as well as the structure of the body wall, and the character of the coelom, are shown in Figure 6-11, which is a transverse section of a female specimen of *Ascaris lumbroicoides*. The body of the worm should be considered as consisting of two tubes, one, the intestine, lying within the other, the body wall; between them is a cavity, the pseudocoel, in which lie the reproductive organs.

The *body wall* is composed of several layers, an outer chitinous cuticle, a thin layer of hypodermis just beneath it, and a thick stratum of longitudinal muscle fibers, mesodermal in origin, lining the pseudocoelom. Thickenings of the hypodermis form the dorsal, ventral, and lateral lines. In each of the last-named lines lies one of the longitudinal excretory tubes. The nerve cords are also embedded in the body wall.

The *intestine* consists of a single layer of columnar cells bearing the microvilli within (Fig. 6-12).

The pseudocoel of *Ascaris* differs from the coelom of the higher animals in several respects. Typically the coelom is a cavity in the mesoderm lined by an epithelium; the excretory organs open into it, and the reproductive cells originate from its walls. In *Ascaris* the pseudocoel is lined only by the mesoderm of the body wall, there being no mesoderm surrounding the intestine. Furthermore, the excretory organs open to the exterior through the excretory pore, and the reproductive cells are not derived from the coelomic epithelium. The body cavity of *Ascaris,* therefore, differs structurally and functionally from that of a true coelom, but nevertheless, is similar in many respects.

Life Cycle

Maturation, fertilization, and the phenomena concerned with the early differentiation of germ cells and somatic cells in *Ascaris* eggs have for many years served to illustrate these processes in animals in general. The eggs of *Ascaris* segment after they leave the body. They are very

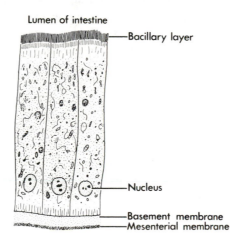

Lumen of intestine

Bacillary layer

Nucleus

Basement membrane
Mesenterial membrane

Figure 6-12 *Ascaris* intestine. Three cells of intestinal wall showing bacillary layer, or "brush border" of cilialike microvilli, lining the intestine. (*Diagrammatic.*)

resistant and may remain alive in the soil for months. Embryos are formed under favorable conditions in about 2 weeks. Infection with *Ascaris* results from ingesting embryonated eggs. The eggs are usually carried to the mouth either with food or water or by accidental transfer of soil containing such ova. They do not regularly hatch in the stomach but pass to the small intestine where they begin to hatch within a few hours after ingestion. Formerly it was believed that the larvae upon hatching settled down in the small intestine and there developed directly into the adult stage. Investigations have shown, however, that the larvae of this parasite leave the intestine immediately after hatching and then follow a definite path of migration through the tissues of the host, afterward returning to the intestine to grow into mature worms.

The newly hatched larvae burrow into the wall of the intestine and enter the lymphatic vessels or the venules. If the larvae enter the lymphatics they are carried to the mesenteric lymph nodes and from there may reach the circulation either by entering the blood capillaries and passing into the portal circulation, or pass into the thoracic duct and from there to the right side of the heart. If the portal circulation is entered the larvae are carried to the liver where they pass from the interlobular veins to the intralobular veins. From the liver they are carried to the right side of the heart and then to the lungs. Within the lungs they break into the alveoli where some further development and growth takes place, after which they pass on to the intestine by way of the trachea, esophagus, and stomach. This journey through the host's tissues requires about 10 days. They become mature worms in the intestine in about 2½ months.

Relations of Ascaris to Humans

Ascarids are pathogenic to humans. When large numbers of larvae pass through the lungs inflammation is set up and generalized pneumonia may result. The adults may be present in the intestine in such large numbers as to produce intestinal obstruction and nervous symptoms may appear as a result of the secretion of toxic substances by the worms. Fortunately several drugs are available which easily remove the worms; the best of these are piperazine citrate and hexylresorcinol. Ascariasis is essentially a children's disease in the United States, but a high percentage of adults in many other countries are infected. Studies in this country have shown that the infestations were particularly severe in families where the children were allowed to pollute the soil near the houses. Under these conditions the soil contains large numbers of embryonated eggs which find their way into the children's mouths on dirty hands. Infection can be prevented easily by enforcing sanitary practices.

The many species of nematodes are thought to number in the hundreds of thousands. The differences between species are in minor features that do not warrant separation at the order level. The tremendous numbers of very similar species have made it difficult, even for specialists on the Nematoda, to arrive at a satisfactory classification. Seventeen superfamilies are noted below. Only the first eight superfamilies contain free-living species.

Free-Living Nematodes

Roundworms that are not parasitic are more abundant and more widely distributed on the earth's surface than most zoologists realize. They live in almost every conceivable type of environment, such as in wet sand or mud, aquatic vegetation, in standing or running water, in the soil, the sea, tap water, fruit juices, and in moist places almost everywhere.

> If all the matter in the universe except the nematodes were swept away, our world would still be dimly recognizable, and if, as disembodied spirits, we could then investigate it, we should find its mountains, hills, vales, rivers, lakes, and oceans represented by a film of nematodes. The location of towns would be decipherable, since for every massing of human beings there would be a corresponding massing of certain nematodes. Trees would still stand in ghostly rows representing our streets and highways. The location of the various plants and animals would still be decipherable, and, had we sufficient knowledge, in many cases even their species could be determined by an examination of their erstwhile nematode parasites (Cobb).

Free-living nematodes are mostly small, a large specimen being only 1 cm in length. The eggs may withstand desiccation, a fact which no doubt is responsible in part for their wide distribution, since they may be carried about by the wind as well as by currents of water, birds and other animals, and by humans. They feed on other nematodes and on microorganisms. Most species are transparent but some have pigment granules in the intestinal cells which give them a yellowish or brownish appearance. Colored eye spots may also be present.

The body is covered by a transparent noncellular coat, the cuticula, beneath which is a cellular layer, the subcuticula or hypoderm, which secretes the cuticula. The cuticula is shed about four times during the growth of the larva and the lining of the mouth, pharynx, and rectum are shed at the same time. Within the cells of the hypoderm are longitudinal contractile fibers. The hypoderm is in many nematodes thickened along the center of the lateral and dorsal and ventral surfaces thus dividing the body wall into four quadrants. The thickenings on the sides are called lateral lines. The digestive system is simple. The

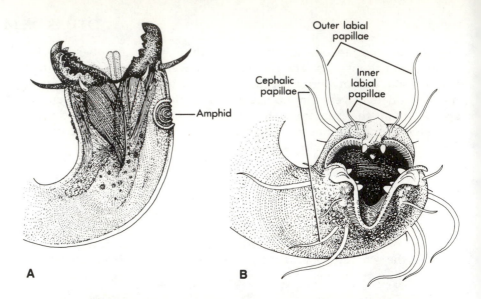

A

B

Cephalic papillae

Outer labial papillae

Inner labial papillae

Amphid

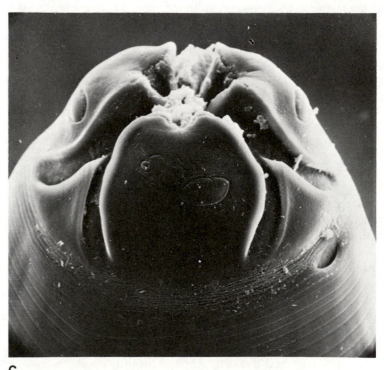

C

Figure 6-13 Nematoda. (A and B) The heads of two free-living species that show the complexity of structure of certain marine nemas. (C) Scanning electron micrograph, showing excretory pore at the base of the ventral interlabium of a fourth stage larva of *Sulcascaris* sp. parasitic in the surf clam. [*A and B, after Cobb; C, from J. R. Lichtenfels, J. W. Bier, and P. A. Madden. 1978. Larval anisakid (Sulcascaris) nematodes from Atlantic molluscs with marine turtles as definitive hosts. Trans. Am. Micros. Soc., 97:199–207. Courtesy of Dr. J. Ralph Lichtenfels.*]

mouth is at the anterior end, usually surrounded by lips or papillae (Fig. 6-13). Teeth may be present for rasping the tissue of plants or animals on which the nematodes feed. The buccal cavity is tubular, funnel-shaped, or expanded into a cup-shaped capsule adapted for sucking. The pharynx is either a thin-walled tube, or thick and muscular, and usually possesses a bulb or valvular apparatus where it joins the intestine. The intestine is a straight tube of columnar epithelial cells. It decreases in size at the posterior end in the female, becoming the rectum. In the male the intestine and genital duct open into a cloaca. The anal opening is on the ventral surface near the posterior end.

The excretory organ is a unicellular gland, the renette, located in the body cavity near the union of pharynx and intestine. A nerve ring surrounds the pharynx. Certain species possess a pair of pigmented eye spots, one on either side of the pharynx. Cephalic, caudal, and terminal setae, probably organs of touch, taste, or smell, occur on free-living but not on the parasitic species. A pair of laterally situated cephalic organs known as *amphids* (Fig. 6-13), probably sensory in nature, also occur on free-living nemas. The free-living stages of certain parasitic nematodes also possess sensory organs. No circulatory system is present, circulation being accomplished by the fluid in the body cavity in which currents are set up by the movements of the worm. The sex organs are, in general, similar to those of *Turbatrix* already described.

1. Superfamily Enoploidea. Marine nematodes with papillae and bristles on the head; a pair of sensory slits on the head in addition to the amphids; pharynx typically a simple straight tube. As with most free-living nematodes, a group of caudal glands are present. These are adhesive in function and are probably homologous to the pedal glands of rotifers and gastrotrichs.

Metoncholaimus pristiurus (Fig. 6-14), found in marine mud of Atlantic coastal regions, is a good example of a free-living marine nematode. It has six lips surrounding the mouth. Each lip bears a papilla. The amphids are lateral and shield-shaped. The male has a ventral row of 10 papillae on the tail as well as numerous bristles and a small preanal papilla. Two slender spicules can be protracted to aid in copulation or retracted. The testes are paired. The female has a single ovary from which eggs pass through the oviduct to the uterus where fertilization occurs. Up to 40 eggs are stored in a single row in the uterus. They are laid before segmentation occurs. A *demanian system*, peculiar to this nematode and others in the same family, occurs in the female as a system of tubes in contact with uterus and intestine and opening through a pore near the anus. It is thought to be an accessory reproductive structure. Also characteristic of the family (Oncholaimidae) is a large mouth cavity containing one to three teeth (Fig. 6-14, B).

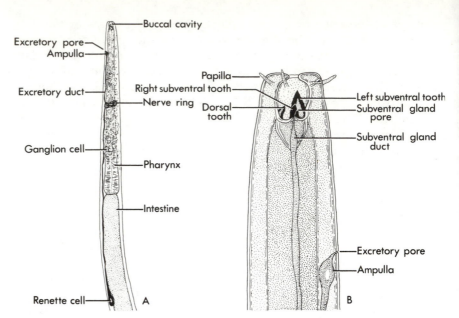

Figure 6-14 Nematoda. *Metoncholaimus pristiurus*. (A) Anterior end. (B) Lateral view of head. (*After Chitwood.*)

2. Superfamily Dorylaimoidea. Buccal cavity containing a hollow protrusible spear (Fig. 6-15, A) mounted on the pharynx and used when sucking juices from plant and animal food organisms; common in soil and fresh water. *Xiphinema index* is a dorylaimid, which has modified cilia in sensory structures (Roggen *et al.*, 1966). The definite presence of ciliary structures in nematodes removes one of the most formidable objections to the use of the Phylum Aschelminthes as a natural, phylogenetic group.

3. Superfamily Mermithoidea. Long, slender juveniles are parasites of insects or other terrestrial and freshwater invertebrates; no buccal cavity; pharynx long and passing through *stichocytes* (cells of the stichosome, a pharyngeal gland); adults do not feed, but live on food stores acquired during the parasitic juvenile stage. The larva has a dorylaimid type of spear on the pharynx. *Agameris decaudata*, a parasite of grasshoppers, leaves its posterior larval half behind when it actively penetrates its young grasshopper host. Other species may infect after the egg stage is ingested.

4. Superfamily Chromadoroidea. Spiral amphids; ornamented cuticle; primarily marine; pharynx with a posterior bulb; usually with teeth in the buccal cavity. The families Draconematidae and Epsilonematidae are marine and have "stilt bristles" near the posterior end of the body.

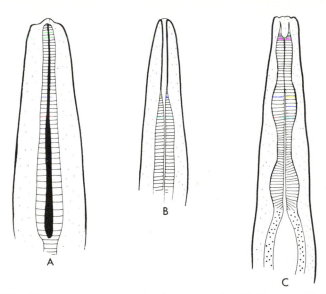

Figure 6-15 Variation in nematode pharynxes. (A) *Xiphinema*. (B) *Cylindrolaimus*. (C) Rhabditid type.

5. Superfamily Araeolaimoidea. Four large cephalic bristles posterior to the labial papillae; loop-shaped or spiral amphids; cuticle without ornamentation.

6. Superfamily Monhysteroidea. Circular amphids; mostly marine. *Cylindrolaimus* is a genus of soil and freshwater nematodes with the buccal capsule in the shape of a long tube (Fig. 6-15, B).

7. Superfamily Demoscolecoidea. All marine; thick-bodied; amphids semicircular; buccal cavity reduced; head distinct from body; four cephalic papillae.

8. Superfamily Rhabditoidea. Pharynx with two expanded regions, or bulbs; amphids reduced; no caudal glands; phasmids present (the only group containing free-living species with phasmids); a large group of mostly terrestrial species. *Turbatrix aceti* has been previously discussed. *Rhabditis* (Fig. 6-16) is a large genus containing both free-living and parasitic forms.

The family Tylenchidae contains species with a buccal stylet. *Heterodera* commonly lives in galls on the roots of plants. *Criconema* (*Iota*) has marked annulations of the cuticle bearing posteriorly projecting scales in eight rows.

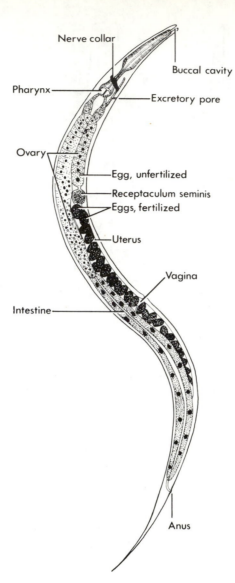

Nerve collar

Buccal cavity

Pharynx

Excretory pore

Ovary

Egg, unfertilized

Receptaculum seminis

Eggs, fertilized

Uterus

Vagina

Intestine

Anus

Figure 6-16 Nematoda. *Rhabditis.* Mature female. (*After Maupas, modified.*)

Parasitic Nematoda

Only two of the preceding superfamilies contained parasites of animals, none of which were vertebrates. The remaining superfamilies all contain parasites of vertebrates. No free-living species occur in them although some may have a developmental stage outside the host. Life cycles require an intermediate host in some species. Phasmids occur in all but the superfamilies Trichuroidea and Dioctophymoidea. *Phasmids* are paired cuticular pits near the caudal end. They are either glandular and excretory or sensory in function. Both types are microscopic. Some of the superfamilies bear a close relationship to free-living superfamilies. This is especially true of the Rhabdiasoidea and Rhabditoidea. In contrast to the free-living nematodes, which are generally microscopic, numerous parasitic species are quite large, some reaching a meter in length.

9. Superfamily Rhabdiasoidea. Parasites of vertebrates; structure much as in the Rhabditoidea; pharyngeal bulb reduced.

Rhabdias bufonis is a common lung nematode in frogs and other amphibians.

Strongyloides stercoralis is a common human parasite especially in the moist tropics. Its life cycle includes a female stage that lives in the human intestine and lays eggs that give rise to male and female larvae. These rhabditiform larvae pass out in the feces and form a free-living generation. They may become infective larvae or develop into sexually mature adults. The offspring of these are filariform larvae, which are infective to people. They may be swallowed but usually penetrate the skin of bare feet. They are carried in the bloodstream to the lungs, break out into the trachea, migrate through the epiglottis into the digestive tract, and become localized in the ileum.

10. Superfamily Oxyuroidea. Parasites of vertebrates, or, in a few cases, of invertebrates; small slender worms; one or two bulbs on pharynx.

This superfamily contains the pinworm of humans, *Enterobius vermicularis* (Fig. 6-17, G), and a similar species in horses. The human pinworm (female) measures from 9 to 12 mm in length, is worldwide in distribution and lives in the adult stage in the upper part of the large intestine. Children are often infected, sometimes over 5,000 worms being present in a single child. The female worms creep out of the rectum, usually at night, causing intense itching. They lay their eggs at once. The infected person scratches the anal region thereby contaminating the hands, especially the fingernails, and the eggs are then conveyed to the mouth and swallowed, thus increasing the infection. Human infections were present 10,000 years ago in North America (Fry and Moore, 1969), as evidenced by eggs in coprolites. Piperazine salts are effective drugs for treatment. *Oxyuris equi* resembles the pinworm of humans; it causes anal pruritus in horses, asses, and mules.

Heterakis contains species that live in birds and mammals. Of particular importance is *Heterakis gallinae* (Fig. 6-17, A), which lives in the cecum of barnyard and wild fowls. Eggs, which are passed in the feces of infected birds, are swallowed and the young that hatch from them in the small intestine move on into the ceca. They do not seriously injure the fowls but are of great economic importance since they carry with them a protozoan parasite, *Histomonas meleagridis,* which is the causative agent of the disease of turkeys known as blackhead.

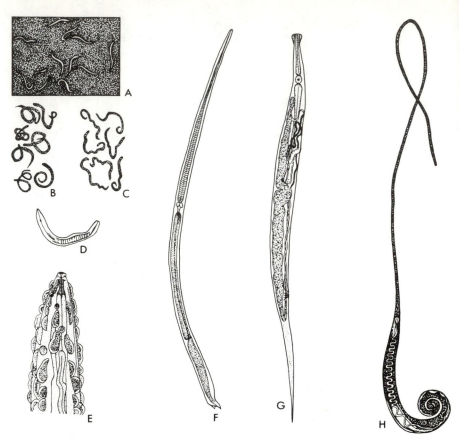

Figure 6-17   Nematoda. (A) *Heterakis gallinae*, from the cecum of fowls; about one-half natural size. (B) *Wuchereria bancrofti*, microfilariae. (C) *Loa loa*, microfilariae. (D) *Trichinella spiralis*, the trichina worm, young larva. (E) *Gongylonema pulchrum*, anterior end. (F) *T. spiralis*, the trichina worm, male. (G) *Enterobius vermicularis*, female; the pinworm of humans (H) *Trichuris ovis*, whipworm, from sheep and cattle. (*After various authors*.)

11. Superfamily Ascaroidea. This group of large nematodes are parasites in the digestive tract of vertebrates. *Ascaris lumbricoides* (Fig. 6-10) from humans has already been described. *Ascaridia galli* (Figs. 6-18 and 6-19) is the common round worm of chickens, living in the small intestine.

Toxocara canis is the common ascarid of dogs being especially prevalent in puppies. Dogs become infected by swallowing the eggs. The larvae migrate through the body as do those of *Ascaris lumbricoides* in people. Dogs acquire an immunity as a result of the infection, and after 3 or 4 months the worms are cast out and susceptibility to further infection is lost.

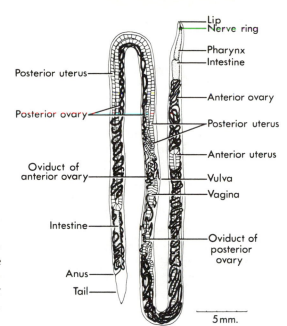

Lip
Nerve ring
Pharynx
Intestine

Posterior uterus

Anterior ovary

Posterior ovary

Posterior uterus

Anterior uterus

Oviduct of
anterior ovary

Vulva
Vagina

Intestine

Oviduct of
posterior
ovary

Anus

Tail

5 mm.

Figure 6-18 Nematoda. The chicken nematode, *Ascaridia galli,* female. Digestive and reproductive organs. The uterus is distinguished from the other organs by crossed lines. (*After Ackert, modified.*)

Figure 6-19 Nematoda. Development of the egg of the chicken nematode, *Ascaridia galli.* (A) Fertilized egg. (B) Two-cell stage. (C) Four-cell stage. (D) Early morula stage. (E, F) Late morula stages. (G) "Tadpole" stage. (H, I) Vermiform stages. (J) Coiled embryo = "embryonated egg." (K, L) Young worm hatching. (*After Ackert.*)

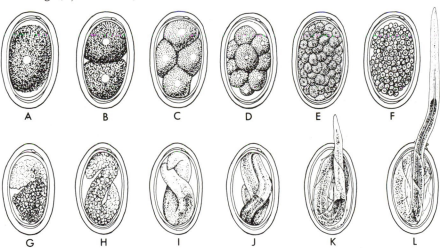

A   B   C   D   E   F

G   H   I   J   K   L

12. Superfamily Strongyloidea. Posterior of the male expanded into a bursa supported by muscular rays. All are parasites of vertebrates, commonly in the intestine. Many interesting and economically important species belong to this superfamily. Five families will be men-

Class V. Nematoda 337

tioned here. The family Strongylidae contains many parasites of domesticated animals. *Strongylus vulgaris* is the most important round worm of horses and other Equidae and is worldwide in distribution but especially prevalent in warm countries. It lives in the cecum or colon attached by the mouth to the mucosa, from which it sucks blood. Loss of blood results in anemia. The eggs of *Strongylus* are deposited in the feces where they give rise to infective larvae. These, when ingested by a horse, migrate to the posterior mesenteric artery where an aneurysm is produced: they then move on to the cecum where they become encysted in the submucosa; and finally they break out into the lumen, attach themselves to the mucosa, and develop into adults.

The nodular worm of sheep and goats, *Esophagostomum columbianum*, also belongs to this family. The young worms encyst in the wall of the intestine forming nodules. Later they break out into the lumen and become adults. Diarrhea and emaciation result from infection and the intestines are not usable for sausage casings.

Syngamus trachea in the family Syngamidae is a slender red worm, which occurs in the trachea of fowls and wild birds and causes gapes. Infection results from the ingestion of infective larvae that hatch from eggs passed in the feces or coughed up by infected birds. The larvae penetrate the wall of the esophagus, migrate to the lungs, where they become mature, and later move into the trachea to the wall of which they attach themselves by the buccal capsule. Fowls suffer from catarrh and from abscesses where the worms are attached.

The hookworms belong to the family Ancylostomidae. The hookworm of the Old World is *Ancylostoma duodenale* (Fig. 6-20). Other species of this genus occur in the intestine of various species of carnivores including the dog, cat, tiger, lion, and wolf. The hookworm of the New World is *Necator americanus* which lives in humans and pigs. Other species of *Necator* have been reported in chimpanzees. The larvae of the hookworm develop in moist earth and usually find their way into the bodies of humans by boring through the skin of the foot. In localities where the hookworm is prevalent, many of the people go barefoot. The larval hookworms enter the veins and pass to the heart; from the heart they reach the lungs, where they make their way through the air passages into the windpipe, and thence into the intestine. The adults attach themselves to the walls of the intestine and feed upon the blood of their host (Fig. 6-21, C). In the case of the dog hookworm and probably also of the human hookworm, blood is continuously being sucked into the body of the worm and expelled from the anus in the form of droplets consisting mainly of red corpuscles. Calculations indicate the possibility that a single worm may withdraw blood from the host at the rate of 0.8 cubic cm in 24 hours (Wells, 1931). When the intestinal wall is punctured, a small amount of poison is poured into the wound by the worm. This poison prevents the

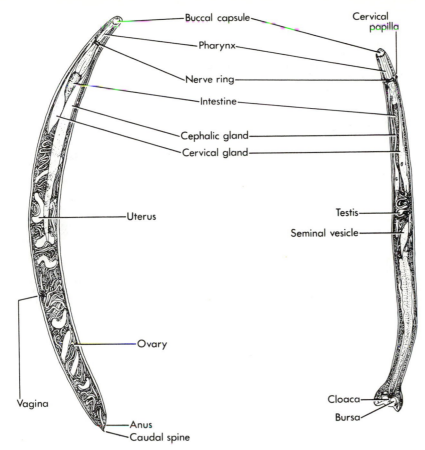

Figure 6-20 *Ancylostoma duodenale,* the Old World hookworm, female (left) and male (right). (*After Loos.*)

blood from coagulating, and therefore results in a considerable loss of blood, even after the worm has left the wound. The victims of the hookworm are anemic, and also subject to tuberculosis because of the injury to the lungs. It is estimated that 2,000,000 persons in 1930 were afflicted by this parasite in the United States. Hookworm disease can be cured by hexylresorcinol or tetrachlorethyl. The most important preventive measure is the disposing of the human feces in rural districts, mines, brickyards, etc., in such a manner as to avoid pollution of the soil, thus giving the eggs of the parasites contained in the feces of infected humans no opportunity to hatch and develop to the infectious larval stage.

Ancylostoma braziliensis is a parasite common in cats and dogs in various countries including the southern United States. Under favorable conditions the infective larvae of this species may penetrate the skin of the person through which it migrates parallel to the surface

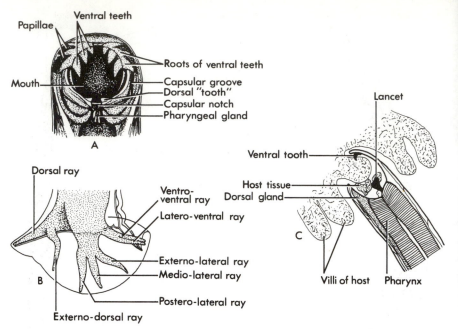

Figure 6-21 Hookworm. (A) Buccal capsule of *Ancylostoma caninum*, the dog hookworm. (B) Bursa of *A. duodenale*, the human hookworm. (*A, from Chandler; B and C, after Loos, modified.*)

forming a tortuous path and causing the condition known as creeping eruption. Ethyl acetate on cotton and carbon dioxide snow have been found to be effective methods of treatment. Another species of hookworm living in cats and dogs is *A. caninum*. It is cosmopolitan in distribution and very pathogenic to puppies.

The species in the family Metastrongylidae are much less important. *Dictyocaulus filaria* is the thread lung worm of sheep, goats, cattle, etc. Its eggs hatch in the lungs and the larvae migrate through the trachea into the alimentary canal from which they escape from the mouth in the saliva or from the anus in the feces. The larvae, which become infective in about 10 days, crawl to the tip of blades of grass with which they may be ingested by herbivorous animals. They establish themselves in the lungs causing a catarrhal condition.

To the family Trichostrongylidae belong several interesting parasites of lower animals. *Haemonchus contortus* is one of the commonest parasites of domestic sheep throughout the world, and also occurs in other ruminants. It lives principally attached to the wall of the fourth stomach. The eggs are deposited in the feces of the host and the infective larvae crawl to the tip of blades of grass where they may be ingested by grazing animals. The infection results in anemia. *Heligmosomum muris* is a species of trichostrongyl parasitic in wild rats in the United States that may be obtained easily for study.

13. Superfamily Spiruroidea. Parasites of vertebrates' digestive tracts, respiratory cavities, and eyes; one or more intermediate arthropod hosts; no pharyngeal bulb; usually with two lateral lips by mouth. *Thelazia* is a parasite in the tear ducts and eyelids of various vertebrates and occasionally humans. *Oxyspirura mansoni,* the "eyeworm" of chickens, uses a cockroach as an intermediate host. Larvae escape from an ingested roach in the chickens crop and migrate to the eye region. *Habronema megastoma* lives in the stomach of the horse and is prevalent in the southern United States. It produces nodules or tumors in the stomach and is transmitted by the house fly. *Adruenna strongylina* is a worm that forms tumors in the stomach wall of pigs in this country. *Gongylonema scutatum* (Fig. 6-17, E) inhabits the mucosa of the esophagus of sheep, cattle, and goats. The intermediate hosts are dung beetles.

14. Superfamily Dracunculoidea. Parasites in the connective tissue or coelom of vertebrates; vulva in posterior half of female, but usually atrophied.

Dracunculus medinensis, the guinea worm, belongs to this superfamily. It is a common human parasite in tropical Africa, Arabia, and India, that has been known for centuries and is probably the "fiery serpent" mentioned by Moses (Num. 21). The adult female, which may reach a length of over 1 m, is located usually in the subcutaneous tissue of the arm, leg, and shoulders. The young larvae are discharged from the worm and escape through an opening in the human skin; if they reach water and chance to encounter the freshwater crustacean *Cyclops* they burrow into it and metamorphose in the body cavity. People become infected by swallowing the *Cyclops* in drinking water. The method of extracting the worm that has been practiced for hundreds of years is to roll it up on a stick gradually, a few turns each day, until the entire worm has been drawn out.

15. Superfamily Filarioidea. Parasites of the body fluids and tissues of vertebrates; transmitted by arthropods; vulva of female near the anterior end; slender worms. *Wuchereria bancrofti* (Fig. 6-17, B) is a species of the Filariidae that lives in humans and is widely spread in tropical countries. The larvae of this species are about ¼ mm long. During the daytime they live in the lungs and larger arteries, but at night they migrate to the blood vessels in the skin. Mosquitoes, which are active at night, suck up these larvae with the blood of the infected person. The larvae develop in the mosquito's body, becoming about 1 mm long, make their way into the mouth parts of the insect, and enter the blood of the mosquito's next victim. From the blood they enter the lymphatics and may cause serious disturbances, probably by obstructing the lymph passages. This results in a disease called elephantiasis. The limbs or other regions of the body swell up to an enormous size, but there is very little pain. Treatment is difficult. The disease may be fatal. Infection with this parasite is common in people in the West Indies, South America, and West and Central Africa, and formerly common in Charleston, S.C., the worms probably having been introduced by the slave trade.

Another interesting species of the family Filariidae is the eye worm, *Loa loa* (Fig. 6-17, C), of West Africa. It migrates around the body through the subdermal connective tissue, sometimes across the eyeball. No severe pathological lesions are produced. The transmitting agents are mango flies of the genus *Chrysops.* Among other species in this family are *Onchocerca volvulus,* which occurs also in West Africa and produces subdermal nodular swellings; *Dirofilaria immitis,* which may be present in tangled masses in the right ventricle of the dog's heart; *Setaria equina,* a frequent parasite of the peritoneal cavity of horses and other Equidae; and species that are common in birds, especially crows and house sparrows, in whose blood large numbers of the microfilariae may be present.

16. Superfamily Trichuroidea. Parasites of vertebrates; slender pharynx embedded in stichosome; anterior usually narrower than posterior; phasmids lacking. Two important human parasites belong to this order. Trichinosis, caused by one, is unusual for a parasitic disease in that it is more common in the United States than in any other region.

Trichinella spiralis (Figs. 6-22 and 6-17, D, F) causes the disease of human beings, pigs, and rats called trichinosis. The parasites enter the human body when inadequately cooked meat from an infected pig is eaten. The larvae soon become mature in the human intestine, and each mature female worm deposits probably about 10,000 young.

Figure 6-22 *Trichinella spiralis.* (A) Larvae among muscle fibers, not yet encysted. (B) A single larva encysted. (C) Piece of pork, natural size, containing many encysted worms. (*After Leuckart.*)

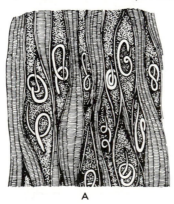

A   B   C

These young Fig. 6-17, D) are either placed directly into the lymphatics by the female worms or burrow through the intestinal wall; they encyst in muscular tissue in various parts of the body (Fig. 6-22). As many as 15,000 encysted parasites have been counted in a single gram of muscle. Pigs acquire the disease by eating offal or infected rats. In a few countries pork is inspected for this and other parasites by government agents.

The whipworm, *Trichuris trichiura,* lives in the cecum and appendix of the human host. Its body is drawn out anteriorly into a long slender whiplike process (Fig. 6-17, H). There is no intermediate host. The eggs escape in the feces and ripen outside of the body. Ripe eggs when swallowed hatch in the intestine and the larvae become located in the cecum. Whipworms injure the host very little. *Capillaria hepatica* occurs in the liver and *Trichosomoides crassicauda* in the urinary bladder of rats. These species may be found of value as examples of round worms for laboratory study.

17. Superfamily Dioctophymoidea. The few worms of this group lack phasmids, are parasites of the digestive tract, kidneys, and body cavity of birds and mammals, and the larvae develop in an intermediate host. *Dioctophyme renale* is the largest known nematode and may reach a length of over 1 meter. It is most frequently encountered in the kidney of the dog, which is gradually consumed by the worm. Renal colic and failure of the infected kidney to function are symptoms. Fish are transport hosts that get the infection from tubificid intermediate hosts.

Class VI. Nematomorpha

The Nematomorpha or Gordiacea resemble the Nematoda in general body features. The muscle cells especially indicate the close relationship. Major differences from the Nematoda are the lack of lateral lines and the presence of only a ventral epidermal chord in most. These opaque-bodied worms are usually referred to as horsehair worms (Fig. 6-23) and are erroneously supposed to arise from horsehairs. Sexes are separate; in both sexes the intestine and genital ducts open into a cloaca; spicules are absent in the male; the body cavity is lined with epithelium; the ova escape into the body cavity and thence into the oviducts; and the alimentary canal is atrophied in the sexually mature worms. The eggs are laid in the water in long strings where the adults live. The larvae that hatch from the eggs penetrate into a young mayfly or some other aquatic insect; they escape in some unknown way from this host and find their way into a second host, usually a beetle, cricket, or grasshopper; in the body cavity of the second host the larvae continue their development eventually passing out into the water where they become sexually mature. Because the adults

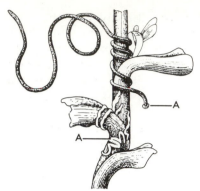

Figure 6-23 *Gordius,* twining around a water plant and laying eggs. *A,* clump and string of eggs. (*After von Linstow.*)

live only in water, those that survive probably emerge from terrestrial insects, which constitute their second intermediate hosts, that chance to become drowned in watering troughs and small pools frequently formed after a rain. *Gordius aquaticus* (Fig. 6-23) is a cosmopolitan species; *Paragordius varius* is common in North America; and *Nectonema agile* lives as a parasite in marine crustaceans at Woods Hole, Mass., and elsewhere.

Two orders are recognized: the Gordioidea with freshwater and terrestrial species; and the Nectonematoidea with the marine genus, *Nectonema.*

Aschelminthes in General

Cell or nuclear constancy is a phenomenon demonstrated in the Aschelminthes. Like organs of adults of the same species contain the same number of cells or nuclei in most instances investigated. The microscopic size of most organisms in the phylum made such determinations possible. This phenomenon is probably correlated with their lack of regenerative ability. Because regenerative abilities are common in other major phyla, it is one indication that considerable divergence has occurred from the ancestral stock from which more advanced phyla developed. The resemblance of a highly modified rotifer, *Trochosphaera* (Fig. 6-6, F), to the trochophore larvae of annelids and mollusks had led some zoologists to think rotifers may be an important ancestral group in the phylogeny of invertebrates. Hyman (1951) states the reasons for denying this assumption.

Axenic culture, or culture in the absence of all other species of organisms, has been achieved with a few nematodes and a rotifer. Thus, the Aschelminthes are one of the few phyla with considerable axenic studies already done. The great economic importance of the nematodes makes further axenic nutritional studies of importance in developing some types of control.

Early investigations were chiefly taxonomic and life-cycle studies. Recent growth of physiological and ultrastructural studies, especially of parasitic nematodes, has emphasized nutrition and basic metabolism. Recent progress is being made in finding evidence of regulation of development and behavior by neurosecretions and other regulation mediated by minute quantities of substances. Some examples include numerous observations by Gilbert, such as the effects of α-tocopherol on development; evidence by Greet of nematode sex pheromones (Bone and Shorey, 1977); and neural and hormonal control of nematode development (Samoiloff, 1973).

Ecology

Many interesting examples of interaction between organism and environment occur in the Aschelminthes. Morphological differences develop in some rotifers under changed environmental conditions. The production of sexual females in rotifers seems to be environmentally controlled.

The presence of nematodes will induce the development of nematode trapping devices on some species of fungi (Pramer, 1964). Attempts to capitalize on this reaction in controlling phytoparasitic nematodes has not yet met with much success.

Similar species from all classes of Aschelminthes except the Gnathostomula and Kinorhyncha can be found widely distributed in fresh waters of the world. The ability to withstand desiccation of eggs or other stages has probably been an important factor in the wide dispersal of freshwater species. Although no fossils are known for the Aschelminthes, the group is undoubtedly quite old and thus time is also a factor contributing to their wide dispersal.

The numerical importance of rotifers in freshwater plankton communities merits increased study of their role as primary consumers in lakes.

The consequences of interstitial habitats and small size have had a major role in selection in the Aschelminthes. The nematodes' loss of cilia and ability to tolerate life in organic-rich material may have preadapted them to life in a host's intestine.

Parasites within a host are known to sometimes time their activities to coincide with external events. Some microfilariae show a daily rhythm of abundance in the peripheral blood that corresponds with activity periods of their mosquito vector. Schad et al. (1973) have shown that hookworms can delay their development so that egg laying corresponds with seasonally favorable conditions.

Parasitic Nematodes

Between the free-living nematodes and those spending all their life cycle in hosts are many degrees and types of parasitism. Some are parasites only as juveniles, others as adults. The plant parasites may be external or internal parasites, are usually of small size, and are the cause of great agricultural loss. Nematodes of animals are usually larger, some reaching many cm in length, and are also the cause of several agricultural losses. Nematodes of humans may do little damage to the host when only a few parasites are present. Any tissue damage done may serve as the avenue of infection by other organisms. Because even asymptomatic infections are dangerous as a source of infection for others, treatment of all infected individuals is desirable. Diagnosis of infection with a particular intestinal nematode parasite of man is based on identification of the eggs or larvae in the feces.

The eggs of roundworms are distinctive in size and shape and of importance in the identification of species, especially of those that live in people (Fig. 6-24). They may be discharged either before or during segmentation, or with the embryo fully developed. In a few species the embryos hatch within the uterus of the female worm and are then brought forth viviparously. The embryonic development is simple and much alike in all species. The larvae, upon hatching, have the main characteristics of nematodes but are not sexually developed. Many parasitic species have certain adaptive larval characters which are subsequently lost. In the course of its development, the worm undergoes about four moults with the adult stage following after the fourth or last moult. Among the parasitic species some of these moults may take place within the egg before hatching, during its free existence, while within the tissue of an intermediate host, or within the tissues of the definitive host.

Of interest is that ascarid embryos can develop in the presence of substances lethal to most other organisms. The protection enabling such development to occur comes, not from the egg shell, but, from a thin lipid membrane.

The Germ Track in Nematodes

A classical study tracing the lineage of the germ cells, or reproductive cells was done with *Parascaris equorum* (*Ascaris megalocephala*). The first cleavage division of the egg results in two daughter cells, each containing two long chromosomes (Fig. 6-25, A). In the second divi-

Figure 6-24   Nematoda. Eggs of species that live in humans: (A) *Ascaris lumbricoides*. (B) *Trichuris trichiura*. (C) *Ancylostoma duodenale*. (D) *Enterobius vermicularis*. (E) *Trichostrongylus orientalis*. (*After Cort.*)

A   B   C   D   E

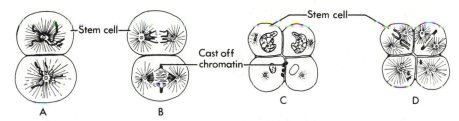

Figure 6-25 The germ track in *Ascaris*. Stages in early cleavage showing the chromatin-diminution process in all cells except the stem cell. (*From Boveri.*)

sion the chromosomes of one cell divide normally and each daughter cell receives one half of each (B). The chromosomes of the other cell behave differently; the thin middle portion of each breaks up into granules (A) which split, half going to each daughter cell, but the swollen ends (B) are cast off into the cytoplasm. In the four-cell state there are consequently two cells with the full amount of chromatin and two with a reduced amount. This inequality in the amount of chromatin results in different-sized nuclei (C); those with entire chromosomes are larger than those that have lost the swollen ends. In the third division one of the two cells with the two entire chromosomes loses the swollen ends of each; the other (D) retains its chromosomes intact. A similar reduction in the amount of chromatin takes place in the fourth and fifth divisions and then ceases. The single cell in the 32-cell stage, that contains the full amount of chromatin has a larger nucleus than the other 31 cells and gives rise to all of the germ cells, wheras the other cells are for the production of somatic cells only. It is likely that any animal having this type of chromatin dominution will have very limited or no regenerative power.

Phylogeny

The Aschelminthes have probably not given rise to other major groups. Dubious relatives might be the Chaetognatha, Priapulida, and Acanthocephala. The ancestral relationship of the Aschelminthes is most likely through ancestral Gnathostomula from ancestral Turbellaria.

Most aspects of phylogeny of the phylum have been discussed under the Gnathostomula. The gnathostomulid–nematode connection is most difficult due to the gross differences superficially. Ultramicroscopic studies show nematodes have ciliary derivatives in their sensory cervical papillae with remnants even in the parasitic forms. The positioning of these is very close to similar ciliary structures forming the gnathostomulid sensorium.

The minute papillae of the nematode posterior that sometimes seems modified into complex bursal structures probably had their forerunner in the generally adhesive surface of gnathostomulids. *Problognathia minima* (Sterrer and Ferris, 1975) shows the oral structures of

Figure 6-26 Scanning electron micrograph of an epizoic rotifer, showing cuticular rings that mimic external segmentation. The rings probably are adaptive for the type of mobility characteristic of rotifers when they are moving over substrates. (*S.E.M. courtesy of Dr. Harvey Blankespoor.*)

gnathostomulids in a close approach to the terminal nematode position.

Further evidence that the differences in developmental features of the Aschelminthes can be adaptive is provided by Gilbert and Wurdak (1978), who showed differences in egg coverings of otherwise closely related rotifers of the same genus. Thus, it is evident that the adaptive features, characteristic of a group, are not necessarily an indication of great phylogenetic distance from groups having a different adaptive aspect for that feature. (See also Fig. 6-26.)

History of the Aschelminthes

The nematodes were separated from other Vermes by Rudolphi in 1808. The Nematomorpha were distinguished from other nematodes in 1866 by Vejdovsky.

Rotifers were part of the Infusoria until separated from the Protozoa by Ehrenberg in 1838. Gastrotrichs were separated from rotifers in 1864. Edmundson and Pennak did much to advance knowledge of the ecology of rotifers.

The kinorhynchs were discovered in 1851 and first allied with nematodes in 1887. The gnathostomulids were reported in 1956 as a new group. Prior to that time Hyman (1951) assembled the diverse groups, along with the Priapulida, into the new phylum Aschelminthes. The phylum has since had a varied acceptance, and further disagreement is anticipated about the opinion that new information, especially of the gnathostomulids, makes the monophyletic nature of the phylum obvious.

The numbers of workers making significant contributions, especially to the Nematoda literature, is so great that only a very few have been cited.

REFERENCES TO LITERATURE ON THE ASCHELMINTHES

BENNETT, L. W. 1979. Experimental analysis of the trophic ecology of *Lepidodermella squammata* (Gastrotricha: Chaetonotida) in mixed culture. *Trans. Am. Micr. Soc.*, **98**:254–260.

BIRKY, C. W., and B. FIELD. 1966. Nuclear number in the rotifer *Asplanchna*: intraclonal variation and environmental control. *Science*, **151**:585–587.

————, and J. J. GILBERT. 1971. Parthenogenesis in rotifers: the control of sexual and asexual reproduction. *Am. Zoologist*, **11**:245–266.

BONE, L. W., and H. H. SHOREY. 1977. Disruption of sex pheromone communication in a nematode. *Science*, **197**:694–695.

BRUNSON, R. B. 1959. Gastrotricha. In W. T. EDMONDSON. *Fresh-Water Biology*. Wiley, New York. 1,248 pp.

————.1963. Aspects of the natural history and ecology of the Gastrotricha. In E. DOUGHERTY. *The Lower Metazoa.* Univ. of California Press, Berkeley, 478 pp.

CHITWOOD, B. G., and D. G. MURPHY. 1964. Observations on two marine monhysterids—their classification, cultivation, and behavior. *Trans. Am. Micr. Soc.*, **83**:311–329.

COBB, N. A. 1915. Nematodes and their relationships. *Yearbook, U.S. Dept. Agric.*, 1914:457–490.

CROLL, N. A., and B. E. MATTHEWS. 1977. *Biology of Nematodes.* Wiley, New York. 193 pp.

CROWE, J. H., and K. A MADIN. 1974. Anhydrobiosis in tardigrades and nematodes. *Trans. Am. Micr. Soc.*, **93**:513–524.

DONNER, J. 1966. *Rotifers.* Warne, London. 80 pp.

FRY, G. F., and J. G. MOORE. 1969. *Enterobius vermicularis:* 10,000-year old human infection. *Science*, **166**:1620.

GAGNE, G. D. 1977. *Dolichodasys elongatus* n. g., n. sp., a new macrodasyid gastrotrich from New England. *Trans. Am. Micr. Soc.*, **96**:19–27.

GILBERT, J. J. 1973. Induction and ecological significance of gigantism in the rotifer *Asplanchna sieboldi. Science*, **181**:63–66.

————. 1974. Dormancy in rotifers. *Trans. Am. Micro. Soc.*, **93**:490–513.

————, and E. S. WURDAK. 1978. Species-specific morphology of resting eggs in the rotifer *Asplanchna. Trans. Am. Micr. Soc.*, **97**:330–339.

GOODEY, T. 1951. *Soil and Freshwater Nematodes.* Wiley, New York. 390 pp.

HIGGINS, R. P. 1965. The homalorhagid Kinorhyncha of Northeastern U.S. coastal waters. *Trans. Am. Micr. Soc.,* **84**:65–72.

———. 1978. *Echinoderes gerardi* n. sp. and *E. riedli* (Kinorhyncha) from the Gulf of Tunis. *Trans. Am. Micr. Soc.,* **97**:171–180.

D'HONDT, J.-L. 1971. Gastrotricha. *Oceanogr. Mar. Biol. Ann. Rev.,* **9**:141–192.

HORN, T. D. 1978. The distribution of *Echinoderes couli* (Kinorhyncha) along an interstitial salinity gradient. *Trans. Am. Micr. Soc.,* **97**:586–589.

HULINGS, N. C. (Ed.). 1971. Proceedings of the First International Conference on Meiofauna. *Smithsonian Contrib. Zool.,* **76**:1–205.

HYMAN, L. H. 1951. *The Invertebrates.* Vol. III. *Acanthocephala, Aschelminthes, and Entoprocta.* McGraw-Hill, New York. 572 pp.

KESSEL, R. G., et al. 1961. Cytological studies on the intestinal epithelial cells of *Ascaris lumbricoides suum. Trans. Am. Micr. Soc.,* **80**:103–118.

LEE, D. L., and H. J. ATKINSON. 1977. *Physiology of Nematodes.* 2nd Ed. Columbia Univ. Press, New York. 215 pp.

LICHTENFELS, J. R., J. W. BIER, and P. A. MADDEN. 1978. Larval anasakid (*Sulcascaris*) nematodes from Atlantic molluscs with marine turtles as definitive hosts. *Trans. Am. Micr. Soc.,* **97**:199–207.

PENNAK, R. W. 1978. *Fresh-Water Invertebrates of the United States.* 2nd Ed. Wiley, New York. 803 pp.

PRAMER, D. 1964. Nematode-trapping fungi. *Science,* **144**:382–388.

PRAY, F. A. 1965. Studies on the early development of the rotifer *Monostyla cornuta* Muller. *Trans. Am. Micr. Soc.,* **84**:210–216.

RIEDL, R. J. 1969. Gnathostomulida from America. *Science,* **163**:445–452.

ROGGEN, D. R., D. J. RASKI, and N. O. JONES. 1966. Cilia in nematode sensory organs. *Science,* **152**:515–516.

SACKS, M. 1964. Life history of an aquatic gastrotrich. *Trans. Am. Micr. Soc.,* **83**:358–362.

SAMOILOFF, M. R. 1973. Nematode morphogenesis: localization of controlling regions by laser microbeam surgery. *Science,* **180**:976–977.

SCHAD, G. A., et al. 1973. Arrested development in human hookworm infections: an adaptation to a seasonally unfavorable external environment. *Science,* **180**:502–504.

SCHMIDT, G. D., and L. S. ROBERTS. 1977. *Foundations of Parasitology.* Mosby, St. Louis. 604 pp.

STERRER, W. 1972. Systematics and evolution within the Gnathostomulida. *Syst. Zool.,* **21**:151–173.

———, and R. A. FERRIS. 1975. *Problognathia minima* n. g., n. sp. representative of a new family of Gnathostomulida, Problognathiidae n. fam. from Bermuda. *Trans. Am. Micr. Soc.,* **94**:357–367.

THORNE, G. 1961. *Principles of Nematology.* McGraw-Hill, New York. 553 pp.

VIGLIERCHIO, D. R., and I. A. SIDDIQUI. 1974. Pigments of the ocelli of Anarctic and Pacific marine nematodes. *Trans. Am. Micr. Soc.,* **93**:338–343.

WELLS, H. S. 1931. Observations on the blood-sucking activities of the hookworm, *Ancylostoma caninum. J. Parasitol.,* **17**:167–182.

WILLIAMS, L. G. 1966. Dominant planktonic rotifers of major waterways of the United States. *Limnol. Oceanogr.,* **11**:83–91.

YAMAGUTI, S. 1962. *Systema Helminthum*. Vol. III. *The Nematodes of Vertebrates*. Wiley, New York. 1,274 pp.

YORKE, W., and P. MAPLESTONE. 1926. *The Nematode Parasites of Vertebrates*. 548 pp. (Reprinted, 1962, by Hafner, New York.)

7

Minor Phyla

The Mesozoa, Ctenophora, Acanthocephala, Rhynchocoela, Priapulida,* Polyzoa, Brachiopoda, Phoronidea, Sipunculida,* Chaetognatha, Pogonophora, Hemichordata, and the invertebrate Chordata discussed in this chapter are phyla that contain relatively few species or are not at present considered important in the phylogeny of major groups of invertebrates. They are of minor economic importance. Some could well be appended to other phyla, but all are distinct groups that differ sufficiently so that recent classifications by many zoologists have listed them as independent phyla. Other groups that could also be considered in this chapter are the classes of Trochelminthes; the Archiannelida,* Echiuroidea, and Myzostomaria, which are treated with the Annelida; and the Onychophora, Pentastomida, Pycnogonida,* and Tardigrada, which are treated as arthropodan groups. All of the preceding have been treated as independent phyla in some classifications, but the relationship to the major phyla with which they are treated here seems clear although they are distinctive groups.

The inclusion of the chordates, whose invertebrate members are of minor importance from all points of view except the phylogenetic, may put them in better perspective. The vertebrates, the dominant form of life for millions of years, are an example of the success of *one* adaptation—an articulated endoskeleton that allowed increased size

---

* The ending -ida is generally an indication of an order name. Two major exceptions allowed in this text are (1) the retention of -ida on phylum and class names of well-established use, and (2) the omission of -ida on some orders, especially in the phylum Arthropoda, where long usage for familiar groups warrants an exception.

and other consequent adaptations. The following chapters will detail phyla that have taken other adaptive routes to size, variety, and complexity that sometimes rivals, or in the case of variety exceeds, those of the chordates.

For an extended discussion of phylogeny and an outline of relationships of the phyla, refer to Chapters 1 and 13.

Phylum Mesozoa

The Mesozoa are small parasites of the kidneys of cephalopods and body spaces of other marine invertebrates. Fewer than 100 species are known.

Morphology

The vermiform body (Fig. 7-1) consists of a layer of ciliated cells surrounding one or more *axial cells*. There are no spaces or organs. The reproductive cells develop from the axial cell. Length is less than 1 cm. The surface or *somatic cells* may be differentiated at the anterior into a *polar cap* of smaller cells where the mesozoan adheres to the host cells.

Reproduction

Both asexual and sexual reproduction occur, although the complete life cycle is not known for the Dicyemida. In the dicyemids the asexual in-

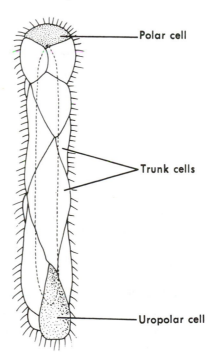

Figure 7-1 Dicyemid mesozoan. *Pseudicyema truncatum*. Position of axial cell indicated by broken line. (*Modified from several authors after Whitman.*)

Polar cell

Trunk cells

Uropolar cell

dividuals are known as *nematogens,* which eventually produce *rhom-bogens,* which have somatic cells bulging with nutrients. The sexual *in-fusorigens* develop within the rhombogen and after self-fertilization ciliated larvae escape from the host. Orthonectids have a vegetative, amoeboid plasmodium within which the sexual individual develops as a ball of cells. As these balls mature as sexual indivduals, they escape from the host and sexual reproduction occurs in the water. The sexual adult resembles the dicyemid rhombogen but has a greater number of cells internally and the somatic cells are more numerous and arranged in rings.

Classification

Two orders of Mesozoa are known. Mesozoa are thought by some zoologists to be degenerate trematodes based largely on life-cycle similarities. Other relationships have been postulated also, including the view, as the name Mesozoa implies, that they represent a stepping stone from the Protozoa to the Metazoa. The complex life cycle and parasitic nature make this seem unlikely. Some authorities think the orders have separate phyletic origins as indicated by quite different life cycles.

Order 1. Dicyemida

Parasites in the kidneys of cephalopods; few axial cells; polar cap; three genera (e.g., *Dicyema typus* from the octopus).

Order 2. Orthonectida

Internal parasites of marine flatworms, annelids, mollusks, and echinoderms; sexual amoeboid stage; ciliated adult with ringed appearance; often many axial cells (e.g., *Rhopalura ophiocomae* from the brittle star).

Phylum Ctenophora

The phylum Ctenophora (Gr. *ktenos,* of a comb; *phoreo,* I bear) includes a small group of free-swimming marine animals that are even more nearly transparent than the coelenterate jellyfishes. They have been placed by many authors under the phylum Coelenterata, but the present tendency is to separate them from that group and rank them as a distinct phylum. They are widely distributed, being especially abundant in warm seas.

Morphology

Ctenophores are commonly called *sea walnuts* because of their shape (Fig. 7-2), or *comb jellies* on account of their jellylike consistency and the comblike locomotor organs arranged in eight rows on the sides of

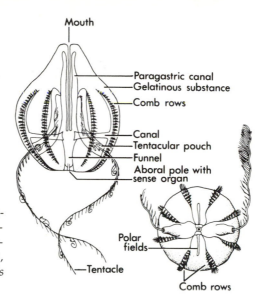

Figure 7-2 Ctenophora. Tentaculata. Left, *Hormiphora plumosa*, side view. Right, *Pleurobrachia pileus*, aboral view. (*Left, after Chun; right, from Lankester's Treatise.*)

the body. A few species have a slender ribbonlike shape and may, like Venus' girdle (Fig. 7-5), reach a length of from 18 cm to over 1 meter.

The general *structure* of a ctenophore is shown in Figure 7-2. It is said to possess *biradial symmetry,* since the parts, though in general radially disposed, lie half on one side and half on the other side of a median longitudinal plane. An aboral view, as in Figure 7-2, illustrates this fact. The *mouth* is situated at one end (*oral*) and a *sense organ* at the opposite or *aboral* end. Extending from near the oral surface to near the aboral end are eight meridional *comb rows* (ciliated bands); these are the *locomotor organs.* Each band has the cilia arranged upon it in transverse rows and fused at the base; each row thus resembles a comb. These are raised and lowered alternately, starting at the aboral end, and cause an appearance like a series of waves traveling from this point toward the mouth. The animal is propelled through the water with the oral end forward. Light is refracted from these moving rows of cilia, and brilliant, changing colors are thus produced. Some species are phosphorescent.

Most ctenophores possess two solid, contractile, large *tentacles* (Fig. 7-2) which emerge from blind pouches, one on either side. With one exception, the tentacles are not provided with nematocysts as are those of the Coelenterata, but are supplied with adhesive cells called *colloblasts* (Fig. 7-3). The colloblasts produce a secretion of use in capturing small animals, which serve as food. The spiral filament in each colloblast is contractile, and acts as a spring, often preventing the struggling prey from tearing the cell away.

The *mouth* (Fig. 7-2) opens into a flattened *stomodaeum*, where most of the food is digested; this leads to the *stomach*, which is flattened at

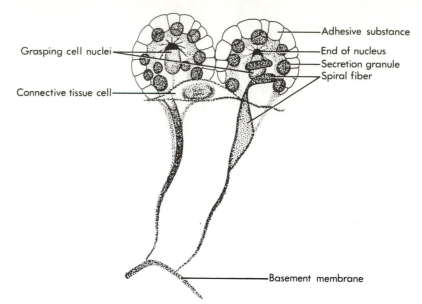

Grasping cell nuclei

Connective tissue cell

Adhesive substance
End of nucleus
Secretion granule
Spiral fiber

Basement membrane

Figure 7-3 *Beroe ovata.* Mature adhesive cells or colloblasts. (*After Schneider, from Dahlgren and Kepner.*)

right angles to the stomodaeum. Six canals arise from the stomach. Two of these, called *anal canals,* open to the exterior near an aboral sense organ; undigested food passes through them, or is ejected through the mouth. The two *paragastric canals* lie parallel to the stomodaeum, ending blindly near the mouth. The two *tentacular canals* pass out toward the pouches of the tentacles, then each gives rise to four branches; these lead into *meridional canals* lying just beneath the comb rows.

Physiology

The *food* of *Mnemiopsis* consists of plankton organisms. A particle of food that is caught in the current produced by the cilia in the auricular grooves is whirled about until it touches one of the small tentacles along the tentacular ridge. The tentacle, often with the aid of several other tentacles, entangles it and then contracts and it is drawn over the labial ridge into the labial trough toward the mouth. Thence the food passes through the mouth into the stomodaeum. Undigested matter is cast out of the mouth or may enter the food canals and pass out through the anus.

The aboral sense organ (Fig. 7-4) is a *statocyst* or organ of equilibrium. It consists of a vesicle of fused cilia enclosing a ball of calcareous granules, the statolith, which is supported by four tufts of fused cilia (balancer cilia). The balancer cilia respond to changed pressure from

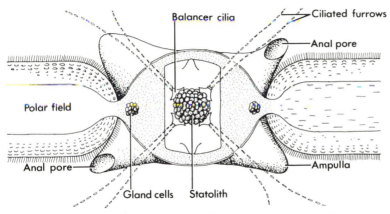

Figure 7-4 *Hormiphora plumosa.* Aboral sense organ viewed from above. (*From Lankester's* Treatise.)

the statolith by modifying the beat of the comb rows they head. The impulses from the balancer cilia that activate the comb plates seem to pass over ciliated axons; inhibitory stimuli for the ciliated system are transmitted by an ectodermal nerve net; a third mechanism of transmission is cell to cell by muscles; a fourth, possibly endodermal, system seems to function for transmitting waves of luminescence (Horridge, 1966). *Pleurobrachia* normally swims upward in response to gravity, but a disturbance causes it to swim downward in response to gravity. Besides such generalized responses, localized responses seem to be under the control of subunits of the nerve net. At least some of the nerve junctions have been proven to be synaptic.

Ctenophores are *hermaphroditic.* The *ova* are formed on one side and the *spermatozoa* on the other side of each meridional canal just beneath the comb rows (Fig. 7-2). The fertilized eggs may escape through the epidermis overlying the meridional canals. The fertilized eggs develop into a larval stage, the *cydippid larva,* similar in general structure to adults such as *Pleurobrachia,* or others in the order Cydippida. This peculiar larval type helps to establish the ctenophore nature of aberrant forms such as *Cestus* and *Coeloplana.*

The cellular layers of ctenophores constitute a very small part of the body, most of it being composed of the transparent jellylike *mesoglea.* The thin ciliated *epidermis* covers the exterior and lines the stomodaeum; and the *gastrodermis,* also ciliated, lines the stomach and the canals to which it gives rise. The muscle fibers that lie just beneath the epidermis and gastrodermis are derived from the *mesenchymal* cells of the embryo. They have more complex organization than coelenterates. Cleavage of the egg is symmetrical and determinate. It is determinate in that each embryonic cell will develop into a particular part even if its position is modified.

Phylum Ctenophora   357

Phylogeny

The Cydippida seem to be nearest the ancestral type of the Ctenophora on the basis of their resemblance to all ctenophore larvae. Little, if any, evidence is present to link them with the polyclad flatworms as some (e.g., Hadzi, 1963) have proposed. The presence of nematocysts in one species, tetramerous symmetry, and a usual planktonic existence support the preferred view that they share a common ancestry with hydrozoan medusae of the Trachylina. The Ctenophora have not given rise to any other invertebrate groups and thus represent an evolutionary dead end.

Classification

Ctenophores are triploblastic animals that exhibit radial combined with bilateral symmetry and possess eight radially arranged comb rows.

The Ctenophora differ from the coelenterates in several important respects besides the presence of a distinct mesoderm. With one exception, ctenophores do not possess nematocysts, and the adhesive cells that take their place are not homologous to nematocysts. Their ciliated bands, aboral sense organs, and pronounced biradial symmetry are peculiarities that warrant placing ctenophores in a phylum by themselves. They are evolved from coelenterate-like ancestors, but can no longer be combined with that phylum. Five orders are listed here.

Order 1. Cydippida

Body spherical, ovoidal, or cylindrical; two long tentacles that may be retracted into aboral tentacular sacs.

Pleurobrachia (Fig. 7-2, B) is ovoidal and sightly compressed laterally; the ciliated bands are eight in number and of equal length. *P. pileus* is about 2 cm long and possesses tentacles about 15 cm long; Long Island northward, Europe. *P. bachei,* California.

Order 2. Lobatida

Body ovate; two large oral lobes; tentacles and tentacular sacs in larva but not in adult.

Mnemiopsis has a pair of long projections (auricles) at the base of each oral lobe upon which the ciliated bands extend. *M. leidyi* is about 10 cm. long; transparent and phosphorescent; Long Island to South Carolina.

Order 3. Cestida

Body shaped like a ribbon; two tentacular sacs and two or more less rudimentary tentacles; many lateral tentacles.

Cestus is a genus containing several species. *C. veneris,* known as Venus' girdle (Fig. 7-5), may be 5 cm wide and nearly 1 meter long; transparent but showing green, blue, and violet colors; tropical seas.

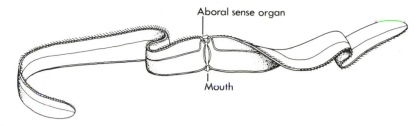

Figure 7-5 Cestida. *Cestus pectenalis*, Venus' girdle. (*After Bigelow.*)

Order 4. Beroida

Tentacles and oral lobes absent. Body conical or ovate and laterally compressed; mouth very large and stomach voluminous; voracious.
Beroe ovata (Fig. 7-6) is about 10 cm long; cosmopolitan.

Order 5. Platyctenida

Creeping ctenophores with two sheathed tentacles. Comb rows in larvae only. Adult flattened. *Coeloplana;* along many coastal regions.

Figure 7-6 Beroida. *Beroe ovata.* (*From Lankester's* Treatise.)

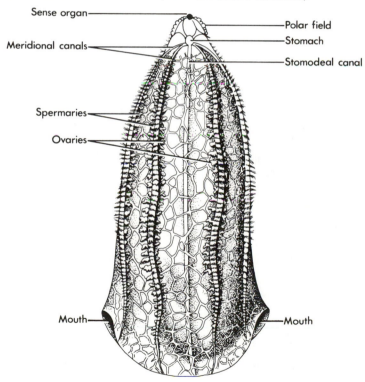

Phylum Acanthocephala

The spinyheaded worms (Fig. 7-7) belong to this phylum. They live in the intestine of vertebrates attached to the wall by a protrusible proboscis usually covered with recurved hooks. The immature stages develop in arthropods such as insects and crustaceans.

Morphology.

The adults vary in length from several millimeters to over 50 cm. An alimentary canal is not present at any stage in their life cycle, although one or two *ligament sacs* running the length of the animal are thought to represent the entoderm derivatives. A thin cuticle, often spiny, covers the syncytial epidermis, which has three layers, the innermost thick and penetrated by lacunae. A layer of circular muscles and an inner layer of longitudinal muscles are between the epidermis and the

Figure 7-7 Acanthocephala. (A) *Macracanthorhynchus hirudinaceus* clinging to the intestinal wall of a pig. The two small worms are males, and the larger is a female. (B) A male specimen of the genus *Acanthocephalus* showing internal organs. (*A, after Brumpt; B, after Van Cleave.*)

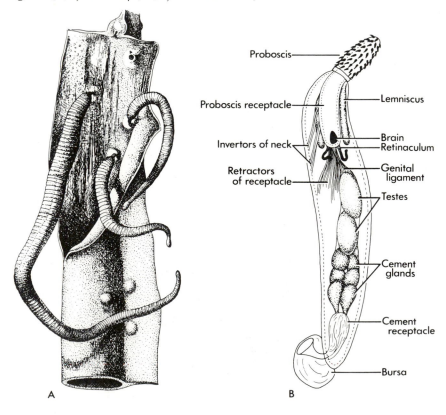

pseudocoelomate body cavity. The anterior spiny *proboscis* is attached by a neck region to a *proboscis receptacle* into which the proboscis can be withdrawn. Special muscles move the proboscis. Two large glandular structures, the *lemnisci*, are thought to provide and receive fluid moved during extension and retraction of the proboscis. The *cerebral ganglion* (brain) is on the ventral surface of the proboscis receptacle. Several nerves connect to the brain. A few acanthocephalans have an excretory system of the flame cell type. Sexes are separate, and the complex reproductive systems are closely associated with the ligament sacs.

The acanthocephalans provide an interesting example illustrating an extreme view of sociobiological theory that the organism is a device designed to perpetuate the genes. The cement glands of the male worm may seal the genital opening of the female after copulation; thus loss of sperm is prevented and so are additional copulations that might bring in outside genes. Some have gone even further and also transfer cement alone to other males during homosexual rape (Abele and Gilchrist, 1977) in a manner that eliminates their competitors ability to successfully inseminate available females.

Life Cycles

After copulation and fertilization a modified spiral cleavage occurs. If ingested by the proper invertebrate host the larva hatches from the egg and penetrates the gut wall to develop into a juvenile worm in the host hemocoel. Some worms are capable of reencapsulating in another host if ingested by one that is not a suitable final host. Thus, more than two hosts may be present in the life cycle. Some species normally have more than two hosts in the life cycle.

Host–Parasite Relationship

Acanthocephalans are most common in fishes and seem to cause few or no symptoms in them. Because the proboscis can penetrate the intestinal lining of the host it is a potential avenue for other infections. One specimen has been reported to be fatal in the robin and a few more in some other birds. One juvenile stage is known to prevent ovarian function in the crustacean host so that an immature stage is maintained in the host. Perhaps this increases the length of life of the intermediate host and the probable success of reaching a final host. Of interest is the apparent dependence of the worm on the host for digestion of lipids, although enzymes useful for the digestion of other substances are present in the worm.

The juvenile worm becomes encapsulated in the intermediate arthropod host. Second intermediate, or transport hosts, are commonly fish or invertebrates in which the juvenile will reencapsulate when the first host is eaten. Ordinarily, no growth occurs while in the transport host.

Phylogeny

Because they are parasitic roundworms covered with a cuticle the spiny-headed worms have sometimes been considered closely related to the nematodes. The type of muscular system present makes relationship to the Aschelminthes doubtful, though not impossible. The muscular system resembles that of the flatworms and the proboscis shows some similarities to the rostellum of the cestodes. However, the pseudocoelomate condition does not support a relationship with the Platyhelminthes. Thus, as with most phyla, the origin of the phylum is not clear.

Sufficient resemblance between all species of the Acanthocephala is present so that no classes and only three orders are recognized.

Order 1. *Archiacanthocephalida*

Small proboscis hooks concentrically arranged; excretory system present; parasites of terrestrial birds and mammals; intermediate hosts are terrestrial arthropods. Examples are *Macracanthorhynchus hirudinaceus* (Fig. 7-7, A) cosmopolitan in pigs, and *Moniliformis moniliformis* in small mammals. Both species have been reported from humans.

Order 2. *Palaeacanthocephalida*

Proboscis hooks alternating in radial rows; no excretory system; parasites of fishes, amphibians, aquatic birds, and marine mammals; first intermediate hosts are crustaceans. *Acanthocephalus* (Fig. 7-2, B) has numerous species in fish and amphibians; *Echinorhynchus salvelini* occurs in lake trout and *Arhythmorhynchus brevis* in the bittern.

Order 3. *Eoacanthocephalida*

Proboscis hooks arranged radially; no excretory system; parasites of fishes and reptiles; first intermediate hosts are aquatic arthropods. *Neoechinorhynchus emydis* is found in turtles in the United States.

Phylum Rhynchocoela

The Rhynchocoela or Nemertea are a group of several hundred mostly marine bottom-dwelling worms. They are flattened and have a ciliated epidermis that contains rhabdites in a few species. The lack of a coelom and the presence of rhabdites add to the evidence that they share a common ancestry with the Platyhelminthes.

Morphology

The most important anatomical features of the nemerteans (Fig. 7-8) are the presence of (1) a long proboscis, which lies in a proboscis sheath just above the digestive tract, and may be everted and used

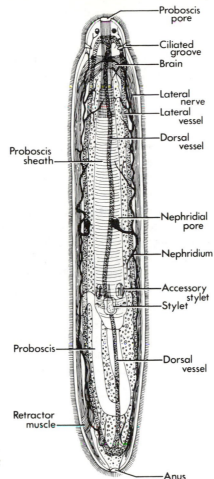

Proboscis
pore

Ciliated
groove

Brain

Lateral
nerve

Lateral
vessel

Dorsal
vessel

Proboscis
sheath

Nephridial
pore

Nephridium

Accessory
stylet

Stylet

Proboscis

Dorsal
vessel

Retractor
muscle

Anus

Figure 7-8 Nemertea. *Tetrastemma.*
General view of internal organs. (*From
Hatschek.*)

as a tactile, protective, and defensive organ; (2) a blood vascular system consisting usually of a median dorsal and two lateral trunks; and (3) an alimentary canal with both mouth and anal openings. The blood vascular system is here encountered for the first time. Nemerteans possess a mesoderm and nervous and excretory systems, which do not differ markedly from those of the flat worms. The proboscis sheath may represent the coelom, but this is not certain.

Ecology

Nemerteans feed on other animals, both dead and alive. They live, as a rule, coiled up in burrows in the mud or sand, or under stones, but some of them frequent patches of seaweed. Locomotion is effected by the cilia which cover the surface of the body, by contractions of the body muscles, or by the attachment of the proboscis and subsequent drawing forward of the body. *Cerebratulus* swims actively like a leech.

A variety of invertebrates have nemerteans that live with them commensally. No truly parasitic species are known. Although the majority are shallow marine-bottom dwellers a few are found in fresh water and a few terrestrial species occur in the tropics and Tasmania. They reach considerable depths in the ocean and a few are pelagic in the ocean.

Reproduction

Sexes are separate in most species of nemerteans. The eggs undergo a determinate type of spiral cleavage. Eventually, a peculiar larval stage called the *pilidium* (Fig. 7-9) is usually passed through. This resembles a helmet with cilia on the surface and a long tuft of cilia at the apex. The adult develops from this larva by the formation of ectodermal invaginations which surround the alimentary canal. This invaginated portion escapes from the *pilidium* and grows into the adult nemertean.

Several species reproduce during the summer by spontaneous fragmentation (autotomy) followed by the regeneration of each of the 2 to 20 or more fragments into complete worms (Fig. 7-10). This type of reproduction is inhibited by cold weather in the autumn at the time when the sex cells are beginning to develop. Nemerteans of these species possess remarkable powers of regeneration since, if, for example, an individual of *Lineus socialis* 100 mm in length is cut into as many as 100 pieces, each piece will regenerate a minute worm within 4 to 5 weeks. These minute worms may again be cut into pieces that regenerate and these may in turn be cut up and so on until miniature worms less than 1/200,000 the volume of the original worm result (Fig. 7-11).

Classification

The Rhynchocoela may be divided into four orders. The first two orders are grouped in the Anopla, the last two in the Enopla, based largely on nervous system variations. Enopla have the main nerve

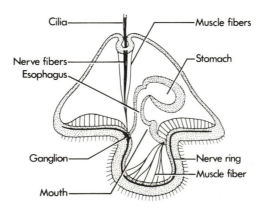

Figure 7-9 Pilidium larva of a nemertine. (*After Salensky.*)

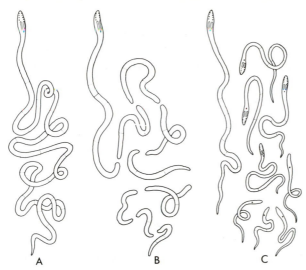

Figure 7-10 *Lineus socialis,* reproduction by fission. (A) Mature worm. (B) Worm divided by fission into nine pieces. (C) Reconstruction of these pieces into complete worms. (*From Coe.*)

tracts internal to all the main muscle layers and the mouth anterior to the brain.

Order 1. *Paleonemertinida*

The species in this order possess a proboscis free from stylets, and cerebral ganglia and lateral nerves in the ectoderm or between the two

Figure 7-11 *Lineus socialis,* regeneration. (A) Normal worm. (B and B′) Section cut from body and its appearance after 12 days. (C) Same, after 30 days, cut in planes 3–3 and 4–4. Successive cuttings and regenerations indicated by arrows. (M and M′) Similar experiments and results on posterior end of body. (*From Coe.*)

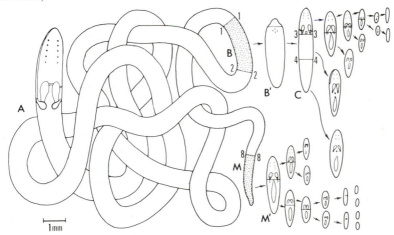

layers of muscles. *Tubulanus pellucida* is a marine species that lives below low water among annelid tubes and has been reported from Long Island and Vineyard Sounds and California. *Cephalothrix linearis* is another marine species living on both our Atlantic and Pacific coasts and in Europe.

Order 2. Heteronemertinida

There are no stylets on the proboscis and the lateral nerve cords lie between layers of circular and longitudinal muscles. *Lineus ruber* (Fig. 7-11) has a long filiform body; is green, brown, or reddish; has a row of from four to eight eyes on either side of the head; and lives under stones in shallow water on our eastern coast, Alaska, and Europe. *Cerebratulus* is a well-known genus with representatives in all seas. *C. lacteus* may reach a length of several meters, is flesh colored, has a white proboscis, and lives in the sand near low-water mark from Florida to Maine. *C. californiensis* is found on the Pacific Coast.

Order 3. Hoplonemertinida

The proboscis in this group bears one or more stylets and the lateral nerves are in the muscle layers. The genus *Tetrastemma* (Fig. 7-8) is widespread and contains many species, mostly unisexual. They are small and possess four eyes. *T. candidum* is slender, about 2 cm long, white, green, or yellow, and lives among algae between tide lines on our eastern coast and in Europe. *Prostoma rubrum* is a freshwater species resembling *Tetrastemma,* but is about 20 mm long, is yellow or reddish, and has three pairs of eyes. *Amphiporus* is represented on both coasts with a number of intertidal species that often have reddish or yellowish nerves.

Order 4. Bdellonemertinida

Commensal nemertines lacking stylets on the proboscis; mouth anterior to brain; only one genus, *Malacobdella. M. grossa* is broad, has a sucker at the posterior end, and lives in the branchial chamber of pelecypod mollusks, such as *Mya* and *Venus,* on both sides of the North Atlantic and on the Pacific Coast of North America.

Phylum Priapulida

This phylum contains only six species of wormlike marine animals of moderate size. The mouth is centered in an eversible spiny proboscis. Bulbs of the flame cells have a single flagellum. The body wall and intestine are both provided with circular and longitudinal muscles. Many other features show striking similarities to the kinorhynchs of the Aschelminthes. They have previously been considered members of that phylum as well as the annelids and echinoderms at different

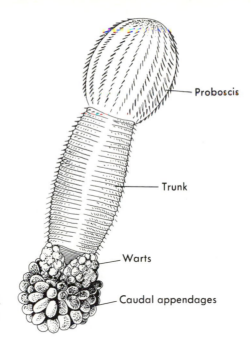

Figure 7-12 Priapulida. General external features of *Priapulus caudatus*.

Labels on figure: Proboscis, Trunk, Warts, Caudal appendages

times. Shapeero (1961) reports that the lining of the coelom is nucleated and thus they differ too much from the aschelminthes to be included with them. *Priapulus caudatus* (Fig. 7-12) occurs in both northern oceans. The pharynx is muscular and provided with numerous teeth, and the anus is located at the posterior end near a caudal appendage of hollow vesicles. The surface of the trunk of the body is warty or spiny. It lives in the mud or sand in shallow water with the anterior end projecting into the water.

Phylum Polyzoa (Bryozoa)

The Polyzoa or Bryozoa are sessile, colonial (an exception occurs in the Entoprocta), lophophore-bearing animals that live in secreted cases. The individuals are barely visible to the unaided eye. The majority of them live in the sea, but a few inhabit fresh water. *Bugula* is a common marine genus which shows the principal characteristics of the group.

Bugula

The soft parts constituting the *polypide* lie within the true *coelomic cavity* bounded by the body wall or *zooecium*. The *mouth* lies in the midst of a crown of *ciliated tentacles* (Fig. 7-13, A) called the *lophophore*, which serve to draw food particles into the body. The U-shaped

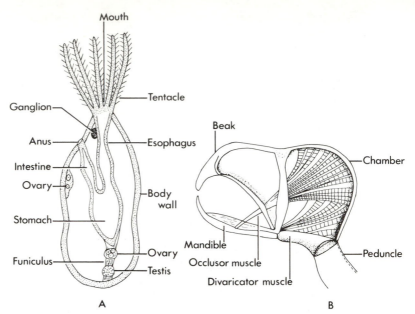

Figure 7-13 Ectoprocta. (A) Diagram showing the structure of a single individual. (B) An avicularium of *Bugula*. (*After Hincks.*)

alimentary canal consists of a ciliated *esophagus,* a *stomach,* and an *intestine,* which opens by means of an anus lying just outside the lophophore. One *retractor muscle* serves to draw the polypide into the zooecium. The *funiculus* is a strand of mesodermal tissue attached to the base of the stomach. There are no circulatory or excretory organs.

Both an *ovary* and a *testis* are present in each individual; they may be found attached to the funiculus or the body wall. The *eggs* are probably fertilized in the coelom and then develop in a modified portion of the zooecium called the *ooecium.* The larvae of some Bryozoa resemble a trochophore.

Certain member of *Bugula* colonies are modified into structures called *avicularia* (Fig. 7-13, B). These have jaws that probably protect the colony from the attacks of small organisms and prevent the larvae of other animals from settling upon it. The latter function is accomplished in some genera by whiplike zooids called *vibracula.*

Other Polyzoa

The coeloms of adjacent zooids are separated in the Stenolaemata and Gymnolaemata of the Ectoprocta. Coordination in lophophore retraction appears to be through nervous impulses conducted over a colonial nervous system at speeds of up to 100 cm per second in *Membranipora membranosa* (Thorpe, Shelton, and Laverack, 1975). Impulses occurred at up to 200 per second. Impulses from the individual zooids are also observed. The muscles retracting the lophophore reached peak contraction speeds equivalent to 20 times their own length per second; this is the fastest comparative speed of muscle contraction known.

Brown body formation occurs in many species by the regression of a few zooids to small dark masses, usually followed by the regeneration from a portion of each zooid of new vigorous replacement zooids. Brown body formation is thought to be a mechanism for ridding the organism of accumulated waste products and/or indigestible residue located mostly in the gut cells.

Classification and Phylogeny

Bryozoa include two widely separated subphyla, the Entoprocta (Endoprocta) and the Ectoprocta. The two subphyla are generally considered phyla in most texts. The lack of a coelom in the Entoprocta may be a consequence of their primitive position or an adaptation for lower mobility, small size, and the special physiology associated with their particular life style. The placement is justified by the opinions of Rogick (1959) and Nielsen (1977) as well as the fact that a coelom is a feature of low complexity and may be highly adaptive, so that it may have many phylogenetic points of origin as well as be obscured by extreme size reduction. The inclusion in one phylum of the pseudocoelomate Entoprocta with the coelomate Ectoprocta is consistent with the inclusion of the acoelomate Gnathostomula with the pseudocoelmate Aschelminthes, where a similar problem is met.

The lophophore is an adaptive feature of a sessile filter feeder that may have multiple origins. Many regard it as an indication of relationship of the phyla having them, perhaps justifying a superphylum Lophophorata. Many of their other similarities may also be convergent due to their sessile, often tube-dwelling life style. How else do you effectively wash away feces without discharging wastes from an anus near the upper part? Why not "forget' about complex genitalia when, as a sessile organism, successful mating may frequently only be done at a distance?

Phylum Polyzoa (Bryozoa)

Sessile, lophophorate, colonial, usually in secreted, and often calcified, zooecia.

SUBPHYLUM ENTOPROCTA

The entoproctan anus opens within a nonretractile lophophore. About 60 species exist; all are pseudocoelomate.

The tentacles of the filter-feeding apparatus, the lophophore, encircle both the mouth and anus of these sessile animals. The cilia-lined digestive tract consists of a mouth, esophagus, stomach, intestine, rectum, and anus. The epidermis secretes a cuticle over much of the body

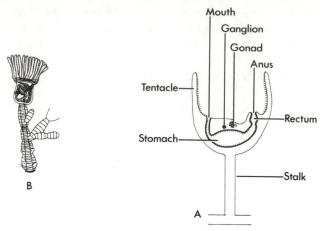

Figure 7-14 Entoprocta. (A) Diagram of arrangement of digestive tract. (B) *Urnatella gracilis.* (*B, after Leidy.*)

or *calyx*. Between the epidermis and the digestive tract the body is filled with a gelatinous substance containing a few cells. The body is attached to the substrate by a *stalk*. The main ganglionic mass, the gonads, and the pair of flame cells are located above the stomach (Fig. 7-14, A). Muscle fibers are very sparse. The tentacles of the lophophore can be bent inward to cover the area within the lophophore.

Some species are dioecious and others hermaphroditic. *Loxosomella davenporti* is a solitary species that bears buds; it is about 2 mm long; has from 22 to 26 tentacles; and possesses a mammary organ in the vestibule to which developing embryos are attached. It is a marine species abundant in Vineyard Sound. *Pedicellina cernua* is a colonial, marine species, the zooids of which arise from a creeping, branching stolon. It attaches itself to shells and algae in shallow water along the Atlantic coast of the United States and Europe. *Urnatella gracilis* (Fig. 7-14, B) is a colonial species that lives on the underside of stones in running fresh water in the eastern and central states. Its zooids, which are few in number, arise from a common disk and form long stalks, which are joined and form branches.

SUBPHYLUM ECTOPROCTA

Members of the classes of Ectoprocta are sessile, lophophore-bearing animals, of small size, living in secreted cases; true coelom; anus opens outside the lophophore; lophophore rectractable; no excretory organs; no special respiratory or circulatory structures; hermaphroditic and unisexual species occur; asexual reproduction by budding is common to all and large branching and encrusting colonies are formed by this means; many species.

Phylactolaemata are freshwater species that have an oval or horseshoe-shaped lophophore, a sort of lip, called the epistome, projecting over the mouth and statoblasts usually provided with marginal air cells which float them on the surface of the water. *Plumatella repens* (Fig. 7-15, A) forms a much-branched, creeping or erect colony and produces elongated statoblasts. *Cristatella mucedo* (Fig. 7-15, D) forms an elongate, creeping gelatinous colony, with zooids on the upper surface, and produces circular statoblasts with two rows of marginal hooks (Fig. 7-15, E). These statoblasts are buds with a hard chitinous shell that arise from the funiculus and give rise to new colonies after the death of the parent or a period of drought. *Pectinatella magnifica* (Fig. 7-15, B) forms rosette-shaped groups of associated colonies which may build up a thick gelatinous base. It has from 60 to 84 tentacles, and black, circular statoblasts with one row of from 10 to 22 marginal hooks. *Fredericella sultana* forms a colony of intertwining branches. It has a lophophore that is oval, and produces dark brown, elliptical

Figure 7-15 Ectoprocta. Phylactolaemata. (A) *Plumatella punctata.* (B) *Pectinatella magnifica.* (C) *Plumatella*, statoblast. (D) *Cristatella mucedo.* (E) *C. mucedo*, statoblast. (*A and B, after Kraepelin; D and E, after Allman.*)

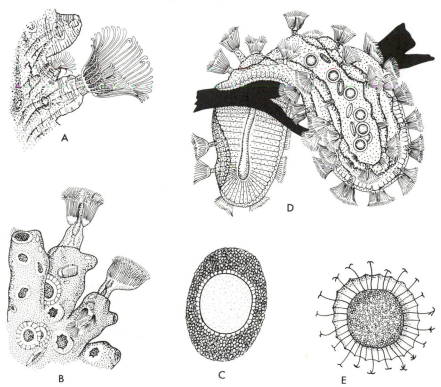

statoblasts without a float. According to their structure and function statoblasts receive various names such as floatoblasts, spinoblasts, and sessoblasts.

Class 2. Stenolaemata

Stenolaemata are marine, with circular lophophore and tubular, calcified zooecium; aperture lacks operculum or membrane; no epistome. Many fossil species in four orders, only one which contains living representatives. The order **Cyclostomida** includes species of *Crisia* (Fig. 7-16, A) and related forms on both coasts.

Class 3. Gymnolaemata

Gymnolaemata are mostly marine species that have a circular lophophore without an epistome.

Order 1. Ctenostomida

Zooecium fleshy or membranous; aperture closed by a folded membrane when the lophophore is retracted; no ooecia or avicularia. *Alcyonidium hirsutum* forms yellowish brown or reddish colonies that are often encrusting, with conical projections on the surface between which are the orifices.

Order 2. Cheilostomida

Zooecium rectangular; aperture closed by an operculum (lid or flap) when the lophophore is retracted; ovicells (ooecia), avicularia, and vibracula may be present; zooecium may be calcified. *Membranipora pilosa* (Fig. 7-16, B) is a common species near Woods Hole. *M. membranacea* occurs in California and on the East Coast, where it may

Figure 7-16 Ectoprocta. Stenolaemata and Gymnolaemata. (A) *Crisia maxima,* zooecium and ovicell. (B) *Membranipora membranacea.* (*From Johnson and Snook.*)

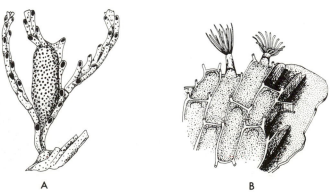

A   B

reach fresh water. *Membranipora* forms a flattened crust without ovicells or avicularia; on kelp, submerged or intertidal objects, and the sediments. *Bugula* also belongs to this order; many widely distributed, shallow-marine species.

Ectoprocta in General

The bryozoans were once though to be plants. The mosslike appearance of *Bulgula* colonies, with their sessile, arborescent habit makes this understandable.

Marine species are numerous and common, their fossil record extends back to the Cambrian. Freshwater species are less numerous and more uniform in structure, although colonies may show considerable variability under different conditions. The cosmopolitan *Plumatella repens* is the most abundant and common freshwater species (Bushnell, 1965). The asexual reproductive bodies, or statoblasts, are an im-

Figure 7-17 Ectoprocta. Phylactolaemata. *Pectinatella magnifica,* expanded polypide. Movement of particles in the coelomic fluid indicates rapid circulation in a specific pattern due to some ciliary patches on the coelomic epithelium. (*Photo by author.*)

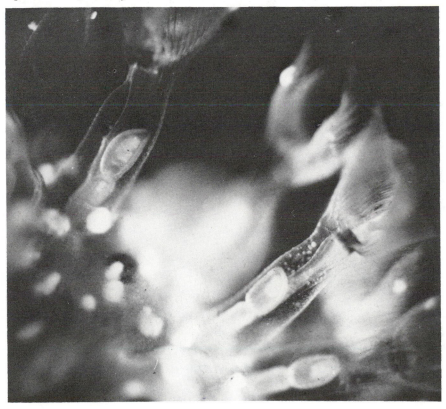

portant aid to the widespread distribution of freshwater species. Statoblasts may survive passage through the digestive tract of birds.

The Polyzoa are of little direct economic importance. They contribute to reef development. Their small size, persistent skeleton, sessile habit, and abundance in the littoral zone make them of good potential for evaluating environmental mishaps. Little is known of their interaction with other species, although at least one species of pycnogonid is specialized to feed upon them (Wyer and King, 1973). A living bryozoan is an animal of unique beauty when its lophophore is fully extended (Fig. 7-17).

Phylum Brachiopoda

The Brachiopoda are marine animals living within a calcareous bivalve *shell* (Fig. 7-19). They are usually attached to some object by a stalk, called the *peduncle* (Fig. 7-18). Because of their shell they were for a long time regarded as mollusks. The valves of the shell, however, are dorsal and ventral instead of lateral as in the bivalve mollusks. Within the shell (Fig. 7-18) is a conspicuous structure called the *lophophore,* which consists of two coiled ridges, called arms; these bear ciliated tentacles. *Food* is drawn into the *mouth* by the lophophore. A true *coelom* is present, within which lie the *stomach, digestive gland,* and the *heart.*

The group Brachiopoda is extremely old, and, although found in all seas today, brachiopods were formerly more numerous in species and of much greater variety in form than at present. Some of them, for example *Lingula* (Fig. 7-19, G), are apparently similar to fossil forms found recently in pre-Cambrian rocks, perhaps a billion years old. Their relationship to the other lophophorates is probably hidden in

Figure 7-18 Brachiopoda. Longitudinal section of *Waldheimia.* (*After Lister, from Shipley and MacBride.*)

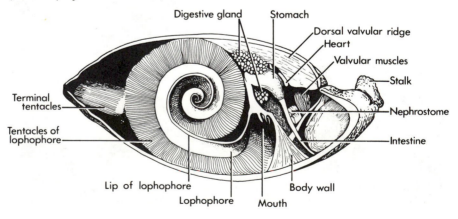

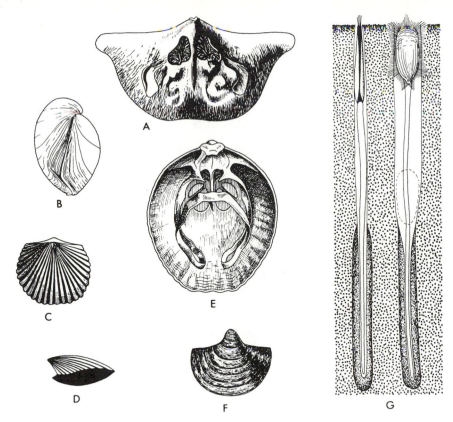

Figure 7-19 Brachiopoda. (A) *Productus giganteus:* Carboniferous limestone; interior of dorsal valve. (B) *Terebratula semiglobosa:* upper Chalk; lateral view. (C and D) *Orthis calligramma:* Ordovician. (E) *Waldheimia:* interior of dorsal valve showing brachial skeleton. (F) *Micromitra labradorica:* Lower Cambrian; ventral valve. (G) *Lingula anatifera* in their tubes in the sand; the dotted line in the right tube indicates the position of the body when retracted. (*A–F, from Woods, after various authors; G, after Francois.*)

unknown recesses of the Precambrian, although molecular data may some day unravel the mystery. Gutmann, Vogel, and Zorn (1978) report indications of a relationship to annelids.

Two classes of brachiopods may be recognized according to whether or not the valves of the shell are joined together by a hinge.

CLASS 1. INARTICULATA OR ECARDINES

An inarticulate brachiopod has a shell without a hinge. There is an alimentary canal with an anus. No internal skeleton is present. Three families with about 32 living and 400 fossil species, mostly Paleozoic, belong to this order. *Lingula* (Fig. 7-19, G) has already been mentioned. *Glottidia albida* occurs on our Pacific coast from low water to a depth of 60 fathoms. It has a smooth, white shell about 30 mm long, and a stout peduncle about 45 mm long. *Crania anomala* lives at a depth of about 100 fathoms in the West Indies. The shell is brown, about 18 mm long and 22 mm broad, and fastened by the ventral valve to a rock. No peduncle is present.

CLASS 2. ARTICULATA OR TESTICARDINES

An articulate brachiopod has a shell with a hinge and an internal skeleton. No anus is present. The three families in this order contain about 80 living and 2,200 fossil species. *Terebratulina septentrionalis* has a yellowish or whitish, thin, semitransparent shell that is broadly oval in shape. *Waldheimia floridana* has a gray or brownish-white shell that is triangular in shape.

Figure 7-20 Phoronidea. *Phoronis,* diagram to show the structure of one half of the lophophore and the main internal organs. (*From Benham.*)

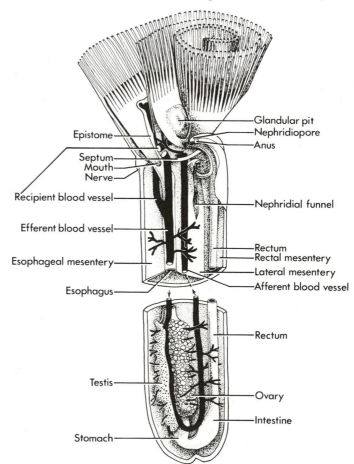

Phylum Phoronidea

The species in this group, about 15 in all, belong to the genera *Phoronis* (Fig. 7-20) and *Phoronopsis*. They are small, marine animals of sedentary habit that live in tubes. The larva, called an actinotroch, is free-swimming and resembles a trochophore. The adults are unsegmented, coelomate, and hermaphroditic. They possess a horseshoe-shaped lophophore, an epistome, two excretory organs, and a vascular system. *Phoronis architecta* occurs in sand flats near the low-water mark in North Carolina. It lives in a straight tube about 13 cm long and 1 mm broad; is flesh colored; and has about 100 tentacles. It and 7 other phoronids live on the Pacific coast of North America.

This phylum and the previous three all possess lophophores. Major differences in other features such as the coelom, excretory system, and other systems are usually considered valid for distinction being made at the phylum level. A further point, clouding the phylogenetic picture for the lophophorates, is the observation of radial cleavage (although an appearance of spiral cleavage is found in some) in the phoronids (Emig, 1977).

Phylum Sipunculida

The Sipunculida (Fig. 7-21) are considered definite relatives of the Annelida on the basis of an annelid type of nervous system, nephridia, and embryology. The complete lack of segmentation indicates a great divergence from the annelids and thus phylum rank is warranted.

The approximately 250 species are all marine-bottom dwellers, although larval stages are sometimes planktonic for as long as a month. Sipunculida are unsegmented with only one pair of nephridia, a large coelom, and an anus on the dorsal surface near the anterior end. They live in the sand or bore into coral rock and are capable of slow, creeping locomotion. The anterior part of the body can be drawn into the larger posterior portion and is therefore called the introvert. Tentacles are usually present at the anterior end. *Phascolosoma gouldi* is about 18 cm long, and has a smooth skin and many tentacles in several rows. *Phascolion strombi* uses a snail shell for protection, closing the entrance with a plug of sand, through a hole in which the introvert may be protruded.

The sipunculids have hemerythrin as their respiratory pigment, as do some brachiopods, polychaetes, and priapulids (Klotz, Klipperstein, and Hendrickson, 1976).

The sexes are separate, with external fertilization. A few species have been found that can reproduce by fission. *Aspidosiphon brocki*

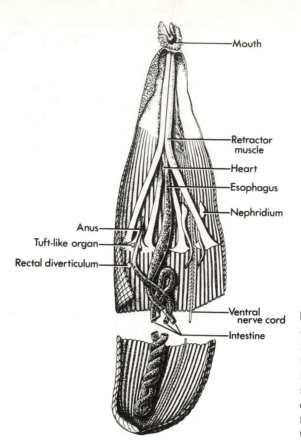

Mouth
Retractor muscle
Heart
Esophagus
Nephridium
Anus
Tuft-like organ
Rectal diverticulum
Ventral nerve cord
Intestine

Figure 7-21 Sipunculida. *Sipunculus nudus,* with introvert and head fully extended, laid open by an incision along the right side to show the internal organs. The spindle-muscle is seen overlying the rectum. (*From Shipley.*)

divides unequally (Rice, 1970), with the smaller posterior portion having a greater job of regenerating the parts not retained.

Phylum Chaetognatha

The Chaetognatha are marine animals which swim about near the surface of the sea. The best known genus is *Sagitta,* the arrowworm. Figure 7-22 shows most of the anatomical features of *Sagitta hexaptera.* The body consists of three regions, head, trunk, and postanal tail. Lateral and caudal fins are present. There is a distinct coelom, an alimentary canal with mouth, intestine, and anus, a well-developed nervous system, two eyes, and other sensory organs. The mouth has a lobe on either side provided with bristles which are used in capturing the minute animals and plants that serve as food. The members of the group are hermaphroditic. There is no circulatory or excretory system in this small phylum of about 50 species. Radial, indeterminate cleavage and type of coelom formation have been used to link this phylum with deuterostomes such as the echinoderms and chordates. Adult anatomy, however, shows considerable resemblance to the aschelminths, hence their phylogenetic position remains problematic.

Sagitta bipunctata is a pelagic species capable of vertical migration through many fathoms. *Eukrohnia hamata* has a slender body with a single lateral fin on either side. *Pterosagitta draco* has a broad body with a single pair of lateral fins and an integumentary expansion along the sides. All these species are cosmopolitan in their distribution.

Sagitta elegans is common in North Atlantic coastal waters and *S. serratodentata* in surface waters of warmer oceans.

Chaetognaths are important members of the plankton of the oceans. Species distribution seems to be determined by temperature.

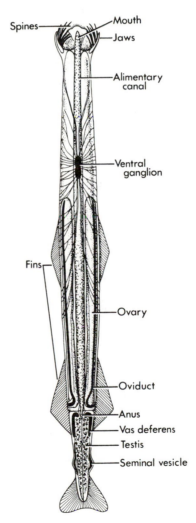

Figure 7-22 Chaetognatha. *Sagitta hexaptera*, ventral view. (*After O. Hertwig, modified.*)

Phylum Pogonophora

The Pogonophora are a recently discovered group of long, slender, solitary, tube-dwelling invertebrates from the deep ocean bottoms. Although discovered this century, they are very widespread in the deeper oceans and recently have been found in somewhat shallow water near the coast of Florida. The most peculiar feature of these free-living animals is the complete lack of a digestive tract.

General Morphology

The long cylindrical body (Fig. 7-23) seldom exceeds 1 mm in diameter but the length may exceed ⅓ meter. They live in a secreted pro-

Figure 7-23 Pogonophora. *Siboglinum fiordicum,* length greatly reduced in indicated areas. (*Modified after Webb.*)

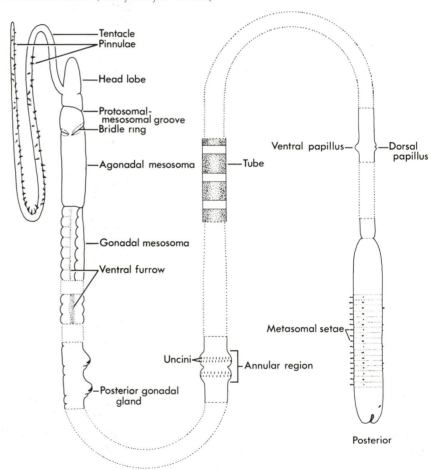

teinaceous and chitinous tube, which usually has many ringlike thickenings evident. The tubes are thought to have a vertical orientation in the bottom mud. Recent discoveries by Webb (1964) indicate that the body regions (Fig. 7-24) described in earlier works are based on incomplete specimens. A short anterior *protosoma* bears one, a few, or many *tentacles*. Beneath this is a short *agonadal mesosomal region* followed by a very long *trunk*. About midway down the trunk is an *annular region* of two raised belts studded with uncini. The *uncini* are minute, toothed, hard structures that probably aid this region as holdfast structures. The long preannular portion has a metameric region at the anterior with paired rows of papillae bounding the *ventral furrow*. A variable number of several types of papillae are distributed over the rest of the trunk. At the extreme lower or posterior end is a setae-bearing region thought to be the true *metasoma*, only recently discovered.

Internally the coelom has several divisions. A concentration of nervous tissue occurs in the protosoma. Two vessels, probably afferent and efferent, from the blood vascular system supply each tentacle. Nephridial structures are present. Sexes are separate, but reports of the position of the gonads vary. In *Siboglinum fiordicum* the oviducts open anterior to the first postgonadal gland and spermatophores leave the genital papillae at the anterior end of the metameric region.

Figure 7-24 Pogonophora body regions. (A) After Ivanov. (B) As revised by Webb following discovery of the posterior extremity. (*Modified after Webb.*)

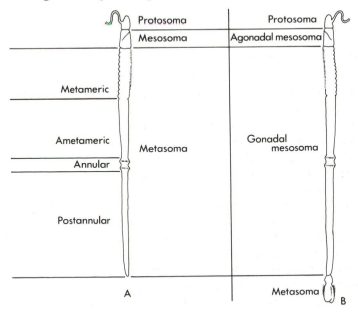

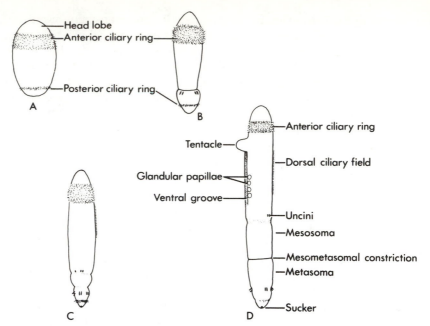

Figure 7-25 Pogonophora. Larval differentiation. (A–D) Early to late larval stages. (*Modified after Webb.*)

Development

Cleavage is holoblastic and unequal. No blastopore is formed. The larva has an anterior and posterior ciliary ring (Fig. 7-25). The great length of the organism is achieved by elongation of the mesosoma.

Pogonophora in General

Many questions remain unanswered about the pogonophorans. The method of nutrition may involve external digestion within the tube or within a chamber formed by the tentacles. Tentacular gland cells support this assumption. A possibility not yet explored is that nutrition is partly saprophytic through the posterior embedded in the organic ooze and that the tentacles also function as gills.

The phylogenetic position is puzzling, although most zoologists seem to assign them a position near the deuterostomes (echinoderms, hemichordates, and chordates). However (Foucart, Gregoire, and Jeuniaux, 1965), the presence of chitin has been reported in their tubes. Chitin is not present in the deuterostomes in general. Several workers think a two tentacled arrangement is primitive and a horseshoe-shaped base has been reported at one stage. Thus, a possible relationship to the lophophorate phyla cannot be eliminated.

It seems possible the pogonophorans are extremely long lived. If the tubes are oriented in a vertical position (tubes are often three times the length of the organism), they may grow upward as the ocean ooze slowly accumulates. It seems unlikely that they are pushed into the ooze as formed, because tubes formed when the animal breaks through the side of the tube with its posterior are thinner and without annulations. Thus, the tubes and their length could indicate extreme ages for invertebrates, probably in the thousands of years (Engemann, 1968). The low food supply, cold temperature, great pressure, and constant conditions are also factors expected to lead to extreme age for many of the deep-sea benthos.

Southward (1971) has reviewed the recent research on the Pogonophora, including their ability to take up nutrients in dilute solution. Discovery (Corliss et al., 1979) of a giant pogonophoran, with tubes up to 338 mm long and to 28 mm in diameter, in an atypical environment (by a deep-sea, thermal spring) is of special interest. Large size in spite of a free-living existence with no digestive tract was apparently due to uptake of nutrients released from the thermal springs that made temperature and nutrient levels above typical deep-sea values.

Classification

Classification as given below does not include changes that may result from study of the giant pogonophorans.

Order 1. Athecanephrida

Nephridiopores lateral; pericardial sac; one or several tentacles. *Siboglinum* (Fig. 7-23) is a genus of about 10 species, each with a single tentacle.

Order 2. Thecanephrida

Nephridiopores medial; no pericardial sac; several to many tentacles, often fused to some extent. In *Lamellisabella* the tentacles are fused into a cylinder, *Spirobrachia* may have over 200 tentacles that are fused and spiraled at the base.

Phylum Hemichordata

This phylum of wormlike animals of shallow ocean bottoms is considered closely related to the chordates. The body (Fig. 7-26) is divided into three regions, an anterior *protosome*, a *mesosome*, and a *metasome*, each with a coelomic compartment. Most species have *gill slits* opening from the pharynx through the body wall of the anterior metasome. A circulatory system is present as well as a complete digestive tract. The brain occurs in the mesosome and the main nerve tracts are in middorsal and midventral positions. A complex of tubules between heart and peritoneum and thought to be excretory structures are in some species. No notochord is found, a diverticulum of the buccal cavity

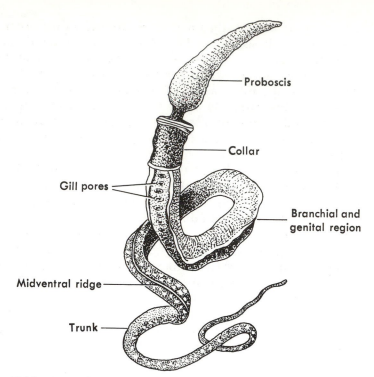

Proboscis

Collar

Gill pores

Branchial and
genital region

Midventral ridge

Trunk

Figure 7-26 Hemichordata. An acorn worm, *Saccoglossus kowalevskii*. (*From Lang, after Spengel.*)

was once erroneously described as one. In species with indirect development a *tornaria larva* is produced that resembles the bipinnaria larvae of some echinoderms. Cleavage is holoblastic and radial.

Classification

Two classes of adult hemichordates are described. A third class, Planctosphaeroida, is known only from larvae.

CLASS I. ENTEROPNEUSTA

Solitary hemichordates; medium length to long worms; mesosome forms a noticeable collar; many gill slits; straight digestive tract; the acorn worms. *Balanoglossus; Saccoglossus,* the most common of several genera; and *Sterobalanus* occur on both coasts. Several other genera also occur on the Pacific Coast.

CLASS II. PTEROBRANCHIA

Colonial; with secreted tubes; small to microscopic; U-shaped digestive tract; with tentacles, bearing resemblance to lophophore. Known mostly from dredgings.

Members of colony connected by a black stolon; two tentacle-bearing arms. Ex. *Rhabdopleura.*

Order 2. Cephalodiscida

Members of colony not connected by a stolon; four or more tentacle-bearing arms. Ex. *Cephalodiscus.*

An extinct class of animals, the Graptolita, were colonial forms secreting tubes resembling the Pterobranchia tubes and thus are often included with the Hemichordata. Some include them with the Hydrozoa.

Phylum Chordata

The phylum Chordata contains invertebrate members in addition to the many thousand species of vertebrates. Members of this phylum have the following characteristics, which set them apart from other bilateral, metameric, coelomate animals, with indeterminate cleavage: (1) a hollow dorsal nerve cord, (2) gill slits, or vestiges of them, and (3) a notochord. The characteristics may not be present in the adult but are present some time during the life cycle. The invertebrate species are all marine. The vertebrate members have been most successful in dominating all habitats of the earth although exceeded numerically by many invertebrate groups from other phyla.

Three subphyla of chordates are recognized. The Hemichordata was formerly considered an additional subphylum but have here been recognized as a separate phylum because of the absence of both metamerism and a true notochord.

SUBPHYLUM I. UROCHORDATA. The tunicates. Notochord absent in adult; sessile or planktonic; with a thick body covering of tunicin, a substance related to cellulose. Ex. *Ciona intestinalis.*

SUBPHYLUM II. CEPHALOCHORDATA. The lancelets. Notochord well developed in the adult. Ex. *Branchiostoma.*

SUBPHYLUM III. VERTEBRATA. The vertebrates. With a backbone.

SUBPHYLUM I. UROCHORDATA

Ciona intestinalis (Fig. 7-27) is a widely distributed, solitary tunicate common on both coasts of this country. The body is more elongate than most tunicates. Opposite the attached end are two prominent openings, the *buccal siphon,* or mouth, and the *atrial siphon.* Water passes in the buccal siphon and is strained through the wall of a very large *pharynx* before passing out the atrial siphon. Planktonic food passing through the openings, or stigmata, in the wall of the pharynx

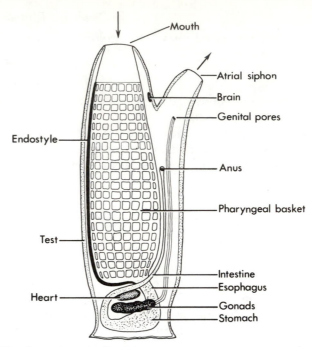

Figure 7-27 *Ciona intestinalis.* Diagrammatic representation of main internal organs. (*Modified after several authors.*)

is moved by cilia on the bars to the *endostyle* where it is trapped in mucus and conveyed to the esophagus. Digestion and absorption occurs in the stomach and intestine, and undigested material is ejected through the anus into the atrium where wastes are carried out the atrial siphon.

The mantle or body covering which secretes the test contains muscles that can contract the tunicate and close the siphons. The coelom is not identifiable. The nervous system is poorly developed. The heart is formed from a fold of the pericardial cavity and has the peculiar fea-

Figure 7-28 Tadpole larva of a urochordate. (*After Van Name.*)

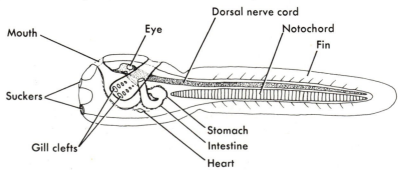

ture of periodically reversing the direction it pumps the blood. Although there are no excretory organs, less soluble wastes are stored by amoebocytic cells which accumulate along the intestine. Although tunicates are usually hermaphroditic and discharge both eggs and sperm into the water, there is apparently no self-fertilization. A larval stage known as the tadpole larva (Fig. 7-28) develops from the egg. The tadpole has a long tail and shows definite chordate characteristics. It becomes sessile by attaching anterior adhesive papillae and then metamorphoses into the adult.

Other Urochordata

Most tunicates are sessile and abundant along marine shores. They may be transparent or of varying colors. A few are pelagic. Three classes are found.

Class 1. Ascidiacea

Sedentary adults that have lost the tail of the tadpole larva. Many produce colonies by asexual budding. They often form dense carpets on wave washed portions of the intertidal zone. When disturbed they contract and discharge two jets of water through the siphons, thus prompting the name *sea squirts*. Ex. *Ciona intestinalis; Halocynthia pyriformis*, the sea peach of the East Coast; *Clavelina huntsmani*, transparent, colonial, common in the tide pools of the West Coast.

Class 2. Thaliacea

Pelagic. The atrium opens posteriorly. The tail of the tadpole is lost in the adult. Ex. *Salpa*.

Class 3. Larvacea

Planktonic. The adult retains the tail of the tadpole larva. Since these are sexually mature in a larval, or larvalike, stage they are said to exhibit neoteny. Ex. *Oikopleura*.

SUBPHYLUM II. CEPHALOCHORDATA

Branchiostoma (Amphioxus) (Figs. 7-29 and 7-30) and other lancelets retain the notochord and hollow dorsal nerve cord in the adult. They look fishlike but have no skeletal container for the brain. The fins are fleshy folds with some connective tissue supports. The muscles of the body wall form V-shaped myotomes. A pharyngeal basket filters out food organisms much as the tunicate's does. The pores in the pharynx are slitlike and called gill slits. They alternate with gill bars. A cecum of the intestine projects anteriorly beside the pharynx. The anus discharges to the outside near the caudal fin. The gonads of the separate sexes discharge into the atrium. Primitive nephridial structures similar to those of some polychaetes are present. Pulsations of the ventral

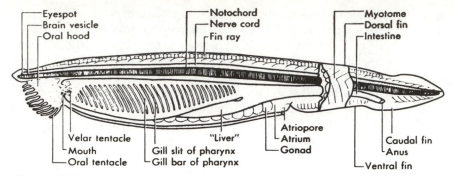

Figure 7-29 Lateral view of an amphioxus with part of the body wall removed from the left side to show the general structure. (*Reprinted with permission of Macmillan Publishing Co., Inc., from* College Zoology, *7th ed., by Robert W. Hegner and Karl A. Stiles. Copyright Macmillan Publishing Co., Inc., 1959.*)

aorta circulates blood through arteries in the gill bars, back through the dorsal aorta, then through the other vessels to the ventral aorta. Coelomic cavities are clearly present. Two closely related families occur in most warm, shallow marine areas. The lancelet spends most of its time partially embedded in the bottom sediments.

CHORDATES IN GENERAL

The invertebrate members of the phylum Chordata are specialized for filter feeding. They are simple organisms of much less complexity than most mollusks, arthropods, and vertebrates. They are of particular interest because they indicate a clearer relationship, of vertebrates with other animals, than would be evident without them. Closest relationship with other phyla appears to be with the Hemichordata and other deuterostomes.

REFERENCES TO LITERATURE ON THE MINOR PHYLA

ABELE, L. G., and S. GILCHRIST. 1977. Homosexual rape and sexual selection in acanthocephalan worms. *Science,* **197**:81–83.

ALVARINO, A. 1965. Chaetognaths. *Oceanogr. Mar. Biol. Ann. Rev.,* **3**:115–194.

BARRINGTON, E. J. W. 1965. *The Biology of Hemichordata and Protochordata.* Freeman, San Francisco. 176 pp.

BUSHNELL, J. H. 1965. On the taxonomy and distribution of freshwater Ectoprocta in Michigan. *Trans. Am. Micr. Soc.,* **84**:231–244, 339–358, 529–548.

CHENG, T. C. 1964. *The Biology of Animal Parasites.* Saunders, Philadelphia. 727 pp.

CORLISS, J. B., et al. 1979. Submarine thermal springs on the Galapagos Rift. *Science,* **203**:1073–1083.

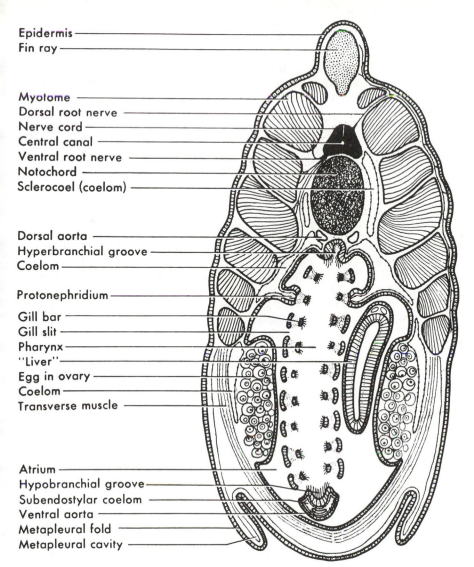

Epidermis
Fin ray

Myotome
Dorsal root nerve
Nerve cord
Central canal
Ventral root nerve
Notochord
Sclerocoel (coelom)

Dorsal aorta
Hyperbranchial groove
Coelom

Protonephridium

Gill bar
Gill slit
Pharynx
"Liver"
Egg in ovary
Coelom
Transverse muscle

Atrium
Hypobranchial groove
Subendostylar coelom
Ventral aorta
Metapleural fold
Metapleural cavity

Figure 7-30 Cross section of an amphioxus in the pharyngeal region, showing various internal structures, including some of the coelomic cavities. (*Reprinted with permission of Macmillan Publishing Co., Inc., from* College Zoology, *7th ed. by Robert W. Hegner and Karl A. Stiles. Copyright Macmillan Publishing Co., Inc., 1959.*)

CROMPTON, D. W. T. 1970. *An Ecological Approach to Acanthocephalan Physiology.* Cambridge Univ. Press, London. 125 pp.

EMIG, C. C. 1977. Embryology of Phoronida. *Am. Zoologist,* **17**:21–37.

ENGEMANN, J. G. 1968. Pogonophora: the oldest living animals? *Pap. Mich. Acad. Sci., Arts Letters,* **53**:105–108.

FOUCART, M. F., S. B. GREGOIRE, and C. JEUNIAUX. 1965. Composition chi-

mique du tube d'un pogonophore (*Siboglinum* sp.) et des formations sque-
lettiques de deux ptérobranches. *Sarsia,* **20:**35–41.

GIBSON, R. 1972. *Nemerteans.* Hutchinson Univ. Library, London. 224 pp.

GRANT, G. C. 1963. Investigations of inner continental shelf waters off lower Chesapeake Bay. Part IV. Descriptions of the Chaetognatha and a key to their identification. *Chesapeake Sci.,* **4:**107–119.

GUTMANN, W. F., K. VOGEL, and H. ZORN. 1978. Brachiopods: biomechanical interdependences governing their origin and phylogeny. *Science,* **199:**890–893.

HADZI, J. 1963. *The Evolution of the Metazoa.* Macmillan, New York. 499 pp.

HEDGPETH, J. W., et al. 1977. Biology of Lophophorates. *Am. Zoologist,* **17:**3–150.

HORRIDGE, G. A. 1965. Relations between nerves and cilia in ctenophores. *Am. Zoologist,* **5:**357–375.

———. 1966. Pathways of co-ordination in ctenophores. Pp. 247–266 in W. J. REES. *The Cnidaria and Their Evolution.* Academic Press, London. 449 pp.

HYMAN, L. H. 1940. *The Invertebrates.* Vol. I. Mesozoa, pp. 233–247; Cteno-phora, pp. 662–696. Vol. II. Rhynchocoela, pp. 459–531. Vol. III. Acanthoce-phala, pp. 1–52; Priapulida, pp. 183–197; Entoprocta, pp. 521–554. Vol. V. Chaetognatha, pp. 1–71; Hemichordata, pp. 72–207; Pogonophora, pp. 208–227; Phoronida, pp. 228–274; Ectoprocta, pp. 275–515; Brachiopoda, pp. 516–609; Sipunculida, pp. 610–696. McGraw-Hill, New York.

IVANOV, A. V. 1963. *Pogonophora.* Consultants Bureau, New York. 479 pp.

KLOTZ, I. M., G. L. KLIPPERSTEIN, and W. A. HENDRICKSON. 1976. Hemerythrin: alternative oxygen carrier. *Science,* **192:**335–344.

KOMAI, T. 1963. A note on the phylogeny of the Ctenophora. Pp. 181–188 in E. DOUGHERTY. *The Lower Metazoa.* Univ. of California Press, Berkeley. 478 pp.

LARWOOD, G. P. (Ed.). 1973. *Living and Fossil Bryozoa.* Academic Press, London. 634 pp.

MAYER, A. G. 1912. *Ctenophores of the Atlantic Coast of North America.* Publ. **162.** Carnegie Inst., Washington.

NIELSEN, C. 1977. Phylogenetic considerations: the prostomian rela-tionships. Pp. 519–534 in R. M. WOOLLACOTT and R. L. ZIMMER. *Biology of Bryozoans.* Academic Press, New York. 566 pp.

RICE, M. E. 1970. Asexual reproduction in a sipunculan worm. *Science,* **167:**1618–1620.

RICHARDSON, J. R., and J. E. WATSON. 1975. Locomotory adaptations in a free-lying brachiopod. *Science,* **189:**381–382.

ROGICK, M. D. 1959. Bryozoa. In W. T. EDMONSON. *Fresh-Water Biology.* Wiley, New York. 1,248 pp.

RUDWICK, M. J. S. 1970. *Living and Fossil Brachiopods.* Hutchinson Univ. Library, London. 199 pp.

RYLAND, J. S. 1970. *Bryozoans.* Hutchinson Univ. Library, London. 175 pp.

SANDBERG, P. A. 1977. Ultrastructure, mineralogy, and development of bryozoan skeletons. Pp. 143–181 in R. M. WOOLLACOTT and R. L. ZIMMER. *Biology of Bryozoans.* Academic Press, New York. 566 pp.

SHAPEERO, W. 1961. Phylogeny of Priapulida. *Science,* **133:**879–880.

————. 1962. The epidermis and cuticle of *Priapulus caudatus* Lamarck. *Trans. Am. Micr. Soc.*, **81**:352–355.

SOUTHWARD, E. C. 1971. Recent researches on the Pogonophora. *Oceanogr. Mar. Biol. Ann. Rev.*, **9**:193–220.

STUNKARD, H. W. 1972. Clarification of taxonomy in the Mesozoa. *Syst. Zool.*, **21**:210–214.

THORPE, J. P., G. A. B. SHELTON, and M. S. LAVERACK. 1975. Colonial nervous control of lophophore retraction in cheilostome Bryozoa. *Science*, **189**:60–61.

TRASON, W. B. 1963. The Life Cycle and Affinities of the Colonial Ascidian *Pycnoclavella stanleyi. Univ. Calif. Publ. Zool.*, **65,** No. 4. 43 pp.

VAN NAME, W. G. 1945. The North and South American Ascidians. *Bull. Am. Mus. Nat. Hist.* **84.** 476 pp. + 31 Pls.

WEBB, M. 1963. *Siboglinum fiordicum* sp. nov. (Pogonophora) from the Raunefjord, Western Norway. *Sarsia*, **13**:33–49.

————. 1964a. The posterior extremity of *Siboglinum fiordicum* (Pogonophora). *Sarsia*, **15**:33–36.

————. 1964b. The larvae of *Siboglinum fiordicum* and a reconsideration of the adult body regions (Pogonophora). *Sarsia*, **15**:57–68.

WOOLLACOTT, R. M., and R. L. ZIMMER (Eds.). 1977. *Biology of Bryozoans.* Academic Press, New York. 566 pp.

WYER, D. W., and P. K. KING. 1973. Relationships between some British littoral and sublittoral bryozoans and pycnogonids. Pp. 199–207 in G. P. LARWOOD. *Living and Fossil Bryozoa.* Academic Press, London. 634 pp.

8

Phylum Annelida

Annelids differ from the other groups of "worms" in possessing the following complex of characteristics: (1) the body is divided into a linear series of similar segments (also called metameres or somites) visible externally because of grooves that encircle the body, and internally because of partitions called septa; (2) the body cavity between the alimentary canal and body wall is a true coelom; (3) there is a single preoral segment, the prostomium; (4) the nervous system consists of a pair of dorsal preoral ganglia, the brain, and a pair of ventral nerve cords with typically a pair of ganglia in each segment; and (5) typically a nonchitinous cuticle on the surface of the body supplied with chitinous bristles or setae.

Four classes may be recognized in the phylum as follows:

CLASS I. POLYCHAETA. Marine annelids with many setae situated on fleshy lateral outgrowths, the parapodia; conspicuous segments; intersegmental septa; a large coelom; usually a well-developed head, bearing appendages; sexes separate; and a free-swimming trochophore larva. Ex. *Nereis virens, Amphitrite ornata, Arenicola cristata.*

CLASS II. OLIGOCHAETA. Mostly terrestrial and freshwater annelids with no parapodia and few setae; conspicuous segments; intersegmental septa; a large coelom; no distinct head with appendages; hermaphroditic; no trochophore larva. Ex. *Lumbricus terrestris, Tubifex tubifex.*

CLASS III. HIRUDINEA. Leeches. Annelids usually dorsoventrally flattened, with a prostomium and 32 body segments; two suckers, one

392

surrounding the mouth and the other at the posterior end; no setae nor parapodia; hermaphroditic; coelom small because of growth of mesenchyme cells. Ex. *Hirudo medicinalis, Macrobdella decora.*

CLASS IV. ECHIUROIDEA. Marine annelids with well-developed prostomium; evidence of segmentation usually absent. Ex. *Echiurus.*

Class I. Polychaeta

NEREIS VIRENS—A CLAMWORM

The clamworm, *Nereis virens,* is a common annelid living in burrows in the sand or mud of the seashore at tide level. The burrows are sometimes 50 cm deep and are kept from collapsing by a lining of mucus, which holds together the grains of sand. By day the clamworm rests in its burrow, but at night it extends its body in search of food, or may leave the burrow entirely.

External Features

The body is flattened dorsoventrally and may reach a length of over 30 cm. The *head* is distinct (Figs. 8-1 and 8-2). Above the mouth is the *prostomium* which bears a pair of terminal *tentacles,* two pairs of simple *eyes,* and, on either side, a thick *palp.* The first true segment is the *peristomium;* from each side of this arise four tentacles or *cirri.* The crustaceans and other small animals that serve as food are captured by a pair of strong chitinous *jaws* which are everted with part of the pharynx when *Nereis* is feeding. Behind the head are a variable number of segments each bearing a fleshy outgrowth on either side, the *parapodia* (Fig. 8-2). These are used as locomotor organs, in addition to undulations of the body that are effective in swimming. New segments are added near the posterior end; here and near the anterior end the parapodia are not as well developed as elsewhere. The peristomium and posterior terminal segment are free from parapodia. Each parapodium consists of a dorsal blade, the *notopodium,* and a ventral blade, the *neuropodium.* The notopodium has a thin, vascular, dorsal lobe, respiratory in function, and bears a dorsal *cirrus* and a bundle of long *setae,* one of which, the *aciculum,* is entirely internal, very thick, and attached to muscles within the body. The neuropodium also bears a cirrus and a similar bundle of setae with an aciculum. The posterior terminal segment bears a pair of ventral cirri that extend posteriorly, and the *anus.*

Body Wall

This (Fig. 8-3) consists of an outer *cuticle,* which is secreted by the cells of the *hypodermis* just beneath it, and several *muscular layers* under the hypodermis. The first are circular muscles; then come two

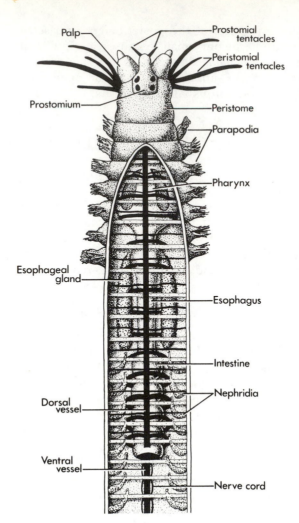

Palp
Prostomial tentacles
Peristomial tentacles
Prostomium
Peristome
Parapodia
Pharynx
Esophageal gland
Esophagus
Intestine
Nephridia
Dorsal vessel
Ventral vessel
Nerve cord

Figure 8-1 *Nereis*, the clamworm. Anterior end of the body with dorsal body wall removed. (*After Parker and Haswell, modified.*)

dorsal and two ventral bands of longitudinal muscles; and also a layer of oblique muscles. These muscles are covered within by a layer of *peritoneal epithelium*.

Coelom

The body cavity between the body wall and the intestine is a coelom (Fig. 8-3). It is divided into chambers by transverse *septa*, which correspond to the limits of the segments. Perforations in the septa beneath the intestine allow fluid to pass from one chamber to the next. The coelom is lined with peritoneal epithelium.

Digestive System

The *mouth* opens into the *pharynx* (Fig. 8-1) which forms a sort of proboscis when protruded (Fig. 8-2, A). This is accomplished by pro-

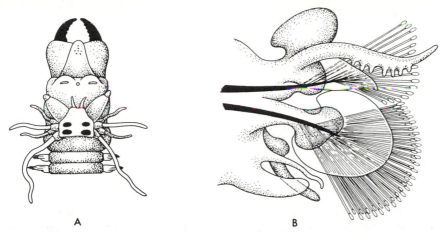

A   B

Figure 8-2 *Nereis,* the clamworm. (A) Head with jaws and pharynx protruded in the position assumed when grasping food. (B) A parapodium showing two acicula (in black), many locomotor setae, and several flattened respiratory gills. (*From Newman, after Leukart.*)

tractor and retractor muscles. The pharynx leads into a slender *esophagus* which has a *digestive gland* on either side opening into it. Following the esophagus is a straight *stomach-intestine* extending to the anus.

Lewis and Whitney (1968) have demonstrated production of cellulase in *Nereis* digestive system shows an increase when the proportion of plant material in the diet is increased.

Figure 8-3 *Nereis.* Diagrammatic transverse section through the body. On the left the chief constituents of the vascular system are represented; on the right side the setae (chaetae) and their muscles, as well as the distribution of the lateral nerve, etc., are shown. (*From Benham.*)

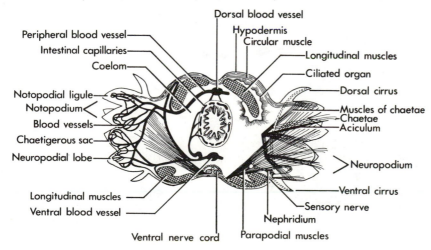

Dorsal blood vessel
Hypodermis
Circular muscle
Longitudinal muscles
Ciliated organ
Dorsal cirrus
Muscles of chaetae
Chaetae
Aciculum
Neuropodium
Ventral cirrus
Sensory nerve
Nephridium
Parapodial muscles
Ventral nerve cord
Ventral blood vessel
Longitudinal muscles
Neuropodial lobe
Chaetigerous sac
Blood vessels
Notopodium
Notopodial ligule
Coelom
Intestinal capillaries
Peripheral blood vessel

Circulatory System

The blood is contained in contractile tubes (Fig. 8-1), the blood vessels. There is a *dorsal vessel*, that lies between the two dorsal longitudinal muscle bands, which carries blood anteriorly, and a *ventral vessel* beneath the intestine which carries blood posteriorly. In each segment the two longitudinal vessels are connected on either side by right and left transverse vessels. From these arise two dorsal and two ventral branchial vessels that form networks of capillaries in the dorsal and ventral lobes of the parapodia. The parapodia serve as respiratory organs.

Excretory System

Every segment except the peristomium and the anal segment contains a pair of *nephridia* (Figs. 8-1 and 8-4, A). Each nephridium opens into the coelomic cavity by means of a *ciliated funnel;* passes posteriorly through the septum into the following segment, where it forms a coiled tube; and opens at the base of a parapodium on the ventral surface through a *nephridiopore.*

Nervous System

Above the pharynx in the head is a pair (Fig. 8-5) of *supra-pharyngeal* ganglia that constitute the *brain.* This is connected with a pair of *subpharyngeal ganglia* by a *circumpharyngeal commissure* on either side that form a ring around the pharynx. Following this is a *ventral nerve cord* with a pair of ganglia in each segment. The brain gives off an optic nerve to each eye, a palpal nerve to each palpus and a tentacular nerve to each group of tentacles. The peristomial tentacles receive nerves from small ganglia connected with the circumpharyngeal commissure. Several dorsal and ventral ganglia around the pharynx, that are connected with one another and with the brain, constitute the *visceral ner-*

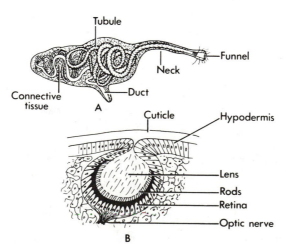

Figure 8-4 *Nereis.* (A) Nephridium. (B) Eye. (*A, after Goodrich; B, after Andrews.*)

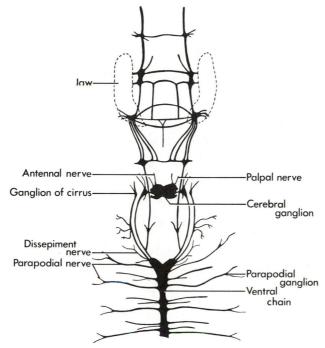

Figure 8-5 *Nereis.* Nervous system, head region. (*After Turnbull, from Petrunkevitch.*)

vous system. From each of the ganglia of the ventral nerve chain arise three pairs of nerves, one pair to the parapodia, one to the anterior segment, and one to the muscles of the segment.

Nereis eyes are light sensitive. The best developed annelid eyes are in rarely collected, transparent, plankton species in the Alciopidae. Several species with well developed eyes have spectral sensitivity that seems to be adaptive to the light at depths they commonly inhabit (Wald and Rayport, 1977).

Reproductive System

The sexes are separate. No well-defined *gonads* are present, but during the breeding season ova or spermatozoa arise from the wall of the coelom in each segment except near the anterior end. At this time the worm is externally differentiated into two regions, an anterior *atoke* and a posterior *epitoke*, which contains the gonads. When thus specialized the worm is called a heteronereid. The gametes leave via the nephridiopores and fertilization occurs in the open water. Trochophore larvae develop from the fertilized eggs.

The principal characteristics of the Polychaeta are exhibited by *Nereis virens*. Many variations from this type occur, some of which are noted below. The orders and a few of the more important families and their commoner representatives are as follows.

Order 1. Errantida

Free-swimming; segments of body similar except at anterior and posterior ends; parapodia with acicula.

FAMILY 1. APHRODITIDAE (Fig. 8-6). Back partly or entirely covered with scales, which usually alternate with slender cirri. *Lepidonotus squamatus* has 12 pairs of scales (elytra); is dark brown and about 3 cm long; and lives under stones near the tide lines along the New England shore.

FAMILY 2. SYLLIDAE. Short, slender worms usually with long slender dorsal cirri; asexual budding common. *Autolytus cornutus* (Fig. 8-7, C) is pinkish and about 15 mm long. Asexual budding results in a linear row of offspring each of which acquires a head before separating from the parent.

Figure 8-6 Polychaeta. *Aphrodite aculeata,* the "sea mouse." (*From Benham, after Regne Animal.*)

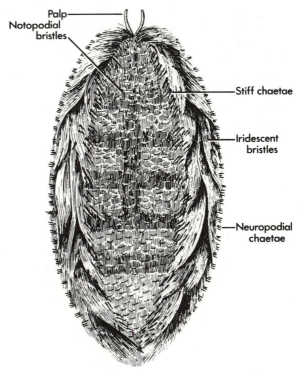

Palp
Notopodial bristles
Stiff chaetae
Iridescent bristles
Neuropodial chaetae

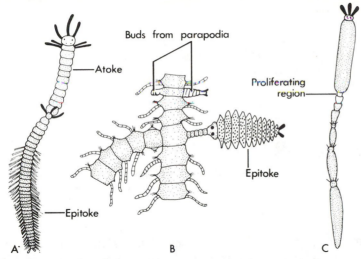

Figure 8-7 Polychaeta. Asexual reproduction. (A) *Syllis*, with the posterior region forming a reproductive individual. (B) *Syllis*, showing branching of the asexual stock and budding of reproductive individuals from parapodia. (C) *Autolytus*, with a chain of reproductive individuals budded off successively from a proliferating region. (*After Borradaile and Potts, modified.*)

FAMILY 3. NEREIDAE. For example, *Nereis virens*, already described.

FAMILY 4. EUNICIDAE. Slender worms of various lengths, with usually branching gills arising from the parapodia near the anterior end; proboscis with complicated jaw apparatus, and generally parchmentlike tube. *Eunice schemacephala* is known as the palolo worm which lives in coral rock in the Gulf of Mexico and the West Indies. Swarming occurs in July within three days of the full moon. At that time the epitokes break free of the atokes and swim to the surface at night. Gametes are freed by ruptures of the epitokes about sunup.

Order 2. Sedentarida

Tube-dwelling; head small or much modified; parapodia simple and without acicula; branchiae localized to a definite region.

FAMILY 1. CHAETOPTERIDAE. Filter feeders, which utilize secreted mucous bags to strain food particles from water circulated through their tube. Typically with a vertical tube and well-developed palps for ejecting large particles (Barnes, 1965). *Chaetopterus* (Fig. 8-8) lives in a U-shaped, parchmentlike tube; the worm may be 15 cm long and has considerable variation in the morphology of the segments and parapodia in different parts of the body; palps are poorly developed.

FAMILY 2. TEREBELLIDAE (Fig. 8-9). Long worms living in burrows or tubes; head with many long tentacular filaments, respiratory in function; usually several pairs of branching gills on segments just back of head; parapodia small. *Amphitrite ornata* is pinkish; up to 30 cm

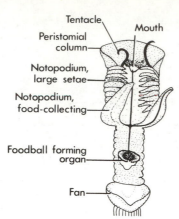

Figure 8-8 Polychaeta. *Chaetopterus*. Dorsal view of anterior end. (*After Borradaile and Potts, modified.*)

long; and lives in a strong tube. *Lanice* has a nonprotein substance that has antitumor activity in mice (Tabrah, Kashiwagi, and Norton, 1970).

FAMILY 3. ARENICOLIDAE. Contains the lugworms. *Arenicola marina* (Figs. 8-10 and 8-11) has a head without tentacles and a proboscis without jaws; the parapodia are rudimentary; branching gills are present above the parapodia of the middle segments of the body. It reaches a length of 20 cm, and lives in deep burrows in the sand.

FAMILY 4. SABELLIDAE. Worms with palps modified into semicircular feathered gills; parapodia rudimentary; living in membranous tubes. *Sabella microphthalma* is greenish yellow and about 5 cm long; the gill filaments are provided with minute eye spots; the tubes often encrust oyster shells.

FAMILY 5. SERPULIDAE. These worms live in calcareous tubes; have prostomial palps modified into semicircular feathered·gills and dorsal gill filaments that serve as an operculum to close the opening to the tube. *Hydroides hexagonus* has usually purplish-brown gills; lives in a

Figure 8-9 Polychaeta. Family Terebellidae. Side view of a young worm removed from its tube. (*After Wilson, modified.*)

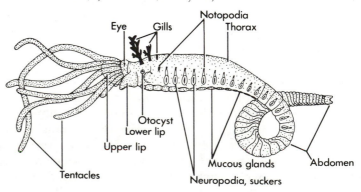

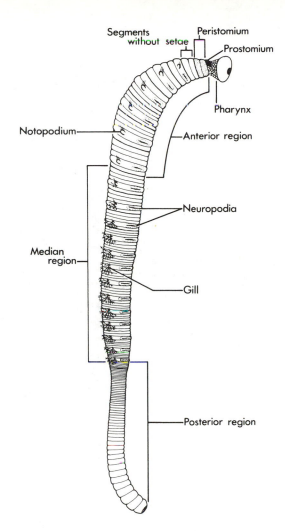

Segments without setae
Peristomium
Prostomium
Pharynx
Notopodium
Anterior region
Neuropodia
Median region
Gill
Posterior region

Figure 8-10 Polychaeta. *Arenicola marina.* Side view. (*After Ashworth, modified.*)

contorted tube encrusted on shells, etc.; and is about 75 mm long. *Spirorbis spirorbis* lives in a coiled tube, often forming a flat spiral, that is encrusted on seaweeds, etc.

Order 3. Myzostomarida

These polychaetes are highly modified as adults for a parasitic existence on or within echinoderms, especially crinoids. The body is externally unsegmented, flat, and oval or discoidal in shape (Fig. 8-12). Five pairs of parapodia armed with acicula and hooks are present on the ventral surface, and around the edge are usually 10 pairs of cirri. Alternating with the parapodia are normally four pairs of suckers.

The digestive system consists of a mouth on the ventral surface near the anterior end, which opens into a retractile pharynx provided with a bulbous musculosus; next comes a short esophagus, then the stom-

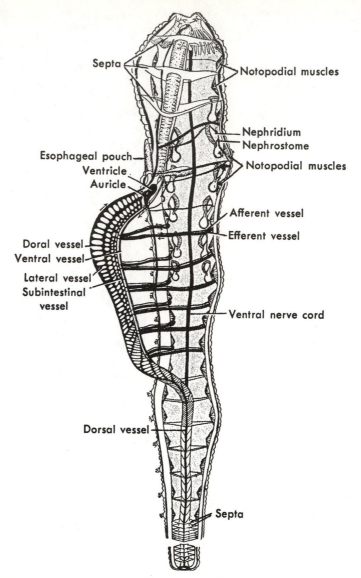

Figure 8-11 Polychaeta. *Arenicola marina.* Dissection to show internal organs. (*After Ashworth, modified.*)

ach, from which a number of ceca arise, and a straight intestine which ends in the anus on the ventral surface near the posterior end. The coelom is greatly reduced by parenchymous tissue. There is no circulatory system except a series of spaces filled with coelomic fluid. No special respiratory system is present. In most species there is one pair of nephridia. The nervous system is well developed and of the ladder type. The worms are normally protandric hermaphrodites, functioning

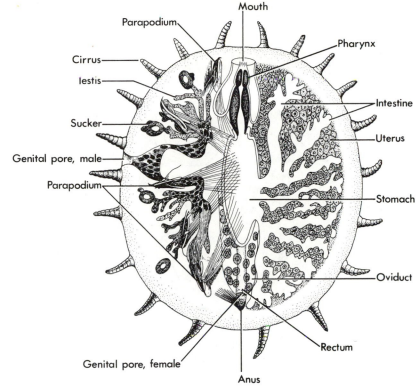

Figure 8-12 *Myzostoma cirriferum.* The organs are supposed to be seen by transparency. On the right side the more dorsal organs are shown and on the left, those lying more ventrally. (*After Land and v. Graff.*)

first as males and then as females. A trochophore larva develops from the egg.

Myzostoma cubanum is a parasite of crinoids in the West Indies. It is about 1.7 mm in diameter and has well-developed parapodia but no suckers. *Myzostoma glabrum* is a species that attaches itself to the oral plates of the crinoid, *Antedon rosacea.* It is nearly circular in outline and about 4 mm long.

Order 4. Archianellida

Small marine polychaetes; parapodia reduced or absent; usually with a tonguelike buccal structure; trochophore larva.

The presence of minute parapodia on *Saccocirrus sonomacus* Martin (1977) supports the inclusion of these worms with the polychaetes. They appear to have their distinctive features as a result of interstitial adaptation.

This order contains six genera of annelids of which *Polygordius* is the best known. *Polygordius* (Fig. 8-13, A) is a marine worm living in the sand. It is about 4 cm long, and only indistinctly segmented exter-

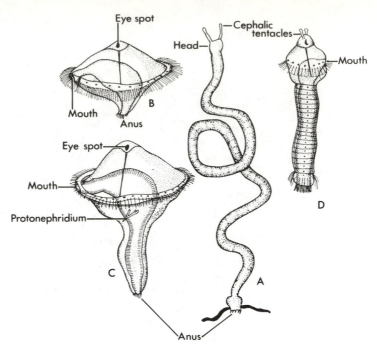

Figure 8-13 Archiannelida. *Polygordius appendiculatus.* (A) Dorsal view. (B) Trochophore larva. (C and D) Stages in development of trochophore into the worm. (*After Fraipont, from Bourne.*)

nally. The prostomium bears a pair of tentacles. The mouth opening is in the ventral part of the first segment, and the anal opening in the last segment. A pair of ciliated pits, one on either side of the prostomium, probably serve as sense organs.

Internally *Polygordius* slightly resembles the earthworm. The coelom is divided into compartments by septa. The internal organs are repeated so that almost every segment possesses coelomic cavities, longitudinal muscles, a pair of nephridia, a pair of gonads, a section of the alimentary canal, and part of the ventral nerve cord. The development of *Polygordius* includes a trochophore stage (Fig. 8-13, B). The adult develops from the trochophore by the growth and elongation of the anal end. This elongation becomes segmented and by continued growth transforms into the adult.

Some authorities think the archiannelids are not primitive, but are specialized in the direction of simplification and reduction, and represent modifications from several groups of polychaetes. Others think they represent remnants of the ancestral annelid stock. In either event, the genera may be arranged in a series beginning with *Polygordius* and ending with forms close to other Polychaeta (e.g., *Nerilla,* Fig. 8-14).

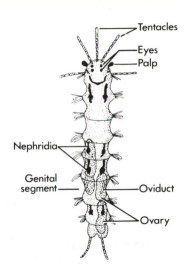

Figure 8-14 Archiannelida. *Nerilla*. Dorsal view of female. (*After Goodrich, modified.*)

Class II. Oligochaeta

LUMBRICUS TERRESTRIS—AN EARTHWORM

The common earthworm, or "night crawler," *Lumbricus terrestris,* or one of its near relatives, is usually used as a type in introductory courses. It illustrates the principal characteristics of the advanced members of the class Oligochaeta so its description will serve for review and reference. Figure 8-15 shows many of the features seen in the usual dorsal dissection.

External Features

The body of *Lumbricus* is cylindroid, and varies in length from about 15 to 30 cm. The *segments,* of which there are over 100, are easily determined externally because of the grooves extending around the body. At the anterior end a fleshy lobe, the *prostomium* (Fig. 8-21), projects over the *mouth;* this is not considered a true segment. It is customary to number the segments with Roman numerals, beginning at the anterior end, because both external and internal structures bear a constant relation to them. Segments XXXI or XXXII to XXXVII are swollen in mature worms, forming a saddle-shaped enlargement, the *clitellum,* of use during reproduction. Every segment except the first and last bears four pairs of chitinous bristles, the *setae* (Fig. 8-16); these may be moved by retractor and protractor muscles, and are renewed if lost. The setae on segment or *somite* XXVI are, in mature worms, modified for reproductive purposes.

The body is covered by a thin, transparent *cuticle* secreted by the cells lying just beneath it. The cuticle protects the body from physical

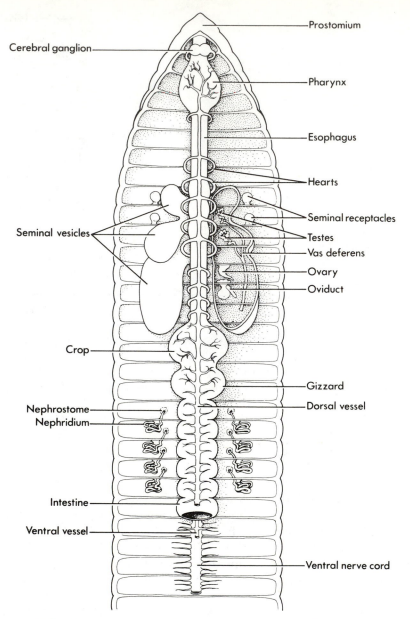

Labels on figure:
- Prostomium
- Cerebral ganglion
- Pharynx
- Esophagus
- Hearts
- Seminal receptacles
- Seminal vesicles
- Testes
- Vas deferens
- Ovary
- Oviduct
- Crop
- Gizzard
- Nephrostome
- Dorsal vessel
- Nephridium
- Intestine
- Ventral vessel
- Ventral nerve cord

Figure 8-15 Earthworm. Anterior part with dorsal body wall removed to show internal organs, diagrammatic. Dorsal wall of seminal vesicles and underlying seminal capsule have been removed on the right side to show gonads. Internal musculature not shown, nephridia are shown in only a few segments.

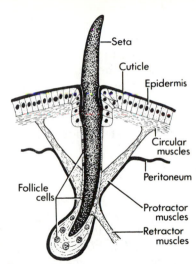

Figure 8-16 Earthworm, *Lumbricus*. Histology of seta and surrounding tissue. (*After Stephenson, modified.*)

and chemical injury; it contains numerous *pores* (Fig. 8-17) to allow the secretions from unicellular glands to pass through, and is marked with fine *striae*, causing the surface to appear iridescent.

A number of *external openings* of various sizes allow the entrance of food into the body, and the exit of feces, excretory products, reproductive cells, etc. (1) The *mouth* is a crescentric opening situated in the ventral half of the first somite (Fig. 8-21); it is overhung by the prostomium. (2) The oval *anal aperture* lies in the last somite. (3) The openings of the sperm ducts or *vasa deferentia* are situated one on either side of somite XV. They have swollen lips; a slight ridge, extends posteriorly from them to the clitellum. (4) The openings of the *oviducts* are small, round pores one on either side of somite XIV; eggs pass out of

Figure 8-17 *Lumbricus*. (A) Superficial view of the cuticle, showing pores and striae. (B) Vertical section of a bit of epidermis, showing four mucous cells in different stages of secretion. Mucus is passing through one of the two pores in the cuticle. (*B, from Dahlgren and Kepner.*)

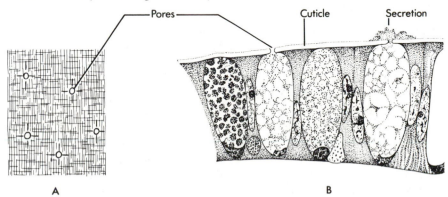

the body through them. (5) The openings of the *seminal receptacles* appear as two pairs of minute pores concealed within the grooves which separate somites IX and X, and X and XI. (6) A pair of *nephridiopores* (Fig. 8-18), the external apertures of the excretory organs, open on every somite except the first three and the last. They are usually situated immediately anterior to the outer seta of the inner pair. (7) The body cavity or *coelom* communicates with the exterior by means of *dorsal pores* located middorsally at the anterior of somites posterior to VII or VIII.

Internal Anatomy

If a specimen is cut open (Fig. 8-15) from the anterior to the posterior end by an incision passing through the body wall a trifle to one side of the middorsal line, a general view of the internal structures may be obtained. The body is essentially a double tube (Fig. 8-18), the body wall constituting the outer, the straight alimentary canal, the inner; between the two is a cavity, the coelom. The external segmentation corresponds to an internal division of the coelomic cavity into compartments by means of partitions called *septa*, which lie beneath the grooves. The alimentary canal passes through the center of the body, and is suspended in the coelom by the partitions. Septa are absent between somites I and II, and incomplete between somites III and IV, and XVII and XVIII. The walls of the coelom are lined with an epithelium, termed the *peritoneum*. The coelomic cavity is filled with a colorless fluid. In somites IX to XVI are the reproductive organs; running along the upper surface of the alimentary canal is the dorsal blood vessel; and just beneath it lie the ventral blood vessels and nerve cord.

The coelomic fluid in the various compartments also serves as a *hydrostatic skeleton* for the earthworm. The opposing action of circular and longitudinal muscles (Fig. 8-18) of the body wall is mediated through hydraulic action of the body fluid. Thus, extension of the worm is accomplished by contraction of the circular muscles. Loss of much body fluid causes a great loss of mobility.

Digestive System

The *alimentary canal* (Fig. 8-15) consists of (1) a *mouth cavity* or *buccal pouch* in somites I to III, (2) a thick muscular *pharynx* lying in somites IV and V, (3) a narrow, straight tube, the *esophagus* which extends through somites VI to XIV, (4) a thin-walled enlargement, the *crop* or *proventriculus,* in somites XV and XVI, (5) a thick muscular-walled *gizzard* in somites XVII and XVIII, and (6) a thin-walled *intestine* extending from somite XIX to the *anal aperture.* The intestine is not a simple cylindrical tube; but its dorsal wall is infolded, forming an internal longitudinal ridge, the *typhlosole* (Fig. 8-18). This increases the digestive surface. Surrounding the alimentary canal and dorsal blood vessel is a layer of *chloragogen cells.* The functions of these cells are somewhat

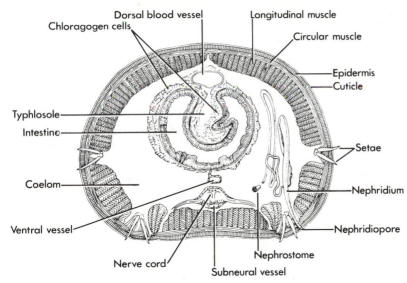

Figure 8-18 *Lumbricus.* Transverse section through the middle region of the body. (*From Woodruff.*)

comparable to those of the liver of higher animals as centers of glycogen and fat synthesis and storage. Three pairs of *calciferous glands* lie at the sides of the esophagus in segments X to XII; they produce calcium carbonate, which they excrete into the digestive tract when the component ions are overly abundant in the body fluids.

The *food* of the earthworm consists principally of pieces of leaves and other vegetation, particles of animal matter, and soil. This material is gathered at night. At this time the worms are active; they crawl out into the air, and, holding fast to the tops of their burrows with their tails, explore the neighborhood. Food particles are drawn into the buccal cavity by suction produced when the pharyngeal cavity is enlarged by the contraction of the muscles which extend from the pharynx to the body wall.

In the pharynx, the food receives a secretion from the pharyngeal glands; it then passes through the esophagus to the crop, where it is stored temporarily. In the meantime the secretion from the calciferous glands in the esophageal walls is added, neutralizing the acids. The gizzard is a grinding organ; in it the food is broken up into minute fragments by being squeezed and rolled about. Solid particles, such as grains of sand, which are frequently swallowed, probably aid in this grinding process. The food then passes on to the intestine, where most of the digestion and absorption takes place.

Circulatory System

The *blood* of the earthworm is contained in a complicated system of tubes which ramify to all parts of the body. A number of these tubes are large and centrally located; these give off branches which likewise branch, finally ending in exceedingly thin tubules, the *capillaries*. The *blood* consists of a plasma in which are suspended a great number of colorless cells, called corpuscles. Its red color is due to a pigment termed *hemoglobin*, which is dissolved in the plasma. In vertebrates the hemoglobin is located in the blood corpuscles.

There are *five longitudinal blood vessels* connected with one another and with various organs by branches, more or less regularly arranged. These are shown in Figure 8-19, and are as follows: (1) the dorsal or

Figure 8-19 *Lumbricus.* Diagrams showing the arrangement of the blood vessels. (A) Longitudinal view of the vessels in somites VIII, IX, and X. (B) Transverse section of same region. (C) Longitudinal view of the vessels in the intestinal region. (D) Transverse section through the intestinal region. (*From Bourne, after Benham.*)

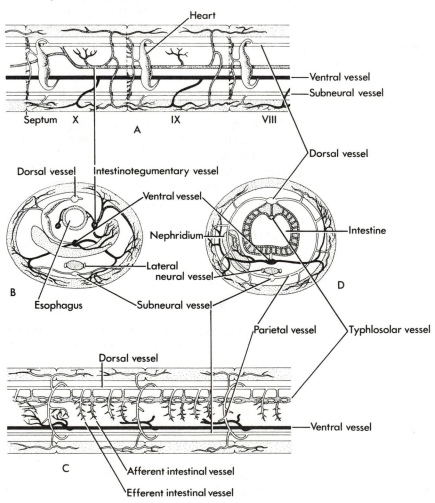

supraintestinal vessel, (2) the ventral or subintestinal trunk, (3) the subneural trunk, (4) two lateral neural trunks, (5) five pairs of hearts in segments VII to XI, (6) two intestinotegumentary vessels (in A and B) arising in segment X and extending to the esophagus, integument, and nephridia in segments X to VI, (7) branches from the ventral trunk to the nephridia and body wall (D), (8) parietal vessels connecting the dorsal and subneural trunks in the intestinal region, (9) branches from the dorsal trunk to the intestine (in C), (10) a typhlosolar vessel connected by branches with the intestine and dorsal trunk, and (11) branches from the ventral vessels to the nephridia and body wall (in D).

The dorsal trunk (Fig. 8-20) and hearts determine the direction of the blood flow, since they furnish the power by means of their muscular walls. Blood is forced forward by wavelike contractions of the dorsal trunk, beginning at the posterior end and traveling quickly anteriorly. These contractions are said to be *peristaltic*, and have been likened to the action of the fingers in the operation of milking a cow. *Valves* in the walls of the dorsal trunk prevent the return of blood from the anterior end. In somites VII to XI the blood passes from the dorsal trunk into the hearts, and is forced by them both forward and backward in the ventral trunk. Valves in the heart also prevent the backward flow. From the ventral trunk the blood passes to the body wall and nephridia. Blood is returned from the body wall to the lateral-neural trunks. The flow in the subneural trunk is toward the posterior end, then upward through the parietal vessels into the dorsal trunk. The anterior region receives blood from the dorsal and ventral trunks. The blood carried to the body wall and integument receives oxygen through the cuticle, and is then returned to the dorsal trunk by way of

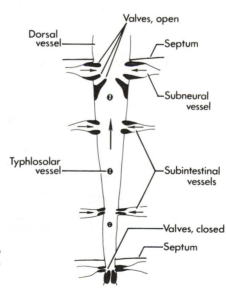

Figure 8-20 Earthworm, *Lumbricus.* Diagram of dorsal blood vessel to show connections and valves. (*After Johnston.*)

the subneural trunk and the intestinal connectives. Because of its proximity to the subneural trunk, the nervous system receives a continuous supply of the freshest blood.

Respiration

The earthworm possesses no respiratory system, but obtains oxygen and gets rid of carbon dioxide through the moist outer membrane. Many capillaries lie just beneath the cuticle, making the exchange of gases easy. The oxygen is combined with the hemoglobin.

Excretory System

Most of the exretory matter is carried outside of the body by a number of coiled tubes, termed *nephridia* (Fig. 8-18), a pair of which are present in every somite except the first three and the last. A nephridium occupies part of two successive somites; in one is a ciliated funnel, the *nephrostome,* which is connected by a thin ciliated tube with the major portion of the structure in the somite posterior to it. Three loops make up the coiled portion of the nephridium. The cilia on the nephrostome and in the nephridium create a current which draws solid waste particles from the coelomic fluid. Glands in the coiled tube take waste matter, principally ammonia and urea, from the blood, and the current in the tube carries it out through the *nephridiopore.* Most salts are removed before the fluid is excreted.

Nervous System

The nervous system is concentrated. There is a bilobed mass of nervous tissue, the *brain* or *suprapharyngeal ganglion,* on the dorsal surface of the pharynx in segment III (Fig. 8-21). This is connected by two *circumpharyngeal connectives* with a pair of *subpharyngeal ganglia* which lie just beneath the pharynx. From the latter the *ventral nerve cord*

Figure 8-21 *Lumbricus.* Diagram of the anterior end to show the arrangement of the nervous system. (*From Shipley and MacBride.*)

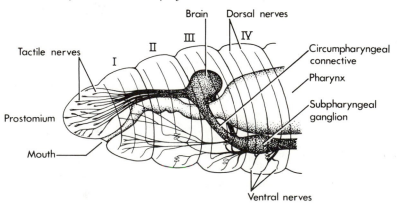

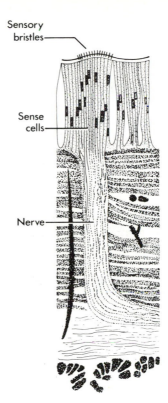

Figure 8-22 *Lumbricus.* Tactile nerve endings in the integument. (*From Dahlgren and Kepner.*)

(Figs. 8-15 and 8-18) extends posteriorly near the ventral body wall. The ventral nerve cord enlarges into a ganglion in each segment and gives off three pairs of nerves in every segment posterior to IV. Each ganglion really consists of two ganglia fused together. Near the dorsal surface of every ganglionic mass are three longitudinal cords, the giant fibers (Fig. 8-23). The brain and nerve cord constitute the *central nervous system;* the nerves that pass from and to them represent the *peripheral nervous system* (Fig. 8-22).

The nerves of the peripheral nervous system are either efferent or afferent. *Efferent* nerve fibers (Fig. 8-23) are extensions from cells in the ganglia of the central nervous system. They pass out to the muscles or other organs, and, because impulses sent along them give rise to movements, the cells of which they are a part are said to be *motor nerve cells.* Some have fibers that exert an excitatory effect whereas other have an inhibitory effect. The *afferent* fibers originate from epidermal nerve cells that are sensory in function, and extend into the ventral cord. The two types may have synaptic connections, or be connected synaptically by *interneurons,* which also send processes to ganglia of neighboring segments.

Neurosecretory substances have been found in the brain of *Lumbricus.* These secretions are thought to function much like hormones.

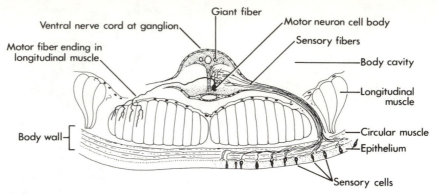

Figure 8-23 *Lumbricus.* Diagram of primary sensory and motor neurons of the ventral nerve cord, showing their connections with the skin and the muscles to form a simple reflex arc. (*After Parker, from Woodruff.*)

Some seem to regulate salt and water balance (Kamemoto, Kato, and Tucker, 1966).

Sense Organs

The sensitiveness of *Lumbricus* to light and other stimuli is due to the presence of a great number of *epidermal sense organs.* These are groups of sense cells connected with the central nervous system by means of nerve fibers and communicating with the outside world through *sense hairs* which penetrate the cuticle. More of these sense organs occur at the anterior and posterior ends than in any other region of the body.

Reproduction

Both male and female sexual organs occur in a single earthworm. Figure 8-15 shows diagrammatically the position and shape of the various structures. The *female system* consists of: (1) a pair of *ovaries* in segment XIII; (2) a pair of *oviducts* which open by a ciliated funnel in segments XIII, enlarge into an *egg sac* in segment XIV, and then open to the exterior; and (3) two pairs of *seminal receptacles* or *spermathecae,* in somites IX and X. The *male organs* are (1) two pairs of glove-shaped *testes* in segments X and XI, (2) two *vasa deferentia* which lead from ciliated funnels to the exterior in segment XV, and (3) three pairs of *seminal vesicles* in segments IX, XI, and XII, and two central reservoirs.

Self-fertilization does not take place, but spermatozoa are transferred from one worm to another during a process called *copulation.* Two worms come together, as shown in Figure 8-24, A; slime tubes are formed, and then a bandlike *cocoon* is secreted about the clitellar region. Eggs and spermatozoa are deposited in the cocoon, but fertilization does not occur until the cocoon is slipped over the head (Fig. 8-24, B).

The *eggs* of the earthworm are *holoblastic,* but cleavage is unequal. A

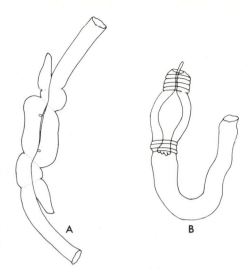

Figure 8-24 *Lumbricus.* (A) The anterior segments of two copulating earthworms. Slime tubes encircle the pair from the 8th to the 33rd segment. (B) Cocoon, freshly deposited, surrounded by one half of a slime tube. (*After Foot.*)

hollow blastula is formed, and a gastrula is produced by invagination. The mesoderm develops from two of the blastula cells, called *mesoblasts*. These cells divide, forming two *mesoblastic bands* which later become the epithelial lining of the coelom. The embryo escapes from the cocoon as a small worm in about 2 to 3 weeks. Stages in the development of an earthworm are shown and explained in Figure 8-25.

Regeneration and Grafting.

Earthworms have considerable powers of regeneration and grafting. Some of the results of experiments along this line are shown in Figure 8-26. A posterior piece may regenerate a head of five segments (A) or in certain cases a tail (B). Such a double-tailed worm slowly starves to death. An anterior piece regenerates a tail (C). Three pieces from several worms may be united so as to make a long worm (D); two pieces may fuse, forming a worm with two tails (E); and an anterior piece may be united with a posterior piece to make a short worm (F). In all these experiments the parts were held together by threads until they became united.

Cooper (1969) found that grafts from different donor species were eventually rejected. Those from the donors in the same family lasted longer before rejection than those from donors in other families.

Ecology

Darwin (1881) has contributed greatly to the understanding of the importance of earthworms in soil formation. More recent information on the subject has been assembled by Rodale (1961) and Edwards and Lofty (1977). They improve the soil by burrowing through it and increasing its porosity while mixing organic and inorganic materials and

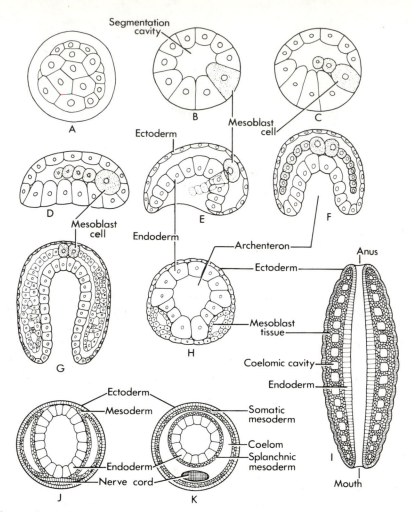

Figure 8-25 *Lumbricus.* Diagrams of stages in development. (A) Blastula (surrounded by a membrane). (B) Section of a blastula, showing blastocoel and one of the primary cells (pole cells) of the mesoderm. (C) Later blastula with developing mesoderm bands. (D) Start of gastrulation. (E) Lateral view of gastrula, showing invagination, which as it proceeds leaves the mesoderm bands on either side of the body as indicated by the cells represented with dotted outline. (F) Section of E, along the line, to show pole cells, mesoderm bands and enteric cavity. (G) Later stage, showing cavities in the mesoderm bands. (H) The same as G, in cross-section. (I) Diagram of a longitudinal section of a young worm after formation of mouth and anus. (J) The same in cross section. (K) Later stage in cross section. (*From Woodruff, after Sedgwick and Wilson.*)

making some of the inorganic materials available to the plants. An interesting side effect of some species voiding their earth containing feces on the surface has been the gradual burial of such things as abandoned Indian cities in Central America.

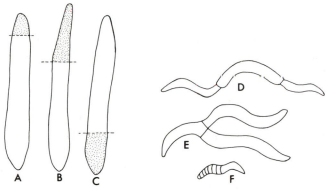

Figure 8-26 *Lumbricus.* (A) Head end of five segments regenerated from the posterior piece of a worm. (B) Tail regenerated from the posterior piece of a worm. (C) Tail regenerated from an anterior piece of a worm. (D) Union of three pieces to make a long worm. (E) Union of two pieces to make a double-tailed worm. (F) Anterior and posterior pieces united to make a short worm. The dotted portions represent regenerated material. (*From Morgan.*)

The accumulation of insecticide residues in earthworms feeding on leaf litter from sprayed trees and the subsequent death of robins feeding on the worms has caused a great rise in public appreciation for ecology following its publicity through Rachael Carson's book, *Silent Spring.*

OTHER OLIGOCHAETA

The principal characteristics of the Oligochaeta are exhibited by *Lumbricus terrestris.* Many variations from the type occur as may be seen from the brief descriptions of some of the more common families.

Classification

Stephenson (1930) classified the oligochaetes into two groups. (1) The Megadrili contained the earthworm families of larger annelids with many segments and sexual reproduction. This group includes the families Lumbricidae, Megascolecidae, Moniligastridae, and Haplotaxidae. (2) The Microdrili included smaller, mostly aquatic, oligochaetes with few segments and frequently asexual reproduction. The microdrile families were Aeolosomatidae, Naididae, Tubificidae, Enchytraeidae, Lumbriculidae, and Branchiobdellidae.

Subsequent classifications have emphasized the positioning of reproductive organs, excretory organs, calciferous glands, and types of setae. Differences in emphasis and interpretation have made the classification unstable.

Brinkhurst and Jamieson (1971) recognize the (1) Lumbriculida and (2) Moniligastrida, each with a single family, and the (3) Haplotaxida, containing the remaining families of oligochaetes, but excluding Aeolosomatidae and Branchiobdellida as closer to other classes. The removal of the Aeolosomatidae because of a simplified gondal condition is not accepted here, although a position as the oligochaetes nearer the polychaetes is acknowledged as very likely.

The Plesiopora–Prosopora–Opisthopora grouping of the last edition is abandoned for a simple listing of some common families. About a dozen families of megadriles, or larger earthworms close to Lumbricidae and Megascolecidae exist but are less likely to be encountered in developed areas.

The variability of morphology in oligochaetes may have occurred in numerous directions and at varying rates. Biochemical and ultrastructural evidence would help to substantiate an acceptable system of order level groupings when one is developed.

FAMILY 1. AEOLOSOMATIDAE. Freshwater worms; microscopic in size; four bundles of setae in each segment; reproduction by asexual division. *Aeolosoma hemprichi* (Fig. 8-27) is 1 mm long and spotted with red oil globules in the integument; lives among algae; and consists of from 7 to 10 segments.

FAMILY 2. NAIDIDAE. Aquatic, mostly in fresh water; small; transparent; two to four bundles of setae on each segment, the ventral setae forked; reproduction principally by transverse division, long chains of offspring formed. *Nais elinguis* (Fig. 8-28) is light brown; 2 to 4 mm long; with 15 to 37 segments; head distinct; often among algae. *Stylaria lacustris* is 25 mm long; has a long, tentaclelike prostomium and 25 segments. *Dero digitata* is reddish; lives in tubes in ponds; has

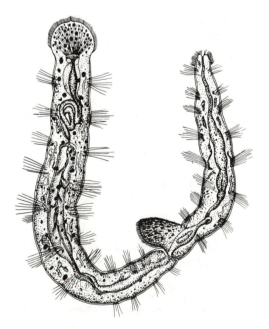

Figure 8-27 Oligochaeta. *Aeolosoma* dividing transversely. (*After Lankester.*)

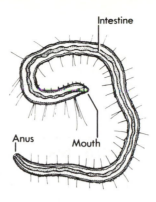

Figure 8-28 Oligochaeta. *Nais.* (*After Leunis.*)

ciliated branchial appendages at the posterior end; and consists of about 48 segments. *D. hylae* is a symbiont in the ureters of frogs. *Chaetogaster limnaei* may be free-living or attached to or in the liver of freshwater snails (*Lymnaea* or *Planorbis*); it is colorless; and has two bundles of ventral hooked setae on each segment.

FAMILY 3. TUBIFICIDAE. Fresh- and brackish-water species living in tubes; slender; with four bundles of setae on each segment; no asexual reproduction by transverse division. *Tubifex tubifex* is reddish and about 4 cm long; it occurs in patches on muddy bottoms where the posterior end of the body, which protrudes from the tube, is waved back and forth. *Limnodrilus hoffmeisteri* resembles *Tubifex* but lacks hair setae in the dorsal bundles of setae. It is probably the most common tubificid worm found in organically polluted waters.

FAMILY 4. ENCHYTRAEIDAE. Both terrestrial and aquatic species; four bundles of hairlike setae in each segment; usually whitish in appearance; up to 25 mm long. *Enchytraeus albidus* is milk-white, slender, and about 25 mm long; it has from 53 to 69 segments, and the setae are nearly straight and of equal length; it lives under débris along the seashore.

FAMILY 5. LUMBRICULIDAE. Free-living aquatic species; four pairs of setae per segment; usually reddish. *Lumbriculus variegatus* has a terminal fork on all the setae. Cook (1975) notes that lumbriculids thrive in oligotrophic situations better than tubificids and thus have been able to invade cave environments.

FAMILY 6. LUMBRICIDAE. Common earthworms. *Lumbricus terrestris* has already been described as the type of this class. *Eisenia foetida* lives in manure and is pink with a dark ring on each segment. *Allolobophora trapezoides* is a common garden worm found in the soil and may be pink, blue, yellow, or gray. These three species are common in such widely separated areas as the United States, England, and New Zealand.

FAMILY 7. MEGASCOLECIDAE. Terrestrial, some species aquatic; many tropical species. *Diplocardia communis* is a North American species that lives in the soil, is flesh colored, and about 30 cm long. The

Asian *Pheretima* has been introduced into the United States. The giant earthworm of Australia, *Megascolides,* is said to exceed 3 meters in length. Other giant species occur in South America and Africa. This family of very large worms has provided excellent material for experimental studies of nerve impulse transmission and other physiological problems.

Class III. Hirudinea

HIRUDO MEDICINALIS—THE MEDICINAL LEECH

The animals included in this class are commonly called *leeches.* They are usually flattened dorsoventrally, but differ externally from the flatworms (Platyhelminthes, Chap. 5) in being distinctly segmented. The external segmentation, does not correspond exactly to the internal segmentation because there are a variable number of external rings or *annuli* (from 1 to 14) to every segment, for example, usually five in the medicinal leech, *Hirudo* (Fig. 8-29), and its allies, and three in *Glossiphonia.* Anatomical features that distinguish the Hirudinea from other Annelida are (1) the presence of a definite number of segments (33 and a prostomium, based on the larval number of ganglia), (2) two suckers, one found around the mouth and the other at the posterior

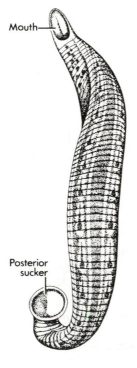

Mouth

Posterior sucker

Figure 8-29. *Hirudo medicinalis,* the medicinal leech. (*From Shipley and MacBride.*)

420   *Phylum Annelida*

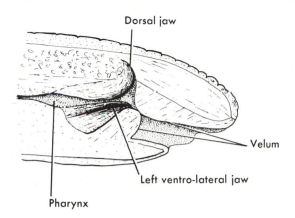

Figure 8-30 Gnathobdel-lida. Dissection showing jaws of leech.

Dorsal jaw

Velum

Left ventro-lateral jaw

Pharynx

end, and (3) the absence of setae (except in one genus). They are her-maphrodites.

Hirudo medicinalis, the medicinal leech (Fig. 8-29), is usually selected as an example of the class. It is about 10 cm long, but is capable of great contractions and longation. The *suckers* are used as organs of at-tachment, and during locomotion are alternately fastened to and re-leased from the substratum, the animal looping along like a measuring worm. Leeches are also able to swim through the water by undulating movements.

The *alimentary tract* (Fig. 8-31) is fitted for the digestion of the blood of vertebrates, which forms the principal food of some leeches. The *mouth* (Fig. 8-30) lies in the anterior sucker and is provided with three *jaws* armed with chitinous teeth for biting. The blood flow caused by the bite of a leech is difficult to stop, since a secretion from glands opening near the jaws prevents coagulation. Blood is sucked up by the dilation of the muscular *pharynx.* The short *esophagus* leads from the pharynx into the *crop,* which has 11 pairs of lateral branches. Here the blood is stored until digested, presumably by the action of symbiotic bacteria. A leech is able to ingest three times its own weight in blood, and, since it may take as long as 9 months to digest this amount, meals are few and far between. The *intestine* leads from the crop and stomach to the rectum which opens through the *anus* on the dorsal surface of the posterior sucker. The absorbed food passes into the *blood vessels* and the *coelomic cavities,* and is carried to all parts of the body. The coelom is usually small because of the development of the *botryoidal tissue* (Fig. 8-32), which is homologous to the chloragogen tissue of the oligochaetes. The spaces in the body which are not filled up by this tissue are called sinuses, and in many species contain a fluid very much like true blood. Reduction of the blood circulatory system with the assumption of much of its function by the coelomic spaces has been a general trend of leech specialization.

Respiration is carried on at the surface of the body, oxygen being

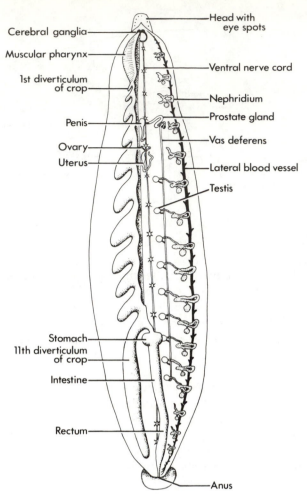

Labels on figure:
Cerebral ganglia
Muscular pharynx
1st diverticulum of crop
Penis
Ovary
Uterus
Head with eye spots
Ventral nerve cord
Nephridium
Prostate gland
Vas deferens
Lateral blood vessel
Testis
Stomach
11th diverticulum of crop
Intestine
Rectum
Anus

Figure 8-31 *Hirudo medicinalis,* internal organs. (*From Shipley and MacBride.*)

taken into, and carbon dixoide given off by, many blood capillaries in the skin. Waste products are extracted from the blood and coelomic fluid by 17 pairs of *nephridia* (Fig. 8-31) which resemble those of the earthworm (Fig. 8-18), but frequently lack the internal opening.

The *ventral nerve cord* is bathed in blood contained in the surrounding *ventral sinus.* Giant fibers are not prominent as in *Lumbricus.* Segmental ganglia of *Hirudo* contain large cell bodies (Retzius cells) of neurons, some of which Lent (1973) has demonstrated function to release mucus. They do not inhibit muscle contraction as had been proposed earlier.

Leeches are *hermaphroditic* (Fig. 8-33), but the eggs of one animal are fertilized by spermatozoa from another leech. The *spermatozoa* arise in the nine pairs of segmentally arranged *testes* (Fig. 8-31): they pass into the *vas deferens,* then into a convoluted tube called the *epididymus,*

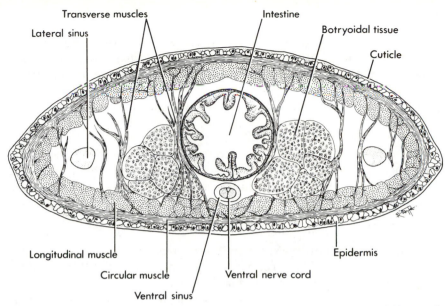

Figure 8-32 Leech. Cross section showing principal features of the mid-region.

where they are fastened into bundles called *spermatophores,* and are finally deposited within the body of another leech by means of the muscular *penis.* The eggs arise in the *ovaries* of which there is a single pair; they pass into the *oviducts,* then into the *uterus,* and finally out through the *genital pore* ventrally situated in segment XI. *Copulation* and the formation of a *cocoon* are similar to these processes in the earthworm.

Figure 8-33 Gnathobdellida. Diagrammatic, lateral view of the main features of the reproductive system. Right members of paired organs are omitted.

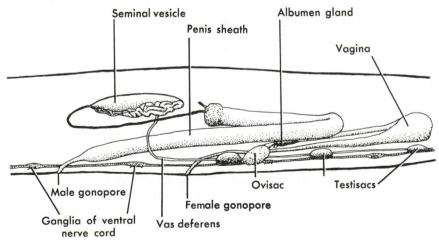

OTHER HIRUDINEA

Hirudo medicinalis (Figs. 8-29 and 8-31) has served to illustrate the principal characteristics of this class. The leeches are also known as "blood suckers," although not all leeches feed on blood. Some are predators on small invertebrates which they swallow whole. Variations from the type described are indicated in the following descriptions.

Order 1. Acanthobdellida

Setae present on anterior end of body; no anterior sucker; parasitic on fish. Includes only the single genus, *Acanthobdella,* which differs so from other leeches it is sometimes separated into its own subclass. Found only in Eurasia.

Order 2. Rhynchobdellida

Both suckers present; no setae; without jaws; proboscis can be protruded from mouth; blood colorless; segments usually of three or four, rarely five, rings; freshwater and saltwater species.

FAMILY 1. PISCICOLIDAE. Mostly marine leeches (Fig. 8-34, A). A few freshwater species, mostly parasites of fish. Body long; more than three annuli on each segment; suckers pedunculate. *Piscicola* contains species parasitic on a variety of fish. *Cystobranchus verrilli* possesses 11 pairs of brownish or purplish papiliform gill-vesicles along the sides of the body, these are larger and more evident on preserved leeches than those of *Piscicola.*

FAMILY 2. GLOSSIPHONIIDAE. Freshwater species; each segment with three annuli; usually very flat and broad near the posterior; anterior sucker fused with the body; posterior sucker distinct. *Placobdella parasitica* occurs under stones or attached to turtles. The body is broad, flat, and greenish or yellowish. *Helobdella stagnalis* is a very active, small species that feeds on snails. It can be recognized by a small brownish plate on the dorsal surface of segment VIII and the general grayish or pinkish color. *Hemiclepsis marginata* attacks frogs and toads. It is greenish with longitudinal stripes. Glossiphonid leeches are very commonly found with eggs or young in a pouchlike area on the ventral surface (Fig. 8-34, B). Undulations by the parent leech ensure an adequate supply of oxygen for the developing leeches.

Order 3. Gnathobdellida

Leeches mostly with jaws; no proboscis present; red blood; freshwater and terrestrial species.

Suborder 1. Gnathobdellae. Jaws with teeth present.

FAMILY 1. HIRUDINIDAE. Each segment with five rings; three toothed jaws; five pairs of eyes. *Hirudo medicinalis* has been described. *Haemopis marmoratis,* the horse leech, is about 10 cm long; it lives in

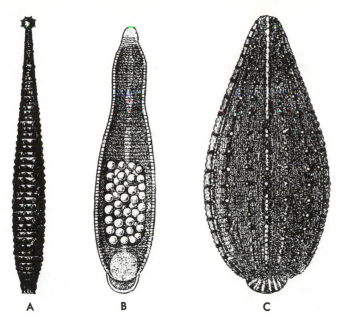

Figure 8-34 Leeches. (A) *Pontobdella muricata*, a green-colored marine species that attacks rays and sharks. (B) *Helobdella algira*, a species that transmits trypanosomes to frogs. Ventral view, showing eggs. (C) *Placobdella catenigera*, a species that transmits hemogregarines to tortoises. (*A, after Bourne; B and C, after Brumpt.*)

mud near fresh water and feeds principally on worms, snails, etc., occasionally sucking blood. *Macrobdella decora* is very common in fresh water and sucks blood avidly from frogs, fish, cattle, humans, etc., although it also eats other small invertebrates. It is easily recognized by its orange ventral surface.

FAMILY 2. HAEMADIPSIDAE. Terrestrial leeches of the tropics and subtropics, especially Southeast Asia; membraneous fold around suckers; blood feeders. Found in moist areas where they will extend from the tips of vegetation to attach to passing mammals by pointing their anterior sucker toward warm objects moving by them.

Suborder 2. Pharyngobdellae. Leeches with three muscular ridges in place of jaws. Represented principally by the family Erpobdellidae, the worm leeches. These leeches lack teeth and are not blood feeders but are carnivores that feed on small invertebrates. *Erpobdella punctata* is common in ponds and streams. It has three pairs of eyes; is about 8 cm long; and is brownish black with four rows of black spots along the dorsal surface.

Order 4. Branchiobdellida

These aberrant forms may be offshoots of primitive leeches. They have fewer segments than other leeches and have been classified as oligochaetes because of somewhat intermediate characteristics. Only one family exists, the family Branchiobdellidae. All species epizoic on crayfishes; with sucker at the posterior end of the body; no setae present; one dorsal and one ventral chitinous jaw. *Cambarincola philadelphica* is 10 mm long; the dorsal and ventral jaws are dissimilar; it lives principally on the ventral surface of the crayfish. Branchiobdellids are thought to be commensals of the crayfish and feed upon algae and other available food particles.

Class IV. Echiuroidea

The class Echiuroidea is sometimes considered as a separate phylum. The members show relationships with the Sipunculida and Annelida. They are included here because of their annelid-type nervous system; nephridia, several pairs in some; circulatory system; setae; and trochophore larvae. The prostomium is well developed as a proboscis which is used for locomotion, for capturing prey, and as an organ of sense. Sexes are separate. All are marine-bottom dwellers. *Urechis caupo,* the "innkeeper," is an interesting west coast species that lives and feeds in a U-shaped burrow. It uses a mucous net to gather food in a manner much as did *Chaetopterus.* When it swallows the net, any lost particles are quickly seized by one of the commensals living with it in the burrow. The scale worm, *Hesperonoe adventor* lives only in association with *Urechis.* Other commensals usually encountered with *Urechis* in-

Figure 8-35 Echiuroidea. *Bonellia viridis.* (A) Female. (B) Male from nephridium of female. (*After Spengler, modified.*)

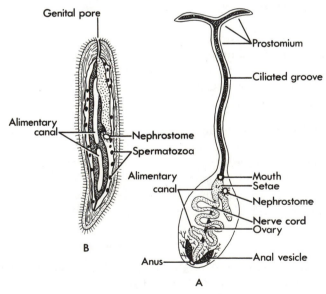

clude a pea crab, goby fish, and a small clam. One reason *Urechis* is a favorable source of material for embryological studies is that eggs and sperm can be removed from them during most of the year. *Urechis* has some peculiarities found also in some distant phyla: a siphon parallels part of the intestine, the posterior portion of the intestine becomes engorged with water taken in through the anus for respiration, and the circulatory system is aided by ciliary circulation of coelomic fluid (as well as by peristaltic movements) although the blood vascular system is reduced.

The other echiuroids possess an annelid-type closed blood vascular system. *Echiurus pallasi* possesses a spoon-shaped proboscis and 22 body rings. In *Thalassema melitta* the proboscis is somewhat pointed. This species often occurs in sand-dollar shells. *Bonellia viridis* (Fig. 8-35) is supplied, in the female, with a very long proboscis which is bifurcated at the end; the male is a small turbellarianlike ciliated organism that lives in the segmental organ of the female.

Annelids in General

Morphology

Three morphological characteristics of the Annelida are especially worthy of notice: (1) metamerism, (2) the coelom, and (3) the trochophore stage in development.

Metamerism. The segmentation of the body as exhibited in annelids is called *metamerism*, and is here demonstrated very clearly. This type of structure is of considerable interest, because the most successful groups in the animal kingdom, the Arthropoda and Vertebrata, have their parts metamerically arranged. How this condition has been brought about is still doubtful, but many theories have been proposed to account for it. According to one view the body of a metameric animal has evolved from that of a nonsegmented animal by transverse fission. The individuals thus produced remained united end to end and gradually became integrated both morphologically and physiologically so that their individualities were united into one complex individuality. Some zoologists maintain that the segmental arrangement of organs such as nephridia, blood vessels, and reproductive organs has been caused by the division of a single ancestral organ, and not by the formation of new organs as the fission theory demands.

True metamerism, as exhibited by annelids, should not be confused with the *pseudometamerism* of the tapeworms (Fig. 5-19). The proglottids of the tapeworms are individuals budded off from the posterior end of the scolex and differ from one another only in the degree of development. The tapeworm may be considered a row of incomplete individuals. Annelids add new segments by budding from the front of the last segment (the *pygidium*) and thus have the oldest segments at the anterior end, as do other segmented animals.

The Coelom. The coelom has already been defined as a cavity in the mesoderm lined by an epithelium; into it the excretory organs open, and from its walls the reproductive cells originate.

Hyman (1951) discusses four theories of coelom development. (1) The *enterocoel theory* views the coelom as a further development of what were gastric pockets in acoelomates. (2) The *gonocoel theory* views the coelom as the cavity of an enlarged gonad. (3) The *nephrocoel theory* of the coelom forming as an expansion of a portion of the nephridium attempts to account for the association of the two. (4) The *schizocoel theory* accepts the coelom as a new formation occurring in the mesoderm. Fauchald (1975) adds adaptation to burrowing as a modification of Clark's *hydrostatic theory* to serve as a hypothesis about the driving force for coelom selection, with the proposal that the coelom has polyphyletic origin.

If we combine the ideas of coelom and segmentation as found in annelids, we note in earthworms a special value in locomotion. We may also suspect that the metamerism may have been derived as a safety mechanism to compartmentalize coelomic fluid so injury is less destructive when localized.

The importance of the coelom should be clearly understood, since it has played a prominent role in the progressive development of complexity of structure. The appearance of this cavity between the digestive tract and body wall brought about great physiological changes and is correlated with the origin of nephridia for transporting waste products out of the body, and of genital ducts for the exit of eggs and spermatozoa. The coelom also affected the distribution of nutritive substances within the body, since it contains a fluid which takes up material absorbed by the alimentary canal and carries it to the tissues. Excretory matter finds its way into the coelomic fluid and thence out of the body through the nephridia.

So important is the coelom considered by most zoologists that the Metazoa are frequently separated into two groups: (1) the Acoelomata without a coelom, and (2) the Coelomata with a coelom. The Porifera, Coelenterata, and Ctenophora are undoubtedly Acoelomata. Likewise, the Annelida, Echinodermata, Arthropoda, Mollusca, and Chordata are certainly Coelomata. The Aschelminthes and a number of other groups with a body cavity having a different development are grouped in the (3) Pseudocoelomata, or those with no epithelium lining the cavity in the mesoderm.

The Trochophore. The term *trochophore* has been applied to the larval stages of a number of marine animals. The description and figures of the development of *Polygordius* (Fig. 8-27) are sufficient to indicate the peculiarities of this larva.

Many other marine annelids pass through a trochophore stage during their life history; those that do not are supposed to have lost this step during the course of evolution.

Because a trochophore also appears in the development of animals belonging to other groups, for example, Mollusca, Nemertea, and Bryozoa, and resembles very closely certain Rotifera, the conclusion has been reached by some embryologists that these groups of animals are all descended from a common hypothetical ancestor, the trochozoan. Strong arguments have been advanced both for and against this theory.

Physiology

Among the interesting modifications encountered among various annelids are a variety of adaptations with respiratory value. These include (1) *gills*, especially in the polychaetes; (2) *behavioral patterns*, such as the undulations of the posterior of tubificid worms under conditions of low dissolved oxygen; (3) presence of *respiratory pigments*, which increase the oxygen carrying capacity of the blood; and (4) the *ability to respire anerobically* for short periods by utilizing stored glycogen. The blood vascular system is also very important in the respiration of the organism.

Locomotion varies considerably. In the polychaetes, parapodia of opposite sides are out of phase and accompanying undulations of the body are lateral in forms such as *Nereis*. In oligochaetes and leeches opposite sides are in phase in their movements. Forms such as *Lumbricus* have waves of alternate contraction and extension of segments which travel the length of the animal. Leeches swim by undulating in a dorsoventral plane or creep by bending and straightening in the same plane while alternately attaching and releasing anterior then posterior suckers. The *hydrostatic skeleton* is esential to locomotion. Also of great importance in coordinating these locomotor movements is the structure and function of the nervous system. The latter topic is thoroughly reviewed by Bullock and Horridge (1965) and is discussed more concisely by Dales (1963). *Aeolosoma* and some others glide smoothly along propelled by cilia.

Neurosecretion is of considerable importance in the regeneration of *Nereis*. Neurosecretory cells are localized in the brain of annelids. Their secretions reach the site of action via the bloodstream and thus function as endocrine substances. Thus, a tail-less, brainless nereid will not regenerate a tail well, but will if brain tissue is implanted although neuronal connections are not reestablished. Neurosecretions have been demonstrated to inhibit metamorphosis in some polychaetes but are required for the development of mature sperm in *Hiru-*

do. They are important in the regulation of salt and water balance in *Lumbricus.*

Ecology

Aside from their role in the food chain and energy cycle, which is considerable in some habitats, annelids have a great role to play in the physical modification of soils and shallow marine bottom deposits. Many species pass large tonnages of substrate through their collective alimentary canals while relocating it within or upon the surface of each acre over a period of a year. The agricultural value of land is usually enhanced in the process. Tube-building polychaetes may stabilize drifting marine bottoms and provide a habitat for a different biological community than would otherwise occur there.

Accumulation and transfer of pesticides via earthworms has been of importance in indicating the hazards that may affect society if we carelessly pollute our environment.

Edwards and Lofty (1977) review much of the current resurgence of interest in earthworm ecology. Although the earthworm is rather tolerant of DDT, chlordane is very toxic to them. Such contaminants may affect the success of composting for solid waste disposal since worms are important in the processes involved.

Pough (1971) has demonstrated what may be the first known case of an invertebrate repellant produced by a vertebrate. A newt had leech-repellent properties for several leech species native to the newt's habitat.

Interrelations and Phylogeny of the Annelids

The polychaetes appear to be the simples of the annelids. Polychaetes show a range in structure of nephridia including those with flame cell elements, their gonads show less specialization in structure, and they are marine animals. The oligochaetes probably arose from the polychaetes or their ancestral stock and adaptation for burrowing resulted in reduced head specialization, loss of parapodia, and reduction of setae. The Oligochaeta and Hirudinea show close relationships in their reproductive structures and habits, and *Acanthobdella* is of somewhat intermediate morphology. The Aeolosomatidae have been considered to be archiannelids by some. The Branchiobdellidae are returned to the leeches after recent classification in the Oligochaeta. The Echiuroidea possess sufficient annelid characteristics to justify their inclusion in this phylum. They seem to have evolved from the Polychaeta but have lost many of the features of this class, probably as a result of their sedentary mode of life.

Based upon general organization, development, and nervous system structure, mollusks and arthropods are generally agreed to be the major phyla most closely related to annelids, although still quite distant. Ultrastructural and biochemical evidence supports a relationship of sipunculans (Goffinet, Voss-Foucart, and Barzin, 1978) with poly-chaetes and pogonophorans.

Pogonophora may be deep-sea adapted derivatives of the Sedentarida.

The History of Our Knowledge of the Annelida

The annelids were by early zoologists included with other worms in the group Vermes but were separated by Cuvier in 1798 from the unsegmented worms and designated by Lamarck the Annelids. This group was sometime later (1820) divided by Savigny into four subdivisions, (1) the Nereideae, (2) the Serpuleae, (3) the Lumbricineae, and (4) the Hirudineae. The terms Polychaeta and Oligochaeta we owe to Grube (1851). Brinkhurst, Hartman, Holt, and Jamieson are among the more active recent researchers in annelid systematics.

REFERENCES TO LITERATURE ON THE ANNELIDA

BARNES, R. D. 1965. Tube-building and feeding in chaetopterid poly-chaetes. *Biol. Bull.*, **129**:217–233.

BELL, A. W. 1962. Enchytraeids (Oligochaeta) from various parts of the world. *Trans. Am. Micr. Soc.*, **81**:158–178.

BRINKHURST, R. O., and B. G. M. JAMIESON. 1971. *Aquatic Oligochaeta of the World.* Univ. of Toronto Press, Toronto. 860 pp.

BULLARD, R. W. 1964. Animals in aquatic environments: annelids and mol-luscs. Pp. 683–695 in *Handbook of Physiology.* Section 4: Adaptation to the Environment. Am. Physiol. Soc., Washington.

BULLOCK, T. H., and G. A. HORRIDGE. 1965. *Structure and Function in the Nervous System of Invertebrates.* Freeman, San Francisco. 2 vols., 1,719 pp.

CARSON, R. 1962. *Silent Spring.* Houghton Mifflin, Boston. 368 pp.

CHAPMAN, G. 1958. The hydrostatic skeleton in the invertebrates. *Biol. Rev.*, **33**:338–371.

Cook, D. G. 1975. Cave-dwelling aquatic Oligochaeta (Annelida) from the eastern United States. *Trans. Am. Micr. Soc.*, **94**:24–37.

COOPER, E. L. 1969. Specific tissue graft rejection in earthworms. *Science*, **166**:1414–1415.

DALES, R. P. 1963. *Annelids.* Hutchinson, London. 200 pp.

DARWIN, C. 1881. *The Formation of Vegetable Mould Through the Action of Worms.* 1898 printing. Appleton, New York. 326 pp.

EDWARDS, C. A., and J. R. LOFTY. 1977. *Biology of Earthworms.* 2nd Ed. Chapman & Hall, London. 333 pp.

FAUCHALD, K. 1975. Polychaete phylogeny: a problem in protostome evolu-tion. *Syst. Zool.*, **23**:493–506.

GOFFINET, G., M. F. VOSS-FOUCART, and S. BARZIN. 1978. Ultrastructure of the cuticle of the sipunculans *Golfingia vulgaris* and *Sipunculus nudus.* *Trans. Am. Micr. Soc.*, **97**:512–523.

GOODNIGHT, C. J. 1959. Oligochaeta. Pp. 522–537 in W. T. EDMONDSON. *Fresh-Water Biology*, 2nd ed. Wiley, New York. 1,248pp.

GORDON, D. C., JR. 1966. The effects of the deposit feeding polychaete *Pec-*

tinaria gouldii on the intertidal sediments of Barnstable Harbor. *Limnol. Oceanogr.,* **11**:327–332.

HARMAN, W. J., and A. R. LAWLER. 1975. *Dero (Allodero) hylae,* an oligochaete symbiont in hylid frogs in Mississippi. *Trans. Am. Micro. Soc.,* **94**:38–42.

HARTMAN, O. 1959. Polychaeta. Pp. 538–541 in W. T. EDMONDSON. *Fresh-Water Biology,* 2nd ed. Wiley, New York.

HYMAN, L. H. 1951. Introduction to the Bilateria. Pp. 1–51 in *The Invertebrates.* Vol. II. McGraw-Hill, New York.

KAMEMOTO, F. I., K. N. KATO, and L. E. TUCKER. 1966. Neurosecretion and salt and water balance in the Annelida and Crustacea. *Am. Zoologist,* **6**:213–219.

LENT, C. M. 1973. Retzius cells: neuroeffectors controlling mucus release by the leech. *Science,* **179**:693–696.

LEWIS, D. B., and P. J. WHITNEY. 1968. Cellulase in *Nereis virens. Nature,* **220**:603–604.

MANN, K. H. 1962. *Leeches (Hirudinea).* Pergamon Press, New York.

MARTIN, G. G. 1977. *Saccocirrus sonomacus* n. sp., a new archiannelid from California. *Trans. Am. Micr. Soc.,* **96**:97–103.

MOORE, J. P. 1959. Hirudinea. Pp. 542–557 in W. T. EDMONDSON. *Fresh-Water Biology,* 2nd ed. Wiley, New York.

PETTIBONE, M. H. 1963. Marine Polychaete Worms of the New England Region. *Bull.* 227, Part 1. U.S. Nat. Mus., Washington. 356 pp.

POUGH, F. H. 1971. Leech-repellent property of Eastern red-spotted newts, *Notophthalmus viridescens. Science,* **174**:1144–1146.

RODALE, R. 1961. *The Challenge of Earthworm Research.* Soil and Health Foundation, Emmaus, Pa. 102 pp.

STEPHENSON, J. 1930. *The Oligochaeta.* Oxford Univ. Press, London. 978 pp.

TABRAH, F. L., M. KASHIWAGI, and T. R. NORTON. 1970. Antitumor activity in mice of tentacles of two tropical sea annelids. *Science,* **170**:181–183.

WALD, G., and S. RAYPORT. 1977. Vision in annelid worms. *Science,* **196**:1434–1439.

9

Phylum Mollusca

The phylum Mollusca includes the snails, slugs, clams, oysters, octopods, and nautili. They exhibit primitive bilateral symmetry, are unsegmented, and many of them possess a shell of calcium carbonate. Mussels, clams, snails, and squids do not appear at first sight to have much in common, but a closer examination reveals several structures possessed by all. One of these is an organ called the *foot*—which the snail (Fig. 9-1, B) uses for creeping over surfaces, the clam (C) generally for plowing through mud, and the squid (D) for seizing prey. In each animal there is a space called the *mantle cavity* between the main body and an enclosing envelope, the *mantle.* The *anus* opens into the mantle cavity.

The mollusks are divided into five classes according to their symmetry and the characters of the foot, shell, mantle, gills, and nervous system.

Definition

Phylum Mollusca: clams, snails, squids, octopods. Triploblastic, bilaterally symmetrical animals; anus and coelom present; no segmentation; shell usually present; the characteristic organ is a ventral muscular foot.

Class I. Amphineura. The chitons, with bilateral symmetry, often a shell of eight transverse calcareous plates, and many pairs of gill filaments.

Class II. Gastropoda. The snails, slugs, whelks, etc., with asymmetry and usually a spirally coiled shell.

433

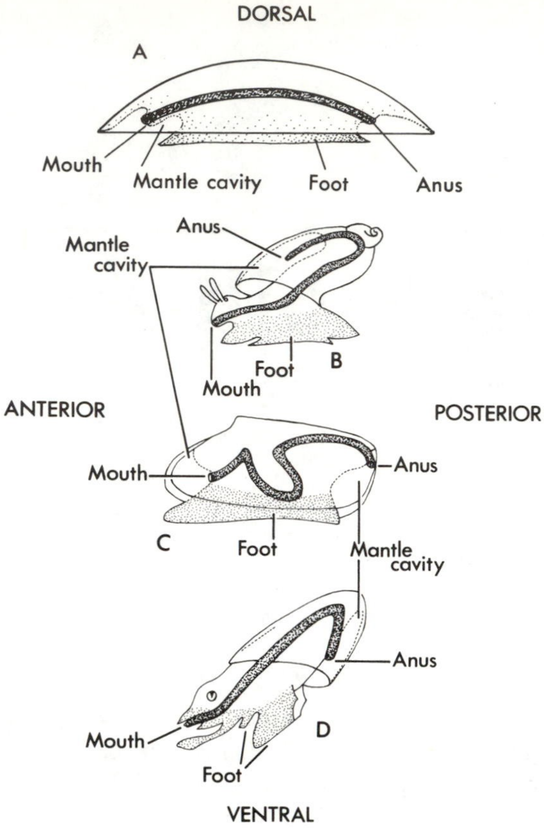

Figure 9-1 Diagrams of four types of mollusks. (A) An amphineuran. (B) A prosobranch gastropod. (C) A pelecypod. (D) A cephalopod. Compare the form of the foot and its regions and the relations of the visceral hump to the anteroposterior and dorsoventral axes. (*B–D, from Shipley and MacBride, after Lankester.*)

CLASS III. PELECYPODA. The clams, mussels, oysters, and scallops, usually with bilateral symmetry a shell of two valves, and a mantle of two lobes.

CLASS IV. SCAPHOPODA. The elephants'-tusk shells, with tubular shell and mantle.

CLASS V. CEPHALOPODA. The squids, cuttlefishes, octopods, and nautili, with bilateral symmetry, a foot divided into arms provided with suckers, and a well-developed nervous system concentrated in the head.

Class I. Amphineura

NEOPILINA GALATHEAE

Neopilina galatheae (Fig. 9-2) is one of very few living members of the Monoplacophora. Fossil members are known from Cambrian rocks. The similarity of the shell to limpet shells made them appear closely related to the gastropods until the soft parts of the living animal were discovered. *Neopilina* is a deep-sea mollusk of special interest, because it illustrates the primitive occurrence of metamerism in mollusks. This strengthens the concept of a close phylogenetic connection between annelids and mollusks that had been suspected prior to the discovery of *Neopilina* in 1952. Metamerism, or the serial repetition of homologous structures, in the genus is probably ultimately derived from a segmented form ancestral to annelids and mollusks.

Unfortunately most zoologists have not accepted the ancestral importance of *Neopilina* because authorities they follow have overlooked the clear evidence (muscle-scar sequences in fossil pelecypods, crystalline style, ventricles paralleling the intestine) toward close relationship to the ancestors of the pelecypods. Feeding, locomotory, nervous, and respiratory evidence of relationship to gastropods and other amphineurans were consequently discounted because of the presumed "aberrant" metamerism. This seeming neglect is only a manifestation of the increasing volume of literature that the speed of modern communications does not yet deal with effectively.

External Features

The single, bilaterally symmetrical *shell* is low, somewhat patelliform, and exogastrically coiled. The largest specimen known is 37 mm long, 35 mm wide, and 13 mm high. *Neopilina* appears somewhat chiton-like from the ventral surface (Fig. 9-2). The central circular *foot* is separated from the encircling mantle by a *pallial groove*, which contains five pairs of lamellated gills with a ciliated epithelium. The head is inconspicuous with two small pre-oral tentacles and two oral tentacle

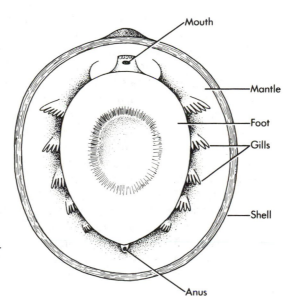

Figure 9-2 *Neopilina.* Ventral view. (*Modified from several authors after Lemche.*)

tufts around the *mouth*. Anterior to the mouth is a velar ridge with two flaps. Unpaired anterior and posterior lips border the mouth. A feeding furrow extends between velar ridge and tentacle tufts. At the opposite, and posterior, end is the *anus*.

Internal Features

Aside from the metameric appearance of some structures, *Neopilina* presents rather generalized molluscan anatomy.

Alimentary Canal. Behind the oral cavity is the *pharynx,* which contains a *radula*. Radulae are usually chitinous, tooth-bearing ribbons borne on a tonguelike structure, the *odontophore*. They are found in all classes of mollusks except the Pelecypoda. In *Neopilina* the radula extends from a sheath to act against a radular plate on the opposite side of the pharynx. The *esophagus* leads from the pharynx to the *stomach*. The stomach contains a *crystalline style.* This peculiar molluscan feature is found in a variety of mollusks and is usually embedded in a ciliated style sack, which rotates the style thus pulling in a mucous food ribbon. The *intestine* contains six coils in contrast to the straight intestine of other amphineurans. A *rectum* leads from intestine to anus.

Systems Showing Metamerism. The *muscular system* is complex. The foot retractors show a segmental type arrangement. The metameric condition of *Neopilina* is also evident in the six pairs of *nephridia,* which are located in the pallial fold. The nephridia open to the coelomic spaces by nephrostomes, and the nephridiopores are near the bases of the gills. Two pairs of nephridia also serve as reproductive ducts. The circulatory system shows some evidence of metamerism in that there are efferent gill vessels from the five metamerically arranged gills. These vessels lead to two pair of *atria,* which fill the single pair of *ventricles.* Other mollusks have only a single ventricle. The ventricles of *Neopilina* are on opposite sides of the rectum. This position makes it likely that the pelecypod heart with its circumintestinal ventricle has arisen from the monoplacophoran type by fusion of the ventricles around the intestine. This tends to substantiate the primitive molluscan nature of *Neopilina*. Blood is pumped from the ventricle via the dorsal aorta to hemocoelic spaces and vascular sinuses.

The *nervous system* is similar to that of the chitons. There is a *circumoral nerve ring*. The main nerves are lateral nerve cords and pedal nerve cords. Paired *statocysts* are located in the body wall behind the tentacle tufts. The *reproductive system* consists primarily of gonads located between intestine and foot. Sexes are separate. The gonads contain cavities in addition to two sets of coelomic cavities. Reproductive products escape via two pair of nephridia.

Neopilina was trawled up from very cold water (2°C) on a soft bottom at a 3,570-meter depth along with a varied catch of fish, echinoderms, crustaceans, and mollusks as well as a few annelids, coelenterates, and sponges. Their food seems to consist of radiolarians and other materials of the bottom.

Although very few specimens of *Neopilina* have been obtained, a very complete account of their anatomy is found in Lemche and Wingstrand (1959) and Purchon (1977).

OTHER AMPHINEURA

This class contains fewer than a thousand species of marine mollusks that usually live on the bottom near the shore. Chitons are the most commonly encountered amphineurans. Their shell consists of eight separate plates (Fig. 9-3). Otherwise there is a general correspondence to *Neopilina* as is shown by comparison with the following general description of a chiton. The body (Fig. 9-3) is elongated and bilaterally symmetrical with the *mouth* at one end and the *anus* at the other end. The *head*, which is sheltered by the mouth, is usually not well developed and lacks both tentacles and eyes. The dorsal surface is occupied by the *mantle*, in which calcareous spicules are embedded and sometimes a continuous covering of plates (Fig. 9-3) with a surrounding scaly girdle. A narrow *mantle cavity* extends along the sides of the body; that part of the mantle that forms this cavity is known as the *girdle*. The *foot* is flat. A *radula* (Fig. 9-4) is usually present. The *nervous system* consists of a circumesophageal ring and two pairs of longi-

Figure 9-3 Polyplacophora. A chiton. (A) Dorsal view. (B) Ventral view.

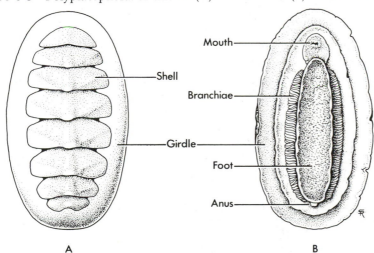

Mouth

Shell

Branchiae

Girdle

Foot

Anus

A

B

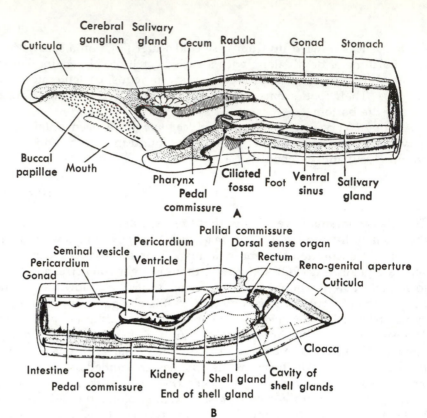

Figure 9-4 Amphineura. *Proneomenia gerlachei.* (A) Right half of anterior part of body, left-side view. (B) Right half of posterior part of body, left-side view. (*From Cooke.*)

tudinal cords, which innervate the foot (pedal) and mantle (pallial). No definite ganglia are present, but the nerve cords are supplied with ganglion cells. In the chitons the valves of the shell are pierced by branching canals through which nerves reach the surface of the mantle where they connect with sense organs called shell eyes.

The *alimentary canal* (Fig. 9-4) extends in a straight line from the mouth to the anus; ducts from mucous, salivary, and hepatic glands open into it. There is a heart near the posterior end, an aorta, and two large sinuses. The blood is, of course, oxygenated in the gills. Two kidneys are present; each possesses a duct that opens outside the body near the posterior end and a duct that opens into the pericardium. The eggs usually give rise to trochophore larvae. Chitons occur as fossils in the Cambrian period.

The Amphineura may be divided into three subclasses.

Subclass 1. Monoplacophora

The only living representatives are *Neopilina*, which has been described, and several related species. A single shell.

Polyplacaphora are known as chitons. They are elliptical in outline, have a flat foot which occupies the entire ventral surface, and a convex dorsal surface characterized by a series of eight transverse overlapping calcareous plates. The mantle groove contains from 4 to 80 pairs of ctenidialike gills. Chitons creep about slowly on the sea bottom giving preference to smooth stones. The overlapping valves of the shell allow them to bend the body and even to roll up into a ball. They feed chiefly on algae. The sexes are separate.

Chaetopleura apiculata is a chiton that is common in shallow water along the Atlantic coast from Cape Cod to Florida. It is oval in shape and about 17 mm long and 10 mm broad. *Chiton tuberculatus* is common in the West Indies. It is about 8 cm long and 0.5 cm broad. *Amicula stelleri,* the giant Pacific Coast chiton, may reach 30 cm.

Subclass 3. Aplacophora

The name Solenogastres is also used for this subclass. They are often wormlike in appearance; the shell is absent; the foot is reduced or lost; the radula may be reduced or absent; the gills are located in the cloaca. Certain species live among corals and hydroids and eat these polyps, and other species feed on protozoans and other minute organisms. This order is believed by some zoologists to be more closely allied with the primitive worms, such as the Archiannelida, than with the mollusks. *Neomenia carinata* is a species living in the North Atlantic; it has a short, thick body about 2.5 cm long; is covered with spicules; but lacks a radula. *Crystallophrisson nitidulum* occurs in the Gulf of Maine and elsewhere; it is slender and wormlike, being over 25 mm long and 2 mm broad.

Class II. Gastropoda

HELIX POMATIA—A TERRESTRIAL SNAIL

External Features

The body of a snail consists of a *head, neck, foot,* and *visceral hump,* as shown in Figure 9-5. The *head* bears two pairs of *tentacles:* (1) a short anterior pair containing the olfactory nerves, and (2) a longer pair containing the *eyes.* The *mouth* is in front and below the tentacles; beneath the mouth is the opening of the pedal *mucous gland.* The *foot* is broad and flat; it is a muscular organ of locomotion with a mucous-secreting integument (Fig. 9-6). Both the foot and head may be withdrawn into the *shell.*

The *spiral shell* encloses the visceral hump, consisting of parts of the digestive, circulatory, respiratory, excretory, and reproductive systems. The *mantle* lines the shell, and is thin except where it joins the

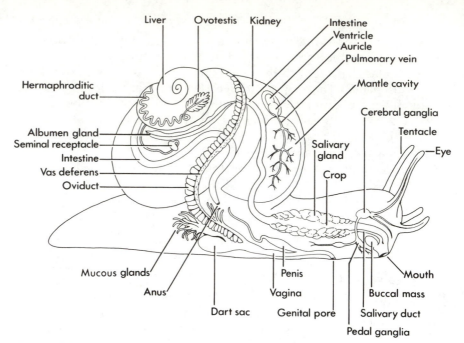

Liver   Ovotestis   Kidney   Intestine
                                    Ventricle
                                    Auricle
                                    Pulmonary vein

Hermaphroditic   Mantle cavity
duct
                                                Cerebral ganglia

Albumen gland   Tentacle
Seminal receptacle   Salivary   Eye
Intestine   gland
Vas deferens   Crop
Oviduct

Mucous glands   Penis   Mouth
Anus   Vagina   Buccal mass
        Dart sac   Genital pore   Salivary duct
                                       Pedal ganglia

Figure 9-5 *Helix.* Side view with shell removed to show internal organs. (*Modified after Root.*)

foot; here it forms a thick collar, which secretes most of the shell. An opening beneath this collar is the *respiratory aperture* leading into the mantle cavity. The *anus* opens just back of this aperture. The *genital pore* is on the side of the head.

Anatomy and Physiology

Digestion. The general anatomy of a snail is shown in Figure 9-5. The *digestive organs* include a *buccal mass, esophagus, salivary glands, crop, stomach, digestive glands, intestine, rectum,* and *anus.*

The food is chiefly, if not entirely, vegetation, such as lettuce. This is scraped up by a horny jaw or *mandible* and devoured after being rasped into fine particles by a band of teeth termed the *radula* (Fig. 9-7). The radula and the cartilages and muscles that move it backward and forward constitute the buccal mass. The *salivary glands,* which lie one on either side of the crop, pour their secretion by means of the *salivary ducts* into the buccal cavity, where it is mixed with the food.

The *esophagus* leads to the *crop,* and from here the food enters the *stomach.* The two *digestive glands* occupy a large part of the visceral hump. They secrete enzymes such as amylase, which converts starch into glucose, and are comparable to the pancreas in vertebrate animals. This secretion enters the *stomach* and aids in digestion. Additional digestive secretions are supplied by cells lining the stomach and

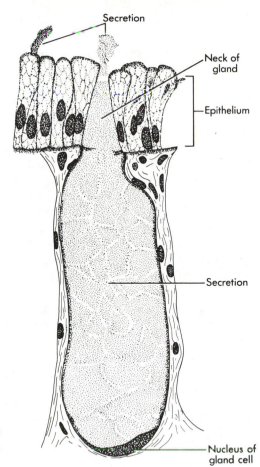

Secretion

Neck of gland

Epithelium

Secretion

Nucleus of gland cell

Figure 9-6 Mucous cell from the edge of the mantle of a terrestrial snail. (*From Dahlgren and Kepner.*)

Figure 9-7 *Helix*, a snail. Vertical longitudinal section through head. (*After Meisenheimer, modified.*)

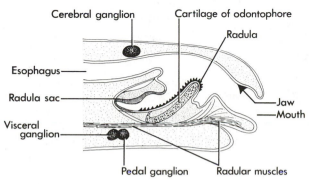

Cerebral ganglion

Cartilage of odontophore

Radula

Esophagus

Radula sac

Visceral ganglion

Jaw

Mouth

Pedal ganglion

Radular muscles

441

intestine and by symbiotic bacteria. The latter seem to be essential for digestion of cellulose and chitin. Absorption takes place chiefly in the *intestine*, and the feces pass out the *anus*.

Circulation and Respiration. The *blood* of the snail consists of a colorless plasma containing corpuscles and serves to transport nutriment, oxygen, and waste products from one part of the body to another. The heart lies in the *pericardial cavity*. The muscular *ventricle* forces the blood through the blood vessels by rhythmical pulsations. One large *aorta* arises at the apex of the ventricle; this gives rise at once to a *posterior branch*, which supplies chiefly the digestive gland, stomach, and ovotestis, and an *anterior branch*, which carries blood to the head and foot. The blood passes from the arterial capillaries into *venous capillaries* and flows through these into *sinuses*. *Veins* lead from these sinuses to the walls of the mantle cavity, where the blood, after taking in oxygen and giving off carbon dioxide, enters the *pulmonary vein* and is carried to the single *auricle* and finally into the ventricle again.

Excretion. The glandular *kidney* lies near the heart. Its duct, the *ureter* or *renal duct*, runs along beside the rectum and opens near the anus.

Nervous System (Figs. 9-8 and 9-9). Most of the nervous tissue of the snail is concentrated just back of the buccal mass and forms a ring about the esophagus. There are five sets of ganglia and four ganglionic swellings. The *supraesophageal* or *cerebral ganglia* are paired and lie above the esophagus. Nerves extend anteriorly from them, ending in the two *buccal ganglia*, the two *eyes*, the two *ocular ganglionic swellings*,

Figure 9-8 Nervous system of Gastropoda as related to torsion. (A) Supposed ancestral condition. (B) Torsion 90°. (C) Torsion 180°. (*After Naef, from Borradaile and Potts, modified.*)

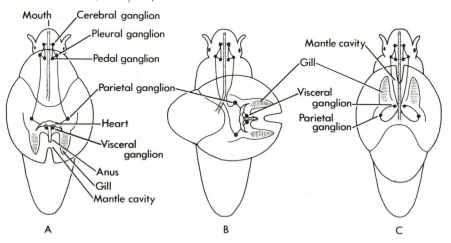

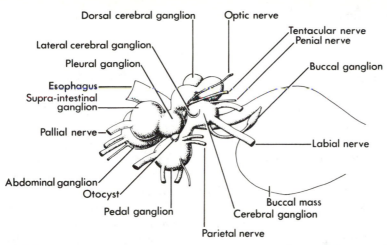

Figure 9-9 *Lymnaea stagnalis;* central nervous system, right-side view. (*From Pelseneer.*)

the two *olfactory ganglionic swellings*, and the mouth. Nerves called commissures connect the supraesophageal ganglia with the ganglia, which lie beneath the esophagus. Here are four pairs of ganglia lying close together—the *pedal, pleural, parietal,* and *visceral.* Nerves pass from them to the visceral hump and the basal parts of the body.

Similar nervous system components occur in the large opisthobranch, *Aplysia* (Fig. 9-16). The large size and spacing of *Aplysia* ganglia, including a small number of giant neuron cell bodies with specific routing of axons and their branches, have made *Aplysia* a favorite experimental animal of many neurophysiologists.

Sense Organs. Both the foot and the tentacles are sensitive to contact, and are liberally supplied with nerves. Each long tentacle bears an *eye.* These eyes are probably not organs of sight, but only sensitive to light of certain intensities. Many snails feed mostly at night, and their eyes may be adapted to dim light.

Snails possess a sense of *smell,* because some of them are able to locate food, which is hidden from sight, at a distance of 18 inches. We are not certain where the sense of smell is located, but investigators are inclined to believe that the small tentacles are the *olfactory organs.* A sense of *taste* is doubtful.

There are two organs of equilibrium (*statocysts*), one on either side of the supraesophageal ganglia. They are minute vesicles containing a fluid in which are suspended small calcaerous bodies (*statoliths*). Nerves connect them with the supraesophageal ganglia.

Locomotion. The snail moves from place to place with a gliding motion. The slime gland that opens just beneath the mouth deposits a film of slime, and on this the animal moves by means of wavelike contractions of the longitudinal muscular fibers of the foot. *Snails* have been observed to travel 5 cm per minute.

Reproduction. Some gastropods are dioecious; others are monoecious. *Helix* is *hermaphroditic,* but the union of two animals is necessary for the fertilization of the eggs, because the spermatozoa of an individual do not unite with the eggs of the same animal. The *spermatozoa* arise in the *ovotestis* (Fig. 9-10); they pass through the coiled *hermaphroditic duct* and into the *sperm duct;* they then enter the vas deferens and are transferred to the *vagina* of another animal by means of a cylindrical *penis,* which is protruded from the *genital pore.*

The *eggs* also arise in the *ovotestis* and are carried through the *hermaphroditic duct;* they receive material from the *albumen gland* and then pass into the *uterine canal;* they move from here down the *oviduct* into the *vagina,* where they are *fertilized* by spermatozoa transferred to the *seminal receptacle* by another snail. In almost all land pulmonates impregnation is mutual, each animal acting during copulation as both male and female.

Helix has a dart sac that secretes a calcareous spicule, which is discharged into a potential mate. It is thought to stimulate effective copulation and thus serves as the equivalent of cupid's arrow.

OTHER GASTROPODA

The gastropods are mollusks that have become modified from the bilaterally symmetrical, unsegmented condition of their ancestors as a result of the coiling of the visceral hump and the rotation of this structure in a counter-clockwise direction through an angle of 180° (Fig. 9-

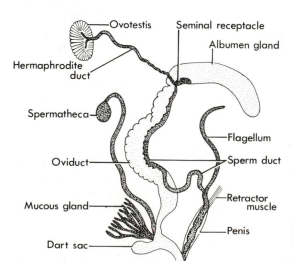

Figure 9-10 *Helix,* a snail. Reproductive organs. (*After Meisenheimer, modified.*)

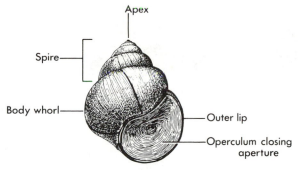

Figure 9-11 *Viviparus*. A dextrally coiled snail showing major features of the shell and the operculum closing the aperture.

8). The foot is flat for creeping; the head is distinct and bears eyes and tentacles; there is usually a shell of one piece (Fig. 9-11); radulae are present; the nervous system has cerebral, pleural, visceral, and usually pedal ganglia and a visceral loop; the visceral hump is often coiled; in some species there is a trochophore larva.

Gastropods are essentially aquatic animals, although some of them live in moist places on land or have means of preventing the escape of moisture from the body such as the epiphragm of certain snails. A few are parasitic on the outside or inside of other animals. Both unisexual and hermaphroditic gastropods occur; most species are oviparous but a few are viviparous. Variations from the type are indicated in the following paragraphs. No classification of this group is accepted by zoologists in general; the arrangement used here includes three subclasses, a number of orders, and some of the more important families.

Subclass 1. Prosobranchia

The large subclass Prosobranchia contains the majority of the gastropods, about 30,000 species, most of which are marine. The gills are in the mantle cavity anterior to the heart (hence Prosobranchia) and the visceral hump is coiled. The visceral connectives are usually twisted into a figure 8. The sexes are separate.

Order 1. *Archaeogastropodida* (*Aspidobranchia*)

Primitive snails that have usually two gills, two auricles, and two nephridia. The gonad opens to the outside through the right nephridium.

FAMILY 1. ACMAEIDAE (Fig. 9-12, A). Limpets. Marine; shell conical, no spiral; one gill; one auricle. Ex. *Acmaea testudinalis*; common on rocks in shallow water from Cape Cod northward. *A. scabra*; common on West Coast.

FAMILY 2. HALIOTIDAE (Fig. 9-12, B). Ear shells. Marine; shell with

A   B   C

Figure 9-12 Archaeogastropoda. (A) *Acmaea,* a limpet. (B) *Haliotus,* an abalone. (C) *Trochus.* (*A and B, from Tryon.*)

flat spiral, very large aperture; spiral series of holes near left margin; two gills, right gill small; two auricles; two nephridia; foot very large, with epipodia projecting through holes in shell. Ex. *Haliotis rufescens:* red abalone, common along coast of California south of Cape Mendocino; abalone shells are made into jewelry, buttons, and buckles, and are used for inlaying; the shells of some tropical abalones are greatly reduced.

FAMILY 3. TROCHIDAE (Fig. 9-12, C). Marine; shell usually conical; horny operculum present; one gill. Ex. *Trochus niloticus:* shell pyramidal, of value as ornament; Indian Ocean. *Margarites obscurus:* shell conical; Long Island Sound northward.

Order 2. Neritacida

Similar to Archaeogastropodida, but with gonoduct extended on to mantle roof and two or more female pores. Mostly marine, few families.

FAMILY HELICINIDAE. Terrestrial, in warm regions; shell conical; operculum oval or triangular; no gills, mantle cavity serving as lung. Ex. *Helicina orbiculata:* in the United States south of Tennessee.

Order 3. Mesogastropodida

Snails that have one auricle, one nephridium, and one gill; gill feathered on one side; gonad opens through a separate duct; from several to many teeth in each row of the radula; shell usually closed with an operculum. Most of the Prosobranchia belong to this suborder of mollusks, which is the largest and also the most varied in morphology and ecological adaptations.

FAMILY 1. CALYPTRAEIDAE (Fig. 9-13, A). Slipper-shell limpets. Shell with inconspicuous or no spiral, but with internal shelf; operculum absent; foot broad; sedentary, attached to rocks. Ex. *Crepidula fornicata:* shell convex, with slight spiral; shelf concave, white; Nova Scotia southward, in shallow water. *C. onyx;* in clusters on West Coast mud flats.

FAMILY 2. LITTORINIDAE. Periwinkles. Salt, brackish, and fresh water;

shell conical and spiral; operculum horny; eyes at base of tentacles. Ex. *Littorina littorea* (Fig. 9-13, B): European edible periwinkle; introduced and common from Delaware Bay northward.

FAMILY 3. VIVIPARIDAE (Fig. 9-13, C). Fresh water; cosmopolitan; shell conical; operculum horny; eyes on short stalks; viviparous. Ex. *Campeloma ponderosum:* shell thick and solid; no umbilicus (cavity in columella); aperture ovate, longer than spire; New York to Illinois southward.

FAMILY 4. PLEUROCERIDAE (Fig. 9-13, D). Fresh water; all in North America, mostly in southern states; shell conical, elongate; operculum present; eyes sessile, at base of tentacles. Ex. *Pleurocera subulare:* long, tapering spire; Great Lakes region.

FAMILY 5. STROMBIDAE (Fig. 9-13, E). Marine; shell usually large, solid, with large lower whorl and expanded lip; foot long, narrow, for springing; eyes at end of stalks; in warm seas. Ex. *Strombus gigas:* conch; largest gastropod in the United States; about 25 cm long; shell

Figure 9-13   Mesogastropoda. (A) *Crepidula.* (B) *Littorina.* (C) *Campeloma.* (D) *Pleurocera.* (E) *Strombus.* (F) *Charonia.* (G) *Vermicularia.* (*A, B, E, and F, from Tryon.*)

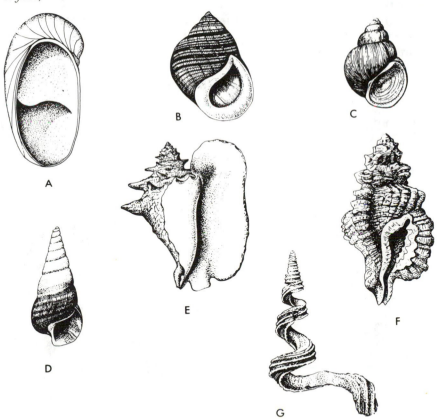

heavy, up to 2 kilograms; flesh edible; shell used as horn and for making ornaments.

FAMILY 6. CYMATIDAE (Fig. 9-13, F). Tritons. Tropical seas. Ex. *Charonia* (*Triton*) *nodifera:* length 45 cm; shell used as war horn or shepherd's horn.

FAMILY 7. EULIMIDAE. Shell elongate; foot long and narrow; proboscis very long; parasitic on echinoderms.

FAMILY 8. CYPRAEIDAE. Cowries. Shell very shiny but covered by mantle in life; developing whorls of shell enclose older whorls; mantle with many sensory papillae; tropical, marine. Ex. *Zonaria spadicea,* the nut brown cowry of the southern California coast.

Order 4. Neogastropodida

Snails with one auricle, one nephridium, and one gill; gill feathered on one side; typically with a pallial siphon; three or fewer teeth in each row of the radula; marine.

FAMILY 1. MURICIDAE (Fig. 9-14, A). Shell rough, with rows of protuberances; central tooth of radula with three cusps; mostly tropical; food mostly other mollusks. Ex. *Urosalpinx cinereus* (Fig. 9-14, B): oyster drill; Massachusetts to Florida and introduced to California; Carriker (1969) determined that they bore a hole in the oyster shell that conforms to the shape of an accessory boring organ whose chemical boring action is aided by mechanical action of the radula. Palmer (1977) demonstrated that the value of shell sculpture projections on *Ceratostoma foliatum* of California includes enabling a dislodged speci-

Figure 9-14 Neogastropoda. (A) *Murex.* (B) *Urosalpinx.* (C) *Busycon.* (D) *Conus.* (*A–C, from Tryon.*)

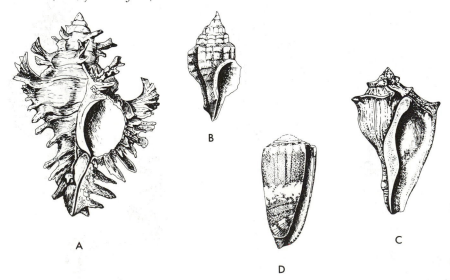

men to land upright more frequently than would occur on a stream-lined specimen without projections.

FAMILY 2. BUCCINIDAE (Fig. 9-14, C). Whelks. Marine. Ex. *Buccinum undatum:* New Jersey northward; used as bait for cod, and in Europe as food for humans. *Busycon canaliculatum;* whelk; marine; large, heavy shell; long siphonal canal; Atlantic Coast.

FAMILY 3. CONIDAE. One genus, *Conus* (Fig. 9-14, D); radula (ultra-structure, Kohn, Nybakken, and Van Mol, 1972) specialized for injection of poison; predators of invertebrates and fish, especially of coral reefs. The cone shells are the most directly dangerous of gastropods. One species, *C. gloriamaris,* has the greatest value to collectors of any shell.

Subclass 2. Opisthobranchia

Sea slugs. All are marine; shell small or absent; gill when present, posterior to the heart; visceral connectives not twisted; hermaphoditic; mostly live among seaweed and under stones near shore, some pelagic; food chiefly animal; four orders.

Order 1. Pleurocoelida

Shell usually present; gills in mantle cavity.

FAMILY 1. AKERIDAE (Fig. 9-15, A). Shell external. Ex. *Haminea solitaria:* Atlantic Coast, Massachusetts to South Carolina.

FAMILY 2. APLYSIIDAE. Sea hares. Shell internal, rudimentary; body sluglike; epipodia over back; brightly colored; cosmopolitan. Ex. *Aplysia protea:* 16 cm long; Florida and West Indies. *A californica,* (Fig. 9-16) a useful experimental organism for behavioral and nervous system studies (for examples, chemoreception, Jahan-Parwar, 1972; DNA in giant neurons, Lasek and Dower, 1971; circadian rhythm, Jacklet,

Figure 9-15 Opisthobranchia. (A) *Haminea.* (B) *Cavolinia.* (C) *Dendronotus.* (D) *Aeolis. (From Tryon.)*

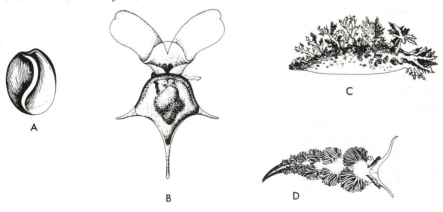

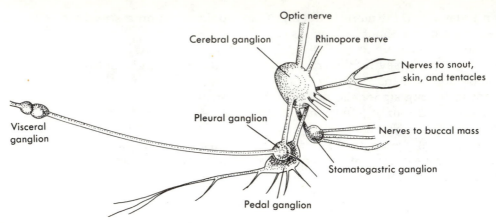

Figure 9-16 *Aplysia californica*. Relationships of major ganglia shown diagrammatically. (*Adapted from Purchon.*)

1969; heart control, Koester, Dieringer, and Mandelbaum, 1979; general anatomy and dissection, Purchon, 1977, pp. 519–526).

Order 2. Pteropodida.

Sea butterflies. Pelagic, with foot modified for swimming.

FAMILY 1. CAVOLINIIDAE (Fig. 9-15, B). Shell present; gill absent; two large epipodia (fins, hence pteropods); pelagic. Ex. *Cavolinia trispinosa;* shell with three long spines; Atlantic Coast.

FAMILY 2. CLIONIDAE. Shell and mantle absent; gill absent; foot modified as a pair of fins; proboscis usually with suckers; pelagic. Ex. *Clione limacina:* Arctic Ocean south to New York; sometimes in enormous schools which color sea; serve as food for whales.

Order 3. Nudibranchida

Sea slugs. Shell absent; gills present or absent.

FAMILY 1. DENDRONOTIDAE. Body long; tentacles branched; cerata (modified gills) branched in two rows on back. Ex. *Dendronotus arborescens* (Fig. 9-15, C): length, 8 cm; about six transparent cerata in each row; Rhode Island northward.

FAMILY 2. DORIDIDAE. Integument with spicules; mantle large; gills retractile; cosmopolitan. Ex. *Doris repanda:* gills in star-shaped group; body covered with small white tubercles; Cape Cod northward.

FAMILY 3. AEOLIDIDAE. No mantle; no spicules in integument; cerata in transverse rows along body; nematocysts in cerata from hydroids used as food. Ex. *Aeolis papillosa* (Fig. 9-15, D): 10 or 12 cerata in a row and 12 to 20 rows on each side; two pairs of cylindrical tentacles; Rhode Island northward.

Shell absent.

FAMILY ELYSIIDAE. No cerata; body ciliated. Ex. *Elysia chlorotica:* 40 mm long; bright green with white and red spots; Massachusetts to New Jersey.

Subclass 3. Pulmonata

Freshwater and terrestrial snails. The characteristics of a member of this order have already been described. No gills, the mantle cavity serves as a lung; the shell is usually a simple, regular spire, sometimes rudimentary or absent; no operculum; shell closed by many land snails by temporary epiphragm of calcified slime; one or two pairs of tentacles; hermaphroditic; mostly oviparous, a few viviparous; mostly terrestrial, many in fresh water, a few marine; mostly vegetarian, a few carnivorous; many species; two orders.

Order 1. Basommatrophorida

Mostly in fresh water, a few terrestrial or marine; one pair of tentacles; eyes at base of tentacles; shell usually with conical spire; cosmopolitan.

FAMILY 1. LYMNAEIDAE (Fig. 9-17, A). Spire of shell acute; tentacles flattened. Ex. *Lymnaea stagnalis:* large shell with six whorls; Northern Hemisphere. There are many similar appearing species of *Lymnaea*.

FAMILY 2. PHYSIDAE (Fig. 9-17, B). Shell with large lower whorl and acute spire; tentacles filiform; cosmopolitan. Ex. *Physa gyrina* (Fig. 9-17, B): shell with five or six whorls; about 22 mm long; central states. *Aplexa hypnorum;* shell thin; body whorl not inflated; northern United States.

FAMILY 3. PLANORBIDAE. Shell discoidal. Ex. *Helisoma trivolvis* (Fig. 9-17, C): shell sinistral and with four whorls; North America.

FAMILY 4. ANCYLIDAE (Fig. 9-17, D). Shell conical, nonspiral, with

Figure 9-17 Freshwater Pulmonata. (A) *Lymnaea.* (B) *Physa.* (C) *Helisoma.* (D) *Ferrissia.*

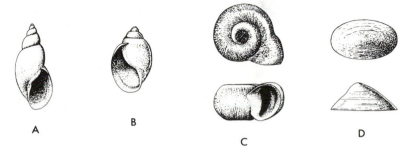

A   B   C   D

oval aperture. Ex. *Ferrissia parallelus:* river limpet; about 5 mm long; New England.

Order 2. *Stylommatophorida*

Terrestrial snails; two pairs of retractile tentacles; eyes at tips of posterior tentacles; shell with conical spire, or rudimentary and concealed, or absent; respiratory pore on right side; cosmopolitan.

FAMILY 1. HELICIDAE. Shell with low, conical spire of five to seven whorls. Ex. *Helix pomatia:* French, edible snail. *Cepaea.*

Cepaea nemoralis (Fig. 9-19) shows great variation in coloration within populations. Jones (1973) has found that populations with yellower shells are associated with higher mean summer temperatures in European locations. Earlier studies have shown that woodland populations are darker than open habitat populations and that thrushes are more effective predators of dark snails in open habitats than light snails in the same habitat. This is an example of the value for adaptability of genetic polymorphism.

FAMILY 2. POLYGYRIDAE (Fig. 9-18, A). A large family of land snails from the temperate regions; similar to the Helicidae; with very low spire; lip usually reflected; aperture often toothed. *Polygyra, Triodopsis, Mesodon,* and *Stenotrema* are genera commonly encountered in woodlands and other moist areas of North America.

FAMILY 3. ENDODONTIDAE. Shell conical or depressed; umbilicus open. Ex. *Anguispira alternata:* shell depressed; gregarious; eastern and Central America. *Helicodiscus parallelus:* shell disk-shaped; diameter about 4 mm; eastern states.

Figure 9-18 Terrestrial Pulmonata. (A) *Triodopsis multilineata.* (B) *Limax.* (C) *Achatina.* (B, from Tryon.)

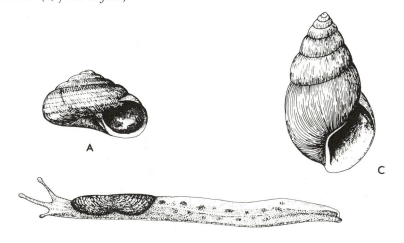

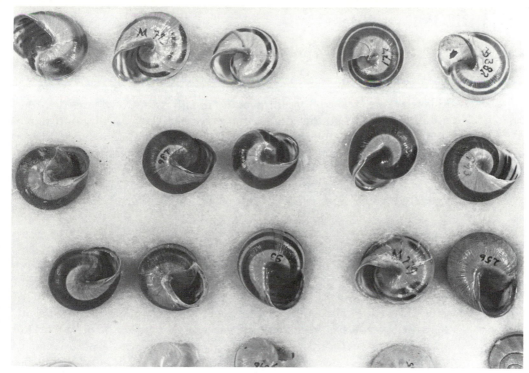

Figure 9-19 *Cepaea nemoralis.* Color varients in one colony. J. L. Howe maintained a colony at Washington and Lee University from which he sent these specimens to C. C. Adams in 1908. (*From the C. C. Adams Collection of Western Michigan University.*)

FAMILY 4. ACHATINIDAE (Fig. 9-18, C). Tropical snails; long, thin shell; several species introduced in southern states. *Achatina fulica,* the giant African snail, may reach 20 cm and is a serious agricultural pest that has been introduced into several countries and Hawaii (Mead, 1961).

FAMILY 5. LIMACIDAE (Fig. 9-18, B). Slugs; shell rudimentary, concealed in mantle; nocturnal; habitat moist; mantle covers less than half of dorsal surface. Ex. *Limax maximus:* about 10 cm long; surface of body with tubercles; European species; eastern states and California. *Derocera laeve:* about 25 mm long; mantle oval; prominent tubercles on dorsal surface; introduced throughout United States.

FAMILY 6. PHILOMYCIDAE. Slugs; shell absent; mantle covering entire dorsal surface. Ex. *Philomycus carolinianus* about 7 cm long; eastern and central states.

Class III. Pelecypoda

ANODONTA—A FRESHWATER MUSSEL

External Features

Mussels, or clams (Fig. 9-20), usually lie almost entirely buried in the muddy or sandy bottom of lakes or streams. They burrow and move from place to place by means of the *foot,* which can be extended from the anterior end of the shell. Water loaded with oxygen and food material is drawn in through a slitlike opening at the posterior end, called the *ventral siphon,* and excretory substances and feces along with deoxygenated water are carried out through a smaller *dorsal siphon.*

The Shell. The shell consists of two parts, called *valves,* which are fastened together at the dorsal surface by an elastic ligamentous hinge. In a related genus, *Unio,* the valves articulate with each other by means of projections called teeth, but these are almost entirely atrophied in *Anodonta.* A number of concentric ridges appear on the outside of each valve; these are called *lines* of *growth,* and, as the name implies, represent the intervals of rest between successive periods of growth. The small area situated dorsally toward the anterior end is called the *umbo;* this is the part of the shell with which the animal was provided at the beginning of its adult stage. The umbo is usually eroded by the carbonic acid in the water.

The *structure* of the *shell* is easily determined (Fig. 9-21). There are three layers: (1) an outer thin, horny layer, the *periostracum,* which is

Figure 9-20 *Anodonta;* external features, soft parts extended from shell, and inner markings of shell. (*From Shipley and MacBride.*)

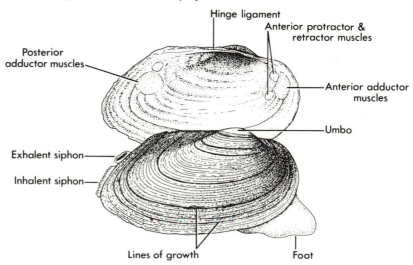

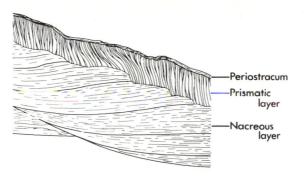

Figure 9-21 *Unio.* Vertical section of shell. The periostracum is formed of the organic material, conchiolin. The other layers are principally calcium carbonate. (*From Woods.*)

Periostracum

Prismatic layer

Nacreous layer

secreted by the edge of the mantle—it serves to protect the underlying layers from the carbonic acid in the water, and gives the exterior of the shell most of its color; (2) a middle portion of crystals of calcium carbonate, called the *prismatic layer,* which is also secreted by the edge of the mantle; and (3) an inner *nacreous layer* (mother-of-pearl), which is made up of many thin lamellae secreted by the entire surface of the mantle, and produces in the light an iridescent sheen.

Anatomy and Physiology

General Account. The valves of the shell are held together by two large *transverse muscles,* which must be cut in order to gain access to the internal organs. These muscles are situated one close to either end near the dorsal surface; they are called *anterior adductors,* and *posterior adductors.* As the shell grows, they migrate outward from a position near the umbo, as indicated by the faint lines in Figure 9-20. When these muscles are cut, or when the animal dies, the shell gapes open, the valves being forced apart by the elasticity of the *ligamentous dorsal hinge,* which is stretched when the shell is closed.

Figure 9-22 *Anodonta mutabilis,* right side, with mantle cut away and right gills folded back. (*After Hatschek and Cori, from Shipley and MacBride.*)

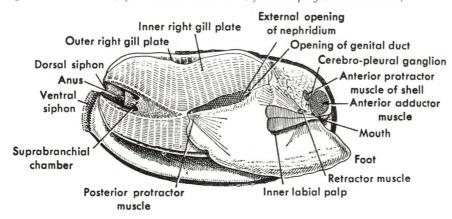

Inner right gill plate

Outer right gill plate

External opening of nephridium

Opening of genital duct

Cerebro-pleural ganglion

Dorsal siphon

Anus

Ventral siphon

Anterior protractor muscle of shell

Anterior adductor muscle

Mouth

Suprabranchial chamber

Foot

Retractor muscle

Posterior protractor muscle

Inner labial palp

Class III. Pelecypoda **455**

The two folds of the dorsal wall of the mussel which line the valves are called the *mantle* or *pallium.* The mantle flaps are attached to the inner surface of the shell along a line shown clearly in Figure 9-20. The space between the mantle flaps containing the two pairs of *gill plates* (Fig. 9-22), the *foot,* and the *visceral mass,* is called the *mantle cavity.*

Digestion. The *food* of the mussel consists of organic material carried into the mantle cavity with the water that flows through the ventral siphon. The *mouth* lies between two pairs of triangular flaps, called *labial palps.* The *cilia* (Fig. 9-26, B) on these palps drive the food particles into the mouth. A short *esophagus* leads from the mouth to the *stomach.* On either side of the stomach is a lobe of a glandular mass called the *digestive gland* or *liver;* a digestive fluid is secreted by the liver and is carried into the stomach by ducts, one for each lobe.

The food is mostly digested and partly absorbed in the stomach; it then passes into the *intestine,* by whose walls it is chiefly absorbed. The intestine coils about in the basal portion of the foot, then passes through the *pericardium,* runs over the posterior adductor muscle, and ends in an *anal papilla.* The feces pass out of the anus and are carried away by the current of water flowing through the dorsal siphon.

Circulation. The circulatory system comprises a heart, blood vessels, and spaces called sinuses. The *heart* (Fig. 9-23) lies in the *pericardium.*

Figure 9-23 Longitudinal section of a freshwater clam with right gills and part of visceral mass removed to show internal organs.

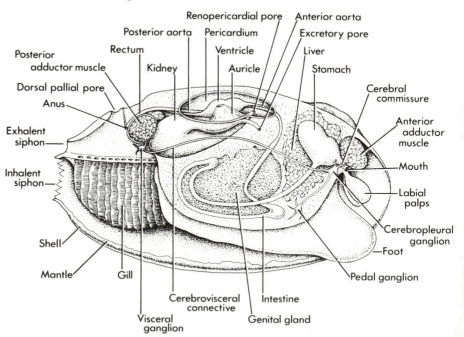

It consists of a *ventricle*, which surrounds part of the intestine, and a pair of *auricles*. By its contractions the ventricle drives the blood forward through the *anterior aorta* and backward through the *posterior aorta*. Part of the blood passes into the mantle, where it is oxygenated, and then returns directly to the heart. The rest of the blood circulates through numerous spaces in the body and is finally collected by a vessel called the *vena cava*, which lies just beneath the pericardium. From here the blood passes into the kidneys, then into the *gills*, and finally through the auricles and into the ventricle. Nutriment and oxygen are carried by the blood to all parts of the body, and carbon dioxide and other waste products of metabolism are transported to the gills and kidneys.

Respiration. The respiratory organs of the mussel are the *gills* or *branchiae* or *ctenidia*. A pair of these hang down into the mantle cavity on either side of the foot (Fig. 9-24).

Each *gill* is made up of two plates or lamellae (Fig. 9-25), which lie side by side and are united at the edges except dorsally. The cavity between the lamellae is divided into vertical *water tubes* by partitions called *interlamellar junctions*. Each lamella consists of a large number of *gill filaments*, each supported by two *chitinous rods*, and covered with *cilia*. Openings, called *ostia*, lie between the gill filaments, and blood vessels are present in the interlamellar junctions and filaments.

Water is drawn through the ostia into the water tubes by the cilia which cover the gill filaments; it flows dorsally into the *suprabranchial chamber* (Fig. 9-24); from here it enters the dorsal mantle cavity and passes out through the dorsal siphon. The blood which circulates through the gills discharges carbon dioxide into the water and takes oxygen from it. Respiration also takes place through the surface of the mantle. There are no respiratory pigments in the blood.

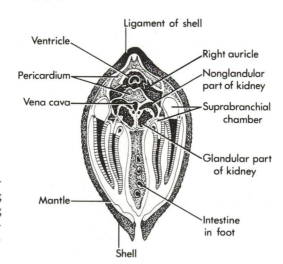

Figure 9-24 *Anodonta.* Diagrammatic section, showing especially the gills hanging down into the mantle cavity. (*After Howes, from Shipley and MacBride.*)

Ligament of shell
Ventricle
Right auricle
Pericardium
Nonglandular part of kidney
Vena cava
Suprabranchial chamber
Glandular part of kidney
Mantle
Intestine in foot
Shell

Class III. Pelecypoda **457**

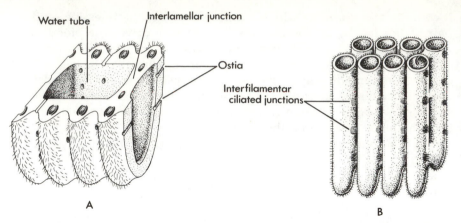

Water tube

Interlamellar junction

Ostia

Interfilamentar ciliated junctions

A

B

Figure 9-25 Gill filaments. (A) Eulamellibranch type of gill with filaments fused laterally and lamellae joined by partitions. (B) Filibranch type of gill with filaments joined loosely by ciliated junctions.

Excretion. The organs of excretion are two U-shaped *kidneys* or *nephridia* lying just beneath the pericardium, one on either side of the vena cava (Fig. 9-23). Each kidney consists of a ventral glandular portion into which the pericardium opens by a ciliated slit and a dorsal thin-walled bladder, which opens to the exterior through the *renal aperture.* Some excretory matter is probably driven into the kidney from the pericardium by cilia, and other excretory matter is taken from the blood by the glandular portion. These waste products of metabolism are carried out of the body through the dorsal siphon.

Nervous System. There are only a few ganglia in the body of the mussel. On each side of the esophagus is a so-called *cerebropleural ganglion* (Fig. 9-23), connected with its fellow by a nerve called the *cerebral commissure,* which passes above the esophagus. From each cerebropleural ganglion a nerve cord passes ventrally, ending in a *pedal ganglion* in the foot. The two pedal ganglia are closely joined together. Each cerebropleural ganglion also gives off a *cerebrovisceral connective,* which may be enclosed by the kidneys and leads to a *visceral ganglion.*

Sensory Organs. Fresh-water mussels are not well provided with sensory organs. A small vesicle, the *statocyst,* containing a calcareous concretion, the *statolith,* lies a short way behind the pedal ganglia. It is an organ of equilibrium (Fig. 9-26, A). A thick patch of yellow epithelial cells covers each visceral ganglion and is known as an *osphradium.* The functions of the osphradia are not certain. They probably test the water that enters the mantle cavity. The edges of the mantle are provided with *sensory cells;* these are especially abundant on the ventral siphon, and are sensitive to contact and light.

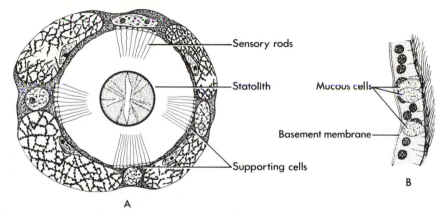

Figure 9-26 (A) Statocyst of *Sphaerium.* (B) Ciliated epithelium of *Mya.* (*From Dahlgren and Kepner.*)

Reproduction. Mussels are usually either male or female; a few are hermaphroditic (van der Schalie, 1969). The *reproductive organs* are situated in the foot (Fig. 9-23). They are paired bunches of tubes and open just in front of the renal aperture on each side. The *spermatozoa* are carried out through the dorsal siphon of the male and in through the ventral siphon of the female. The eggs pass out of the genital aperture and come to lie in various parts of the gills according to the species. The spermatozoa enter the gill of the female with the water and fertilize the eggs. That portion of the gill in which the eggs develop is termed the *marsupium.*

The *eggs* undergo complete but unequal segmentation. Blastula and gastrula stages are passed through, and then a peculiar larva known as a *glochidium* is produced (Fig. 9-27). The glochidium has a shell con-

Figure 9-27 (A) Young mussel or glochidium. (B) Gills of a fish in which are embedded many young mussels forming "blackheads." (*A, after Balfour; B, after Lefever and Curtis.*)

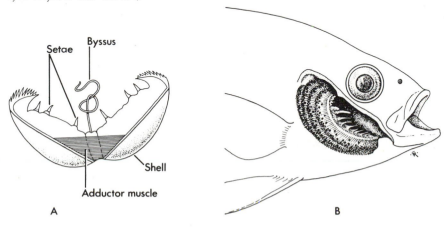

sisting of two valves, which are hooked in some species; these may be closed by a muscle when a proper stimulus is applied. A long, sticky thread called the *byssus* extends from the center of the larva, and bunches of *setae* are also present.

In *Anodonta* the eggs are usually fertilized in August, and the glochidia that develop from them remain in the gills of the mother all winter. In the following spring they are discharged, and, if they chance to come in contact with the external parts of a fish, this *contact stimulus* causes them to seize hold of it by closing the valves of their shell. The valves cut through the soft tissues of the host, but merely clasp the firmer, deeper organs, such as blood vessels and fin rays. Cells of the host lying near the glochidium migrate toward and completely enclose the partly buried organism. The glochidium probably chemically stimulates the skin of the fish to grow around it, forming the well-known "worms" or "blackheads." While thus embedded the glochidium undergoes a stage of development (metamorphosis), during which the foot, muscles, and other parts of the adult are formed. The glochidium is nourished by tissue severed from the host at the time of attachment, and by the disintegrated larval muscle and larval mantle. Other nutriment consists of transuding tissue juices absorbed from the host. After a *parasitic life* within the tissues of the fish from 3 to 12 weeks the young mussel is liberated and takes up a free existence. The liberation of the glochidium is due largely to its own efforts. The injured host tissue is rapidly repaired. Some fishes possess a natural, or racial immunity to infestation with glochidia; others are known to acquire immunity. In the former, many glochidia are destroyed by cytolysis and invasion by the cells of the host. In both cases glochidia are shed, prematurely, at about the second day.

In most other mussels the eggs are fertilized during the late spring and summer, and the glochidia are discharged before the middle of September. The glochidium of *Unio* is smaller than that of *Anodonta* and is usually hookless. It does not as rule become permanently attached to the fins, operculum, or mouth as in *Anodonta,* but usually lodges on the gill filaments of the fish.

One result of the parasitic habit of larval mussels is the *dispersal of the species* through the migrations of the fish. Only in this way can we account for the rapid colonization of certain streams by mussels, because the adult plows its way through the muddy bottom very slowly.

OTHER PELECYPODA

This class, for which the term Lamellibranchia is also used, contains the bivalve mollusks, such as the clams, mussels, oysters, scallops, cockles, and ship worms. About four fifths of the more than 10,000 species are marine; the rest live in fresh water. The body typically exhibits bilateral symmetry, is compressed laterally, and is completely covered by a bilobed mantle. The mantle secretes the shell, which usually consists of two valves hinged together dorsally and separated ventrally.

Trueman (1966) discusses the importance of body fluids and shell adduction in the hydraulic mechanism of foot extension. Thus, the adductor muscles are important for locomotion. Water forced from between the valves can also loosen the substrate, making penetration easier; rocking motions, due to alternating contractions of upper and lower foot retractor muscles, are especially useful in movement through the substrate for some with shell sculpturing. The sculpturing is useful, rather than an impediment, in displacing materials for easier movement.

Anodonta was observed, moving in aquarium gravel in our laboratory, to expand its foot with terminal bulges enlarging in synchrony with contractions of the ventricle, visible through a "window" cut in the umbonal region of the shell.

Striated adductor muscles of the Pectinidae adduct valves with such force and speed that water trapped by the velar folds of the mantle is expelled in jets through the wings by the hinge, enabling the scallop to swim jerkily. The smooth, "catch muscle" portion of the adductor muscle provides sustained contraction to hold the valves in a closed position, apparently with very little energy expenditure.

The head is absent as well as tentacles and radula. The foot is usually wedge-shaped and adapted for burrowing. There are usually two gills on either side of the mantle cavity; these are covered with cilia, which help carry minute particles of food to the mouth as well as create a current of fresh water through the mantle cavity. The sexes are usually separate and there are trochophore and veliger larval stages in marine species. The Pelecypoda usually are divided into subclasses according to the characteristics of their gills and their feeding habits. The mussel used as an example of the class has served to illustrate many of their characteristics; variations from this are indicated in the following account of the subclasses and a few of the families.

Subclass 1. Protobranchia

Marine. The gills are not lamellar but consist of two rows of short, flat leaflets projecting into the mantle cavity; the foot has a creeping surface with sometimes a fringed margin; sediment feeders.

FAMILY 1. NUCULIDAE. Valves of shell equal in size, oval or triangular; hinge with many teeth; oral palps large; cosmopolitan. Ex. *Nucula proxima*: shell 10 mm long; in shallow water; South Carolina northward. *Yoldia limatula*: shell about 5 cm long; North Carolina northward, Pacific Coast, Europe.

FAMILY 2. SOLEMYIDAE. Valves of shell equal in size, long, semicy-lindrical, open at ends; foot long and slender; oral palps long. Ex. *Solemya velum:* about 25 mm long; North Carolina northward.

Subclass 2. Polysyringia.

Usually adapted for filter feeding, many families placed in two groups for convenience.

Group 1. Filibranchia

Marine. The gills consist of two rows of long filamentous leaflets hanging down into the mantle cavity on either side; the two lamellae of each gill are usually not joined by partitions; the foot is poorly developed; a byssus is present consisting of tough strings secreted by the foot and of use for purposes of attachment.

FAMILY 1. ARCIDAE (Fig. 9-28, A). Shell oval, with radial corrugations; valves equal. Ex. *Arca pexata:* shell thick; umbo directed obliquely forward; about 35 corrugations; Florida northward.

FAMILY 2. MYTILIDAE. Mussels; valves equal, long; umbo near or at anterior end; foot cylindrical; attached by byssus. Ex. *Mytilus edulis:* edible mussel; umbo at anterior end; North Carolina and San Francisco northward. *Modiolus modiolus:* horse mussel; umbo near anterior end; burrow in gravel; New Jersey northward and San Francisco northward.

FAMILY 3. PECTINIDAE (Figs. 9-28, B and 9-29). Scallops; valves with a wing on either side of umbo, and with radiating ribs; muscle near center of body. Ex. *Aequipecten irradians:* common scallop; wings large; 20 radiating ribs; Cape Cod southward. *Pecten aequisulcatus* of California similar to *A. irradians. Placopecten magellanicus:* giant scallop; New Jersey northward. Scallops have numerous eyes and the ability to swim short distances.

FAMILY 4. OSTREIDAE. Oysters; valves unequal, irregular and variable; one adductor muscle; adults sessile; marine. *Ostrea lurida* is the West Coast oyster. *Crassostrea virginica* (Fig. 9-28, H) is the common oyster of the East Coast. It is found in shallow, brackish-water areas. Oyster fishermen return empty shells to the oyster beds so that the planktonic larva will have a suitable place to settle and develop. The adult shell has calcium carbonate in the form of calcite as its principal component whereas most mollusk shells have the calcium carbonate as aragonite.

Group 2. Eulamellibranchia

Fresh water and marine. The two gills on each side have the filamentous leaflets connected so as to form two lamellae (Fig. 9-25); siphons present; foot large; byssus small or absent. This group contains most of the pelecypods.

Figure 9-28 Pelecypoda. Types of shells: (A–B, H) Filibranchia; (C–F) Eulamellibranchia; (G) Septibranchia. (A) *Arca*. (B) *Pecten*. (C) *Sphaerium*. (D) *Mactra*. (E) *Ensis*. (F) *Teredo*. (G) *Cuspidaria*. (H) *Crassostrea*. (A–G, from Tryon.)

FAMILY 1. UNIONIDAE. Freshwater clams or mussels; shell large; valves equal; umbo anterior to center; eggs usually carried in outer gill of each pair; glochidia develop from eggs; about 1,000 species. Ex. *Lampsilis ventricosa*: pocketbook clam; shell used for making buttons. *Anodonto grandis*: central states. *Elliptio dilatatus*: Mississippi Valley, etc. *Amblema costata*: blue foot clam; shell used for making buttons; Mississippi Valley, etc.

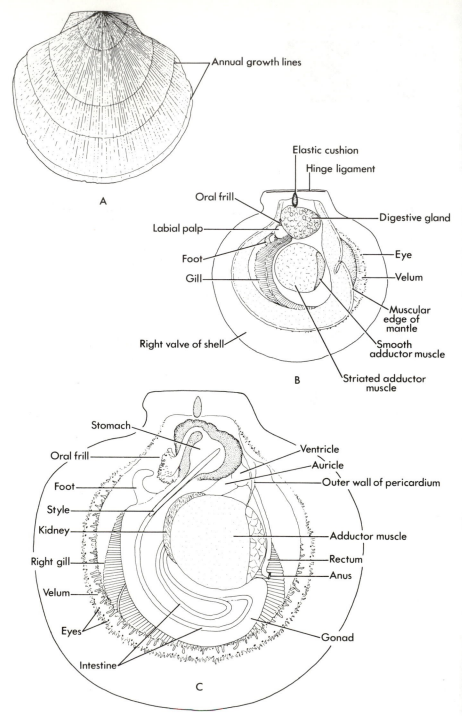

Figure 9-29 *Pecten.* Shell and soft parts of the scallop. (A) Outer surface of left valve. (B) View with left valve removed. (C) Internal anatomy as seen in section through visceral mass with left shell, mantle, and gill removed.

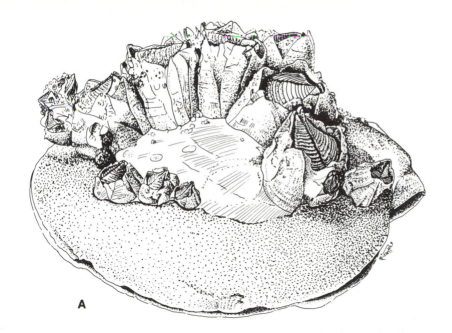

A

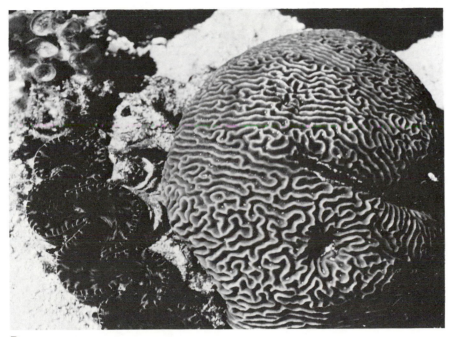

B

Figure 9-30 Some symbionts of pelecypods. (A) Valve of *Pecten*, with numerous epizoic barnacles, polychaetes, and ectoproctans. (B) Photo of *Tridacna* boring in coral rock beside a colony of brain coral. *Tridacna* is exposing its mantle with enclosed photosynthetic zooxanthellae to light. Although the giant clam is anchored hinge down, the soft parts have rotated to the position normal to clams with the hinge uppermost.

Family 2. Sphaeriidae (Fig. 9-28, C). Freshwater or brackish water; shell small, oval; foot elongate, flattened; eggs hatch in inner gill of each pair; many species. Ex. *Sphaerium occidentale:* umbo median; about 7 mm long. *Pisidium abditum:* umbo posterior to middle of shell; about 4 mm long.

Family 3. Corbiculidae. This family of freshwater clams has a general resemblance to the Sphaeriidae, but most are larger and native to Asia. *Corbicula* has been introduced into some streams of this country where it has become a pest because it has a planktonic veliger stage not found in our native freshwater bivalves. The larva may settle in water intake pipes in large numbers and, with growth, clog the pipes.

Family 4. Tridacnidae. Marine, the giant clams of coral reefs. *Tridacna* (Fig. 9-30), the largest of pelecypods, lives hinge down in a hole it has bored in coral. It is a plankton feeder but presumably gets considerable nutriment from zooxanthellae growing in its exposed mantle. It may reach a length of almost 2 meters and weigh over 100 kg.

Family 5. Mactridae (Fig. 9-28, D). Marine; shell large, triangular. Ex. *Spisula solidissima:* surf clam; 15 cm long; edible.

Family 6. Veneridae. Marine, shell thick. *Mercenaria mercenaria* (Fig. 9-31): hard-shell clam or quahog, 11 cm long, edible, source of wampum of the Indians. *Tivela stultorum,* the Pismo clam, California.

Figure 9-31 *Mercenaria mercenaria,* the hard-shell clam or quahog. Dissection showing organs revealed when left valve, mantle, and gills are removed. (*From Woodruff.*)

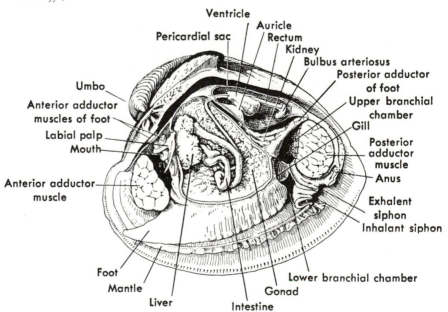

FAMILY 7. MYIDAE. Marine, shell not pearly, burrowing habit. Ex. *Mya arenaria:* soft-shell clam, 20 cm long, edible.

FAMILY 8. SOLENIDAE (Fig. 9-28, E). Marine; shell thin, long, narrow; foot large, cylindrical. Ex. *Ensis directus:* razor clam, 15 cm long and 25 mm high.

FAMILY 9. TEREDIDAE (Fig. 9-28, F). Marine, vermiform, burrow in wood or clay. Ex. *Teredo navalis:* ship worm, tube about 25 cm long and 8 mm in diameter; shell 6 mm long and 2 mm broad. At Woods Hole, Massachusetts, *Teredo navalis* breeds from May 10 to October 10. The eggs remain in the mother's gills for about 3 weeks during cleavage and early larval development. A free-swimming veliger results. The trochophore is nonmotile.

Subclass 3. Septibranchia

Marine. The gills are absent and a perforated septum divides the mantle cavity horizontally into two chambers.

FAMILY CUSPIDARIIDAE (Fig. 9-28, G). Valves of shell small and almost equal in size; siphons short, fused together. Ex. *Cuspidaria pellucida:* right valve smaller than left, about 12 mm long; umbo anterior to center; deep water; Massachusetts northward.

Class IV. Scaphopoda

The Scaphopoda, or tooth shells (Fig. 9-32), are all marine mollusks of which about 12 genera and 200 living species are known; nearly 300 fossil species have also been described. The body exhibits bilateral symmetry and is enclosed in a tubular shell open at both ends. The foot is small and used for burrowing. The head is supplied with many tentacles (captacula) and contains a radula. No gills are present and the circulatory system is rudimentary. Excretory organs are paired and the duct of the right one serves as the outlet for the gametes. Sexes are separate. Development includes a trochophore, then a veliger stage. The mantle fuses ventrally to form an open cylinder.

Scaphopods are typically found in sandy ocean bottoms from one mile deep to just below low tide level. They are usually small. *Dentalium pretiosum* was used by West Coast Indians for wampum. *D. entale* is about 5 cm long by 5 mm broad and occurs from Cape Cod northward. *Dentalium* has a trilobate foot and a tapered shell. The shell of *Cadulus* is broadest near the midregion of the shell. *Siphonodentalium lobatum* has an elongate vermiform foot and is only 10 mm long. It occurs in Arctic seas from New England to Europe.

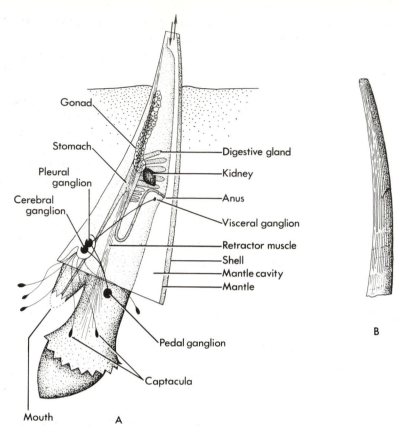

Gonad

Stomach

Pleural
ganglion

Cerebral
ganglion

Digestive gland

Kidney

Anus

Visceral ganglion

Retractor muscle

Shell

Mantle cavity

Mantle

Pedal ganglion

Captacula

Mouth

A

B

Figure 9-32 Scaphopoda. *Dentalium* (A) Structure of specimen buried in sand, except for the end through which inhalant and exhalant currents of water pass. (B) Shell. (*A, after Naef, modified from Borradaile and Potts.*)

Class V. Cephalopoda

LOLIGO PEALII—A SQUID

Loligo pealii (Fig. 9-33) is one of the common squids found along the eastern coast of North America from Maine to South Carolina. It probably lives in deep water during the winter, but about May 1 it enters shallow water in large schools to lay its eggs. Squids are of some economic importance, because they are used as food by Chinese and Italians, and as bait for line and trawl fishing. They feed on small fish, Crustacea, and other squids, and in turn furnish food for cod and other large fish.

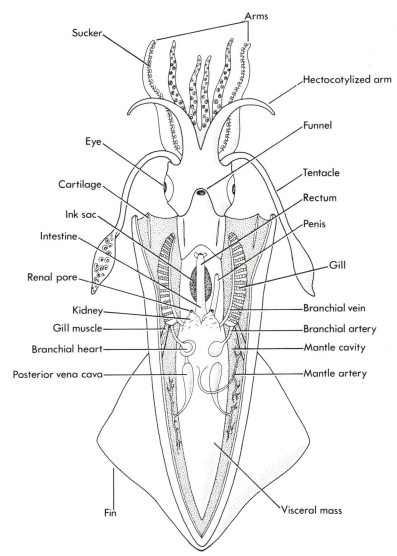

Figure 9-33 *Loligo,* with ventral portion of mantle removed to show internal anatomy. (*Modified after Root.*)

Anatomy and Physiology

The body of *Loligo* is spindle-shaped. When swimming through the water the foot is anterior instead of ventral as in other classes of mollusks. The *skin* may change color rapidly; sometimes it is bluish white, at others, mottled red or brown. The color changes are produced, under nervous control, by changes in the size of *chromatophores.* The chromatophores of *Loligo* are elastic sacs filled with red or yellow pigment and changes in the amount of pigment visible result from contraction or relaxation of muscles. Squids are more heavily pigmented on the functionally dorsal surface. In that respect they resemble many vertebrates which show such obliterative countershading that they blend with a uniform background when they are illuminated from above.

The *foot* consists of 10 *lobes* and a *funnel*. Eight of the lobes are *arms* and two are long *tentacles*. The inner surfaces of both arms and tentacles are provided with *suckers*. The arms are pressed together and used for steering when the squid swims, but when capturing prey the tentacles are extended, seize the victim with their suckers, and draw it back to the arms, which hold it firmly to the mouth. The *funnel* is a muscular tube extending out beyond the edge of the *mantle collar* beneath the *head*. Water is expelled from the *mantle cavity* through it. The funnel is the principal steering organ; if it is directed forward, the jet of water passed through it propels the animal backward; if directed backward, the animal is propelled forward.

A thick muscular *mantle* encloses the visceral mass and mantle cavity. It terminates anteriorly in a collar, which articulates with the visceral mass and funnel by three pairs of interlocking surfaces. Water is drawn into the mantle cavity at the edge of the collar by the expansion of the mantle and forced out through the funnel by the contraction of the mantle. On each side of the animal is a triangular finlike projection of the mantle; these *fins* may propel the squid slowly forward or backward by their undulatory movements, or may change the direction of the squid's progress by strong upward or downward strokes.

The *head* is the short region between the arms and the mantle collar; it contains two large eyes.

The *skeletal system* consists of the pen and several cartilages. Some of the cartilages form articulations for funnel and mantle, others protect the ganglia (Fig. 9-35) or support the eye (Fig. 9-36). The *pen* is homologous with the shell of other mollusks. In *Loligo* the pen is a thin, chitinous, feather-shaped plate concealed beneath the dorsal region of the mantle.

The *muscular system* comprises the major part of mantle, foot, funnel, and fins. Large retractor muscles of the head and funnel run through the visceral mass to their origin under the pen. The mantle is mostly circular muscle.

The *respiratory* structures consist of a pair of *gills* in the mantle cavity. The gill filaments are well supplied with blood capillaries. The blood contains the respiratory pigment *hemocyanin*.

Circulation of the blood is through a closed vascular system (Fig. 9-34). Arterial blood is forced by a muscular *systemic heart* to all parts of the body by three *aortae:* (1) anterior, (2) posterior, and (3) genital. It passes from *arterial capillaries* into *venous capillaries*, and thence into the large *veins*. From these it enters the right and left *branchial hearts*, and then is forced into the *gills* through the *branchial arteries*. In the gills the blood is oxygenated and is finally carried by the *branchial veins* back to the systemic heart.

The two *nephridia* or *kidneys* (Fig. 9-34) are white triangular bodies extending forward from the region of the branchial hearts and opening on either side of the intestine at the ends of small papillae.

The *digestive system* is found in the head and visceral mass. There are two powerful chitinous *jaws* in the *buccal mass;* the jaws resemble a parrot's beak and they are moved by strong muscles. A *radula* is present. Two *salivary glands* lie in the buccal mass. The third salivary gland is near the anterior end of the liver and secretes a poison, which passes through a duct to the region of the jaws. The *esophagus* leads from the buccal mass through the liver and into the stomach. The digestive gland is in two parts, the anterior liver and posterior *pan-*

Figure 9-34 *Loligo pealii,* male. Dissection revealing the circulatory and renal organs. (*From Brooks, modified.*)

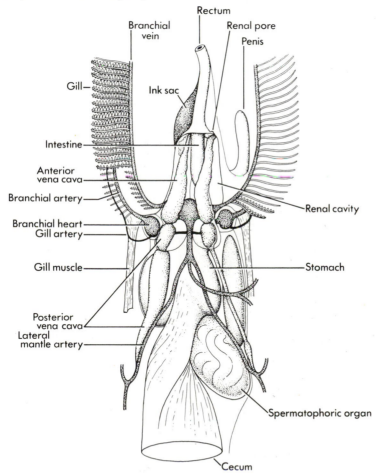

creas. Both pass their secretions via a common duct, to the cecum. The *stomach* is muscular and mixes food and secretions from the digestive gland. Material from the stomach may pass into the intestine or into the cecum. The *cecum* is a thin-walled sac, that absorbs the digested nutrients from the food. Near its anterior is a ciliary sorting area, which removes particulate matter for elimination through the *intestine* and *rectum*. The cecum may fill up much of the visceral mass in the apex of the mantle. Because specimens are usually collected during the reproductive season, the cecum is frequently crowded into a much smaller space by the reproductive products. There is a flap on both sides of the rectum near the anus. A slim, silvery-appearing rectal gland, the *ink sac,* found above the rectum contains a dark pigment, which may be discharged from an opening just inside the anus. The ink may protect the animal in two ways. When discharged in a small cloud about the size of the cephalopod, it may divert the attention of a predator while the squid escapes. It has been found to protect the octopus from its principal enemy, the moray eel, by deadening the olfactory sense the eel uses to locate the octopus.

Figure 9-35 *Loligo pealii.* Longitudinal section through the head, showing ganglia and cartilages. (*After Williams, from Petrunkevitch.*)

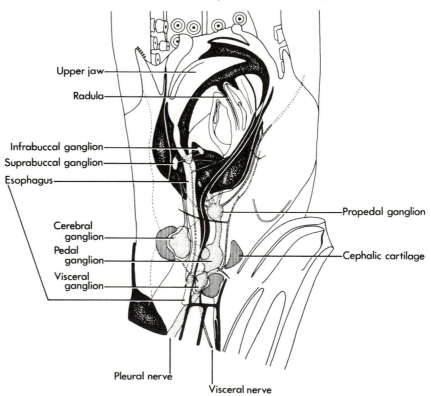

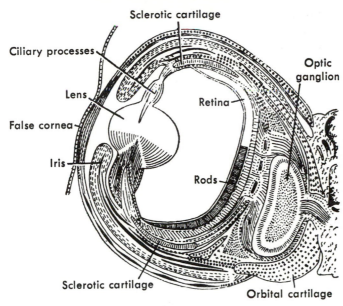

Figure 9-36 Section through the eye of a cephalopod. (*After Hensen, from Parker and Haswell.*)

The *nervous system* consists of ganglia and nerves. The cerebral, pedal, visceral, suprabuccal, infrabuccal, and optic ganglia are in the head (Figs. 9-35 and 9-36). The stellate ganglia are easily seen on both sides near the anterior junction of visceral mass and mantle. Giant nerve fibers passing from neurons in these ganglia to the circular muscles of the mantle cause simultaneous contraction of all parts of the mantle when their neurons are stimulated by giant fibers from neurons in the visceral ganglia. This is possible because larger fibers run to the more distant parts and conduction is more rapid in the larger fibers. Giant fibers from larger species of squids were the material studied for learning much of the basic information now known about nerve impulse transmission.

The *sensory organs* are two very highly developed *eyes,* two *statocysts,* and probably an *olfactory organ.* The *statocysts* are two vesicles lying side by side in the head; each contains a concretion, the *statolith,* and is probably an organ of equilibrium. The *eyes* (Fig. 9-36) are large and somewhat similar superficially to those of vertebrates. Just behind the eye, a fold projects backward under the collar and is probably olfactory.

Squids are either male or female. The *male* has a large white *testis* dorsal to the cecum. Released sperm are picked up by the *vas deferens* and transferred to the *spermatophoric* organ, a compact convoluted organ that packages the sperm into *spermatophores.* These are long slender structures, which may be seen in the thin-walled *spermato-*

phoric sac in which they are then stored until ejection by the *penis*. In the *female* the *ovary* and *oviduct* may be packed with eggs and fill much of the apical region of the viscera. Part of the oviduct is expanded into a thick-walled *oviducal gland*. Before eggs are layed they are coated with secretions from the large *nidamental glands* located very conspicuously in the ventral part of the visceral mass. The function of the smaller accessory nidamental glands is not known. The horseshoe organ is located on the buccal membrane of the female and may serve as a seminal receptacle. The lower left arm of the male (hectocotylized arm) is used to remove spermatophores from the penis for transfer to the female. The eggs are deposited in clusters on the bottom and develop directly into small squids.

OTHER CEPHALOPODA

The cephalopods are marine mollusks known as squids, devilfish, cuttlefish, octopods, and nautili. About 150 genera contain living representatives; most of the species in the class are known only as fossils. The most striking differences between cephalopods and other mollusks are their ability to move about rapidly and their aggressive, carnivorous habits. The body is bilaterally symmetrical. The head is well developed and contains a radula and large, often complex, eyes. Part of the foot has grown around the head and gives rise to mobile, prehensile tentacles, and the other part forms the muscular funnel, or siphon through which water is expelled from the mantle cavity. The mouth lies in the midst of the tentacles. One or two pairs of gills are present in the mantle cavity. The typical cephalopod possesses a chambered shell and lives in the last and largest, chamber, but most of the living representatives are either without a shell or have one that is reduced and internal.

Certain species of squids possess luminous organs that contain symbiotic, luminous bacteria; these are not transmitted from mother to offspring but enter the organs from outside each generation. *Abraliopsis* sp., a squid from dimly lighted midwater depths of the ocean, illuminates photophores on its ventral surface when subjected to overhead lighting (Young and Roper, 1976) and may thus match its background to escape visual detection from below. Person (1969) found cartilaginous dermal scales on the squid *Cranchia scabra*.

Many fossil cephalopods are now extinct. One group, the Ammonoidea, resembled *Nautilus* but had complex sutures where the septa joined the outer shell (Fig. 9-37, C). Another group, the Belemnoidea, had internal shells. The belemnites are thought to be ancestral to all living cephalopods except *Nautilus*.

The living Cephalopoda may be classified in two subclasses as follows.

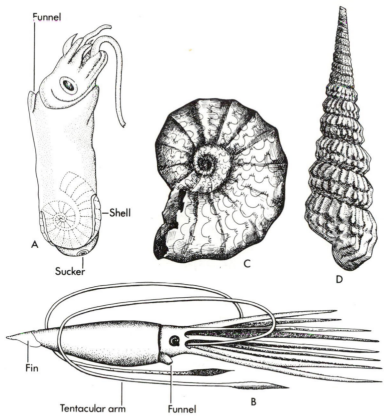

Figure 9-37 Cephalopoda; living and fossil types. (A) Decapoda. *Spirula peronii,* a living species containing a spiral, chambered shell almost completely covered by the mantle. (B) Decapoda. *Architeuthis princeps;* the largest living invertebrate. (C) Tetrabranchia. *Ceritites nodosus,* a Triassic ammonite with shell removed to reveal the sutures. (D) Tetrabranchia. *Turrilites costatus,* an uncoiled ammonite from the Lower Chalk. (*A, after Owen, from Cooke; B, after Verrill; C and D, from Woods.*)

Subclass 1. Nautiloidea (Tetrabranchia)

Calcareous shell, closely coiled; tentacles numerous and without suckers; eyes simple; no chromatophores; no ink sac; two pairs of gills and two pairs of nephridia. The only family that contains living representatives is the Nautilidae with one genus. Ex. *Nautilus pompilius:* pearly nautilus (Fig. 9-38); shell coiled in flat spiral; many chambers formed by curved septa; a slender tube, the siphuncle, passes through chambers near center; shell up to 25 cm in diameter; edible; Pacific and Indian Oceans.

Marked *Nautilus* were found to move as much as 150 km in 332 days,

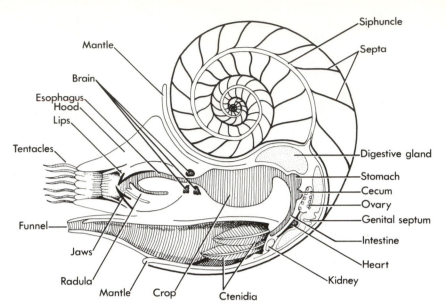

Labels on figure: Mantle, Brain, Esophagus, Hood, Lips, Tentacles, Funnel, Jaws, Radula, Mantle, Crop, Ctenidia, Siphuncle, Septa, Digestive gland, Stomach, Cecum, Ovary, Genital septum, Intestine, Heart, Kidney

Figure 9-38 *Nautilus pompilius.* Dissection of the "pearly nautilus," showing chambered shell and anatomy of the animal. (*From Borradaile and Potts.*)

although some were in the same area nearly a year later (Saunders and Spinosa, 1979). Shells of dead individuals drifted as much as 1,000 km in 5 months. Their habitat is found at depths greater than 90 meters with temperatures less than 24°C around the fringe reefs.

Subclass 2. Coleoidea (Dibranchia)

Shell absent or reduced and internal, calcareous or horny, not coiled; one pair of gills; one pair of nephridia; from 8 to 10 tentacles, with suckers or hooks; eyes complex; chromatophores and ink sac present; two orders.

Order 1. Decapoda

Shell present, chitinous or calcareous, internal, reduced; 10 tentacles, one pair long with suckers near distal end, four pairs short with suckers along entire length.

FAMILY 1. SPIRULIDAE. An exception to other members of order, because shell is partly external and loosely coiled. Ex. *Spirula peroni* (Fig. 9-37, A): white shell; 25 mm in diameter; Caribbean Sea.

FAMILY 2. SEPIIDAE. Shell calcareous; body oval; one pair narrow fins; cosmopolitan. Ex. *Sepia officinalis:* about 20 cm long; Europe; edible.

FAMILY 3. SEPIOLIDAE. Shell chitinous, straight (called a pen); body short and broad; fins large, rounded; eyes with cornea. Ex. *Rossia sublevis:* about 46 mm long; pinkish, spotted; Cape Cod northward.

FAMILY 4. LOLIGINIDAE. Body long; fins near posterior end; eyes with cornea. Ex. *Loligo pealii:* common squid of the East Coast; *L. opalescens:* common squid of the West Coast.

FAMILY 5. ARCHITEUTHIDAE. Body long, cylindrical; fins near posterior end; eyes without cornea; cosmopolitan. Ex. *Architeuthis princeps* (Fig. 9-37, B): giant squid; body up to 6 meters long; total length, including arms, over 17 meters; largest living invertebrate; deep sea.

Order 2. Octopoda

Shell absent (except in *Argonauta*); eight arms.

FAMILY 1. ARGONAUTIDAE (Fig. 9-39). Shell of female thin, spiral, without septa, serves as an egg case; male small. Ex. *Argonauta argo:* paper nautilus; Atlantic Ocean.

FAMILY 2. OCTOPODIDAE (Fig. 9-40). Devilfishes; body globose; head very large. Ex. *Octopus bairdi:* sea polyp; body about 75 mm long; arms about 10 cm long; Cape Cod northward. *O. apollyon* of the Pacific Coast attains a weight of over 45 kg. Octopuses can swim by jet propulsion as does the squid, although they typically live in crevices on the bottom.

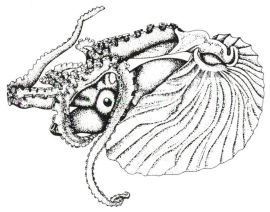

Figure 9-39 Octopoda. *Argonauta pacifica,* a paper nautilus, with its egg case. (*From Johnson and Snook.*)

Figure 9-40 Octopoda. An octopus.

Class V. Cephalopoda **477**

Mollusks in General

The mollusks are a large, successful group of invertebrates with many marine, freshwater, and terrestrial species. Most are free-living filter feeders or grazers, but numerous predatory and parasitic mollusks exist. They are an economically important group supplying food and some other materials, damaging crops, and serving as intermediate hosts for trematodes. They show considerable diversity and include the largest invertebrate, the squid, and other cephalopods which are among the most complex of invertebrates.

Mollusks and chordates both probably arose from their ancestral stocks in Precambrian times with forms much simpler than exist today. Thus, parallels in functional features of the nervous system demonstrate the selective values of a highly developed nervous system are of sufficient importance to cause a functional convergence in this feature of the two phyla. There is evidence that the octopus has both a long-term and a short-term memory. Memory seems more closely related to the amount of brain tissue left after surgery than on the presence of specific cells (Boycott, 1965). Both characteristics are also thought to apply to man. The general similarity of the eyes has already been cited. That eyes of such complexity are new developments in both groups is indicated by the different position of nerve fibers serving the retinal cells in the two phyla.

Interrelations and Phylogeny of the Mollusca

The Amphineura appear to be the most primitive class in the phylum and to have changed the least from the ancestral condition. The modifications of the other classes have taken place in various directions. The Gastropoda became a short, creeping type with a spiral, visceral hump revolved through an angle of 180°. The Pelecypoda seem to be derived from the Monoplacophora as indicated by the Ordovician fossil, *Babinka* (Fig. 9-41, A); they became flattened laterally and developed a large bilobed mantle that secreted a shell of two valves, a large mantle cavity containing gills, and a burrowing foot in place of the creeping type. The Cephalopoda have become free-swimming animals, with the foot modified into prehensile tentacles, and with the brain and sense organs highly developed. Trochophore larvae among both mollusks and annelids indicate that these may be derived from the same ancestral type. This is supported by the presence of metamerism in *Neopilina*. The general relationships of the mollusks are indicated in Figure 9-41 and by the table, Ocurrence of Molluskan structures.

The value of internal metamerism is lost when animals live in molluskan shells, especially when the locomotory adaptation is not dependent on metamerism in those whose ancestors retained some me-

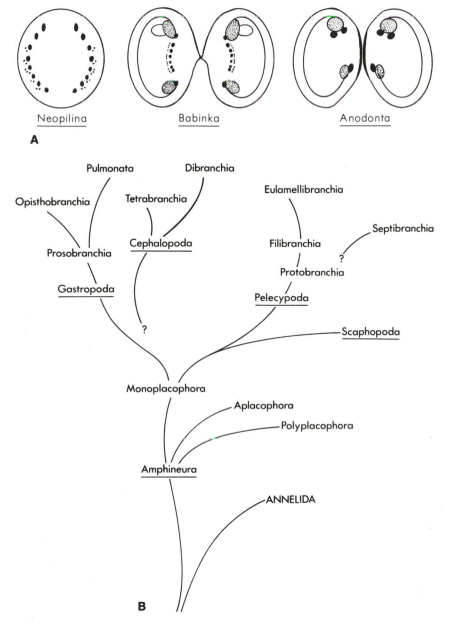

Figure 9-41 Mollusk phylogeny. (A) Diagram illustrating transition from amphineuran to pelecypod muscle scars. (B) Diagram indicating one view of the presumed relationships of the main groups of mollusks. (*A, adapted from McAlester.*)

Occurrence of Molluskan Structures

	Monoplacophora	*Gastropoda*	*Pelecypoda*	*Cephalopoda*
Metamerism	×			
Gill repetition	×	×	×	×
Muscle scar repetition	×		×	
Ganglia repetition	?	×	×	×
Shell, univalved	×	×	lateral valves	×
Mantle cavity	×	×	×	×
Visceral mass	×	×	×	×
Muscular foot	×	×	×	×
Creeping-type sole	×	×		
Radula	×	×		×
Style sac, style	×		×	ciliary sorting area
Ventricle(s) lateral to intestine	×		×	? branchial hearts
Veligerlike larvae		×	×	
Sexes usually separate	×		×	×

tamerism. The loss is so complete and the adaptation so successful that metamerism has not returned in advanced shell-less mollusks.

Neopilina, as with all living "fossils," has also had the opportunity to change since the mollusks diverged from metameric stem forms. This may be most notable in their lack of ganglia remote from the paired cerebral and buccal ganglia. Nerve cell bodies are present in some nerves remote from the brain of *Neopilina.* Rather than the suggested departure from metamerism beginning with *Neopilina,* it could indicate that the *Neopilina* type of nervous system is closer to the beginnings of metamerism, before the advanced metamerism of groups with prominent ganglia connected by nerves of only neuronal processes.

Reproduction

Sexes are usually separate, although hemaphroditism is common. A number of unusual reproductive phenomena have evolved in different groups. The peculiar dart sac of *Helix* has already been noted. The hectocotylized arm of the squid derives its name from the fragment of the male arm found in the mantle cavity of the female in some species following its separation when transferring sperm. The fragment was first thought to be a parasite and given the name *Hectocotylus.* The cephalopods package their sperm in complex packets, the spermatophores. In some pelecypods the sperm are released in *Volvox*-like balls. Gastropods may form spermatozeugma in which small sperm which will fertilize the eggs are carried on the surface of a large motile cell which may also give them some nourishment. In some species the eggs hatch within the female, others lay them in distinctive masses or shed them into the water.

Embryology

Mollusks and annelids correspond remarkably closely in their embryonic development. As in the flatworm, *Hoploplana* (Fig. 5-23), and in representatives of other phyla, a definite cell lineage has been worked out in a number of species. One of the most interesting results obtained is the fact that the quartets that are destined to give rise to definite parts of the embryo and adult have similar values in the flatworms, annelids, and mollusks. There are many variations of a minor type but the first three quartets or micromeres given off by the macromeres form the entire ectoblast of the larva. The first quartet in annelids and mollusks give rise to the pretrochal ectoblast and the second and third to the posttrochal ectoblast. In some cases, as in the Turbellaria, the ectoblastic quartets may give rise to mesoblast cells. Three of the macromeres (4A, 4B, and 4C), after three quartets have been given off, are entirely endoblastic; the fourth (4D) is the primary endoblast. It divides into two equal teloblasts from which arise the me-

Figure 9-42 *Patella coerulea.* Sections of young larvae. (A) Sagittal section of early stage. (B) Frontal section of veliger larva. (*After Patten.*)

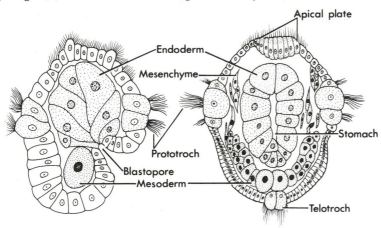

soblastic bands that develop into the coelomic wall and much of the muscular system (Fig. 9-42). Development typically includes a trochophore stage followed by a veliger larval stage.

Ecology

The ecology of gastropods has had special attention because of their tremendous importance as intermediate hosts of trematodes. Concern has particularly been raised about the increased dangers from schistosomiasis in developing countries with ambitious programs of dam-building and irrigation. Simultaneous programs to combat the problem are vital.

Many species of *Conus* coexist, with overlapping ranges and food resources in Hawaii (Kohn, 1959). Numerous species of mollusks occupy the intertidal zone with partial overlap of ranges. A typical rocky shore situation might show the sequence Littorinidae, Acmaeidae, Mytilidae, Ostreidae, *Turbo,* chitons, and Trochidae as the more conspicuous forms of mollusks from the splash zone to seldom-exposed low water levels. Barnacles, polychaetes, kelps, and green algae are interspersed abundantly on solid substrates.

Their usually sedentary ways, abundance, size, and filter-feeding habit make pelecypods potential biological monitors of environmental conditions. Lee, Sauerheber, and Benson, (1972) found that *Mytilus* absorbed petroleum hydrocarbons to which they were exposed. They lost 80 to 90 percent of the petroleum hydrocarbons in 1 to 3 days after exposure to them ceased.

Population fluctuations and range changes have been documented for mollusks. *Rangia* has become abundant 1,000 miles from its recent range limits (Hopkins and Andrews, 1970). Mussels in fresh water have undergone reduction in species diversity in many streams over historic times. Boss (1974) discusses the periods of inactivity natural to mollusks that they may have used to enable them to survive unfavorable periods of natural origin in the past.

Snails grazed periphyton in a small stream at rates approximating the rate of production (Elwood and Nelson, 1972).

The opisthobranchs have been very successful at using organelles of other organisms within their own tissues. Both nematocysts of coelenterates and algal chloroplasts are known to be used. This provides additional evidence of the symbiotic theory of chloroplast and mitochondrial origin since it shows cytocompatibility in widely divergent groups.

Reduced diversity is found in bivalve populations as they are found in increasingly poleward positions. The effect is more pronounced in Recent than in Miocene populations. The higher-latitude populations also represented more ancient taxonomic groups (Hecht and Agan, 1972). Clarke (1978) notes greater variation in arctic and deep-sea species.

Many leave a record of annual growth on their shells, where heavy dark lines on the periostracum indicate the slow growth of winter. From this it has been established that some freshwater clams reach an age of 100 years. The giant *Tridacna* reaches its large size by rapid growth, as much as by long life, reaching a length of ½ meter in 9 years (Bonham, 1965). Nothing is known of the age reached by the giant squid, the largest living invertebrate, although the extremely rare sporadic abundance of it may indicate a very long slow growth if, as in some cephalopods, egg laying is followed by death.

The History of Our Knowledge of the Mollusca

Mollusks were well known to the ancients because of their conspicuous shells, and constituted an important article of diet of primitive people. At first the group included a number of types of animals that possessed a superficial resemblance to the mollusks because of their shells, for example, the Brachiopoda, Tunicata, and Cirripedia. When the larval stages of the Cirripedia were studied it was found that they really belonged to the Crustacea. Investigations of the Tunicata revealed their chordate nature and vertebrate affinities. Finally, the Brachiopoda were excluded but placed near the Mollusca in the group Molluscoidea. Cuvier (in 1795) recognized three subdivisions of the Mollusca: (1) Gastropoda, (2) Cephalopoda, and (3) Acephala. They were included by Linnaeus as one of the four divisions of the Vermes. Many workers have made significant contributions to our knowledge of the mollusks in modern times; among these may be mentioned Thiele, Yonge, Lemche, Clench, van der Schalie, Baker, Tryon, Pilsbry, Purchon, and Wilbur.

REFERENCES TO LITERATURE ON THE MOLLUSCA

ARNOLD, J. M. 1965. Normal embryonic stages of the squid, *Loligo pealii* (Lesueur). *Biol. Bull.,* **128**:24–32.

BONHAM, K. 1965. Growth rate of giant clam *Tridacna gigas* at Bikini Atoll as revealed by radioautography. *Science,* **149**:300–302.

BOSS, K. J. 1974. Oblomovism in the Mollusca. *Trans. Am. Micr. Soc.,* **93**:460–481.

BOYCOTT, B. 1965. Learning in the octopus. *Sci. Am.,* **212**:42–50.

BROWN, F. A., H. M. WEBB, and F. H. BARNWELL. 1964. A compass directional phenomenon in mud-snails and its relation to magnetism. *Biol. Bull.,* **127**:206–220.

BURCH, J. B. 1962. *How to Know the Eastern Land Snails.* Brown, Dubuque. 214 pp.

CAMPBELL, J. L. 1965. The structure and function of the alimentary canal of the black abalone, *Haliotis cracherodii* Leach. *Trans. Am. Micr. Soc.,* **84**:376–395.

CARRIKER, M. R. 1969. Excavation of boreholes by the gastropod, *Urosal-*

pinx: an analysis by light and scanning microscopy. *Am. Zoologist,* **9**:917–933.

CLARK, K. B., H. M. STIRTS, and K. R. JENSEN. 1978. Functional symbiotic chloroplasts in non-elysiid Opisthobranchia. *Am. Zoologist,* **18**:665.

CLARKE, A. H. 1978. Polymorphism in marine mollusks and biome development. *Smithsonian Contrib. Zool.,* No. 274, pp. 1–14.

CLENCH, W. J. 1959. Mollusca. Pp. 1117–1160 in W. T. EDMONDSON. *Fresh-Water Biology,* 2nd ed. Wiley, New York, 1248 pp.

DAZO, B. C. 1965. The morphology and natural history of *Pleurocera acuta* and *Goniobasis livescens* (Gastropoda: Cerithiacea: Pleuroceridae). *Malacologia,* **3**:1–80.

EDGAR, A. L. 1965. Observations on the sperm of the pelecypod *Anodontoides ferussacianus* (Lea). *Trans. Am. Micr. Soc.,* **84**:228–230.

ELWOOD, J. W., and D. J. NELSON. 1972. Periphyton production and grazing rates in a stream measured with a ^{32}P material balance method. *Oikos,* **23**:295–303.

GALTSOFF, P. S. 1964. The American Oyster. *Fish. Bull.* **64.** U.S. Dept. Interior. 480 pp.

GHISELIN, M. T. 1965. Reproductive function and the phylogeny of opisthobranch gastropods. *Malacologia,* **3**:327–378.

HECHT, A. D., and B. AGAN. 1972. Diversity and age relationships in Recent and Miocene bivalves. *Syst. Zool.,* **21**:308–312.

HERRINGTON, H. B. 1962. A revision of the Sphaeriidae of North America (Mollusca: Pelecypoda). *Misc. Publ. Mus. Zool.,* No. 118. Univ. Mich. 74 pp. + 7 Pl.

HOPKINS, S. H., and J. D. ANDREWS. 1970. *Rangia cuneata* on the East Coast: thousand mile range extension or resurgence? *Science,* **167**:868–869.

HUNTER, R. D. 1975. Growth, fecundity and bioenergetics in three populations of *Lymnaea palustris* in upstate New York. *Ecology,* **56**:50–63.

HYMAN, L. H. 1967. *The Invertebrates: Mollusca I* (Vol. VI). McGraw-Hill, New York. 792 pp.

JACKLET, J. W. 1969. Circadian rhythm of optic nerve impulses recorded in darkness from isolated eye of *Aplysia. Science,* **164**:562–563.

JAHAN-PARWAR, B. 1972. Behavioral and electrophysiological studies on chemoreception in *Aplysia. Am. Zoologist,* **12**:525–537.

JONES, E. C. 1963. *Tremoctopus violaceus* uses *Physalia* tentacles as weapons. *Science,* **139**:764–766.

JONES, J. S. 1973. Ecological genetics and natural selection in molluscs. *Science,* **182**:546–552.

KEEN, A. M. 1963. *Marine Molluscan Genera of Western North America: An Illustrated Key.* Stanford Univ. Press, Stanford. 126 pp.

KOESTER, J., N. DIERINGER, and D. E. MANDELBAUM. 1979. Cellular neuronal control of molluscan heart. *Am. Zoologist,* **19**:103–116.

KOHN, A. J. 1959. The ecology of *Conus* in Hawaii. *Ecol. Monogr.,* **29**:47–90.

———, J. W. NYBAKKEN, and J.-J. VAN MOL. 1972. Radula tooth structure of the gastropod *Conus imperialis* elucidated by scanning electron microscopy. *Science,* **176**:49–51.

LANE, C. E. 1961. The Teredo. *Sci. Am.,* **204**:132–142.

LANE, F. W. 1960. *Kingdom of the Octopus.* Sheridan House, New York, 300 pp.

Lasek, R. J., and W. J. Dower. 1971. *Aplysia californica*: analysis of nuclear DNA in individual nuclei of giant neurons. *Science,* **172**:278–280.

Lee, R. F., R. Sauerheber, and A. A. Benson. 1972. Petroleum hydrocarbons: uptake and discharge by the marine mussel *Mytilus edulis. Science,* **177**:344–346.

Lemche, H., and K. G. Wingstrand. 1959. The anatomy of *Neopilina galatheae* Lemche, 1957. *Galathea Report,* **3**:9–72, +56 Pl.

McAlester, A. L. 1964. Transitional Ordovician bivalve with both monoplacophoran and lucinacean affinities. *Science,* **146**:1293–1294.

McCraw, B. M. 1961. Life history and growth of the snail, *Lymnaea humilis* Say. *Trans. Am. Micr. Soc.,* **80**:16–27.

Mead, A. R. 1961. *The Giant African Snail: A Problem in Economic Malacology.* Univ. of Chicago Press, Chicago. 257 pp.

Morris, P. A. 1973. *A Field Guide to Shells of the Atlantic and Gulf Coasts and the West Indies.* Edited by W. J. Clench. Houghton Mifflin, Boston. 330 pp.

Morse, D. E., N. Hooker, H. Duncan, and L. Jensen. 1979. γ-aminobutyric acid, a neurotransmitter, induces planktonic abalone larvae to settle and begin metamorphosis. *Science,* **204**:407–410.

Morton, J. E. 1958. *Molluscs.* Hutchinson, London. 232 pp.

Mpitsos, G. J., and S. D. Collins. 1975. Learning: rapid aversive conditioning in the gastropod mollusk *Pleurobranchaea. Science,* **188**:954–957.

Muscatine, L. 1967. Glycerol excretion by symbiotic algae from corals and *Tridacna* and its control by the host. *Science,* **156**:516–519.

Nixon, M., and J. B. Messenger (Eds.). 1977. *The Biology of Cephalopods.* Academic Press, London. 614 pp.

Palmer, A. R. 1977. Function of shell sculpture in marine gastropods: hydrodynamic destabilization in *Ceratostoma foliatum. Science,* **197**:1293–1295.

Person, P. 1969. Cartilaginous dermal scales in cephalopods. *Science,* **164**:1404–1405.

Purchon, R. D. 1977. *The Biology of the Mollusca.* 2nd Ed. Pergamon Press, Oxford. 560 pp.

Raven, C. P. 1958. *Morphogenesis: The Analysis of Molluscan Development.* Pergamon Press, New York. 311 pp.

Runnegar, B., and J. Pojeta, Jr. 1974. Molluscan phylogeny: the paleontological viewpoint. *Science,* **186**:311–317.

Saunders, W. B., and C. Spinosa. 1979. *Nautilus* movement and distribution in Palau, Western Caroline Islands. *Science,* **205**:1199–1201.

Stenzel, H. B. 1964. Oysters: composition of the larval shell. *Science,* **145**:155–156.

Tasaki, I., and M. Luxoro. 1964. Intracellular perfusion of Chilean giant squid axons. *Science,* **145**:1313–1314.

Thompson, T. E., and I. Bennett. 1969. *Physalia* nematocysts: utilized by mollusks for defense. *Science,* **166**:1532–1533.

Trueman, E. R. 1966. Bivalve mollusks: fluid dynamics of burrowing. *Science,* **152**:523–525.

van der Schalie, H. 1969. Two unusual unionid hermaphrodites. *Science,* **163**:1333–1334.

Vernberg, W. B., F. J. Vernberg, and F. W. Beckerdite, Jr. 1969. Larval trematodes: double infections in common mud-flat snail. *Science,* **164**:1287–1288.

WELLS, M. J.   1978.   *Octopus*. Chapman & Hall, London. 417 pp.

WILBUR, K. M., and C. M. YONGE.   1964.   *Physiology of Mollusca*. Vol. I. Academic Press, New York. 473 pp.

YONGE, C. M., and T. E. THOMPSON.   1976.   *Living Marine Molluscs*. Collins, London. 288 pp.

YOUNG, J. Z.   1961.   Learning and discrimination in the octopus. *Biol. Rev.*, **36**:32–96.

YOUNG, R. E., and C. F. E. ROPER.   1976.   Bioluminescent countershading in midwater animals: evidence from living squid. *Science*, **191**:1046–1048.

10

Phylum Arthropoda

The arthropods comprise the largest phylum in the animal kingdom, surpassing in number of species all the other phyla combined. They may be said to be the dominant animals on the earth at the present time, if the numbers of different species and numbers of individuals are accepted as criteria of dominance. To the phylum belong the lobsters, crabs, water fleas, barnacles, millipedes, centipedes, scorpions, spiders, mites, and insects. The most important characteristic common to all of these is the possession of jointed appendages. They exhibit bilateral symmetry; consist of a linear series of segments, on all or some of which is a pair of appendages; are covered with a chitinous exoskeleton, which is flexible at intervals to provide movable joints; possess a nervous sytem of the annelid type; have a coelom, which is small in the adult, the body cavity being a hemocoel filled with blood; and are free from cilia.

Regional specialization into groups of segments (*tagmata*) with a common function such as those forming the head or thorax or abdomen have occurred in arthropods. The jointed, chitinous exoskeleton has been an important factor in the success of this phylum by contributing to the protection and efficient locomotion of its members. Problems of growth and respiration caused by an effective exoskeleton have been averted by molting and the development of trachea, gills, and other modifications in various groups.

A brief classification of the phylum is presented below. Smaller divisions are given under each class.

Excluded from the classification are an extinct group of paleozic fossils, the trilobites. They may be considered as a separate subphylum, the Trilobitomorpha. Cisne (1974) suggests that the trilobites are a link between the Crustacea and Chelicerata.

SUBPHYLUM 1. MANDIBULATA. Typically with mandibles and antennae.
> CLASS 1. CRUSTACEA. Ex. Lobster, crab, barnacle, sow bug.
> CLASS 2. INSECTA. Ex. Grasshopper, honeybee, fruit fly.
> CLASS 3. CHILOPODA. Ex. Centipede.
> CLASS 4. DIPLOPODA. Ex. Millipede.
> CLASS 5. PAUROPODA. Ex. *Pauropus.*
> CLASS 6. SYMPHYLA. Ex. *Scutigerella.*

SUBPHYLUM 2. ONYCHOPHORA. Only slight development of oral appendages. With several annelidlike characteristics.
> CLASS 1. ONYCHOPHORA. Ex. *Peripatus.*

SUBPHYLUM 3. CHELICERATA. Typically with chelicerae and pedipalps as head appendages; no antennae; usually a cephalothorax and abdomen; four pairs of legs on most species.
> CLASS 1. MEROSTOMATA. Ex. *Xiphosura.*
> CLASS 2. ARACHNOIDEA. Ex. Spider, scorpion, mite.
> CLASS 3. PYCNOGONOIDEA. Ex. Sea spiders.
> CLASS 4. TARDIGRADA. Ex. *Macrobiotus.*
> CLASS 5. PENTASTOMOIDEA. Ex. *Linguatula.*

Class I. Crustacea

THE CRAYFISH

Crayfishes inhabit freshwater lakes, ponds, and streams. The genera *Procambarus, Cambarus,* and *Orconectes* are common in the central and eastern United States. Pacific Coast crayfishes are in the genus *Pacifastacus.* The lobster is so nearly like the crayfish in structure that the anatomical portion of this chapter may be applied also in large part to this animal. In Europe the most common crayfish is *Astacus fluviatilis,* a species made famous by Huxley's classical work *The Crayfish.* Crocker and Barr (1968) review the biology of crayfishes in general, as well as discuss those of the Ontario area.

External Features

Exoskeleton. The outside of the body of the crayfish is covered by an extremely hard chitinous *cuticle* inpregnated with lime salts. This exoskeleton is thinner and flexible at the joints, allowing movement.

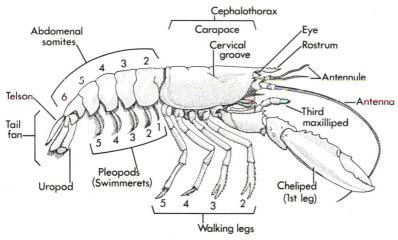

Figure 10-1 External features of a lobster. (*After Calman.*)

Regions of the Body. The body of the crayfish consists of two distinct regions, an anterior rigid portion, the *cephalothorax,* and a posterior series of segments, the *abdomen.* The entire body is segmented, but the joints have been obliterated on the dorsal surface of the cephalothorax (Fig. 10-1).

Structure of a Segment. A typical segment is shown in cross section in Figure 10-5. It consists of a convex dorsal plate, the *tergum,* a ventral transverse bar, the *sternum,* plates projecting down at the sides, the *pleura,* and smaller plates between the pleura and the basis of the limb, the *epimera.*

Cephalothorax. The cephalothorax consists of segments I to XIII, which are enclosed dorsally and laterally by a cuticular shield, the *carapace.* An indentation, termed the *cervical groove,* runs across the middorsal region of the carapace, and obliquely forward on either side, separating the *cephalic* or head region from the posterior *thoracic* portion. The anterior pointed extension of the carapace is known as the *rostrum.* Beneath this on either side is an *eye* at the end of a movable peduncle. The *mouth* is situated on the ventral surface near the posterior end of the head region. It is partly obscured by the neighboring appendages. The carapace of the thorax is separated by *branchiocardiac grooves* into three parts, a median dorsal longitudinal strip, the *areola,* and two large convex flaps, one on either side, the *branchiostegites,* which protect the gills beneath them.

Abdomen. In the abdomen there are six segments, and a terminal extension, the *telson,* bearing on its ventral surface the longitudinal anal opening. Whether or not the telson is a true segment is still in dispute; we shall adopt the view that it is not. The first abdominal segment (XIV) is smaller than the others and lacks the pleura. Segments XV to XIX are like the type described above.

Appendages. With the possible exception of the first abdominal segment in the female, every segment of the body bears a pair of jointed appendages. These are all variations of a common type, consisting of a basal segment, the *protopodite,* which bears two branches, an inner *endopodite,* and an outer *exopodite.* Beginning at the anterior end, the appendages are arranged as follows (Fig. 10-2). In front of the mouth are the *antennules,* and the *antennae;* the mouth possesses a pair of mandibles, behind which are the first, and the second *maxillae;* the thoracic region bears the first, the second, and the third *maxillipeds,* the *pinchers* or *chelipeds,* and four other pairs of *walking legs;* beneath the abdomen are six pairs of *swimmerets,* some of which are much modi-

Figure 10-2 Appendages of crayfish as seen from the ventral side. (A) First antenna. (B) Second antenna. (C) Mandible. (D) First maxilla. (E) Second maxilla. (F) First maxilliped. (G) Second maxilliped. (H) Third maxilliped. (I) Fourth leg. (*From Kerr.*)

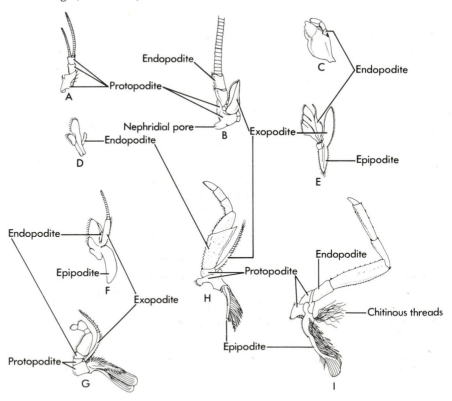

fied. The accompanying Descriptive Table of the Appendages of the Crayfish gives brief descriptions of the different appendages, and shows the modifications due to differences in function.

Three kinds of appendages can be distinguished in the adult crayfish; (1) the *foliaceous,* e.g., the second maxilla, (2) the *biramous,* e.g., the swimmerets, and (3) the *uniramous,* e.g., the walking legs. All of these appendages have doubtless been derived from a single type, the modifications being due to the functions performed by them. The biramous type probably represents the condition from which the other types developed. The uniramous walking legs, for example, pass through a biramous stage during their embryological development. Again, the biramous embryonic maxillipeds are converted into the foliaceous type by the expansion of their basal segments.

General Internal Anatomy

The body of the crayfish (Fig. 10-3) contains all of the important systems of organs characteristic of the higher animals. The coelom is not a large cavity but is restricted to the cavities of the reproductive organs. Certain of the organs are metamerically arranged (e.g., the *nervous system*); others (e.g., the *excretory organs*) are concentrated into a small space. The systems of organs and their functions will be presented in the following order: (1) digestive, (2) vascular, (3) respiratory, (4) excretory, (5) nervous, (6) sense organs, (7) muscular, and (8) reproductive.

Digestive System

The alimentary canal of a crayfish consists of the following parts.

1. The *mouth* opens on the ventral surface between the jaws.
2. The *esophagus* is a short tube leading from the mouth to the stomach.
3. The *stomach* is a large cavity divided by a constriction into an anterior *cardiac* chamber and a smaller posterior *pyloric* chamber. In the stomach are a number of chitinous ossicles of use in chewing the food, and collectively known as the *gastric mill.* The most important of these are (a) a median cardiac ossicle, (b) a median urocardiac ossicle, (c) two lateral pterocardiac ossicles, (d) a pair of lateral zygocardiac ossicles, (e) a pyloric ossicle, and (f) a prepyloric ossicle. The ossicles are able to move one upon another, and, being connected with powerful muscles, are effective in grinding up the food. On either side of the pyloric chamber enters a duct of the digestive glands, and above is the opening of the small cecum.

At certain times two calcareous bodies, known as *gastroliths,* are present in the lateral walls of the cardiac chamber of the stomach. Their function is not certain, but is probably the storage of the calcareous matter used in hardening the exoskeleton.

Descriptive Table of the Appendages of the Crayfish

Appendage	Protopodite	Exopodite	Endopodite	Function
I. Antennule	3 segments: statocyst in basal segment	many-jointed filament	many-jointed filament	tactile; chemical; equilibration
II. Antenna	2 segments; excretory pore in basal segment	broad, thin, dagger-like lateral projection	a long many-jointed "feeler"	tactile; chemical
III. Mandible	2 segments; a heavy jaw and basal segment of palp	absent	small; 2 distal segments of palp	crushing food
IV. 1st Maxilla	2 thin lamellae extending inward	absent	a small outer lamella	
V. 2d Maxilla	2 bilobed lamellae	dorsal half of plate, the scaphognathite	1 segment; small, pointed	creates current of water in gill chamber
VI. 1st Maxilliped	2 thin segments extending inward; a broad plate, the epipodite extending outward	a long basal segment bearing a many-jointed filament	small; 2 segments	chemical; tactile; holds food
VII. 2d Maxilliped	2 segments; a basal coxopodite bearing a gill, and a basipodite bearing the exopodite and endopodite	similar to VI	5 segments; the basal one long and fused with the basipodite	similar to VI
VIII. 3d Maxilliped	similar to VII	similar to VI	similar to VII, but larger	similar to VI
IX. 1st Walking Leg (Chela, Cheliped, or Pincher)	2 segments; coxopodite, and basipodite	absent	5 segments, the terminal two forming a powerful pincher	offense and defense; aids in walking; tactile
X. 2d Walking Leg (Pereiopod)	similar to IX	absent	as in IX, but not so heavy	walking; prehension; toilet implements

XI. 3d Walking Leg	similar to IX; coxopodite of female contains genital pore	absent	similar to X	similar to X
XII. 4th Walking Leg	similar to IX	absent	similar to X, but no pincher at end	walking
XIII. 5th Walking Leg	similar to IX; coxopodite of male bears genital pore	absent	similar to XII	walking; cleaning abdomen and eggs
XIV. 1st Abdominal (1st Pleopod or Swimmeret)				reduced in female; in male, protopodite and endopodite fused together forming an organ for transferring sperm
XV. 2d Abdominal (2d Pleopod or Swimmeret)	in female 2 segments	in female many-jointed filament	in female like exopodite but longer	in female as in XVI, in male modified for transferring sperm to female
XVI. 3d Abdominal (3d Pleopod or Swimmeret)	2 segments	many-jointed filament	like exopodite but longer	creates current of water; in female used for attachment of eggs and young
XVII. 4th Abdominal (4th Pleopod or Swimmeret)	2 segments	as in XVI	as in XVI	as in XVI
XVIII. 5th Abdominal (5th Pleopod or Swimmeret)	as in XVI	as in XVI	as in XVI	as in XVI
XIX. 6th Abdominal (Uropod)	1 short, broad segment	flat oval plate divided by transverse groove into two parts	flat oval plate	swimming

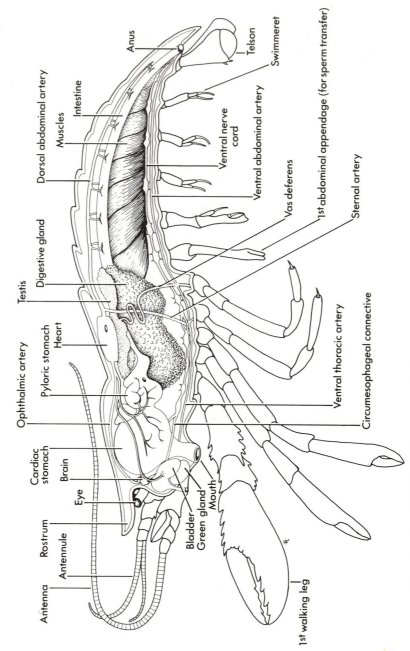

Figure 10-3 Longitudinal section of a lobster showing internal organs. (*After Root modified.*)

494

4. The *intestine* is a small tube passing posteriorly near the dorsal wall of the abdomen, and opening to the outside through the *anus* on the ventral surface of the telson.

5. The *digestive glands,* or "liver," are situated in the thorax, one on either side. Each consists of three lobes composed of a great number of small tubules. The glandular epithelium lining these tubules produces a secretion which passes into the *hepatic ducts* and thence into the pyloric chamber of the stomach.

Nutrition

Food. The food of the crayfish is made up principally of living animals such as snails, tadpoles, insect larvae, small fish, and frogs; but decaying organic matter is also eaten. Not infrequently crayfishes prey upon others of their kind. They feed at night, being more active at dusk and daybreak than at any other time. Their method of feeding may be observed in the laboratory if a little fresh meat is offered to them. The maxillipeds and maxillae hold the food while it is being crushed into small pieces by the mandibles. It then passes through the esophagus into the stomach. The coarser parts are ejected through the mouth.

Digestion. In the cardiac chamber of the stomach, the food is ground up by the ossicles of the gastric mill. When fine enough, it passes through the *strainer,* which lies between the two divisions of the stomach. The strainer consists of two lateral and a median ventral fold which bear hairlike setae, and allow the passage of only liquids or very fine particles. In the pyloric chamber, the food is mixed with the secretion from the digestive glands brought in by way of the hepatic ducts. From the pyloric chamber the disolved food passes into the intestine by the walls of which the nutritive fluids are absorbed. Undigested particles pass on into the posterior end of the intestine, where they are gathered together into feces, and egested through the anus.

Circulatory System

The Blood. The blood into which the absorbed food passes is an almost colorless liquid in which are suspended a number of amoeboid cells, the *blood corpuscles* or amoebocytes. The principal *functions* of the blood are the transportation of food materials from one part of the body to another, of oxygen from the gills to the various tissues, of carbon dioxide to the gills, and of urea to the excretory organs. If a crayfish is wounded, the blood, on coming in contact with the air, thickens, forming a clot. It is said to *coagulate.* This clogs the opening and prevents loss of blood. The chelipeds and other walking legs of the crayfish have a breaking point near their bases. When one is injured the animal may break the limb at this point and lessen the blood flow, since only a small space is present in the appendage at this particular spot, and coagulation, therefore, takes place very quickly.

Blood Vessels (Figs. 10-3, 10-4, and 10-5). The principal blood vessels are a heart, seven main arteries, and a number of spaces called sinuses.

Heart. The *heart* is a muscular-walled, saddle-shaped sac lying in the *pericardial sinus* in the median dorsal part of the thorax. It may be considered as a dilation of a dorsal vessel resembling that of the earthworm. Six elastic *ligaments,* two anterior, two posterior, and two running along the ventral border of each lateral surface, fasten it to the walls of the pericardial sinus. Three pairs of valvular apertures, called *ostia,* one dorsal and two lateral, allow the blood to enter from the surrounding sinus.

Arteries. Five arteries arise from the anterior end of the heart:

1. The *ophthalmic* artery is a median dorsal tube passing forward over the stomach, and supplying the cardiac portion, the esophagus, and head.

2,3. The two *antennary* arteries arise one on each side of the ophthalmic artery, pass forward, outward, and downward, and branch, sending a gastric artery to the cardiac part of the stomach, arteries to the antennae, to the excretory organs, and to the muscles and other cephalic tissues.

4,5. The two *hepatic* arteries leave the heart below the antennary arteries. They lead directly to the digestive glands.

A single dorsal abdominal artery arises from the posterior end of the heart.

Figure 10-4 Circulatory system of a lobster. (*After Gegenbaur.*)

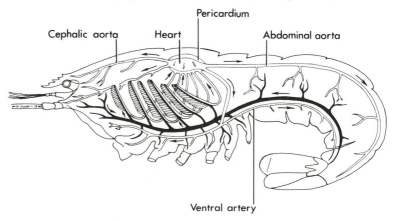

Cephalic aorta   Heart   Pericardium   Abdominal aorta

Ventral artery

6. The *dorsal abdominal artery* is a median tube leading backward from the ventral part of the heart, and supplying the dorsal region of the abdomen. It branches near its point of origin, giving rise to the *sternal* artery; this leads directly downward, and, passing between the nerve cords connecting the fourth and fifth pairs of thoracic ganglia (see Fig. 10-3), divides into two arteries. One of these, the *ventral thoracic* artery, runs forward beneath the nerve chain, and sends branches to the ventral thoracic region, and to appendages III to XIII; the other, the *ventral abdominal* artery, runs backward beneath the nerve chain, and sends branches to the ventral abdominal region and to the abdominal appendages.

Sinuses. The blood passes from the arteries into spaces lying in the midst of the tissues, called sinuses. The pericardial sinus has already been mentioned. The thorax contains a large ventral blood space, the *sternal sinus,* and a number of *branchiocardiac canals* extending from the bases of the gills to the pericardial sinus. A *pervisceral sinus* surrounds the alimentary canal in the cephalothorax.

Blood Flow (Fig. 10-5). The heart, by means of rhythmical contractions, forces the blood through the arteries to all parts of the body. Valves are present in every artery where it leaves the heart; they prevent the blood from flowing back. The finest branches of these arteries, the *capillaries,* open into spaces between the tissues, and the blood eventually reaches the sternal sinus. From here it passes into the *afferent channels* of the gills and into the gill filaments, where the carbonic acid in solution is exchanged for oxygen from the water in the branchial chambers. It then returns by way of the *efferent gill channels,* passes into the branchiocardiac sinuses, thence to the pericardial sinus, and finally through the ostia into the heart. The valves of the ostia allow the blood to enter the heart, but prevent it from flowing back into the pericardial sinus.

Repiratory System

Between the branchiostegites and the body wall (Fig. 10-5) are the branchial chambers containing the gills. At the anterior end of the branchial chamber is a channel in which the *scaphognathite* of the second maxilla (Fig. 10-2) moves back and forth, forcing the water out through the anterior opening. Water flows in through the posterior opening of the branchial chamber.

Gills. There are two rows of gills, named according to their points of attachment. The outermost, the *podobranchiae,* are fastened to the coxopodites of certain appendages and the inner double row, the *arthrobranchiae,* arise from the membranes at the bases of these appendages. In *Astacus* there is a third row, the *pleurobranchiae,* attached to the

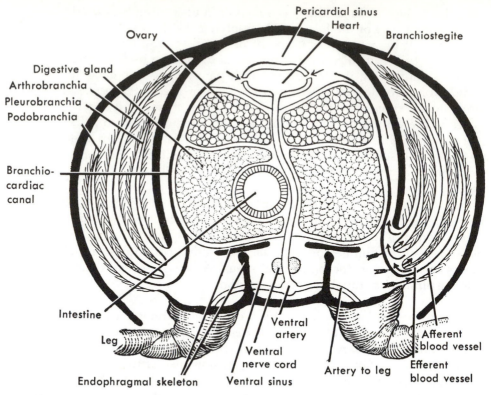

Figure 10-5 Diagrammatic cross section through the thorax of a crayfish at the level of the sternal artery. The arrows indicate the direction of blood flow. (*From R. W. Hegner,* College Zoology, *5th ed., Macmillan Publishing Co., Inc., 1947.*)

walls of the thorax. The number and arrangement of these gills are shown in the accompanying table. The podobranchiae consist of a basal plate covered with delicate setae and a central axis bearing a thin, longitudinally folded corrugated plate on its distal end, and a featherlike group of *branchial filaments.* The arthrobranchiae have a central stem on either side of which extends a number of filaments, causing the entire structure to resemble a plume. Attached to the base of the first maxilliped is a broad thin plate, the *epipodite,* which has lost its branchial filaments.

Excretory System

The excretory organs are a pair of rather large bodies, the *"green glands"* (Fig. 10-3), situated in the ventral part of the head anterior to the esophagus. Each green gland consists of a green glandular portion, a thin-walled dilation, the bladder, and a duct opening to the exterior through a pore at the top of the papilla on the basal segment of the antenna.

The Number and Position of the Gills of the Crayfish

| Segment | Podobranchiae | Arthrobranchiae | | Total Number |
		Anterior	Posterior	
VI	0 (ep.)	0	0	0 (ep.)
VII	1	1	0	2
VIII	1	1	1	3
IX	1	1	1	3
X	1	1	1	3
XI	1	1	1	3
XII	1	1	1	3
	6 (ep.)	6	5	17 (ep.)

Nervous System

The morphology of the nervous system (Fig. 10-3) of the crayfish is in many respects similar to that of the earthworm. The *central nervous system* includes a dorsal ganglionic mass, the *brain*, in the head, and two *circumesophageal connectives* passing to the *ventral nerve cord*, which lies near the median ventral surface of the body.

Brain. The brain is a compact mass larger than that of the earthworm, and supplies the eyes, antennules, and antennae with nerves.

Ventral Nerve Cord. The ganglia and connectives of the ventral nerve cord are more intimately fused than in the earthworm, and it is difficult to make out the double nature of the connectives, except between segments XI and XII, where the sternal artery passes through. Each segment posterior to VII possesses a ganglionic mass, which sends nerves to the surrounding tissues. The large subesophageal ganglion in segment VII consists of the ganglia of segments III to VII fused together. It sends nerves to the mandibles, maxillae, and first and second maxillipeds.

The *visceral nervous system* consists of an anterior visceral nerve which arises from the ventral surface of the brain, is joined by a nerve from each circumesophageal connective, and, passing back, branches upon the dorsal wall of the pyloric part of the stomach, sending a *lateral nerve* on each side to unite with an *inferolateral nerve* from the *stomatogastric ganglion.*

Schrameck (1970) found that giant fibers of the nerve cord are not essential for operation of the tail flips used for escape responses, although they normally are the route of initiation and may be used for segmental synchronization of the muscles involved.

Sense Organs

Eyes. The eyes of the crayfish are situated at the end of movable stalks which extend out, one from each side of the rostrum (Fig. 10-3). The external convex surface of the eye is covered by a modified portion of the transparent cuticle, called the *cornea*. This cornea is divided by a large number of fine lines into four-sided areas, termed *facets*. Each facet is but the external part of a long visual rod known as an *ommatidium*.

Sections (Fig. 10-6) show the compound eye to be made up of similar ommatidia lying side by side, but separated from one another by a layer of dark pigment cells. The average number of ommatidia in a single eye is 2,500.

Two ommatidia are shown in Figure 10-7. Beginning at the outer surface, each ommatidium consists of the following parts: (a) a *corneal facet*; (b) two *corneagen* cells which secrete the cornea; (c) a *crystalline cone* formed by four cone cells, or *vitrellae*; (d) two *retinular cells* surrounding the cone; (e) seven retinular cells which form a *rhabdom* consisting of four *rhabdomeres*; and a number of *pigment cells*. Fibers from the *optic nerve* enter at the base of the ommatidium and communicate with the inner ends of the retinular cells.

Vision. The *eyes* of the crayfish are supposed to produce an *erect mosaic* or *"apposition image"*; this is illustrated in Figure 10-8, where the ommatidia are represented by *a–e* and the fibers from the optic nerve by *a'–e'*. The rays of light from any point *a, b,* or *c,* will all encounter the dark pigment cells surrounding the ommatidia and be absorbed, except the ray, which passes directly through the center of the cornea as *d* or *e;* the ray will penetrate to the retinulae, and thence to the fibers from the optic nerve. The sum of similar events involving all the ommatidia presumably provides a mosaic image. This method of

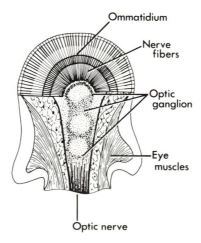

Ommatidium

Nerve fibers

Optic ganglion

Eye muscles

Optic nerve

Figure 10-6 Crayfish. Eye in longitudinal section. (*After Borradaile and Potts, modified.*)

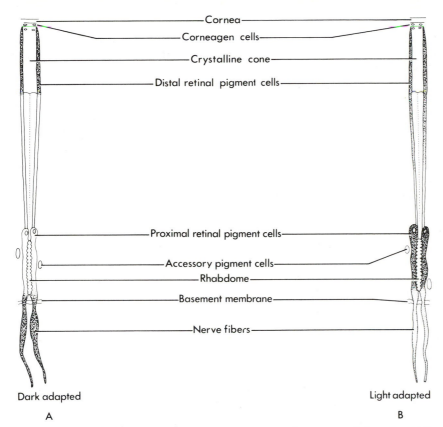

Cornea
Corneagen cells
Crystalline cone
Distal retinal pigment cells

Proximal retinal pigment cells

Accessory pigment cells
Rhabdome
Basement membrane

Nerve fibers

Dark adapted

A

Light adapted

B

Figure 10-7 *Cambarus.* Longitudinal sections of two crayfish ommatidia, showing the arrangement of retinal pigment under light-adapted (B) and dark-adapted (A) conditions. (*After Bennitt and Merrick.*)

image formation is especially well adapted for recording motion, since any change in the position of a large object affects the entire 2,500 ommatidia.

When the pigment surrounds the ommatidia (Fig. 10-7, B), vision is as described above; but it has been found that in dim light the pigment migrates partly toward the outer and partly toward the basal end

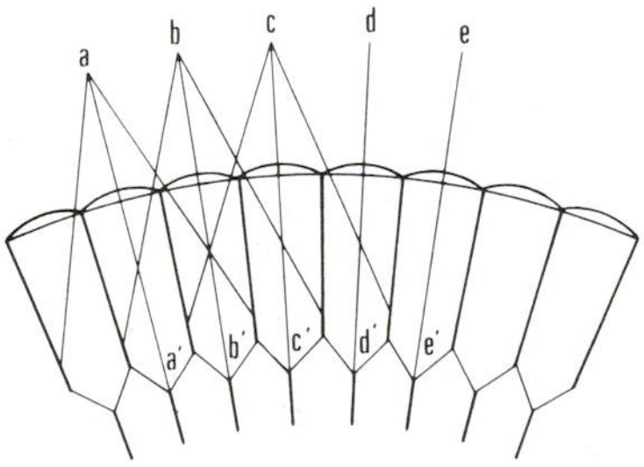

Figure 10-8 Diagram of part of a compound eye which serves to explain mosaic vision. (*After Lubbock.*)

of the ommatidia (Fig. 10-7, A). When this occurs the ommatidia no longer act separately, but a combined image is thrown on the retinular layer. The erect *"superposition image"* thus formed is probably a strong image with little detail.

A reflecting pigment is present in the eyes of certain crustaceans which consists of amorphous guanin and forms a reflecting layer below the rhabdomes. This layer increases the efficiency of the eye at low light intensities.

Hormones have been obtained from the eye stalk of many crustaceans. One brings about contraction of chromatophores in Crustacea. It produces melanophore expansion in frog tadpoles and melanophore contraction in fishes.

Statocysts. The statocysts of the crayfish are chitinous-walled sacs situated one in the basal segment of each antennule. In the base of the statocyst is a ridge, called the *sensory cushion,* and three sets of *hairs,* over 200 in all, each innervated by a single nerve fiber (Fig. 10-9). Among these hairs are a number of large grains of sand, the *statoliths,* which are placed there by the crayfish. Beneath the sensory cushion are *glands* which secrete a substance for the attachment of the statoliths to the hairs.

The statocyst for many years was considered an auditory organ, and it may possibly function as such, though later investigations have proven that it is primarily an organ of *equilibration.* The contact of the statoliths with the statocyst hairs determines the orientation of the body while swimming, because any change in the position of the animal causes a change in the position of the statoliths under the influence of gravity. When the crayfish changes its exoskeleton in the process of molting, the statocyst is also shed. Individuals that have just molted, or have had their statocysts removed, lose much of their powers of orientation. Perhaps the most convincing proof of the function of equilibration is that furnished by the experiments of Kreidl.

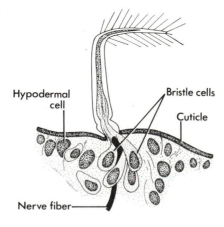

Hypodermal cell

Bristle cells

Cuticle

Nerve fiber

Figure 10-9 Sensory hair from the statocyst of a shrimp, *Palaemonetes.* (*After Prentiss, from Dahlgren and Kepner.*)

This investigator placed shrimps, which had just molted and were therefore without statoliths, in filtered water. When supplied with iron filings, the animals filled their statocysts with them. A strong electromagnet was then held near the statocyst, and the shrimp took up a position corresponding to the resultant of the two pulls, that of gravity and of the magnet.

Muscular System

The principal muscles in the body of the crayfish are situated in the abdomen, and are used to bend that part of the animal forward upon the ventral surface of the thorax, thus producing backward locomotion in swimming. Other muscles extend the abdomen in preparation for another stroke. Figure 10-3 shows a longitudinal section of a crayfish. The powerful flexor muscles of the abdomen are shown. In the dorsal region are the less powerful extensor muscles. Other muscles of considerable size are situated within the tubular appendages, especially the chelipeds. A comparison of the skeleton and muscles of the crayfish with those of humans is interesting. The skeleton of the crayfish is external and tubular, except in the ventral part of the thorax (Fig. 10-5). The muscles are internal, and attached to the inner surface of the skeleton. In humans, on the other hand, the skeleton is internal and the muscles are external.

Reproductive System

Crayfishes are normally *dioecious*, there being only a few cases on record where both male and female reproductive organs were found in a single specimen.

Male Reproductive Organs. The male organs (Fig. 10-10, A) consist of a *testis* and two *vasa deferentia* which open through the coxopodites of the fifth pair of walking legs. The testis lies just beneath the pericardial sinus (Fig. 10-3). It is a soft white body possessing two anterior lobes and a median posterior extension. The vasa deferentia are long coiled tubes.

Spermatogenesis. The primitive germ cells within the testis pass through two maturation divisions, and then metamorphose into spermatozoa. These are flattened spheroidal bodies when enclosed within the testis or vas deferens, but if examined in water or some other liquid they are seen to uncoil, finally becoming star-shaped.

The spermatozoa remain in the testis and vasa deferentia until copulation takes place. As many as 2 million spermatozoa are contained in the vasa deferentia of a single specimen.

Female Reproductive Organs. The *ovary* (Fig. 10-10, B) resembles the testis in form, and is similarly located in the body (Fig. 10-3). A short

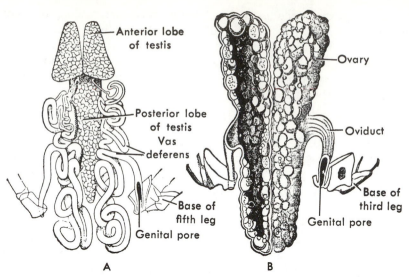

Figure 10-10 Reproductive organs of the crayfish. (A) Male. (B) Female. (*From Shipley and MacBride.*)

oviduct leads from near the center of each side of the ovary to the external aperture in the coxopodite of the third walking leg.

Oogenesis. The primitive germ cells in the walls of the ovary grow in size, become surrounded by a layer of small cells, the follicle, which eventually break down, allowing the eggs to escape into the central cavity of the ovary. At the time of laying, the ova pass out through the oviduct.

Breeding Habits

The details of copulation, egg laying, and the larval stages of crayfishes have been derived from observations of several different species.

Ameyaw-Akumfi and Hazlett (1975) provide evidence that sex recognition among crayfish is behavioral and pheromonal. Pheromones, which are probably carbohydrates in this case, are perceived on the inner branch of the antennule.

The principal events in the reproduction of crayfishes may be enumerated as follows:

(1) Copulation, during which the spermatozoa are transferred from the vasa deferentia of the male to the seminal receptacle of the female; (2) egg laying; (3) the embryonic development of the eggs; (4) hatching; and (5) the growth of the young crayfishes.

Copulation. Copulation in crayfishes, in most cases, takes place in September, October, or November of the first year of their lives, that is, when they are about 4 months old. A second copulating season is passed through at the end of the second summer, when the animals are about 17 months old, and a third copulating season occurs at the end of the third summer. At these times a male approaches a female, grasps her by her cephalic appendages, and, after a struggle, turns her over on her back. He then stands over her and transfers spermatozoa to the seminal receptacle. During this process, the spermatozoa flow out of the openings of the vasa deferentia, pass along the grooves on the first abdominal appendages of the male, and enter the seminal receptacle. Here they are stored during the winter. The *seminal receptacle* is a cavity in a fold of cuticle lying between the sterna of the segments bearing the fourth and fifth pairs of walking legs.

Egg Laying. The eggs are laid at night during the month of April. First the ventral side of the abdomen is thoroughly cleaned of all dirt by the hooks and comblike bristles near the end of the fifth pair of walking legs. A clear slime or *glair* is secreted by *cement glands* situated chiefly on the basal parts of the uropods, and on the endopodites of the other abdominal appendages. This milky glair gradually covers the swimmerets. The female then lies on her back, and an apronlike film of glair is constructed between the ends of the uropods and telson, and the bases of the second pair of walking legs. The eggs emerge from the openings of the oviducts in the bases of the third pair of walking legs, flow posteriorly, and become attached to the hairs on the swimmerets by strings of a substance no doubt secreted by the cement glands. This is brought about by the turning of the animal first on one side and then on the other a number of times. From 100 to over 600 greenish eggs are laid by a single female, depending upon the size and age of the animal. After the eggs are laid the crayfish rights herself, the apron of glair breaks down, and the abdomen is extended.

Fertilization. The method of fertilization has never been discovered. It is supposed that as the eggs are laid they pass over the opening of the seminal receptacle, and are then penetrated by the spermatozoa which were placed there by the male the preceding autumn.

Embryology

While the eggs are developing they are protected by the abdomen of the female, and are *aerated* and kept free from dirt by the waving of the swimmerets back and forth.

As already noted, the eggs of the crayfish are attached to the swimmerets of the mother by a glutinous secretion. They die if removed, although those of the lobster do not. All the eggs carried by one crayfish are in the same stage of development at the same time. Embryonic development is slow, 2 weeks to several months being required. Most of the egg consists of *yolk,* which greatly interferes with cleavage (Fig.

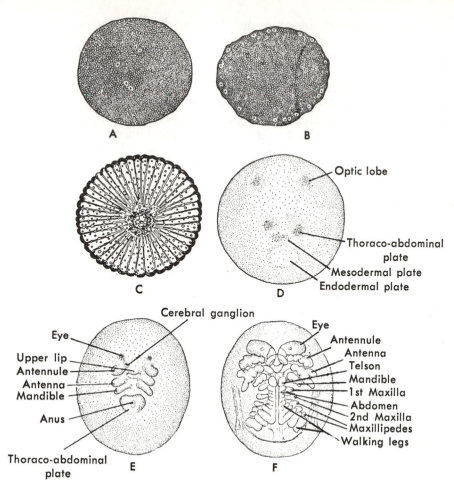

Figure 10-11 Stages in the embryology of the crayfish. (A) Egg with a few cleavage nuclei. (B) Nuclei arranged near the surface. (C) Egg divided into yolk pyramids. (D) Embryo beginning to form. (E) Embryo in nauplius stage. (F) Older embryo. (*After Reichenbach.*)

10-11, A). The *fusion nucleus* lies within the yolk in a little island of cytoplasm. The nuclei that arise from it migrate to the periphery (B) and the egg then divides into incomplete columnar *blastomeres* called *primary yolk pyramids,* each with a single nucleus at the outer end (C). This *blastula* is not hollow but has a *blastocoel* filled with yolk globules, and a single layer of cells at the periphery which constitute the *blastoderm*. On one side of the egg the blastoderm cells increase in number and become columnar, thus forming a *ventral plate*. Five thickenings of the ventral plate are recognizable (D); these are two *cephalic lobes* from which the eyes and cerebral ganglia will develop, two *thoraco-abdominal rudiments,* and one *endodermal disk* at the anterior end of which mesoderm cells are destined to arise. An invagination at the anterior end of the endodermal disk indicates the beginning of *gastrulation.*

Further changes in the ventral plate result in what is known as the *Nauplius* stage as illustrated in Figure 10-11, E. At this time the following rudiments are recognizable: two cephalic lobes with cerebral rudiments in the center; three pairs of outgrowths, which develop into the first and second antennae and the mandibles; the upper lip or labium; the crescent-shaped mouth; and the anus in the center of the thoracoabdominal disk.

As development continues the thoracoabdominal disk grows out into a long forked process, which bends underneath the egg toward the anterior end as shown in Figure 10-11, F. The following pairs of rudiments may now be seen: optic lobes, optic ganglia, cerebral ganglia, antennal ganglia, first antennae, second antennae, mandibles, first and second maxillae, first, second, and third maxillipeds, and first to fifth walking legs. The abdominal segments with the telson at the end are less distinct; the labium is a triangular outgrowth; and the fold that becomes the carapace is now visible.

The time of the various stages in development is as follows: first week, cleavage; second week, embryo formation; third through fifth weeks, nauplius stage and further development prior to hatching. Hatching time is from the fifth to eighth week, depending upon temperature.

Hatching. In hatching, the egg capsule splits and the larva emerges head foremost. The helpless young crayfish would drop away from the mother at once but for a thread extending from its telson to the inner surface of the egg capsule.

Soon the larva possesses strength enough to grasp the egg string with its claws. The telson thread then breaks. After about 48 hours the larva passes into a *second stage.* This is inaugurated by the shedding of the first larval cuticular covering, a process known as *molting* or *ecdysis.* This casting off of the covering the body is not peculiar to the young, but occurs in adult crayfishes as well as in young and adults of many other animals. In the larval crayfish the cuticle of the first stage becomes loosened and drops off. In the meantime the hypodermal cells have secreted a new covering. Ecdysis is necessary before growth can proceed, since the chitin of which the exoskeleton is composed does not allow expansion. In adults it is also a means of getting rid of an old worn-out coat and acquiring a new one.

In the *second larval stage* the young crayfish is again supported immediately after casting off its covering by a thread extending from the new to the old telson. When the larva becomes strong enough, it grasps the old larval skin or swimmerets, and the telson thread drops off. The duration of the second larval stage is about 6 days.

No telson thread is present after the molt which ushers in the *third larval stage,* but the young is able at once to cling to the old cuticle. In about a week the third larvae become entirely independent of the mother, although they at first always return to her when separated. From this time on the larvae shift for themselves, growing rapidly, and molting at least four more times during the first summer. The first winter no growth nor molts occur. There are four or five molts the second summer, three or four in the third summer, and perhaps one or two in the fourth summer. The life of a single individual extends over a period of about 3 or more years.

Regeneration

The crayfish and many other crustaceans have the power of regenerating lost parts, but to a much more limited extent than such animals as *Hydra* and the earthworm. Experiments have been performed upon almost everyone of the appendages as well as the eye. The second and third maxillipeds, the walking legs, the swimmerets, and the eye have all been injured or extirpated at various times and subsequently renewed the lost parts. Many species of crayfish of various ages have been used for these experiments. The growth of regenerated tissue is more frequent and rapid in young specimens than in adults.

The new structure is not always like that of the one removed. For example, when the annulus containing the sperm receptacle of an adult is extirpated, another is regenerated, but, although this is as large as that of the adult, it is comparable in complexity to that of an early larval stage. A more remarkable phenomenon is the regeneration of an apparently functional (tactile) antenna-like organ in place of a degenerate eye which was removed from the blind crayfish, *Orconectes pellucidus testii.* In this case a nonfunctional organ was replaced by a functional one of a different character. The regeneration of a new part which differs from the part removed is termed *heteromorphosis.*

Autotomy

Perhaps the most interesting morphological structure connected with the regenerative process in the crayfish is the definite breaking point near the bases of the walking legs. If the chelae are injured, they are broken off by the crayfish at the breaking point. The other walking legs, if injured, may be thrown off at the free joint between the second and third segments. A new leg, typically a smaller leg, develops from the end of the stump remaining. This breaking off of the legs at a definite point is known as *autotomy,* a phenomenon that also occurs in a number of other animals. The breaking point in decapod crustaceans is in the base of the ischium. The leg is flexed by the autotomizer muscle until the ischium comes in contact with the coxa. Continued pull of the autotomizer muscle separates the leg at the breaking point.

The muscles are not damaged and the valve and diaphragm prevent hemorrhage. Autotomy results when the leg is stimulated by some harmful agent or may be brought about in certain instances by injury to a ventral ganglion or by a strong general stimulus. Immediately after the leg has been thrown off, a membrane of ectoderm cells covers the canal through which the nerve and blood passed; five days later regeneration begins by an outward growth of the ectoderm cells which lined the exoskeleton. An interesting point in this new growth is that the muscles of the regenerated part are probably produced by ectoderm cells, whereas in the embryonic development of the crayfish the muscles are supposed to arise from the mesoderm.

The power of autotomy is of advantage to the crayfish, since the wound closes more quickly if the leg is lost at the breaking point.

As in the earthworm, the *rate of regeneration* depends upon the amount of tissue removed. If one chela is amputated, a new chela regenerates less rapidly than if both chelae and some of the other walking legs are removed.

Behavior

When at rest, the crayfish usually faces the entrance to its place of concealment, and extends its antennae. It is thus in a position to learn the nature of any approaching object without being detected. Activity at this time is reduced to the movements of a few of the appendages and the gills; the scaphognathites of the second maxillae move back and forth baling water out of the forward end of the gill chambers; the swimmerets are in constant motion creating a current of water; the maxillipeds are likewise kept moving; and the antennae and eye stalks bend from place to place.

Crayfishes are more active at nightfall and at daybreak than during the remainder of the day. At these times they venture out of their hiding places in search of food, their movements being apparently all utilitarian and not for spontaneous play or exercise.

Locomotion. Locomotion is effected in two ways, walking and swimming. Crayfishes are able to walk in any direction, forward usually, but also sidewise, obliquely, or backward. In walking, the fourth pair of legs are most effective and bear nearly all of the weight of the animal; the fifth pair serve as props, and to push the body forward; the second and third pairs are less efficient for walking, since they are modified to serve as prehensile organs, and as toilet implements. Swimming is not resorted to under ordinary conditions, but only when the animal is frightened or shocked. In such a case the crayfish extends the abdomen, spreads out the uropods and telson, and, by sudden contractions of the bundles of flexor abdominal muscles, bends the abdomen and darts backward. The swimming reaction apparently is not voluntary, but is almost entirely reflex.

Equilibration. The crayfish either at rest or in motion is in a state of unstable equilibrium, and must maintain its body in the normal position by its own efforts. The force of gravity tends to turn the body over. From a large number of experiments it has been proven that the statocysts are the organs of equilibration. The structure of these organs has been described. The contact of the statoliths with the statocyst hairs furnishes the stimulus which causes the animal to maintain an upright position.

When placed on its back, the crayfish has some difficulty in righting itself. Two methods of regaining its normal position are employed. The usual method is that of raising itself on one side and allowing the body to tip over by the force of gravity. The second method is that of contracting the flexor abdominal muscles which causes a quick backward flop, bringing the body right side up.

Senses and Their Location. The sense of *touch* in crayfishes is perhaps the most valuable, since it aids them in finding food, avoiding obstacles, and in many other ways. It is located in specialized hairs on various parts of the body. *Vision* in crayfishes is probably of minor importance, since the compound eyes are almost useless in recognizing form, and are of real value only in detecting moving objects. No reactions to *sound* have ever been observed in crayfishes, and apparently they do not hear, although tactile reflexes may simulate reaction to sound. In aquatic animals it is so difficult to distinguish between reactions of *taste* and *smell* that these senses are both included in the term *chemical sense*. The end organs of this sense are distributed all over the body.

Reactions to Stimuli. *Contact.* The crayfish has tactile receptors on all surface areas. Receptors are concentrated on extremities, ventral surface, and mouth parts. The antennae are usually considered the special organs of touch, but experiments seem to prove that they are not so sensitive as other parts of the body. The tactile hairs are plumed, and supplied by a single nerve. *Positive* reactions to contact are exhibited by crayfishes to a marked degree, the animals seeking to place their bodies in contact with a solid object, if possible. The normal position of the crayfish when at rest under a stone is such as to bring its side or dorsal surface in contact with the walls of its hiding place. This, no doubt, is of distinct advantage, since it forces the animal into a place of safety.

Light. Light of various intensities in the majority of cases causes the crayfish to retreat. Individuals prefer colored lights to white, having a special liking for red. Negative reactions to light play an important role in the animal's life, since they influence it to seek a dark place where it is concealed from its enemies.

Chemicals. The reactions of the crayfish to food are due in part to a chemical sense. All surfaces seem sensitive to chemicals. The anterior appendages, however, are the most sensitive, especially the outer ramus of the antennule. *Positive reactions* result from the application of food substances. For example, if meat juice is placed in the water near an animal, the antennae move slightly and the mouth parts perform vigorous chewing movements. The meat causes activity and, if the food is touched, localization of the food. Acids, salts, sugar, and other chemicals produce a sort of *negative reaction* indicated by scratching the carapace, rubbing the chelae, or pulling at the part stimulated.

Habit Formation. It has been shown by certain simple experiments that crayfishes are able to learn habits and to modify them. They learn by experience, and modify their behavior slowly or quickly, depending upon their familiarity with the situation. Proprioceptors as well as external sensory receptors are factors in habit formation.

A large male *Orconectes* was collected during a trip when inappropriate equipment was available. A hemispherical sieve about 15 cm in diameter was used for temporary transport. The crayfish was placed on its back in the sieve and its legs gyrated wildly with little effect until after about 10 minutes of trying it righted itself. Returned to the original position, it achieved the righting much more rapidly. After several more experiences it was righting itself immediately with reasonably well coordinated movements.

Ecology of the Crayfish

Crayfish are important scavengers. They also feed upon living invertebrates and are potential control organisms for light-shelled gastropod vectors of trematodes. *Cambarus* may go overland on moist nights and often builds burrows in meadows with obvious "chimneys" at the openings.

They are not collected by dredging in lakes in proportion to their importance. David Flanagan ran some 1- by 10-meter SCUBA transects in Gull Lake, Michigan, and found the following distribution of burrows by depth on June 16, 1979. An area of fairly uniform, silty sand with few aquatic weeds was chosen.

Depth of transect (meters)	3	5	7	9	11
Number of crayfish excavations	2	3	6	3	0

Large dead individuals, perhaps postreproductive mortalities, were found in the 7- to 11-meter depths outside the transects. A few live ones, including an egg-bearing female, were present outside of excavations.

Species show definite substrate preferences (Crocker and Barr,

1968). Many use stones and other obstructions as roofs for their burrows.

ASELLUS

The water sowbug or water slater, *Asellus* (Fig. 10-12), is a common member of the order Isopoda in many shallow streams and ponds of the northern hemisphere. Comparison with the crayfish will show a form that has a similar number of segments organized in different functional groupings.

External Anatomy

Asellus tapers to a narrow small head with sessile eyes and two pair of uniramous antennae. The first pair of antennae bear a number of olfactory setae (Fig. 10-13, E). The mandibles bear a three-jointed palp, cutting teeth, and a molar process. A spine row prevents escape of food being ground by the molar process (Fig. 10-15). Two pair of maxillae posterior to the mandibles aid in food handling.

There are eight thoracic appendages as in the crayfish. Only the first forms a food-handling maxilliped that can be considered a mouthpart. The segment it is borne on is fused with the head. The following seven thoracic segments are separate and each bears a pair of legs.

The first pair of legs is more specialized than the rest. It has a terminal segment (dactyl) that can fold against the segment bearing it (pro-

Figure 10-12 *Asellus communis.* (A) Dorsal view of male. (B) Ventral view of female. (*A, from Van Name, after Harger; B, from Van Name.*)

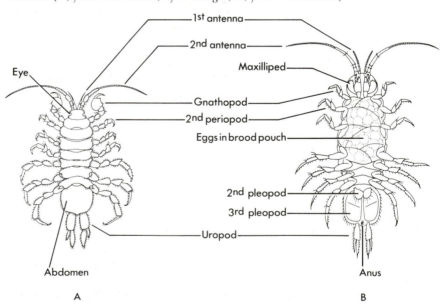

1st antenna
2nd antenna
Maxilliped
Eye
Gnathopod
2nd periopod
Eggs in brood pouch
2nd pleopod
3rd pleopod
Uropod
Abdomen
Anus

A

B

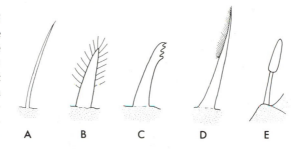

Figure 10-13 Setae of *Asellus*. (A) Simple tactile seta from leg. (B) Plumose seta from first maxilla. (C) Spinose seta from first maxilla. (D) Pectinate seta from maxilliped. (E) Olfactory seta from first antenna. (*Not to same scale.*)

A   B   C   D   E

pod). In some related species the side of the propod is expanded to form a more efficient grasping organ. Comparison of these types with the chelae of a crayfish show how the pincer has probably formed by even greater growth of the lateral surface of the propod (Fig. 10-14). This simple illustration, although it involves different thoracic appendages, indicates how a simple modification can change the functional ability of an appendage. Thus, part of the success of the arthropods has been due to the presence of many segments with similar appendages, which have undergone individual modifications selected to fill the needs of their diverse ways of life.

The abdominal segments form a unit bearing five pair of pleopods and a terminal pair of branched uropods. The female has only four pair of pleopods and the first pair is greatly reduced. The second pair of pleopods in the male is used for sperm transfer. The third pair bears

Figure 10-14 Terminal three segments of crustacean appendages, showing varying levels of chela development. (A) Fourth leg of female *Asellus communis*, no chelate function. (B) Gnathopod of female *A. communis*. (C) Gnathopod of female *Colubotelson thomsoni*. (D) Cheliped on the crab, *Planes minuta*. (E) Cheliped of *Portunus pelagicus*. (*D and E, modified from Hale.*)

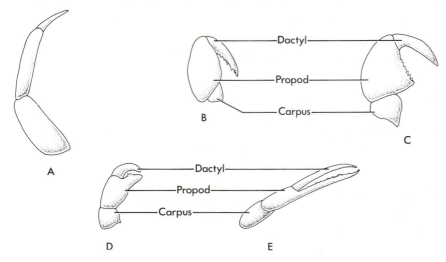

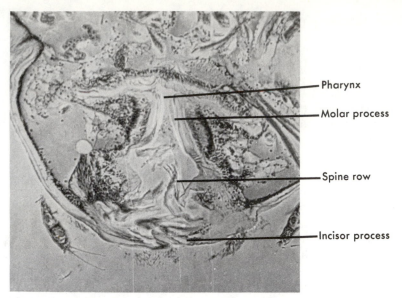

Pharynx

Molar process

Spine row

Incisor process

Figure 10-15 Section through mandibles of an aquatic isopod, showing opposing action of right and left cutting (incisor) and grinding (molar) processes. An inwardly directed spine row prevents escape of food materials being ground by the molar processes.

a protective cover for the remaining pair. The uropods seem to have a sensory function and are held poised like a pair of posterior antennae.

Internal Anatomy

The stomach contains a gastric mill and receives two pairs of long digestive glands. The intestine is straight and tubular.

The heart lies in a blood sinus in the posterior thoracic region. In the head are a prominent pair of maxillary glands which are thought to be excretory organs. As with other arthropods there is a supraesophageal ganglion with circumesophageal connectives to the subesophageal ganglion. Smaller ganglia are located along the ventral nerve cord.

Reproduction

Asellus belongs to a large group of crustaceans that carry their eggs in a temporary brood pouch. The pouch is formed by platelike expansions (*oostegites*) from the coxal segments of the first four pairs of legs (Fig. 10-12, B). The male carries the female with his fourth legs prior to fertilization. Fertilization occurs as soon as the posterior half of the female exoskeleton is shed. After fertilization the anterior half of the exoskeleton is shed and the brood pouch is fully developed. Eggs are rapidly deposited in it.

A peculiarity of the embryology of *Asellus* eggs is the growth of an

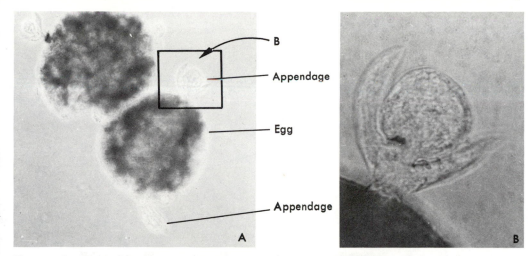

Figure 10-16 *Asellus* egg, showing the appendage, which is an early developmental feature that is lost after hatching of the embryo.

appendage (Fig. 10-16) that seems an aid for respiration by the embryo. The appendage disappears from its dorsal, thoracic attachment shortly after hatching. The young isopod resembles the adult but has one less pair of legs. *Asellus,* with the egg appendage, hatches in about two weeks. This contrasts with a six months incubation period in a southern hemisphere isopod (Fig. 10-17), which lacks the appendage. It seems that the appendage is one of several factors enabling *Asellus* to develop quite rapidly.

Ecology

Asellus is a detritus feeder. They reach greatest abundance in mildly polluted waters. Sometimes large aggregations of them occur. The latter phenomenon led W. C. Allee, one of the leading early ecologists in this country, to investigate many facets of their behavior. The isopod forms an important item in the diet of many game fish. Since it feeds on the abundant decaying organic matter, or detritus, found in its habitat and can be fed upon by useful food fishes it serves as an efficient link in the food chain.

OTHER CRUSTACEA

The Crustacea are arthropods most of which live in the water and breathe by means of gills. The body is divided into head, thorax, and abdomen, or the head and thorax may be fused, forming a cephalothorax. The head usually consists of five segments fused together; it bears two pairs of antennae (feelers), one pair of mandibles (jaws), and two pairs of maxillae. The thorax bears a variable number of appendages, some of which are usually locomotory. The abdominal seg-

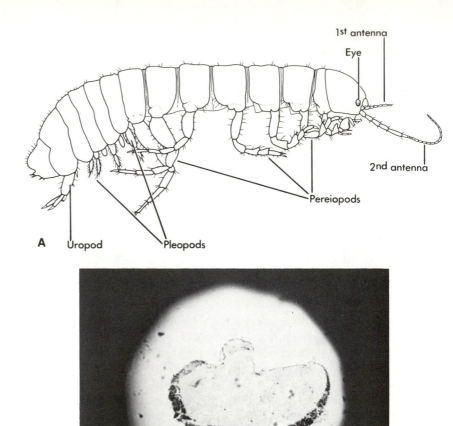

Labels on figure A: 1st antenna, Eye, 2nd antenna, Pereiopods, Uropod, Pleopods

Figure 10-17 A, *Colubotelson thomsoni*, a Tasmanian representative of the phreatoicids, a group of freshwater isopods that are nearly the ecological equivalent of *Asellus* in the southern hemisphere. B, *Mesamphisopus capensis*, a South African phreatoicid embryo sectioned to show the homologue of the dorsal appendage of *Asellus;* note the typical superficial cleavage of the yolky arthropod egg. (*A, from Engemann, 1963; B, from a slide prepared in 1913 by Dr. K. H. Barnard, courtesy of the South African Museum, Capetown.*)

ments are generally narrow and more mobile than those of the head and thorax; they bear appendages which are often reduced in size.

Hormones in crustaceans have been investigated in a number of the *Decapoda.* Secretions originating in nerve cells of the brain and X-organ move along neurons and accumulate in the sinus gland of the eyestalk. The secretions, or hormones, are involved in regulation of such things as color changes, molting and related physiological changes. A Y-organ in one of the head appendages is also involved in regulation of molting.

516 *Phylum Arthropoda*

Circadian rhythms have been observed most extensively in *Uca,* the fiddler crab. Circadian rhythms are cyclic changes that approximate a day in length. Because *Uca* has such cyclic changes in coloration that persist when it is placed in a constant environment it is said to have a "biological clock." Such changes presumably involve cyclic physiological or metabolic rhythms that develop in phase with cyclic events in nature and are somewhat regulated by the external environment. Studies of such rhythms have become increasingly important with the increased speed of transportation. Perhaps studies of such invertebrates will help develop safe methods of resetting "biological clocks" of statesmen flown to different time zones so they can negotiate at peak efficiency during resting hours of their homeland.

A special type of such cyclic rhythms especially common in planktonic and pelagic crustaceans are the *vertical migrations* normally regulated by light. Forms exhibiting such activity include cladocerans, copepods, euphausids, and decapods. The movement is typically up into or toward surface waters in the evening hours and back down into deeper waters at other times except for some upward movement around dawn. In shallow freshwater lakes it may involve a vertical movement of only a few meters whereas some larger crustaceans in the ocean undertake daily vertical migrations of nearly 1,000 meters.

Ecology

Crustaceans are very important members of aquatic ecosystems. The food value of shrimps, crabs, and lobsters is well known. Of even greater importance are such planktonic genera (Fig. 10-18) as *Calanus, Daphnia,* and *Cyclops.* They harvest smaller plankton organisms, including the photosynthetic forms, and serve as food organisms for many larger animals. The marine copepod, *Calanus finmarchicus,* is perhaps the most abundant animal in the world. Other crustaceans are important as scavengers and predators.

Vermeij (1977) has found that crab claws of species in the tropical Pacific and Indian oceans are relatively heavier than those of the Atlantic and eastern Pacific. Their intertidal gastropod prey have a similar pattern of heaviness for their shells.

Waxes are important food stores and may be flotation aids as well for marine copepods, although the copepods sink when not swimming.

Euphausia superba is an extremely important source of food for whales in southern oceans. Its life cycle is timed to fit the ocean currents it occupies. Adults in surface currents drift north in warm months while they grow rapidly and spawn. Eggs sink and are slowly carried south in deeper currents as they hatch and develop. The young euphausids continue an upward movement until they reach surface waters at about 2 years of age near the place their parents had surfaced as young adults.

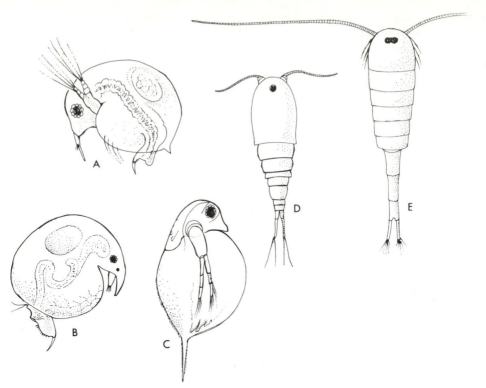

Figure 10-18 Some common genera of planktonic Cladocera and Copepoda. (A) Cladocera, *Bosmina*. (B) Cladocera, *Chydorus*. (C) Cladocera, *Daphnia*. (D) Copepoda, *Cyclops*. (E) Copepoda, *Limnocalanus*. (*Adapted from Edmondson, Pennak, and others.*)

Some marine crustacea are reported to have digestive tracts normally devoid of biota due to antibiotics in their algal food.

A marine isopod has had considerable effect in halting the seaward incursion of mangroves in Florida (Rehm and Humm, 1973). They destroy mangrove prop roots up to the mean high tide level.

The ecology of Cladocera has been reviewed by Frey et al. (1971).

Isopods are the only crustaceans to achieve much success on land. They are frequently the dominant invertebrate detritus feeder in the litter layer of forests and grasslands. The terrestrial adaptations in Crustacea are treated in a symposium arranged by Bliss and Mantel (1968).

Classification

The diversity of the crustaceans necessitates a somewhat complex classification. The more important orders are briefly described in the following pages. The term Entomostraca is no longer used in classification, although it is still useful to designate an odd mixture of subclasses including the small crustaceans once thought to be insectlike.

Subclass I. Branchiopoda

Branchiopoda are free-swimming crustaceans with flattened, leaflike thoracic appendages that are respiratory in function. There is usually a carapace.

Order 1. Cephalocarida

An order of few species first reported in 1955. The Cephalocarida show considerable departure from the other Branchiopoda because of interstitial adaptations. They are often placed in a separate subclass. Hermaphroditic, marine mud-dwellers; carapace covers the head region; two ovisacs, each with one egg, carried on the genital segment; adbomen of 10 segments. Sanders (1957) proposes these as the closest living relatives of the primitive crustacean stock. Ex. *Hutchinsoniella macracantha* (Fig. 10-19) from Long Island Sound, *Lightiella serendipita* from San Francisco Bay.

Order 2. Anostracida

Body elongate; eyes stalked; thoracic segments many, no carapace. Ex. *Eubranchipus vernalis* (Fig 10–20, D): fairy shrimp; pinkish; freshwater

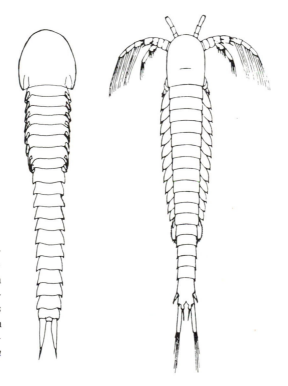

Figure 10-19 Cephalocarida. *Hutchinsoniella macracantha* with the nearest known relatives of the Cephalocarida, *Lepidocaris rhyniensis* (with antennae showing), a Devonian fossil. (*From Sanders, 1957;* Lepidocaris, *from Sanders, after Scourfield.*)

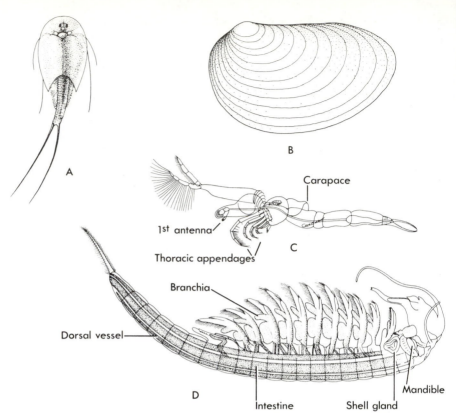

Figure 10-20 Branchiopoda. (A) Notostracida. (B) Conchostracida. *Estheria.* (C) Cladocera. *Leptodora.* (D) Anostracida. *Branchipus. (B, after Sars; C, after Smith; D, after Sedgwick.)*

pools, especially in spring; eastern North America. *Artemia salina:* brine shrimp; western United States.

Order 3. Notostracida

Body elongate; eyes sessile (Fig. 10-20, A); thoracic segments many; carapace broad, shield-shaped. Ex. *Apus lucasanus:* about 4 cm long; in fresh water; western United States.

Order 4. Conchostracida

Body enclosed in carapace in the form of a bivalve shell; eyes sessile; antennae large, biramous. Ex. *Estheria morsei* (Fig. 10-20, B); shell about 12 mm long; central and western United States.

Order 5. Cladocera

Bivalve carapace usually covers the body but not the head; eyes sessile; antennae large, biramous; four to six thoracic segments. Ex. *Daphnia pulex* (Fig. 10-21): water flea; reddish; sharp caudal spine;

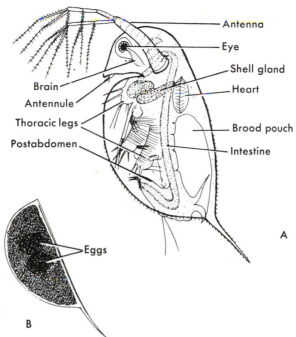

Figure 10-21 Branchiopoda. *Daphnia.* (A) Female. (B) Ephippium containing fertilized eggs. (*A, after Claus.*)

ventral beak large; America. *Simocephalus vetulus:* no caudal spine; America. *Leptodora hyalina* (Fig. 10-20, C): shell small; abdomen terminated by two claws; antennae large in male.

Subclass II. Ostracoda

Free-swimming crustaceans with a bivalve carapace; compound eyes present or absent; two or less thoracic, nonphyllopod appendages; mandibular palp, usually biramous.

The two major orders listed contain individuals with two pairs of legs on the trunk. Two additional orders have few species; the Cladocopida, with no legs on the trunk; and the Platycopida, with one pair of legs on the trunk.

Order 1. Myodocopida

Marine; second antennae biramous, one branch large, the other small; deep notch in anterior margin of shell. Ex. *Sarsiella zostericola:* three eyes present; notch in shell (antennal sinus) absent in female; eastern United States.

Order 2. Podocopida

Mostly in fresh water; second antennae uniramous (Fig. 10-22); no notch in anterior margin of shell. Ex. *Eucypris virens:* in fresh water; shell covered with short hairs; one median eye; second antennae with

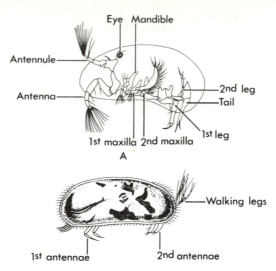

Figure 10-22 Ostracoda. (A) *Cypris.* (B) *Candona.* (*A, after Zender; B, after Baird.*)

five natatory bristles; parthenogenetic; no males known; cosmopolitan. *Loxoconcha impressa:* marine; Vineyard Sound.

Subclass III. Mystacocarida

Microscopic; marine; in the interstitial water of sandy beaches; four pair of thoracic legs. The single order Mystacocarida contains only one genus, *Derocheilocaris.* Considered close relatives of the Copepoda.

Subclass IV. Copepoda

Free-swimming or parasitic; compound eyes and carapace absent; typically with six pairs of thoracic appendages; marine and fresh water.

Several orders contain parasites of fish and marine invertebrates. A Cretaceous fossil copepod parasite of a euryhaline fish is the oldest known copepod (Cressy and Patterson, 1973). The copepod was a caligid and shows mouth cones indicating the parasites of the order Caligoida are derived from the Cyclopoida. The Caligoida, Lernaeoida, Monstrilloida (with parasitic larvae in snails or polychaetes), and Notodelphoida (in tunicates) are the orders of parasitic copepods. Free-living species are in the first three orders below.

Order 1. Calanoida

Body distinctly divided into an anterior metasome and a narrow posterior urosome; first antennae approximately as long as body; females with a single egg sac. Ex. *Calanus finmarchicus:* marine; yellow or red;

pelagic in North Atlantic; important as fish food. *Diaptomus sanguineus:* in fresh water; bright red; about 2 mm long; in pools in spring; North America.

Order 2. Cyclopoida

Body tapers to a narrow urosome; first antennae no longer than metasome; two egg sacs carried laterally by the female. Ex. *Cyclops viridis* (Fig. 10-23, A): usually greenish; small ponds; United States and Europe. *Ergasilus versicolor:* parasitic on gills of bullhead and catfish.

Order 3. Harpacticoida

Urosome not distinctly narrower than metasome; first antennae no longer than head segment; female with one medial egg sac; typically

Figure 10-23 Copepoda. (A) *Cyclops,* dorsal view of female, showing cyclopoid body type. (B) Calanoid body type, diagrammatic. (C) Harpacticoid body type, diagrammatic. (D) *Lernaea,* parasite of fish (haddock). (*A, after Hartog; D, after Scott.*)

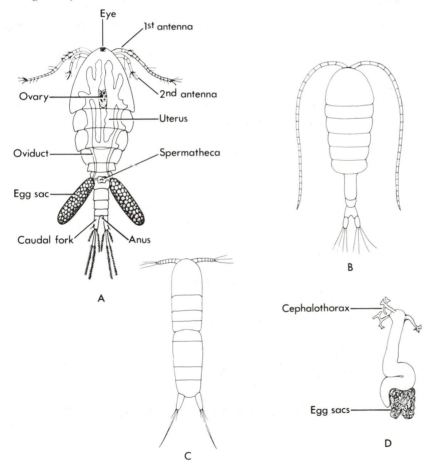

from surface of bottom sediments in shallow water. Ex. *Canthocamptus assimilis;* ponds and ditches; North America.

Order 4. Lernaeoida

Parasitic; segments usually not evident. Ex *Lernaea branchialis* (Fig. 10-23, D): body of fertilized female, wormlike, twisted; two convoluted egg sacs; on gills of codfish, etc.

Subclass V. Branchiura

Parasitic on marine and freshwater fishes; no egg sacs; body flattened; five pair of thoracic appendages. Sometimes considered an order of the Copepoda. Order of the same name as the subclass. Ex. *Argulus laticauda* (Fig. 10-24): fish louse; suckers flank a hollow spine used to penetrate host tissues; about 6 mm long; on eel, flounder, etc.

Subclass VI. Cirripedia

Sessile as adults; no compound eyes in adult; carapace enclosing body; mostly hermaphroditic.

Order 1. Thoracida

Barnacles; marine; body surrounded by calcareous shell; six pairs of biramous thoracic legs. Ex. *Lepas anatifera* (Fig. 10-25, A and B): goose barnacle; body attached by short stalk; shell of five thin plates; cosmopolitan. *Balanus balanoides* (Fig. 10-25, C and D): rock barnacle; shell thick, closed by two hinged plates; stalk absent; both coasts of United States.

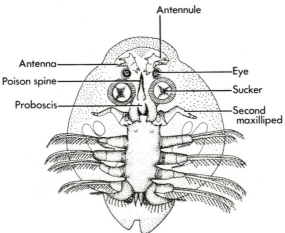

Figure 10-24 Brachiura. *Argulus*, ventral view of female. (*After Wilson.*)

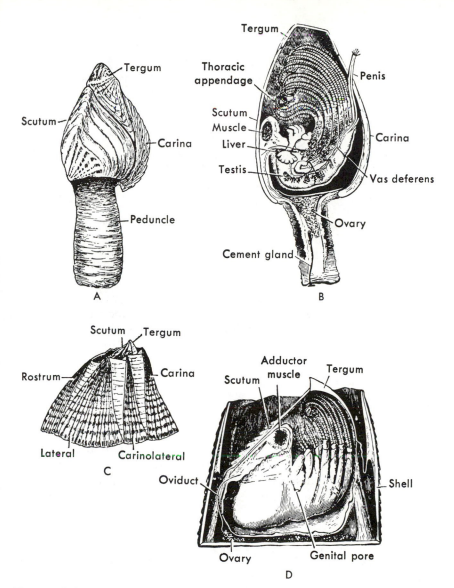

Figure 10-25 Cirripedia. Thoracida. (A) *Lepas,* external view. (B) *Lepas,* internal view. (C) *Balanus,* external view. (D) *Balanus,* internal view. (*A and B, after Darwin; C and D, after Claus.*)

Order 2. Acrothoracida

Parasitic; no calcareous shell; body covered by large mantle. Ex. *Alcippe lampas:* males, small, legless, and attached to females; bores into *Natica* shells containing hermit crabs: Woods Hole, Mass.

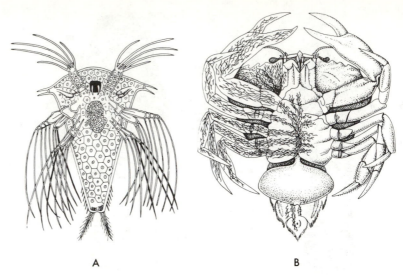

A   B

Figure 10-26 Cirripedia. Rhizocephalida. *Sacculina.* (A) Nauplius larva. (B) Adult attached like a sac to the undersurface of the abdomen of a crab and sending "roots" throughout the crab's body for extracting nourishment. (*After Leuckart.*)

Order 3. Ascothoracida

Parasites on or in echinoderms and in corals.

Order 4. Rhizocephalida

Parasitic; no appendages, alimentary canal nor segmentation in adult; attached by stalk with roots penetrating tissue of host. Ex. *Sacculina carcini* (Fig. 10-26): parasitic on decapod crustaceans degenerating into a sac lying on the central surface between the thorax and abdomen.

Subclass VII. Malacostraca

Lobsters, crayfish, crabs, etc.: mostly large crustaceans; usually five segments in the head, eight in the thorax, and six in the abdomen; gastric mill in the stomach. Twelve orders are usually placed in five divisions or superorders. Two orders, the Spelaeogriphida and Thermosbaenida, are known only from one or a few rare species from caverns and hot springs, mostly in Africa; they are omitted from the following account. The division Phyllocaridea contains the Nebalida, forms with a large carapace. The division Syncaridea contains the Anaspidida and a few dozen species from subterranean waters, the Bathynellida. The division Peracaridea is very large and contains the Mysida, Cumacida, Tanaiida, Isopodida, and Amphipodida. The Peracaridea have little or no carapace and are the only crustaceans with oostegites forming a brood pouch. The division Eucaridea is also very large; it contains only the Euphausida and Decapodida, which have the carapace covering the thorax. The division Hoplocaridea contains only the Stomatopodida.

Order 1. Nebalida

Primitive marine Malacostraca (Fig. 10-27, A) with seven abdominal segments; stalked eyes; thorax with eight pairs of leaflike gills. Ex. *Ne-*

Figure 10-27 Malacostraca. (A) Nebalida; *Nebalia geoffroyi.* (B) Anaspidida; *Anaspides tasmaniae.* (C) Mysida; *Mysis oculata.* (*A, after Claus; B, after Woodward; C, after Sars.*)

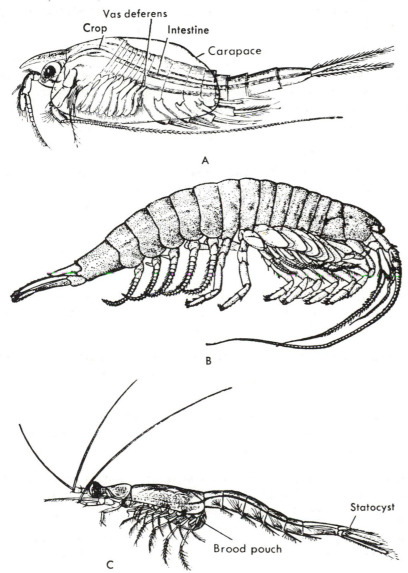

balia bipes: body slender, compressed; eggs carried by setae on the thoracic appendages of female.

Order 2. Anaspidida

Fresh water; no carapace; a tail fan. Ex. *Anaspides tasmaniae* (Fig. 10-27, B): from ponds in Tasmania.

Order 3. Mysida

Carapace covers almost entire thorax (Fig. 10-27, C); eyes stalked; thoracic appendages all biramous. Ex. *Mysis stenolepis:* body elongate, cylindrical; abdomen bends between first and second segments; Atlantic Coast of North America.

Order 4. Cumacida

Carapace does not cover entire thorax (Fig. 10-28, A); eyes sessile; no tail fan. Ex. *Diastylis quadrispinosus:* a spine on each side of carapace; New Jersey northward. Cumaceans inhabit the bottom sand and mud; all are marine.

Order 5. Tanaiida

Small, marine, bottom-dwelling crustaceans. Usually in burrows or tubes. Isopodlike but carapace covers first two thoracic segments; strongly developed chelae on first leg. Ex. *Apseudes* (Fig. 10-28, C).

Order 6. Isopodida

Marine, fresh water, terrestrial, or parasitic; eyes sessile; body usually flattened dorsoventrally.

Suborder 1. Flabellifera. Uropods flattened and forming part of a tail fan; all marine. Ex. *Limnoria lignorum:* bores in submerged wooden structures causing considerable damage.

Suborder 2. Valvifera. Marine, uropods form a protective cover for the pleopods. Ex. *Idothea rectilinea* (Fig. 10-28, D): about 2 cm; California.

Suborder 3. Asellota. Uropods styliform, abdominal segments usually fused. Ex. *Asellus communis* (Fig. 10-12): common among vegetation in fresh water of the United States.

Suborder 4. Phreatoicidea. Uropods styliform; abdominal segments separate; fresh water in the southern hemisphere. Ex. *Colubotelson thomsoni* (Fig. 10-17): montane pools; Tasmania.

Suborder 5. Epicaridea. Uropods styliform; parasitic on other crustaceans; mostly marine. Ex. *Probopyrus pandalicola:* parasitic in the gill

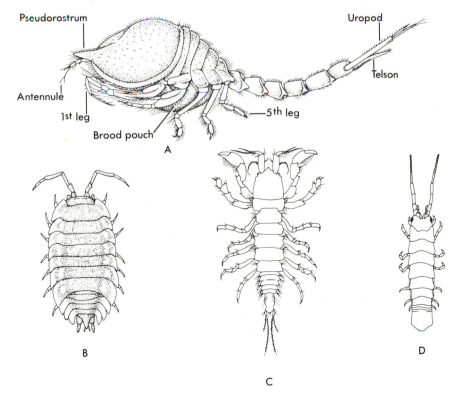

Figure 10-28 Malacostraca. (A) Cumacida; *Diastylis goodsiri*. (B) Isopodida; *Oniscus asellus*. (C) Tanaiida; *Apseudes australis*. (D) Isopodida; *Idothea rectilinea*. (*A, after Sars; B, after Smith; C, after Hale; D, after Johnson and Snook.*)

chamber of *Palaemonetes vulgaris,* a shrimp that sometimes moves up into streams of the United States.

Suborder 6. Oniscoidea. Pleopods adapted for air breathing; most species terrestrial. Ex. *Oniscus asellus* (Fig. 10-28, B): sow bug; under stones, logs, bark etc.; introduced into the northern United States from Europe. *Ligia exotica* (Fig. 10-29): one of many similar species common on marine beaches of the world.

Order 7. Amphipodida

Mostly marine; usually laterally compressed; eyes sessile; no carapace; no tail fan. Ex. *Talorchestia longicornis:* beach flea; body large; two large eyes; on sand beaches; Cape Cod to New Jersey. *Gammarus fasciatus* (Fig. 10-30, A): in freshwater ponds and streams; eastern United States. *Caprella geometrica* (Fig. 10-30, B): body elongate, slender; on seaweeds and hydroids; Cape Cod to Virginia.

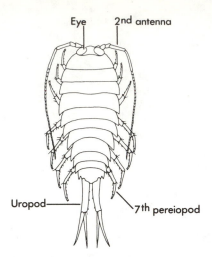

Figure 10-29 Isopodida. *Ligia exotica*. Dorsal view of female. (*After Van Name*.)

Order 8. Euphausida

Filter-feeding; pelagic; marine crustaceans. None of the thoracic appendages are developed as maxillipeds. Ex. *Euphausia pacifica:* 25 mm; common off southern California.

Order 9. Decapodida

Lobsters, crayfish, shrimps, crabs. This order contains over 8,000 species of comparatively large and familiar crustaceans. The carapace covers the entire thorax; three pairs of maxillipeds; five pairs of thoracic walking legs, the first pair usually larger and with pinching claws (chelae); mostly marine.

Suborder 1. Natantia. Shrimps and prawns. Pleopods adapted for swimming; rostrum long; body typically long and round or laterally compressed. Ex. *Crago septemspinosus:* edible shrimp; eastern coast. *C. nigricauda:* blacktail shrimp; Pacific Coast. *Palaemonetes vulgaris* (Fig.

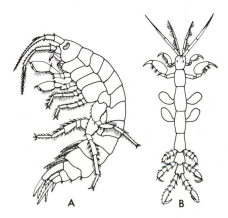

Figure 10-30 Malacostraca. Amphipodida. (A) *Gammarus fasciatus*. (B) *Caprella geometrica*. (*From Paulmier*.)

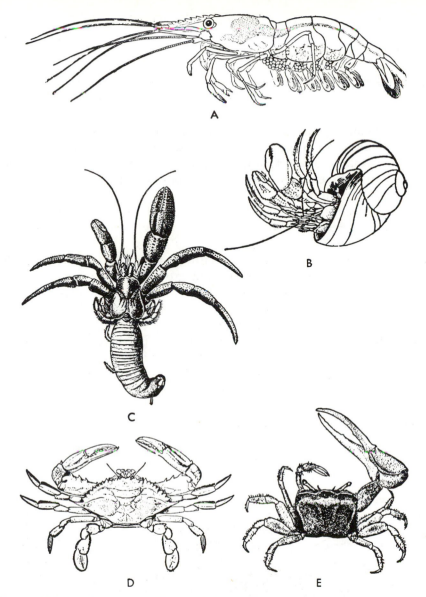

Figure 10-31 Malacostraca. Decapodida. (A) *Palaemonetes vulgaris*, the prawn. (B) *Pagurus*, the hermit crab, in snail shell. (C) *Pagurus*, removed from snail shell. (D) *Callinectes sapidus*, the edible or blue crab. (E) *Uca crenulata*, the fiddler crab. (*A, from Davenport; B, after Emerton; C, after Cuvier; D, after Rathbun; E, after Johnson and Snook.*)

10-31, A): prawn; eastern coast. *Alpheus*, the pistol prawn or snapping shrimp; smooth matching disks on the dactyl and propus do not separate until the closer muscles have built up sufficient tension to close the claw with an audible snap in *A. californiensis* (Ritzmann, 1973).

Shrimps predominate in the over $500,000,000 retail value of crustaceans marketed each year in the United States; the major portion of the supply is imported.

Suborder 2. Reptantia. Lobsters, crayfish, and crabs. Usually dorsoventrally flattened; first pair of legs usually chelipeds; rostrum not markedly keeled if present. Divided into four tribes.

Tribe Palinura contains the spiny lobster and the other small-clawed lobster relatives with well developed and extended abdomens. *Panulirus argus* forms long single-file queues when migrating or grouped without protective cover; tactile sensations are the cues needed for queuing (Herrnkind, 1969).

Tribe Astacura contains the large clawed species with a well-developed, extended abdomen. Ex. *Homarus americanus:* lobster; marine. *Orconectes virilis:* crayfish; central states.

Tribe Anomura contains species with the abdomen bearing uropods and flexed under the thorax or soft and extended asymmetrically. Ex. *Emerita talpoida:* sand crab; both coasts. *Pagurus longicarpus* (Fig. 10-31, B, C): small hermit crab; East Coast. *P. hirsutiusculus:* blue and white banding on legs; West Coast.

Tribe Brachyura contains the true crabs. Uropods are usually lacking on the abdomen which is reduced and tightly flexed under the thorax. Ex. *Libinia emarginata:* spider crab; Atlantic Coast. *Cancer irroratus:* rock crab; edible; Northeast Coast. *C. magister;* edible crab; Pacific

Figure 10-32 *Callinectes sapidus,* carapace removed to show arrangement of internal organs.

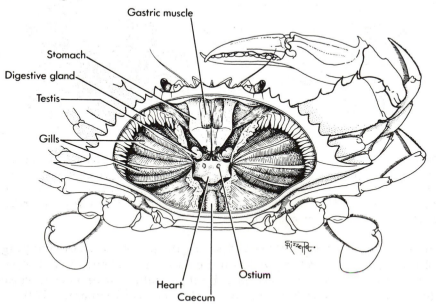

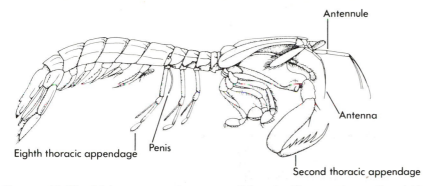

Antennule

Antenna

Second thoracic appendage

Eighth thoracic appendage

Penis

Figure 10-33 Malacostraca. Stomatopodida. *Squilla mantis,* male. (*After Calman.*)

Coast. *Callinectes sapidus* (Fig. 10-31, D; 10-32): blue or edible crab; Southeast Coast. *Uca pugnax:* fiddler crab; East Coast. *U. crenulata:* fiddler crab (Fig. 10-31, E); California.

Order 10. Stomatopodida

Marine; body elongate (Fig. 10-33) with broad abdomen; carapace covering three of seven thoracic segments; stalked eyes; gills on first five pairs of abdominal appendages; second pair of maxillipeds very large and subchelate. Ex. *Squilla empusa:* body up to 25 cm long; in burrows in mud near shore; edible; Eastern Coast. *Pseudosquilla biglowi:* similar to *Squilla;* south California coast.

Class II. Insecta

The insects are treated separately in the next chapter. This is done because of the large size of the class, to emphasize their importance, and as a convenience for varied course coverage modified for various curricular systems. Their general features are (a) three body regions— *head, thorax,* and *abdomen;* (b) one pair of antennae; (c) mouth parts— mandibles, maxillae, and labium; (d) usually two pair of wings; and (e) three pair of thoracic legs.

Class III. Chilopoda

The Chilopoda are called centipedes (Fig. 10-35). The body is flattened dorsoventrally, and consists of from 15 to 173 segments, each of which bears one pair of legs except the last two and the one just back of the head. The latter bears a pair of poison claws called maxillipeds, with which insects, worms, mollusks, and other small animals are killed for food. The antennae are long consisting of at least 12 segments.

The internal anatomy of a common centipede is shown in Figure 10-

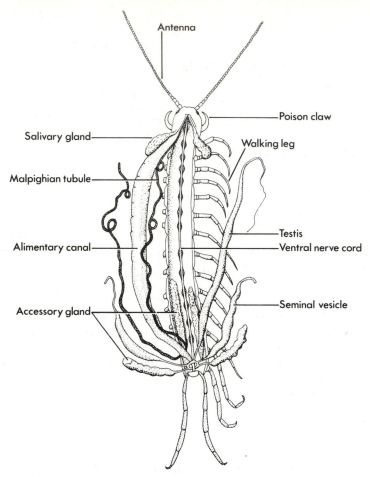

Antenna

Poison claw

Salivary gland

Walking leg

Malpighian tubule

Alimentary canal

Testis

Ventral nerve cord

Accessory gland

Seminal vesicle

Figure 10-34 Chilopoda. *Lithobius forficatus*, dissected to show the internal organs. (*After Vogt and Jung, after Sedgwick*.)

34. The alimentary canal is simple; into it open the excretory organs—two Malpighian tubules. The tracheae are branched as in insects and open by a pair of stigmata in almost every segment. The reproductive organs are connected with several accessory glands. Eggs are usually laid. Those of *Lithobius* are laid singly and covered with earth.

The centipedes are swift-moving creatures. Many of them live under the bark of logs, or under stones. The genera *Lithobius, Geophilus,* and *Scutigera* are common. The poisonous centipedes of tropical countries belong to the genus *Scolopendra*. They may reach 25 cm in length, and their bite is painful and even dangerous to humans. Fossil Chilopoda have been found in amber of oligocene age. Four of the families in this class are as follows:

FAMILY 1. GEOPHILIDAE (Fig. 10-35, A). Body long, with over 31 segments; eyes absent; antennae with 14 segments; young leave egg with full number of segments and legs. Ex. *Geophilus rubens:* about 45 mm long; 47 to 53 pairs of legs; eastern and central states.

FAMILY 2. SCOLOPENDRIDAE. Body long, with 21 to 23 segments; eyes present or absent; antennae with 17 to 31 segments; young leave egg with full number of segments and legs. Ex. *Scolopendra morsitans* (Fig. 10-35, B): 21 pairs of legs; eyes present; cosmopolitan. *Scolopocrytops sexspinosa:* 23 pairs of legs; eyes absent; United States.

FAMILY 3. LITHOBIIDAE. Body with legs on 15 segments; maxillary palp with three segments; young leave egg with seven pairs of legs. Ex. *Lithobius forficatus* (Fig. 10-34): body about 3 mm long; antennae with 33 to 43 segments; America, Europe.

FAMILY 4. SCUTIGERIDAE. Body short, with about 15 segments; 15 pairs very long legs, last pair longest; antennae very long. Ex. *Scutigera forceps* (Fig. 10-35, C): about 25 mm long; last pair of legs about 50 mm long; United States

Figure 10-35 Chilopoda. (A) Geophilidae; *Geophilus longicornis.* (B) Scolopendridae; *Scolopendra morsitans.* (C) Scutigeridae; *Scutigera forceps.* (*B, from Claus; C, from Herrick.*)

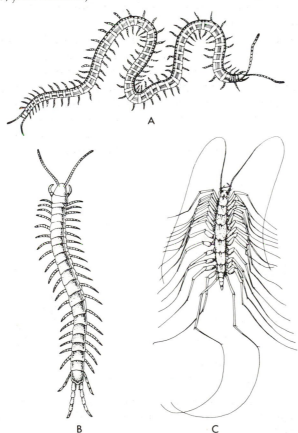

Class IV. Diplopoda

The Diplopoda are called millipedes (Fig. 10-37). The body is subcylindrical, and consists of from about 25 to more than 100 segments, according to the species. Almost every segment bears two pairs of appendages (Fig. 10-36), and has probably arisen by the fusion of two segments. One or both pairs of legs on the seventh segment of the male are modified as copulatory organs. The mouth parts are a pair of mandibles and a pair of maxillae. One pair of short antennae and either simple or aggregated eyes are usually present. There are olfactory hairs on the antennae and a pair of scent glands in each segment, opening laterally. The scent glands, or *repugnatorial glands,* of some have a toxic secretion containing hydrocyanic acid. Some can discharge the secretion nearly a meter. The secretions of some tropical species have caused blindness in children. The breathing tubes (tracheae) are usually unbranched; they arise in tufts from pouches which open just in front of the legs. The heart is a dorsal vessel with lateral ostia; it gives rise to arteries in the head. The two or four excretory organs are threadlike tubes (Malpighian tubules), which pour their excretions into the intestine.

The millipedes move very slowly in spite of their numerous legs. Some of them are able to roll themselves into a spiral or ball. They live in dark, moist places and feed principally on decaying vegetable matter but sometimes on living plants and may thus be destructive. The sexes are separate, and the eggs are laid in damp earth. The young have few segments and only three pairs of legs when they hatch, and resemble apterous insects. Other segments are added just in front of the anal segment. Fossil millipedes have been discovered in geological

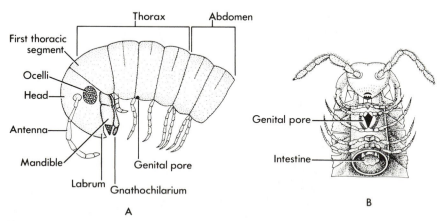

Figure 10-36 Diplopoda. (A) *Julus terrestris,* side view of anterior end. (B) *Polydesmus complanatus,* ventral view of anterior end. (*A, after Borradaile and Potts; B, after Latzel.*)

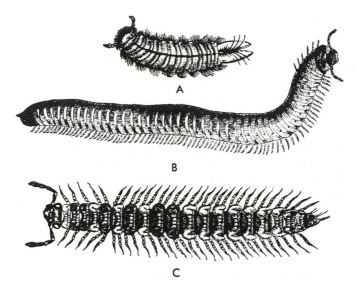

Figure 10-37 Diplopoda. (A) Polyxenidae; *Polyxenus lagurus.* (B) Julidae; *Julus nemorensis.* (C) Polydesmidae; *Polydesmus collaris.* (*From Koch.*)

formations as far back as the Devonian and Upper Silurian. Three of the families in this class are as follows:

FAMILY 1. POLYXENIDAE (Fig. 10-37, A). Body small; integument soft; each segment with tuft of bristles on either side; maxillae leglike; no copulatory feet in male. Ex. *Polyxenus fasciculatus:* about 2.5 mm long; 13 pairs of legs; Long Island and southern states.

FAMILY 2. JULIDAE (Fig. 10-37, B). Integument hard; maxillae form a plate; two pairs of copulatory feet on seventh segment of male. Ex. *Julus virgatus:* body with 30 to 35 segments; 50 to 60 pairs legs; no legs on third segment; body about 12 mm long.

FAMILY 3. POLYDESMIDAE (Fig. 10-37, C). Body with 19 to 22 segments; first pair of legs on seventh segment of male, copulatory. Ex. *Polydesmus serratus:* body about 37 mm long; male with 30, female with 31 pairs of legs; eastern and central states.

Class V. Pauropoda

The Pauropoda, Symphyla, Diplopoda, and Chilopoda are sometimes grouped together in one class, the Myriapoda. However, one large group, the Chilopoda, resembles the insects more than the other large group, the Diplopoda; and the two smaller groups the Pauropoda and Symphyla have certain characteristics in common with the Diplopoda. It seems advisable, therefore, to consider these groups as separate classes. The first three (Pauropoda, Symphyla, and Diplopoda) are sometimes combined into a superclass, the Progoneata, characterized by reproductive organs that open to the outside near the anterior end of the body, and the Chilopoda and Insecta into a superclass, the Opisthogoneata, with reproductive organs that open near the posterior end.

The Pauropoda are less than 2 mm in length. They prey on microscopic animals or eat decaying animal and vegetable matter. They are without eyes, heart, and special respiratory organs, and evidently breathe through the general surface of the body, as in the earthworm. The head is distinct, and the body contains 12 segments and bears nine pairs of legs. The young possess three pairs of legs when hatched. The two families in this class are the Pauropidae and Eurypauropidae. Ex. *Pauropus huxleyi* (Fig. 10-38, C): body elongate and cylindrical; 1.3 mm long; eastern and central United States, Europe. *Eurypauropus spinosus:* body broad and flat; 1.25 mm long; eastern and central United States, Europe.

Class VI. Symphyla

The Symphyla are small arthropods with 12 pairs of legs. The head bears antennae, mandibles, maxillulae, maxillae, and a labium. Fewer than 100 species belong to the class. They resemble certain wingless insects in habits and appearance, but have a greater number of legs. They live in moist places and avoid light. Their food probably consists of decaying plants. Ex. *Scutigerella immaculata* (Fig. 10-38, D): head distinct; first pair of legs well developed; about 6 mm long; eastern United States, Europe. *Scolopendrella texana:* head not distinct; first pair of legs rudimentary; 2.8 mm long; Austin, Texas.

Class VII. Onychophora

The Onychophora are frequently treated as a separate phylum. They share enough characteristics with the Arthropoda to be included here as the only class in a subphylum of the same name, Onychophora. The 10 genera and 70 odd species belonging to this class are all usually referred to as *Peripatus* (Fig. 10-39). They are especially interesting because they evidently separated from the main arthropodan stock at an early period and obviously exhibit both arthropod and annelid characteristics as well as peculiarities of their own. Unfortunately, they are very rare and hence not available for laboratory study.

Genera thus far described and their geographical distribution include the following: (1) *Peripatus,* tropical America; (2) *Oroperipatus,* Pacific watershed of tropical America; (3) *Metaperipatus,* Chile; (4) *Paraperipatus,* New Britain, New Guinea, and Ceram; (5) *Mesoperipatus,* west-central Africa; (6) *Peripatopsis,* (7) *Opisthopatus,* southern

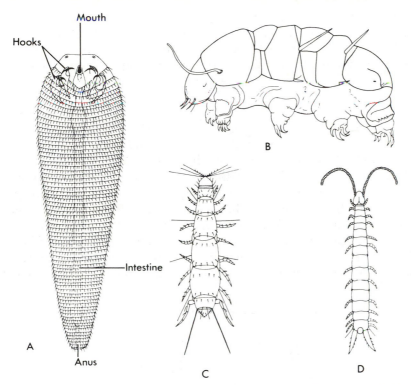

Figure 10-38 (A) Pentastomoidea; *Linguatula serrata*. (B) Tardigrada; *Echiniscus spinulosus*. (C) Pauropoda; *Pauropus huxleyi*. (D) Symphyla; *Scutigerella immaculata*. (*A, from Sedgwick; B, after Doyere; C and D, after Latzel.*)

Africa; (8) *Peripatoides*, Australia, Tasmania, and New Zealand; (9) *Eoperipatus*. Sumatra and Malay Peninsula; (10) *Typhloperipatus*, Tibet.

The group furnishes an excellent example of discontinuous distribution. Even in the area where a species occurs, specimens are present only in a few of the many available habitats. This seems to indicate that the class had once a continuous distribution but has disappeared throughout most of its range and is on the road to extinction.

Peripatus lives in crevices of rock, under bark and stones, and in other dark moist places and is active only at night. As the animal moves slowly from place to place by means of its legs, the two extremely sensitive antennae test the ground over which it is to travel, while the eyes, one at the base of each antenna, enable it to avoid the light. When irritated, *Peripatus* often ejects slime, sometimes to the

Figure 10-39 Onychophora. *Peripatus capensis*. (*After Sedgwick.*)

distance of almost 30 cm, from a pair of glands which open on the oral papillae. This slime sticks to everything but the body of the animal itself; it is used principally to capture flies, woodlice, termites, and other small animals, and in addition is probably a weapon of defense. A pair of modified appendages serve as jaws and tear the food to pieces.

Most species of *Peripatus* are viviparous, and a single large female may produce 30 to 40 young in a year. These young resemble the adults when born, differing chiefly in size and color.

Figure 10-40 shows the principal internal organs of a male specimen. The head (Fig. 10-41, B) bears three pairs of appendages, the *antennae, oral papillae,* and *jaws,* a pair of simple *eyes,* and a ventrally placed *mouth.* The fleshly *legs* (Fig. 10-41, C) number from seventeen pairs to over forty pairs in different species; each bears two *claws.* The *anus* is at the posterior end (Fig. 10-41, A); the *genital pore* is situated between the last pair of legs; and a *nephridiopore* lies at the base of each leg. The *skin* is covered with *papillae,* each bearing a *spine;* these papillae are

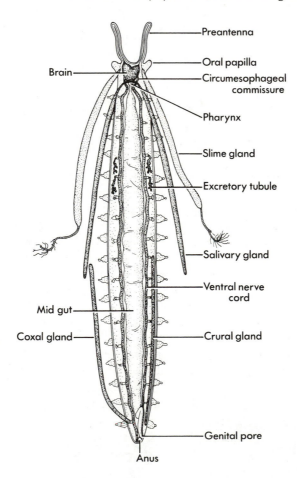

Figure 10-40 Onychophora. *Peripatus capensis.* Dissection of male. (*After Balfour, modified.*)

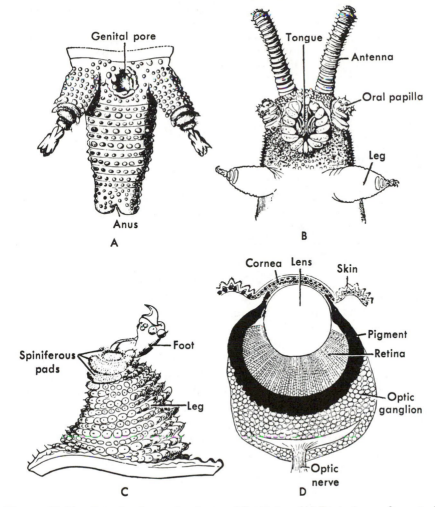

Figure 10-41 Onychophora. Anatomy of *Peripatus*. (A) Posterior end, ventral view. (B) Anterior end, ventral view. (C) Leg, front view. (D) Eye, longitudinal section. (*From Sedgwick; D, after Balfour.*)

especially numerous on the antennae, lips, and oral papillae, and are probably *tactile*. The external rings of the body are more numerous than the internal segments.

The *digestive system* (Fig. 10-40) is very simple, consisting of a muscular *pharynx,* a short *esophagus,* a long, saccular *stomach,* and a short *intestine.* The pair of *salivary glands,* which open into the mouth cavity, are modified nephridia. The *heart* is the only blood vessel; it is a dorsal tube with paired ostia connecting it with the pericardial cavity in which it lies. The *body cavity* is a blood space (i.e., a hemocoel). The breathing organs are air tubes, called *tracheae,* which open by means of pores on various parts of the body. The excretory organs are

Arthropod Characteristics	Annelid Characteristics	Structures Peculiar to Peripatus
appendages modified as jaws	paired segmentally arranged nephridia	number and diffusion of tracheal apertures
a hemocoelic body cavity	cilia in reproductive organs	single pair of jaws
no coelom around alimentary canal	chief systems of organs arranged as in annelids	distribution of reproductive organs
tracheae present	cholinergic neurons	texture of skin
character of the cuticle		simplicity and similarity of segments behind the head

nephridia, one at the base of each leg. The vesicular end of the nephridium is part of the coelom. The nervous system consists of a *brain,* dorsally situated in the head, and a pair of *ventral nerve cords,* which are connected by many transverse nerves. The sexes are separate, and the cavities of the reproductive organs are coelomic.

Peripatus is of special interest because its body exhibits certain structures characteristic of annelids and other structures found only in arthropods. It is, however, undoubtedly an arthropod. The accompanying table presents briefly these characteristics and shows in what respects it differs from other arthropods. Hackman and Goldberg (1975) have extended the similarity noted for *Peripatus* cuticle and arthropod exoskeleton. The cuticle is about 8 percent chitin and gives no evidence of collagen.

Class VIII. Merostomata

This class contains arthropods of the subphylum Chelicerata that have a cephalothorax (prosoma) consisting of six segments, six pair of laminate, nonlocomotor body appendages, and a terminal segment without a caudal fin but with a postanal spine or plate.

Five living species exist. They are the king, or horseshoe, crabs. All belong in one order of the subclass Xiphosura. The living king crab, *Xiphosura (Limulus) polyphemus* (Fig. 10-42) is a marine animal occurring along the Atlantic Coast from Maine to Yucatan. It possesses gills but no Malpighian tubules. *Xiphosura* is a burrowing animal and lives in the sand. It may be active at night, moving by "short swimming hops, the respiratory appendages giving the necessary impetus, whilst between each two short flights the animal balances itself for a

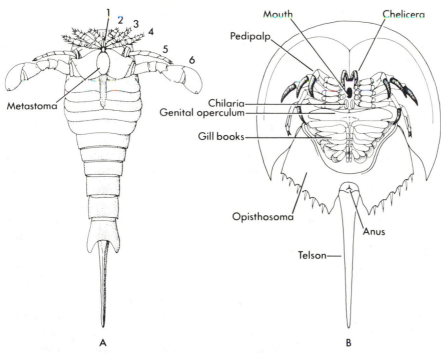

Figure 10-42 (A) Eurypteridea; *Eurypterus fischeri*. Restoration of ventral surface from an Upper Silurian fossil. 1–6, appendages of the prosoma. (B) Xiphosura; *Xiphorsura polyphemus* (*Limulus polyphemus*) the king or horseshoe "crab" ventral view. (C) Trilobita; *Phacops*, a Devonian fossil. (*A, after Holm, from Woods; B, after Shipley and MacBride modified; C, photo from a specimen courtesy of Mr. James Hodgson.*)

543

moment on the tip of its tail." The food of *Xiphosura* consists chiefly of worms, such as *Nereis,* and mollusks. These are caught while burrowing through the sand, are held by the chelicerae, and are chewed by the bases of the walking legs. In the spring the king crabs come near the shore to spawn. In some respects the Xiphosura resemble fossil trilobites and the latter are placed by some students near the king crabs in the arthropod series.

The extinct trilobites (Fig. 10-42, C) are a class of arthropods placed in their own subphylum, the Trilobitomorpha. They were marine animals, especially abundant in the Cambrian and Silurian. One pair of antennae was present. Each segment posterior to the antennary somite bore a pair of biramous appendages each with a gnathobase. Trilobites are associated in the strata of the earth's crust with the remains of crinoids, brachiopods, and cephalopods. The best known species, *Triarthrus becki,* occurs in the Utica shale (Lower Silurian) of New York State.

Included in the Merostomata is an extinct subclass, the Eurypteridea (Fig. 10-42, A), known only from fossils from Paleozoic strata. They were scorpionlike in appearance with a small cephalothorax and an abdomen of 12 segments, the anterior six of which were provided with unbranched, platelike appendages. The largest arthropods known belonged to this order, having reached a length of about 2 meters. Ex. *Eurypterus, Pterygotus, Stylomurus.*

Class IX. Arachnoidea

The Arachnoidea include the spiders, scorpions, mites, etc. These animals have no antennae nor true jaws; the first pair of appendages are chelicerae; and a cephalothorax, and abdomen are evident. Eleven orders are described here following a more detailed account of the spider.

THE SPIDER

External Features

The body of the spider consists of a *cephalothorax* which is undivided, and an *abdomen* which is usually soft, rounded, and unsegmented.

There are six pairs of *appendages* attached to the cephalothorax. Antennae are absent; their sensory functions are in part performed by the walking legs. The first pair of appendages are called *chelicerae*. They are in many species composed of two parts, a basal "mandible," and a terminal claw. *Poison glands* are situated in the chelicerae. The poison they secrete passes through a duct and out of the end of the chelicera; it is strong enough to kill insects and to injure large animals. The second pair of appendages are the *pedipalps;* their bases, called "maxillae," are used as jaws to press or chew the food. The pedipalps of the male are used as copulatory organs.

Following the pedipalps are four pairs of *walking legs*. This number easily distinguishes spiders from insects, since the latter possess only three pairs. Each leg consists of seven joints—(1) coxa, (2) trochanter, (3) femur, (4) patella, (5) tibia, (6) metatarsus, (7) tarsus—and is terminated by two toothed claws (Fig. 10-45, B) and often a pad of hairs which enables the spider to run on ceilings and walls. The bases of certain of the legs sometimes serve as jaws.

The *sternum* lies between the legs, and a *"labium"* is situated between the "maxillae." The *eyes*, usually eight in number, are on the front of the head (Fig. 10-45, C). The *mouth* (Fig. 10-43) is a minute opening between the bases of the pedipalps (maxillae); it serves for the ingestion of juices only, since spiders do not eat solid food.

The *abdomen* is connected by a slender waist (*pedicel*) with the cephalothorax. Near the anterior end of the abdomen on the ventral surface is the *genital opening*, protected by a pair of appendages which have fused together to form a plate called the *epigynum*. On either side of the epigynum is the slitlike opening of the respiratory organs or *book lungs*. Some spiders also possess *tracheae*, which open to the outside near the posterior end of the ventral surface (Fig. 10-43). Just back of the tracheal opening are three pairs of tubercles or *spinnerets* (Fig. 10-45, A), used for spinning threads. The *anus* lies posterior to the spinnerets.

Internal Anatomy and Physiology

The *food* of the spider (Fig. 10-43) consists of juices sucked from the bodies of other animals, principally insects. Suction is produced by the enlargement of the sucking stomach, due to the contraction of muscles attached to its dorsal surface and to the chitinous covering of the cephalothorax. The *true stomach*, which follows the sucking stomach, gives off five pairs of *ceca* or blind tubes in the cephalothorax. The *intestine* passes almost straight through the abdomen; it is enlarged at a point where ducts bring into it a digestive fluid from the *"liver,"* and again near the posterior end, where it forms a sac, the "stercoral pocket." Tubes, called *Malphighian tubes*, enter the intestine near the posterior end. The alimentary canal is surrounded in the abdomen by a large digestive gland or "liver." This gland secretes a fluid resembling pancreatic juice and pours it into the intestine through ducts.

The *circulatory system* consists of a heart, arteries, veins, and a number of spaces or sinuses. The *heart* is situated in the abdomen and is surrounded by the digestive glands. It is a muscular, contractile tube lying in a sheath, the pericardium, into which it opens by three pairs of ostia. It gives off posteriorly a caudal artery, anteriorly an aorta which branches and supplies the tissues in the cephalothorax,

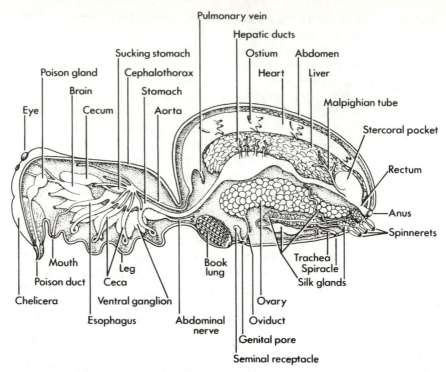

Pulmonary vein

Hepatic ducts

Sucking stomach · Ostium · Abdomen

Poison gland · Cephalothorax · Heart · Liver

Brain · Stomach

Eye · Cecum · Aorta · Malpighian tube

Stercoral pocket

Rectum

Anus

Spinnerets

Mouth · Leg · Book lung · Trachea · Spiracle

Poison duct · Ceca · Silk glands

Chelicera · Ventral ganglion · Ovary

Esophagus · Abdominal nerve · Oviduct

Genital pore

Seminal receptacle

Figure 10-43 Longitudinal section of a spider, showing internal organs. (*After Root.*)

and three pairs of abdominal arteries. The *blood,* which is colorless and contains mostly amoeboid corpuscles, passes from the arteries into sinuses and is carried to the book lungs, where it is aerated; it then passes to the pericardium by way of the pulmonary veins, and finally enters the heart through the ostia.

Respiration is carried on by tracheae and book lungs; the latter are peculiar to arachnids. The *book lungs* (Fig. 10-44), of which there are usually two, are sacs, each containing generally from 15 to 20 leaflike horizontal shelves through which the blood circulates. Air entering through the external openings is thus brought into close relationship with the blood. Tracheae are also usually present, but do not ramify to all parts of the body as in the insects.

The *excretory organs* are the *Malphighian tubules,* which open into the intestine, and two *coxal glands* in the cephalothorax. The coxal glands are sometimes degenerate, and their openings are difficult to find; they are homologous with the green glands of the crayfish.

The *nervous system* consists of a bilobed ganglion above the esophagus, a subesophageal ganglionic mass and the nerves which arise from them. There are *sensory hairs* on the pedipalps and probably on the walking legs, but the principal sense organs are the eyes. There

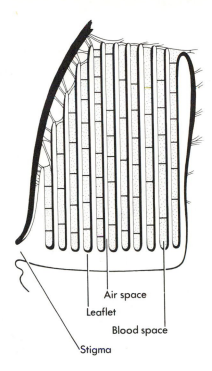

Figure 10-44 Spider. Book lung, longitudinal section. (*After MacLeod, modified.*)

Air space

Leaflet

Blood space

Stigma

are usually eight *eyes* (Fig. 10-45, C), and these differ in size and arrangement in different species. Spiders apparently can see objects distinctly only at a distance of about 12 cm.

The sexes are separate, and the *testes* or *ovaries* form a network of tubes in the abdomen. The spermatozoa are transferred by the pedipalps of the male to the female, and fertilize the eggs within her body. The eggs are laid in a silk cocoon, which is attached to the web or to a plant, or carried about by the female. The young leave the cocoon as soon after hatching as they can run about.

The *spinning organs* of spiders are three pairs of appendages called *spinnerets* (Fig. 10-45, A). The spinnerets are pierced by hundreds of microscopic tubes through which a fluid secreted by a number of abdominal *silk glands* (Fig. 10-43), passes to the outside and hardens in the air, forming a thread. These threads are used to build nests, form cocoon, spin webs, and for many other purposes. An *orb web*, such as is shown in Figure 10-46, is spun in the following manner. A thread is stretched across the space selected for the web; then from a point on this thread other threads are drawn out and attached in radiating lines. These threads have their strength as soon as drawn from the silk gland, apparently due to the orientation of the molecules as the silk is drawn out. Next a wide spiral may be placed upon the radii to temporarily hold them. A fine spiral of silk, covered with sticky droplets (Fig. 10-45, D), is then added to replace the first spiral as it is cut

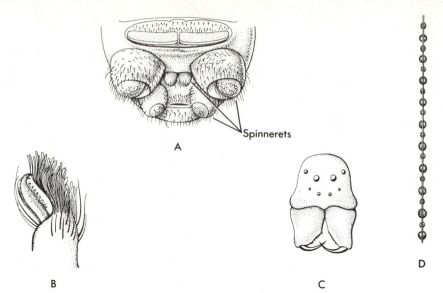

Figure 10-45 Parts of a spider's body. (A) Ventral view of posterior end of abdomen showing three pairs of spinnerets. (B) Foot, showing claws and bristles. (C) Front of head, showing eyes and jaws. (D) A thread from a spider's web, showing adhesive droplets. (*From Warburton.*)

away. Various species of orb weavers make characteristic modifications of this basic pattern. The spider stands in the center of the web or retires to a nest at one side and waits for an insect to become entangled in the sticky thread; it then rushes out and spins threads about its prey to restrain it prior to feeding on it.

Silk is used very conservatively, with threads reinforced only where strength is needed, and is often eaten when use is completed, providing efficient recycling of web protein.

Many spiders do not spin webs, but wander about capturing insects, or lie in wait for them in some place of concealment. In this group belong the crab-spiders (Thomisidae, Fig. 10-53, D), jumping-spiders (Attidae, G, J), ground-spiders (Drassidae), and running spiders (Lycosidae, H). The cobweb spiders spin various kinds of nets for capturing insects. The tube-weavers (Agelenidae) build platforms on the grass and hide in a tube at one side.

OTHER ARACHNOIDEA

The 11 orders of living arachnids are described below.

Order 1. Scorpionida

The scorpions (Fig. 10-47) are rapacious arachnids measuring from 1 to 20 cm in length. They hide in crevices or in pits in the sand during the

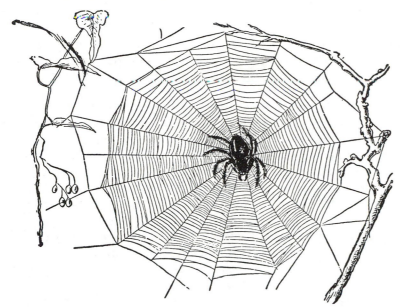

Figure 10-46 Web of the garden spider. (*After Blanchard.*)

Figure 10-47 Scorpionida. Scorpion. (A) Ventral view. (B) Dorsal view.

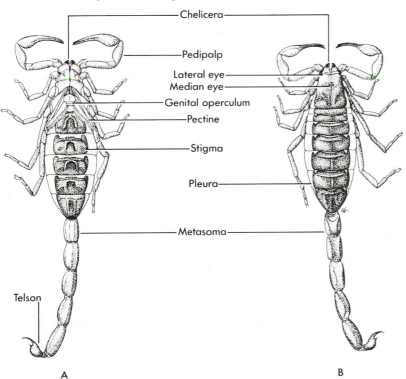

549

daytime, but run about actively at night. They capture insects and spiders with their pedipalpi, tear them apart with their chelicerae, and devour the pieces. Larger animals are paralyzed by the sting on the end of the tail. This sting does not serve as a weapon of defense unless the scorpion is hard pressed. Most American scorpions are relatively harmless, although *Centruroides sculpturatus* can give lethal stings.

The scorpion's body is more obviously segmented than that of most of the other arachnids. There is a *cephalothorax* (prosoma), and an abdomen of two parts—a thick anterior portion (mesosoma), and a slender tail (metasoma), which is held over the back when the animal walks. The dorsal shield of the cephalothorax bears a pair of median *eyes* and three lateral eyes on each side. The sense of sight is, however, poorly developed. On the ventral surface of the second abdominal segment are two comblike appendages called pectines; these are probably special tactile organs. *Tactile hairs* are distributed over the body, and the sense of touch is quite delicate. There are four pairs of *book lungs* opening by means of *stigmata* on the undersurface of abdominal segments III–VI.

The mating activities of scorpions are very curious, and include a sort of promenade. Scorpions are viviparous. The young ride about upon the back of the female for about a week, and then shift for themselves. They reach maturity in 1 to 5 years.

Order 2. Uropygida (Thelyphonida)

Whip scorpions; with long slender tail, but no sting; pedipalps very large; first pair of legs long and slender. The one American species, found in the southern United States, is the largest of the order; *Mastigoproctus giganteus;* vinegarroon; about 13 cm long; emits odor of vinegar from posterior glands, which secrete acetic and caprylic acids.

Order 3. Palpigradida

The order Palpigradida contains a single family, Koeneniidae, of small arachnids. They possess a long segmented tail and long, slender pedipalps and legs. A few species found in a few scattered localities. Ex. *Koenenia wheeleri* (Fig. 10-48, A) about 2.5 mm long; in moist places under stones; Texas.

Order 4. Pseudoscorpionida

The pseudoscorpions (Fig. 10-48, B) resemble scorpions in general shape and in the possession of scorpionlike pedipalps but a tail with a caudal sting at the end is lacking. Silk glands are present in the cephalothorax, opening near the end of the chelicerae. The silk is used to construct nests in which they molt or pass the winter. They live under stones, in moss, under bark, in the nests of termites, ants, and bees, and in houses, and feed on small insects and mites. Large insects to

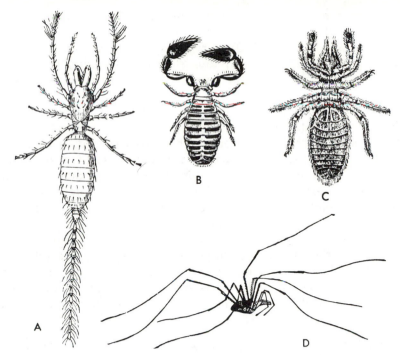

Figure 10-48 Arachnoidea. (A) Palpigradida; *Koenenia mirabilis.* (B) Pseudoscorpionida; *Chelifer cyrneus.* (C) Solpugida; *Rhagodes.* (D) Phalangida; *Phalangium opilio.* (*After various authors.*)

which they cling often carry them about—a fact that probably accounts for their wide distribution.

FAMILY 1. CHELIFERIDAE. Ex. *Chelifer cancroides:* book scorpion; about 3 mm long; in books, furniture, and clothing; cosmopolitan.

FAMILY 2. CHTHONIIDAE. Ex *Chthonius pennsylvanicus:* about 1.9 mm long; eastern United States.

Order 5. Schizomida

A small group of small arachnids sometimes united with the Uropygida. They differ from them by their small size, shorter tail, several tergal plates on the cephalothorax, and presence of only one pair of book lungs. The pedipalps are not chelate. Mostly tropical in distribution. Ex. *Trithyreus pentapeltis:* about 1 cm long; southern California.

Order 6. Acarida

The order Acarida includes the ticks and mites, most of which are of small size. The head, thorax, and abdomen are fused together forming an unsegmented body. At the anterior end, a small part of the head region is usually segmented off and hinged to the body proper to serve as a movable base for the mouth parts. In some mites the chitinous integument is membranous throughout, in others some portions of the integument are thickened into protective plates or shields. The body and legs usually bear regularly arranged hairs or bristles. Acarida usually pass through four stages, egg, larva, nymph, and adult. The larva has only three pairs of legs. When it molts to become a nymph it acquires a fourth pair of legs but still lacks a genital aperture. The nymph, in turn, molts and becomes an adult, with four pairs of legs and a genital aperture.

Mites are extremely numerous both in species and in individuals and are found in all sorts of habitats. Some are ectoparasitic, others endoparasitic on or in all sorts of animals and plants. Some are terrestrial, others live in fresh or salt water. Some feed on decaying animal or vegetable matter, others on stored foods, others on plants, others on minute animals, and the parasitic forms on blood.

There are about 166 families; a few follow.

FAMILY 1. ARGASIDAE (Fig. 10-49, A, B). Soft ticks. The ticks of this family have much the same habits as bedbugs, hiding in cracks or crevices in houses or in the nests of their hosts and coming out at night to feed on the blood of the host for a short period, usually less

Figure 10-49   Acarida. (A) Argasidae; *Argas reflexus*, female, ventral view. (B) Argasidae; *Ornithodorous moubata*, male, dorsal view; the vector of tick fever. (C) Ixodidae; *Dermacentor andersoni*, male, dorsal view; the vector of certain fevers of humans. (D) Hydrachnidae; *Atax alticola*, a freshwater mite. (E) Demodecidae; *Demodex follicularum*, ventral view; the causative agent of some "blackheads" in the skin of the human face. (*Mostly after Brumpt.*)

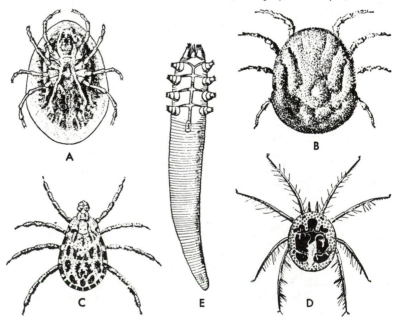

than ½ hour. Ex. *Argas persicus:* parasitic on fowls; vector of fowl spirochaetosis; tropics and subtropics. *Ornithodorus moubata* (Fig. 10-49, B): vector of African relapsing fever or tick fever.

FAMILY 2. IXODIDAE (Fig. 10-49, C). Hard ticks. In the hard ticks the two sexes are often very dissimilar in appearance because the hard dorsal scutum, which both sexes possess, covers the entire dorsal surface in the male but extends over a small part of the anterior portion of the dorsum of the female. The ticks of this family attach themselves firmly to their hosts and remain upon them, sucking blood, for days or even weeks. The lesion resulting from a tick bite is due largely to inflammatory responses of the host, although tick secretions do cause some vascular trauma (Tatchell and Moorhouse, 1970). Ex. *Boophilus annulatus:* vector of Texas cattle fever. *Haemaphysalis leporispalustris;* rabbit tick. *Dermacentor andersoni* (Fig. 10-49, C); vector of Rocky Mountain spotted fever.

FAMILY 3. ERIOPHYIDAE. Gall mites. These mites feed on plant juices and stimulate the formation of galls, the cavities of which open to the outside. *Phyllocoptes pyri:* pear-leaf blister mite.

FAMILY 4. DEMODECIDAE. Mites that live in the sebaceous glands and hair follicles of humans and domestic animals. Ex. *Demodex folliculorum* (Fig. 10-49, E): parasite of cattle, hogs, and people causes "blackheads" in skin of human face.

FAMILY 5. DERMANYSSIDAE. A number of parasitic species belong to this family. Ex. *Dermanyssus gallinae* (Fig. 10-50, D): chicken mite; causes dermatitis in humans.

FAMILY 6. SARCOPTIDAE. Itch mites. Many parasites of people, mammals, and birds. Ex. *Sarcoptes scabei* (Fig. 10-50, B): human itch mite. *Psoroptes communis ovis* (A): sheep scab mite.

FAMILY 7. TROMBIDIIDAE (Fig. 10-50, C): Harvest mites and chiggers. Ex. *Trombicula* spp. North American chigger or red bug. (See also Siphonaptera.)

FAMILY 8. TETRANYCHIDAE. Red spiders. Ex. *Tetranychus telarius:* lives on house plants, fruit trees, cotton plants, etc.

FAMILY 9. HYDRACHNIDAE (Fig. 10-49, D). Water mites. Ex. *Hydrachna geographica:* adults, dark red; larvae parasitic on aquatic insects; cosmopolitan.

Order 7. Phalangida

The harvestmen or daddy longlegs may be distinguished from spiders by their extremely long legs (Fig. 10-48, D), the absence of a waist, and their segmented abdomen (Fig. 10-51). Tracheae are present and a single pair of spiracles. There are no silk glands, and hence no web or nest is constructed. Harvestmen are able to run rapidly over leaves and grass. Their food consists of small living insects and the juices of fruit and vegetables. The adults die in the autumn, in the northern part of their range, but their eggs live over winter and hatch in the

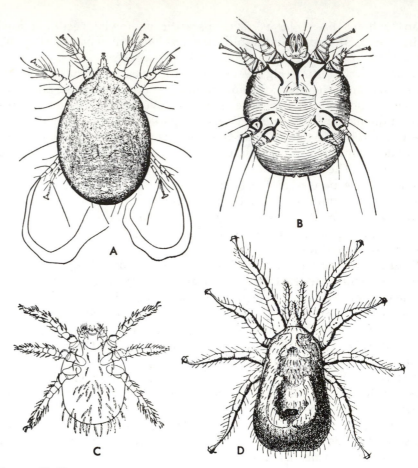

Figure 10-50 Acarida. (A) Sarcoptidae; *Psoroptes communis;* sheep scab mite. (B) *Sarcoptes scabei,* female ventral view; human itch mite. (C) Trombidiidae; *Trombicula akamushi,* larva with six legs; harvest mite. (D) Dermanyssidae; *Dermanyssus gallinae,* female; chicken mite. (*A, after Salmon and Stiles; B, after Gudden; C and D, from Brumpt.*)

spring. Ex. *Leiobunum vittatum:* body about 9 mm long; legs from 42 to 90 mm long; in fields and woods; eastern and central America.

Order 8. Solpugida (Solifugae)

The Solpugida (Fig. 10-48, C) live in warm countries in sandy regions. Their large chelate mandibles give them a dangerous appearance but they are not poisonous. Tracheae are present opening through a pair of spiracles on the thorax and three pairs on the abdomen. Mostly tropical. Ex. *Eremobates pallipes:* about 13 mm long; southern states west of Mississippi.

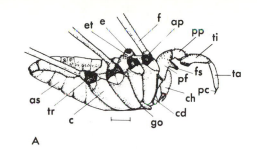

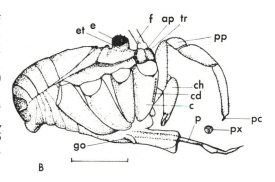

Figure 10-51 Phalangids. (A) Lateral view of *Leiobunum calcar*, male. (B) Lateral view of *L. politum*, male, with penis exerted. *as,* abdominal segment; *ap,* autotomy plane; *c,* coxa; *cd,* coxal denticle; *ch,* chelicera; *e,* eye; *et,* eye tubercle; *f,* femur; *fs,* spur of femur; *go,* genital operculum; *p,* penis; *pc,* palpal claw; *pf,* femur of palpus; *pp,* patella of palpus; *px,* penis, cross-section, showing sclerotized shaft or nonalate type; *tr,* trochanter. Scale lines are 1.0 mm in length. (*From A. L. Edgar. Phalangida of the Great Lakes Region. Am. Midl. Nat.,* **75:***347–366. Copyright 1966 by the American Midland Naturalist.*)

Order 9. *Amblypygida*

Very large nonchelate pedipalps; first leg long with many segments (Fig. 10-52); segmentation evident on abdomen; no spinnerets; no tail. About 60 species, mostly tropical. Ex. *Tarantula whitei:* about 2 cm long; Texas to California.

Order 10. *Araneida (Araneae)*

The spiders; with spinnerets; abdomen unsegmented and joined to cephalothorax by a narrow pedicel; poison glands open on chelicerae.

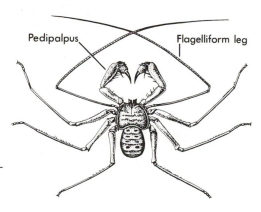

Figure 10-52 Amblypygida. *Admetus pumilio.* (*From Sedgwick.*)

Kaston (1978) lists 89 families. Some commonly encountered spiders are in the following families.

FAMILY 1. AVICULARIIDAE (Fig. 10-53, A). Tarantulas and trapdoor spiders; two pairs of lungs; eight eyes; chelicerae move up and down. Ex. *Eurypelma hentzi*: tarantula; large; southwestern states. *Pachy-*

Figure 10-53 Araneida. Species belonging to various families. (A) Ariculariidae; *Chilobrachys stridulans.* (B) Theridiidae; *Latrodectus mactans.* (C) Theridiidae; *Trithena tricuspidata.* (D) Thomisidae; *Thomisus*, a crab spider. (E) Argiopidae; *Epeira angulata.* (F) Argiopidae; *Argiope aurelia.* (G) Salticidae; *Attus*, jumping spider. (H) Lycosidae; *Lycosa fabrilis.* (I) Theridiidae; *Theridion tepidariorum.* (J) Salticidae; *Salticus scenicus. (From Warburton.)*

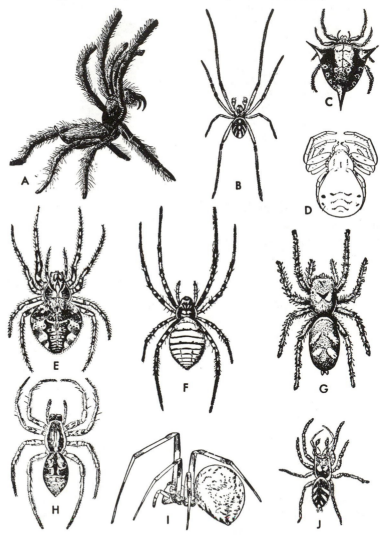

lomerus audonini: trap-door spider; lives in cylindrical hole in ground that can be closed by hinged earthen door.

FAMILY 2. AGELENIDAE. Funnel-web spiders; legs long; feet with three claws; posterior spinnerets long; eight eyes in two rows. Ex. *Agelena naevia:* grass spider; web concave with funnel-shaped tube at side. *Argyroneta aquatica:* European aquatic spider; lives at bottom of ponds among plants; air carried down from surface.

FAMILY 3. THERIDIIDAE (Fig. 10-53, B, C, I). Comb-footed weavers; tarsi of posterior legs with comb of curved, toothed setae. Ex. *Theridion tepidariorum* (Fig. 10-53, I): common house spider; irregular web in corners, etc. *Latrodectus mactans* (Fig. 10-53, B): black-widow spider.

FAMILY 4. ARGIOPIDAE (Fig. 10-53, E, F). Orb weavers; tarsi clothed with hairs; three claws on feet; eight eyes; web (Fig. 10-46). Ex. *Epeira foliata:* common on vegetation and fences.

FAMILY 5. THOMISIDAE (Fig. 10-53, D). Crab spiders; short, flat, and crab-like; two claws on feet; eight eyes in two rows. Ex. *Misumena vatia:* frequents flowers; no web; color may change from white to yellow according to color of flower.

FAMILY 6. LYCOSIDAE (Fig. 10-53, H). Running or wolf spiders; body long, hairy; eight eyes in three rows, larger eyes on posterior row; legs long and stout. Ex. *Lycosa helluo:* no web; cocoon attached to female; young carried on mother's back.

FAMILY 7. SALTICIDAE (Fig. 10-53, G, J). Jumping spiders; body short; legs stout; no web; can run or jump forward, backward, and sideways; large eyes. Ex. *Salticus scenicus* (Fig. 10-53, J): common on fences and buildings.

Order 11. Ricinulei

A rare order, less than 500 living individuals have been collected of about 14 species with a general resemblance to spiders. There are no spinnerets; the mouth and chelicerae can be covered by a hood; eyeless. Ex. *Ricinoides sjostedti* from the Cameroons.

Class X. Pycnogonoidea

The Pycnogonoidea (Fig. 10-54), order Pycnogonida, are marine arthropods, sometimes called sea spiders, with very small bodies and disproportionally long legs. They live among algae and hydroids, on which they feed, and range from one millimeter to several centimeters in length. They are of uncertain systematic position sometimes being placed in a separate phylum. The cephalothorax is segmented and the abdomen much reduced; the appendages consist of a pair of long chelate mandibles, a pair of slender pedipalps, a pair of slender ovigerous legs and four pairs of very long legs usually with nine seg-

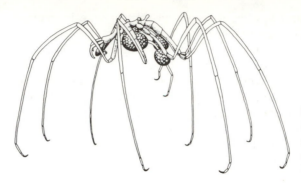

Figure 10-54 Pycnogon-oidea. *Nymphon stromii;* male carrying egg masses on his ovigerous legs. (*From Thompson.*)

ments each; four eyes are present; long diverticulae extend from the intestine into the legs; the sexes are separate; the eggs are carried on the ovigerous legs of the males and give rise to larvae with three pairs of legs.

The special bryozoan-feeding adaptation of *Achelia echinata* was noted in Chapter 7. Increased interest in marine groups is reflected in an entire issue of the Linnaean Society journal dealing with pycnogonids; the lead-off article is by Manton (1978). King and Hedgpeth are workers who have given special attention to the pycnogonids. Psycnogonids are cosmopolitan, benthic, marine arthropods. Ex *Pycnogonum, Pallene, Nymphon.*

Class XI. Tardigrada

The Tardigrada, or water bears (Figs. 10-55 and 10-38, B), are minute animals that are often included with the arachnids, although they may not even be arthropods. Here they are given the value of a class for the sake of convenience. They live in damp moss or sand and in fresh water or salt water. The head is more or less distinct and the body consists of four segments fused together. Each body segment bears a pair of short, thick legs, which are unsegmented but terminated by from four to nine sharp claws. No circulatory, respiratory, and excretory systems are present. The nervous system, however, is well developed. The sexes are separate. The eggs are large and the young that hatch from them sometimes possess only three pairs of legs. Tardigrades are able to live for several years in a dessicated condition.

Two orders are common. A third order contains one species from Japanese hot springs. Slight differences between orders perhaps justify placing all tardigrades in the same suborder, rather than into orders as here, or classes as elsewhere. Tardigrades are free-living, although I have found one species on several occasions in the umbilicus of *Viviparus* snails.

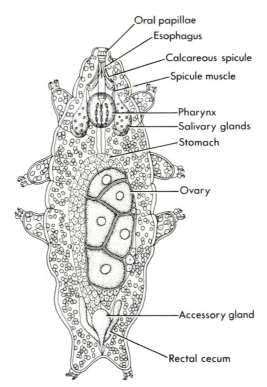

Oral papillae
Esophagus
Calcareous spicule
Spicule muscle
Pharynx
Salivary glands
Stomach
Ovary
Accessory gland
Rectal cecum

Figure 10-55 Tardigrada. *Macrobiotus schultzei*, showing internal anatomy. *(From Shipley, after Greeff.)*

Order 1. Eutardigrada

Head without cirri; legs not retractile telescopically; claws united into one branched pair. Ex. *Macrobiotus hufelandi*: 0.7 mm long; one pair strong teeth; freshwater ponds. *Hypsibius, Milnesium.*

Order 2. Heterotardigrada

Head with two pairs of cirri; legs somewhat retractile telescopically; claws separate. Ex. *Echiniscus testudo*: 0.3 mm long; one pair long teeth; two red eyes; body reddish, opaque; frequent in damp moss on roofs.

Class XII. Pentastomoidea

The wormlike animals in this subclass are all parasitic. They were formerly grouped with the worms but the morphology of the adults and young indicate they may be arthropods allied to the arachnids. The body is unsegmented although the body wall consists of many rings. No circulatory, respiratory, and excretory systems are present. The alimentary canal is straight and close to the mouth is a pair of horny hooks. The sexes are separate. The larvae have two pairs of legs and are mitelike in shape. Ex. *Reighardia sternae:* genital opening anterior; hooks posterior to mouth; parasitic in gulls and terns. *Linguatula serrata (Pentastomum taeniodes,* Fig. 10-38, A): female about 8 cm long; male 2 cm long; adult in nasal cavities of dog, wolf, and fox; eggs discharged with mucus, when swallowed by rabbits, domesticated animals, or humans, hatch in stomach, migrate to lungs, kidneys, etc., and encyst; excyst if host devoured by a carnivore and migrate to nose. *Porocephalus (Armillifer) armillatus:* adult in lungs of snakes; eggs swallowed by various animals, hatch and larvae migrate into the liver and body cavity.

Arthropods in General

The characteristics of each group of arthropods have been presented in the preceding pages, hence it will suffice here to point out in what respects these various groups differ from or resemble each other.

Anatomy

Divisions of the Body. The ancestors of the arthropods were probably annelidlike, as in *Peripatus,* which has a muscular body wall and a body that is not divided into well-defined regions. In the Crustacea, Insecta, Chilopoda, and Diplopoda there may be three distinct divisions, head, thorax, and abdomen, or the head and thorax may be united into a cephalothorax. The Chelicerata usually possess a cephalothorax or prosoma and an abdomen or opisthosoma consisting of the mesosoma and metasoma combined. The size of and number of segments in each of these divisions of the body differ widely within the group and are correlated with the environment and activities of each species.

Appendages. The paired, jointed appendages, from which the term Arthropoda is derived, are of almost infinite variety. Typically each segment bears one pair of appendages but in many species appendages are lacking on many segments. Those that are present were probably derived from a primitive type which has become modified for the numerous functions performed by them. The accompanying table, Segments and Appendages of some Adult Arthropods, indicates in general the appendages of the larger divisions of the phylum.

Cuticle. The exoskeleton of arthropods consists of chitin often hardened by lime salts. The chitin is inelastic and hence increase in size is impossible except during the period when the exoskeleton is cast off (molting or ecdysis). The new chitin secreted by the hypodermis then allows a certain amount of expansion. The chitin is rigid except between certain segments and joints where it is thin and allows the movement of the adjoining parts. The chemical structure of chitin is shown in Figure 1-2.

Digestive Tract. The alimentary canal consists of three divisions united end to end. The foregut or stomodaeum and the hindgut or proctodaeum are of ectodermal origin, and are lined with chitin which is shed when molting occurs. The midgut may be derived from mesoderm since it lacks a chitinous lining. The length, diameter, and sections into which the various parts of the alimentary canal are divided are correlated with the food habits of the species and vary greatly in the different groups.

Circulatory System. There is, among the arthopods, no complex series of large blood vessels nor fine capillaries that come into direct contact with the tissues; but the circulatory system is of the so-called open type. The blood is contained largely in sinuses and bathes the tissues

Segments and Appendages of Some Adult Arthropods

Segment	Malacostracan	Insect	Centipede	Scorpion
3rd preoral	(-)	(-)	(-)	
2nd preoral	1st antennae	antennae	antennae*	chelicerae
1st preoral	2nd antennae			pedipalps*
1st postoral	mandible	mandible	mandible	1st legs
2nd postoral	1st maxillae	1st maxillae	1st maxillae	2nd legs
3rd postoral	2nd maxillae	2nd maxillae	2nd maxillae	3rd legs
4th postoral	1st maxillipeds	1st legs	maxillipeds	4th legs
5th postoral	2nd thoracic	2nd legs + wings	1st legs	(not developed)
6th postoral	3rd thoracic	3rd legs + wings	2nd legs	♀♂ genital append.
7th postoral	4th thoracic	(1st abdom.)	3rd legs	pectines
8th postoral	5th thoracic	(2nd abdom.)	4th legs	(book lungs)
9th postoral	♀ 6th thoracic	(3rd abdom.)	5th legs	(book lungs)
10th postoral	7th thoracic	(4th abdom.)	6th legs	(book lungs)
11th postoral	♂ 8th thoracic	(5th abdom.)	7th legs	(book lungs)
12th postoral	1st abdominal	(6th abdom.)	8th legs	(7th abdom.)
13th postoral	2nd abdominal	(7th abdom.)	9th legs	(1st metasomal)
14th postoral	3rd abdominal	♀ (8th abdom.)	10th legs	(2nd metasomal)
15th postoral	4th abdominal	♂ (9th abdom.)	11th legs	(3rd metasomal)
16th postoral	5th abdominal	(10th abdom.)	12th legs +++	(4th metasomal)
17th postoral	last uropod	cerci	♀♂ genital	(5th metasomal)
postsegmental	(telson)		(telson)	(telson, sting)

Modified primarily from Borradaile et al. (1955).
* = postoral embryonic origin;   ♀ = segment bears female opening;
( ) = not appendages;   ♂ = segment bears male opening;
+++ = 3 up to 158 additional leg-bearing segments.

directly. Blood enters the heart through ostia and is pumped through arteries to the sinuses.

Respiratory System. This differs widely in the several divisions of the phylum. In aquatic species respiratory exchanges occur through the general body wall or through well-developed gills or branchiae. Terrestrial species breathe by means of tracheae, or, as in the arachnids, by book lungs or both book lungs and tracheae.

Coelom. The "body cavity" of arthropods is not a coelom but is filled with blood and known as a hemocoel. The coelom is represented in the embryo by cavities in the mesodermal segments and in the adult is restricted to the cavities of the reproductive organs and certain excretory organs.

Excretory System. There are no true nephridia among the arthropods. Malpighian tubules are present in insects, chilopods, diplopods, and arachnids and some arthropods possess coelomoducts.

Nervous System. This is probably derived from an ancestral condition in which a double ventral nerve cord with a pair of ganglia in each segment was present. In the arthropods of today there is a brain and more or less shortened ventral nerve chain. The brain consists of several pairs of ganglia fused together, and may exhibit several divisions known as the forebrain, midbrain, and hindbrain (Fig. 11-13). The ventral nerve cord likewise usually consists of a reduced number of masses of nervous tissue each comprising several pairs of ganglia.

Eyes. In *Peripatus* the eyes are a pair of simple, closed vesicles with a lens derived from the wall and a thickened, pigmented posterior complex. In all other arthropods the lens is derived from the cuticle; beneath the lens are cells known as retinulae, in each of which is a vertical rod called the rhabdom; and in and around the retinulae is pigment which may migrate toward or away from the surface according to the intensity of the light. The simple eyes of insects (ocelli) consist of a single vesicle with several retinulae. In chilopods and diplopods a number of vesicles lie close together in a group. The compound eyes of crustaceans and insects are very complex structures, the units of which are known as ommatidia. The arachnids possess eyes that resemble the ocelli of insects but may be degenerate compound eyes.

Phylogeny and Interrelations

The arthropods no doubt evolved from the same ancestors as the polychaete annelids. The development of a rigid exoskeleton with joints brought about a change in the distribution of the muscles from the continuous type forming a muscular body wall, as in the annelids, to the discontinuous type in which the muscles are separately and diversely developed for the movement of special segments.

The Onychophora resemble most closely the ancestral condition, with a thin cuticle, a continuous and muscular body wall, no joints, one pair of jaws, appendages on the first segment, and a series of coelomoducts of which one pair are cilated oviducts.

The Crustacea, Insecta, Chilopoda, and Diplopoda have become modified in many ways. The Crustacea differ from the others in the possession of a second pair of antennae and in their aquatic habit. The Insecta have only three pairs of legs and usually wings. Both Crustacea and Insecta possess compound eyes. The Arachnoidea have anterior appendages modified into chelicerae and lack true compound eyes.

The rigid exoskeleton has made a fit between male and female copulatory structures important for successful fertilization. Very slight morphological changes may result in effective genetic isolation of previously close populations having this type of lock-and-key genitalia. These reproductive features have been an important factor in the development of many specialized species of arthropods.

Gross changes in genitalia and other structures are often considered an indication of long divergence and very distant taxonomic relationship. Husband (1979, and unpublished) has studied mites of one family, Podapolipodidae, and found a sequence including some species with conventionally placed genitalia and others ranging to a species with dorsally positioned genitalia that enable copulation to occur while male and female remain attached to opposing surfaces formed by the folded wings and body of their beetle host (Fig. 10-56). Such

Figure 10-56 Aedeagal position in the Podapolipodidae. These mites are six-legged parasites under the elytra of certain beetles and of some other insects. The ancestral type was presumably some family such as the Pyemotidae (A) with eight legs and posterior aedeagus. Intermediate forms such as *Eutarsopolipus* (B) and *Dorsipes* (C) occur with the extreme reached by genera such as *Rhynchopolipus* Husband and Flechtmann, 1972 (D, and Figure 10-58) which shows a dorsal anterior aedeagus on the male that presumably enables fertilization of the female inverted on the elytron above. (*Based on unpublished illustrations courtesy of Dr. Robert Husband, Adrian College.*)

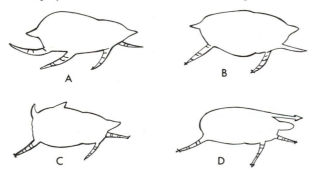

observations, and the fact that considerable variation occurs throughout the arthropods in position of genital openings when classes are compared, indicate that reproductive structures are sometimes one of the more variable arthropod characteristics.

History of Our Knowledge of the Arthropoda

The term Arthropoda was proposed by von Siebold in 1845 for a group of animals including the Crustacea, Insecta, and Arachnoidea. This group resembled the Articulata of Cuvier except that the Annelida were removed and placed with the Vermes. Linnaeus employed Insecta as a term for all of the arthropods; but separate classes were recognized later, the Myriapoda by Latrielle in 1796, the Crustacea by Cuvier in 1800, and the Arachnoidea by Lamarck in 1881. Many changes of comparatively minor importance have taken place since that time and are still being made, a condition that indicates both our lack of knowledge of certain sections of the phylum and the fact that we are making progress.

Many recent and current workers have made important contributions to this enormous group. A few and the general area of their contribution are: Snodgrass and Manton, anatomy and phylogeny; Sars, Stebbing, Calman, crustaceans; Savory, Levi, Kaston, spiders; Wharton, mites; and Eisner, defensive adaptations.

Embryology and the Biogenetic Law

Studies of the embryology of arthropods has furnished some of our best evidence in favor of what is known as the biogenetic law or the *recapitulation theory*. Organic evolution, that is, the evolution of one organism from another, is accepted as an established fact by practically all zoologists at the present time. Evolutionists do not claim that the more complex forms have evolved directly from the simpler animals, but that their ancestors were related. Beginning with the simplest animals we find that a single cell performs all the necessary processes of life (e.g., *Amoeba*). Within the lowest phylum, the Protozoa, there are animals consisting of a number of cells more or less intimately bound together in a hollow spherical colony (e.g., *Volvox*). Passing to the next higher group of organisms, we are introduced to animals that possess two layers of cells, surrounding a single cavity (e.g., *Hydra*). All animals above the coelenterates have three layers of cells forming their body walls (i.e., they are triploblastic). Four stages in the evolution of animals are represented in the groups just mentioned—(1) the single cell, (2) a ball of cells, (3) a two-layered sac, and (4) a three-layered organism.

Early in the past century it was noticed that these stages correspond to the early stages in the embryology of the Metazoa; in other words, that the development of the individual recapitulates the stages in the evolution of the race, or *ontogeny recapitulates phylogeny*. These stages contrasted appear as follows:

Phylogenetic Stage	Ontogenetic Stage
1. Single-cell animal	egg cell
2. Ball of cells	blastula
3. Two-layered sac	gastrula
4. Triploblastic animal	three-layered embryo

Later, other zoologists became interested in the recapitulation theory, and enlarged upon it. Of these Fritz Müller and Ernst Haeckel are especially worthy of mention. The latter expressed the facts as he saw them in his "fundamental law of biogenesis." The ancestor of the many-celled animals was conceived by him as a two-layered sac something like a gastrula, which he called a *Gastraea*. The coelenterates were considered to be gastraea slightly modified.

Fritz Müller derived strong arguments in favor of biogenesis from a study of certain Crustacea belonging to the Malacostraca. Many members of this group do not emerge from the egg so nearly like the adult as does the crayfish. The lobster, for example, upon hatching resembles a less specialized prawnlike crustacean, called *Mysis* (Fig. 10-27, C), and is said to be in the *Mysis* stage. The shrimp, *Penaeus*, passes through a number of interesting stages before the adult condition is attained. It hatches as a larva, termed a *Nauplius* (Fig. 10-57, A), possessing a frontal eye and three pairs of appendages; this Nauplius molts and grows into a *Protozoea* stage (B) which bears three more pairs of appendages and the rudiments of segments III–VIII. The *Protozoea* stage grows into the *Zoea* stage (C). The cephalothorax and abdomen are distinct at this time; eight pairs of appendages are present (I–VIII) and six more are developing. The *Zoea* grows and molts and becomes a *Mysis* (D) with eight pairs of appendages (I–VIII) on the cephalothorax. Finally, the *Mysis* passes into the adult shrimp, which possesses the characteristic number of appendages (I–XIX) each modified to perform its particular function. The *Nauplius* of *Penaeus* resembles the larvae of many simple crustaceans; the *Zoea* is somewhat similar to the condition of an adult *Cyclops* (Fig. 10-23, A); the *Mysis* is like the adult *Mysis* (Fig. 10-27, C); and finally the adult *Penaeus* is more specialized than any of its larval stages, and belongs among the higher crustacea. These facts have convinced some zoologists that *Penaeus* recapitulates in its larval development the progress of the race; that the lobster has lost many of these stages, retaining only the *Mysis;* and that the crayfish hatches in practically the adult condition. The *Nauplius* stage of the latter is supposed to be represented by a certain embryonic phase (Fig. 10-11, E).

The law of biogenesis should not be taken too seriously, since it has been criticized severely by many prominent zoologists, but it has fur-

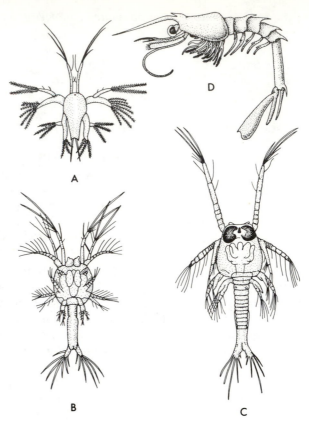

Figure 10-57 Four stages in the development of *Penaeus*, a shrimp. (A) Nauplius. (B) Protozoea. (C) Zoea. (D) Mysis. (*After Muller and Claus.*)

nished a hypothesis, which has concentrated the attention of scientists upon fundamental embryological processes, and has therefore had a great influence upon zoological progress.

Ecology of the Arthropoda

The remarkable success of arthropods as a group is largely an effect of the articulated exoskeleton, giving them mobility, protection, and an effective reproduction isolating mechanism, as was discussed under phylogeny.

The long-lasting success is demonstrated by the largest animals, the baleen whales, adaptation to feeding on myriads of small crustaceans of very few species (Pivorunas, 1979). Such finely tuned habits required very long presence of the association. This is also demonstrated by the barnacles, whose odd adaptations have enabled them to dominate the most energy intense zones of rocky coasts (Paine, 1979).

Examples of the mechanisms of success include structural features such as the aedeagus of *Rhynchopolipus* (Fig. 10-58) (Husband, 1979);

the specialized hairs (Rovner, Higashi, and Foelix, 1973) that enable the female wolf spider to carry its spiderlings; the amazing webs constructed by other spiders via the complex spinning apparatus; behavioral adaptations (Potter, Wrensch, and Johnston, 1976); behavioral combined with chemical adaptations (Eisner et al., 1971); biochemical adaptations (Clegg, 1974; Lester et al., 1975); hormonal regulation (Wright, 1969); and pheromonal communication (Dahl, Emanuelsson, and von Mecklenburg, 1970; Ameyaw-Akumfi and Hazlett, 1975).

The success of the exoskeleton is buffered by some limitations. Arthropods do not generally achieve great mass, perhaps never more than 20 kg and with the innermost tissue never more than 10 cm from the surface. The nearly 3 meters spanned by the giant spider crab is at the expense of a rather emaciated appearance.

Growth is usually more rapid at higher temperatures (Hughes, Sullivan, and Shleser, 1972), so the occurrence of enormous amphipods (Hessler, Issacs, and Mills, 1972), several times larger than any others known, at abyssal depths with perpetual cold and little food is an

Figure 10-58 *Rhynchopolipus rhynchophori.* Dorsal view of male, showing spearlike aedeagus directed forward over head region used for mating with larviform female attached to the elytron of the palm weevil while male is on the opposing surface. (*Adapted from S.E.M. in Husband, 1979, and drawing in Husband and Flechtmann, 1972.*)

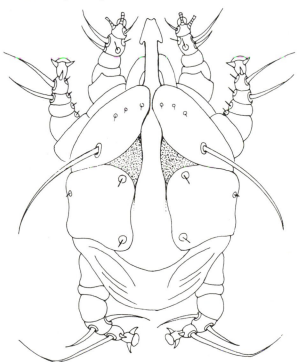

anomaly that will be explored in the last chapter. The largest isopods are also deep-sea species.

As crustaceans pervade aquatic communities, the insects pervade terrestrial and many benthic freshwater communities. The insects in the soil and litter are exceeded numerically by mites. Biomass of isopods produced is sometimes greatest for animals in soil communities. As expected, predators do not usually dominate numerically nor in biomass, but as predators the spiders are ubiquitous terrestrial animals.

Arthropods are to the community what ATP is to the organism.

REFERENCES TO LITERATURE ON THE ARTHROPODA

AGRAWAL, V. P. 1965. Feeding appendages and the digestive system of *Gammarus pulex. Acta Zool.,* **46:**67–81.

AMEYAW-AKUMFI, C., and B. A. HAZLETT. 1975. Sex recognition in the crayfish *Procambarus clarkii. Science,* **190:**1225–1226.

BAKER, E. W., and G. W. WHARTON. 1952. *An Introduction to Acarology.* Macmillan, New York, 465 pp.

BARNARD, J. L., R. J. MENZIES, and N. BACESCU. 1962. *Abyssal Crustacea.* Columbia Univ. Press, New York. 223 pp.

BARNWELL, F. H. 1966. Daily and tidal patterns of activity in individual fiddler crab (genus *Uca*) from the Woods Hole Region. *Biol. Bull.,* **130:**1–17.

BLISS, D. E., and L. MANTEL. (Eds.). 1968. Terrestrial adaptations in Crustacea. *Am. Zoologist,* **8:**307–685.

BLOWER, J. G. (Ed.). 1974. *Myriapoda.* Academic Press, New York. 714 pp.

BRINTON, E. 1962. The Distribution of Pacific Euphausids. *Bull. Scripps Inst. Oceanogr.,* **8,** No. 2. 219 pp.

BROOKS, J. L. 1957. The Systematics of North American Daphnia. *Mem. Conn. Acad. Arts Sci.* **XIII,** 180 pp.

———, and S. I. DODSON. 1965. Predation, body size, and composition of plankton. *Science,* **150:**28–35.

BROOKS, M. A. 1963. Symbiosis and aposymbiosis in arthropods. Pp. 200–231 in *Symbiotic Associations.* 13th Symp., Soc. Gen. Microbiol. Cambridge Univ. Press, New York. 356 pp.

BUTT, F. H. 1960. Head development in arthropods. *Biol. Rev.,* **35:**43–91.

CISNE, J. L. 1974. Trilobites and the origin of arthropods. *Science,* **186:**13–18.

Clegg, J. S. 1974. Biochemical adaptations associated with the embryonic dormancy of *Artemia salina. Trans. Am. Micr. Soc.,* **93:**481–490.

CLOUDSLEY-THOMPSON, J. L. 1958. *Spiders, Scorpions, Centipedes, and Mites.* Pergamon Press, New York. 228 pp.

———. 1964. Terrestrial animals in dry heat: arthropods. Pp. 451–465 in *Handbook of Physiology.* Section 4: Adaptation to the Environment. Am. Physiol. Soc., Washington.

COOPER, W. E. 1965. Dynamics and production of a natural population of a fresh-water amphipod, *Hyalella azteca. Ecol. Monogr.* **35:**377–394.

CRESSEY, R., and C. PATTERSON. 1973. Fossil parasitic copepods from Lower Cretaceous fish. *Science,* **180:**1283–1285.

CROCKER, D. W., and D. W. BARR. 1968. *Handbook of the Crayfishes of Ontario*. Univ. of Toronto Press, Toronto. 158 pp.

CURTIS, H. 1965. Spirals, spiders, and spinnerets. *Am. Scientist*, **53**:52–58.

DAHL, E., H. EMANUELSSON, and C. VON MECKLENBURG. 1970. Pheromone transport and reception in an amphipod. *Science*, **170**:739–740.

EDGAR, A. L. 1966. Phalangida of the Great Lakes region. *Am. Midl. Nat.*, **75**:347–366.

EISNER, T., H. E. EISNER, J. J. HURST, F. C. KAFATOS, and J. MEINWALD. 1963. Cyanogenic glandular apparatus of a millipede. *Science*, **139**:1218–1220.

———, A. F. KLUGE, J. E. CARREL, and J. MEINWALD. 1971. Defense of phalangid: liquid repellent administered by leg dabbing. *Science*, **173**:650–652.

FLOREY, E., and E. FLOREY. 1965. Cholinergic neurons in the Onychophora (*Opisthopatus costei*): A comparative study. *Comp. Biochem. Physiol.*, **15**:125–136.

FREY, D. G., et al. 1971. Symposium of ecology of the Cladocera. *Trans. Am. Micr. Soc.*, **90**:100–121.

GALUN, R., and S. H. KINDLER. 1965. Glutathione as an inducer of feeding in ticks. *Science*, **147**:166–167.

GREEN, J. 1961. *A Biology of the Crustacea*. Witherby, London. 180 pp.

GRICE, G. D., and A. D. HART. 1962. The abundance, seasonal occurrence and distribution of the epizooplankton between New York and Bermuda. *Ecol. Monogr.*, **32**:287–309.

GUPTA, A. P. 1979. *Arthropod Phylogeny*. Van Nostrand Reinhold, New York. 762 pp.

HACKMAN, R. H., and M. GOLDBERG. 1975. *Peripatus*: its affinities and its cuticle. *Science*, **190**:582–583.

HARTLAND-ROWE, R. 1965. The Anostraca and Notostraca of Canada with some new distribution records. *Can. Field-Nat.*, **79**:185–189.

HERRNKIND, W. 1969. Queuing behavior of spiny lobsters. *Science*, **164**:1425–1427.

HESSLER, R. H. 1966. Life cycle of the interstitial crustacean *Derocheilocaris typicus* Pennak & Zinn. *Am. Zoologist*, **6**:322.

———, J. D. ISSACS, and E. L. MILLS. 1972. Giant amphipod from the abyssal Pacific Ocean. *Science*, **175**:636–637.

———, and H. L. SANDERS. 1966. *Derocheilocaris typicus* Pennak & Zinn (Mystacocarida) revisited. *Crustaceana*, **11**:141–155.

HOBBS, H. H., JR. 1975. New crayfishes (Decapoda: Cambaridae) from the Southern United States and Mexico. *Smithsonian Contrib. Zool.*, No. 201. 34 pp.

HUGHES, J. T., J. J. SULLIVAN, and R. SHLESER. 1972. Enhancement of lobster growth. *Science*, **177**:1110–1111.

HUSBAND, R. W. 1979. The reproductive anatomy of male *Rhynchopolipus rhynchophori*, parasitic mite of the palm weevil. *Micron*, **10**:165–166.

———, and C. H. FLECHTMANN. 1972. A new genus of mite, *Rhynchopolipus*, associated with the palm weevil in Central and South America (Acarina, Podapolipodidae). *Rev. Brasil. Biol.*, **32**:519–522.

JUDD, W. W. 1965. Terrestrial sowbugs (Crustacea: Isopoda) in the vicinity of London, Ontario. *Can Field-Nat.*, **79**:197–202.

Kaston, B. J. 1965. Some little known aspects of spider behavior. *Am. Midl. Nat.,* **73**:336–356.

———. 1978. *How to Know the Spiders.* 3rd Ed. Brown, Dubuque. 272 pp.

King, P. E. 1974. *Pycnogonids.* St. Martin's Press, New York. 144 pp.

Kinne, O. 1964. Animals in aquatic environments: crustaceans. Pp. 669–682 in *Handbook of Physiology.* Section 4: Adaptation to the Environment. Am. Physiol. Soc., Washington.

Krantz, G. E., R. R. Colwell, and E. Lovelace. 1969. *Vibrio parahaemolyticus* from the blue crab *Callinectes sapidus* in Chesapeake Bay. *Science,* **164**:1286–1287.

Lankester, Sir R. 1906 (1964 reprint). *A Treatise on Zoology.* Part VII, 3rd Fasc. Crustacea.

Lee, R. F., J. C. Neuenzel, and G.-A. Paffenhöfer. 1970. Wax esters in marine copepods. *Science,* **167**:1510–1511.

Lester, R., M. C. Carey, J. M. Little, L. A. Cooperstein, and S. R. Dowd. 1975. Crustacean intestinal detergent promotes sterol solubilization. *Science,* **189**:1098–1100.

Maguire, B. 1965. *Monodella texana* n. sp., an extension of the range of the crustacean order Thermosbaenacea to the Western Hemisphere. *Crustaceana,* **9**:149–154.

Manton, S. M. 1960. Concerning head development in the arthropods. *Biol. Rev.,* **35**:265–282.

———. 1978. Habits, functional morphology and the evolution of pycnogonids. *Zool. J. Linn. Soc.,* **63**:1–22.

Marshall, S. M., and A. P. Orr. 1972 reprint. *The Biology of a Marine Copepod.* Springer-Verlag, New York. 195 pp.

Massaro, E. J. 1970. Horseshoe crab lactate dehydrogenase: tissue distribution and molecular weight. *Science,* **167**:994–996.

Paine, R. T. 1979. Disaster, catastrophe, and local persistence of the sea palm *Postelsia palmaeformis. Science,* **205**:685–687.

Paris, O. H., and F. A. Pitelka. 1962. Population characteristics of the terrestrial isopod *Armadillidium vulgare* in California grassland. *Ecology,* **43**:231–248.

Peakall D. B. 1971. Conservation of web proteins in the spider, *Araneus diadematus. J. Exp. Zool.,* **176**:257–264.

Pivorunas, A. 1979. The feeding mechanisms of baleen whales. *Am. Scientist,* **67**:432–440.

Potter, D. A., D. L. Wrensch, and D. E. Johnston. 1976. Aggression and mating success in male spider mites. *Science,* **193**:160–161.

Puglia, C. R. 1964. Some tardigrades from Illinois. *Trans. Am. Micr. Soc.,* **83**:300–311.

Rehm, A., and H. J. Humm. 1973. *Sphaeroma terebrans:* a threat to the mangroves of southwestern Florida. *Science,* **182**:173–174.

Ritzmann, R. 1973. Snapping behavior of the shrimp *Alpheus californiensis. Science,* **181**:459–460.

Rovner, J. S., G. A. Higashi, and R. F. Foelix. 1973. Maternal behavior in wolf spiders: the role of abdominal hairs. *Science,* **182**:1153–1155.

Sanders, H. L. 1957. The Cephalocarida and crustacean phylogeny. *Syst. Zool.,* **6**:112–129.

Savory, T. H. 1960. Spider webs. *Sci. Am.,* **202**(4):114–124.

————. 1962. Daddy longlegs. *Sci. Am.*, **207** (Oct.):119–128.

————. 1964. *Arachnida*. Academic Press, New York. 291 pp.

SCHRAMECK, J. E. 1970. Crayfish swimming: alternating motor output and giant fiber activity. *Science*, **169**:698–700.

SHAROV, A. G. 1966. *Basic Arthropodan Stock*. Pergamon Press, Oxford. 271 pp.

SNODGRASS, R. E. 1952. *A Textbook of Anthropod Anatomy*. Cornell Univ. Press, Ithaca.

TAPPA, D. W. 1965. The dynamics of the association of six limnetic species of *Daphnia* in Aziscoos Lake, Maine. *Ecol. Monogr.* **35**:395–423.

TATCHELL, R. J., and D. E. MOORHOUSE. 1970. Neutrophils: their role in the formation of a tick feeding lesion. *Science*, **167**:1002–1003.

VERMEIJ, G. J. 1977. Patterns in crab claw size: the geography of crushing. *Syst. Zool.*, **26**:138–151.

WARNER, G. F. 1977. *The Biology of Crabs*. Von Nostrand Reinhold, New York. 202 pp.

WATERMAN, T. H. 1960–1961. *The Physiology of Crustacea*. Vols. I and II. Academic Press, New York.

WHARTON, G. W. 1970. Mites and commercial extracts of house dust. *Science*, **167**:1382–1383.

WITT, P. N., and C. F. REED. 1965. Spider-web building. *Science*, **149**:1190–1197.

WRIGHT, J. E. 1969. Hormonal termination of larval diapause in *Dermacentor albipictus*. *Science*, **163**:390–391.

11

Class Insecta

Insects form the most important class of invertebrates in many ways. The class contains a greater number of species than all other living species combined. They are an extremely abundant group in terrestrial and freshwater environments as well as having the only winged invertebrates. Although many are harmful insects, many others are useful in control of pests and pollination of plants.

One group of insects, the mosquitoes, are humanity's worst enemy, carrying diseases that cause more deaths than any other single cause (Gillett, 1973). Another insect, the honey bee, is perhaps our most valuable friend, providing honey, wax, and pollinating many important food crops.

Some distinctive features are partly responsible for the terrestrial success of insects. They include (1) their protective exoskeleton that allows mobility and flight; (2) a tracheal respiratory system that reduces water loss; (3) Malpighian tubules that eliminate nitrogenous wastes as uric acid with feces containing little water; (4) behavioral, biochemical, and anatomical adaptations radiating into many specialized groups; and (5) their reproductive mechanism, including a high biotic potential.

The remarkable success of insects is moderated by their failure to produce many marine species and their restriction to small and moderate size.

Complex developmental changes may be of value to the advanced insect orders for the reduction of intraspecific competition between larvae and adults.

DISSOSTEIRA CAROLINA—A GRASSHOPPER

The grasshopper (Figs. 11-1 and 11-8) is a favorable representative of the class Insecta, because it is less specialized then many of the other common insects and, therefore, exhibits insect characteristics in their more generalized condition.

External Features

Exoskeleton. As in the crayfish, the grasshopper is covered by an exoskeleton that protects the delicate systems of organs within. This exoskeleton is the *cuticule,* which consists of chitin and is divided into a linear row of *segments.* The cuticule is mostly hard and rigid; some of the segments are firmly fastened together at *sutures,* but it is softer between other segments, thus allowing movements of the abdomen, wings, legs, antennae, etc. These softer regions are also known as *sutures.* The body wall consists of the cuticule beneath which is a layer of cells, the *hypodermis,* which secretes it and under this a *basement membrane.* Each segment is made up of separate pieces known as *sclerites;* usually some of the sclerites of a typical segment cannot be distinguished and the sutures are therefore said to be obsolete. In the grasshopper the body is divided into three groups of segments that constitute the head, thorax, and abdomen.

Head. Six segments are fused together to form the head (Fig. 11-2). These are not visible in the adult but may be observed in the embryo and are indicated by the paired appendages of the adult. They are as follows: *preoral, antennal, intercalary, mandibular, maxillary,* and *labial.*

Figure 11-1 Diagram of a female grasshopper with wings removed on left side. (*After Walden, from Wellhouse.*)

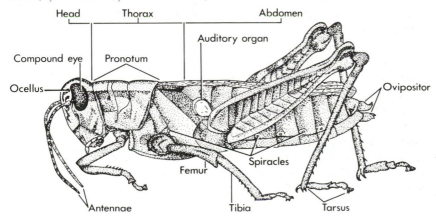

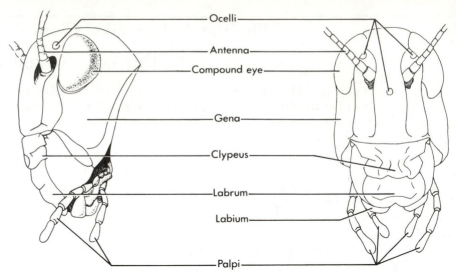

Ocelli
Antenna
Compound eye
Gena
Clypeus
Labrum
Labium
Palpi

Figure 11-2 Head of a grasshopper in side and front views. (*From Herms, after Folsom.*)

The exoskeleton of the head is known as the *epicranium;* the front of this is the *frons,* and the sides are the cheeks or *genae.* The rectangular sclerite beneath the front of the epicranium is the *clypeus.* On either side of the head is a *compound eye,* and on top of the head and near the inner edge of each compound eye is a simple eye or *ocellus.* The mouth parts are attached to the ventral side of the epicranium.

Mouth Parts (Fig. 11-3). The grasshopper is a biting insect, and its mouth parts are therefore *mandibulate.* There is a *labrum* or upper lip attached to the ventral edge of the *clypeus.* Beneath this is the membranous tonguelike organ, the *hypopharynx.* On either side is a single, hard jaw or *mandible* with a toothed surface fitted for grinding. Beneath the mandible are a pair of *maxillae;* each of these consists of the following parts: a basal *cardo,* central *stipes,* a long curved *lacinia,* a long, rounded *galea,* and a *maxillary palp* which arises from a *palpifer.* The *labium* or lower lip comprises a basal *submentum,* a central *mentum,* two movable flaps, the *ligulae,* and a *labial palp* on either side. The labium of the insects appears to have evolved from the lateral union of two appendages resembling the biramous limbs of a crustacean. The *maxillae* have obviously also arisen from this type of appendage. The labrum and labium serve to hold food between the mandibles and maxillae which move laterally and grind it. The maxillary and labial palps are supplied with sense organs that no doubt serve to distinguish different kinds of food.

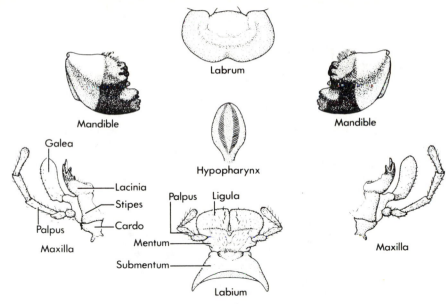

Figure 11-3 Mouth parts of a grasshopper. (*From Herms.*)

Ocellus (Fig. 11-4, B). The simple eyes consist of a group of visual cells, the *retinulae,* in the center of which is an optic rod, the *rhabdome,* and a transparent lenslike modification of the cuticule.

Compound Eye (Fig. 11-4, A). The compound eye is covered by a transparent part of the cuticule, the *cornea,* which is divided into a large number of hexagonal pieces, the *facets.* Each facet is the outer end of a unit known as an *ommatidium.* Such a structure gives mosaic vision as described in the crayfish (Fig. 10-8). Some insects, possibly the grasshopper, are able to distinguish colors.

Antennae. The antennae are filiform in shape and consist of many joints. Sensory bristles, probably olfactory in nature, are present on them and this, combined with the ability of the insect to bend them about, makes them efficient sense organs.

Thorax. The thoracic portion of the body is separated from the head and abdomen by flexible joints and consists of three segments: an anterior prothorax, a middle mesothorax, and a posterior metathorax. Each segment bears a pair of legs, and the mesothorax and metathorax each bear a pair of wings. On either side of the mesothorax and metathorax is a spiracle, an opening into the respiratory system.

Prothorax. A typical segment includes ten sclerites. The dorsal *tergum* (called the *pronotum* in this segment) consists of four sclerites in a row,

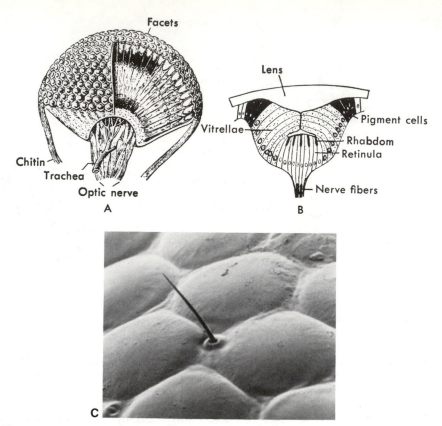

Figure 11-4 Insect eyes. (A) Diagram of compound eye. (B) Diagram of simple eye or ocellus. (C) Scanning electron micrograph of some facets of a cricket eye; 1,500×. (*A, from Schmeil; B, from Borradaile and Potts, modified; C, courtesy D. Buthala and R. Ulrich, Western Michigan University.*)

an anterior *prescutum* followed by the *scutum, scutellum,* and *postscutellum.* The lateral *pleura* consist of three sclerites, the *episternum, epimeron,* and *parapteron.* The single ventral sclerite is the *sternum.* The *pronotum* of the prothorax is large and extends down on either side; its four sclerites are indicated by transverse grooves. The sternum bears a spine.

Mesothorax. In this segment the tergum (*mesonotum*) is small, but the sclerites of the pleuron are distinct.

Metathorax. The metathorax resembles the mesothorax.

Legs (Fig. 11-5). Each leg consists of a linear series of five segments as follows: the *coxa* articulates with the body; then come the small *trochanter* fused with the *femur,* the *tibia,* and the *tarsus.* The femur of the

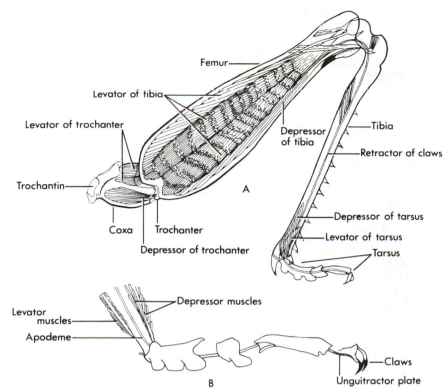

Femur

Levator of tibia

Levator of trochanter

Depressor
of tibia

Tibia

Retractor of claws

Trochantin

A

Coxa | Trochanter

Depressor of trochanter

Depressor of tarsus

Levator of tarsus

Tarsus

Levator
muscles

Apodeme

Depressor muscles

Claws

B

Unguitractor plate

Figure 11-5 Carolina locust, *Dissosteira carolina*. (A) Musculature of left, metathoracic leg. (B) Tarsus of mesothoracic leg disjointed, showing levator and depressor muscles and tendon-like apodeme of retractor of claws attached to unguitractor plate and extending through tarsus. (*After Snodgrass.*)

metathoracic legs are enlarged to contain the muscles used in jumping. The tarsus (Fig. 11-5, B) of each leg consists of three visible segments; the one adjoining the tibia has three pads on the ventral surface, and the terminal segment bears a pair of *claws* between which is a fleshy lobe, the *pulvillus*.

Wings (Figs. 11-6 and 11-7). The wings of insects arise from the region between the tergum and pleuron as a double layer of hypodermis, which secretes upper and lower cuticular surfaces. Between these are tracheae around which spaces occur and the cuticule thickens; they later become the longitudinal wing veins. The *veins* are of value in strengthening the wings. They differ in number and arrangement in different species of insects but are so constant in individuals of any one species that they are very useful for purposes of classification. The mesothoracic wings of the grasshopper, the *tegmina*, are leathery and not folded; they serve as covers for the metathoracic wings which lie beneath them. The latter are thin and folded like a fan.

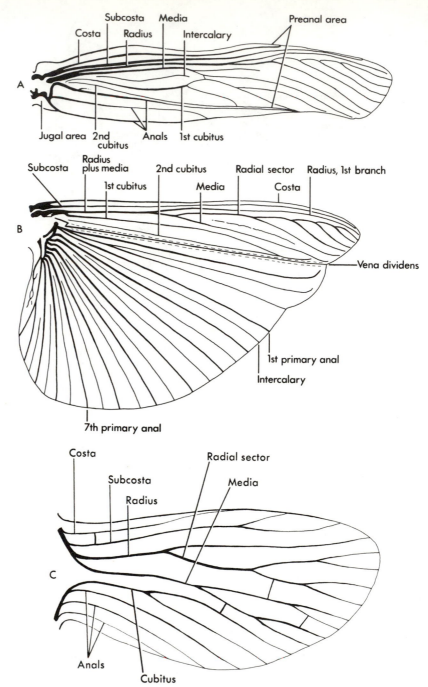

Figure 11-6 Principal wing veins. (A) Fore wing of the Carolina locust, *Dissosteira carolina*. (B) Hind wing of *D. carolina*. (C) Hypothetical wing of the presumed ancestral insect from which other winged forms are derived. (*A and B, after Snodgrass.*)

578

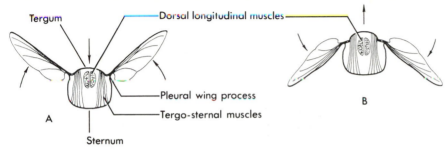

Tergum

Dorsal longitudinal muscles

Pleural wing process

Tergo-sternal muscles

A

Sternum

B

Figure 11-7 Movement of wings in flight. (A) The wings are elevated on the pleural wing processes by the depression of the tergum due to the contraction of the tergosternal muscles. The hind margins of the wings are deflected. (B) The wings are lowered by the elevation of the tergum due to the contraction of the dorsal, longitudinal muscles. Hind margins of wings elevated. (*After Snodgrass.*)

Abdomen. The number of segments in the abdomen of the embryo insect is 11 and each of these bears a pair of rudimentary appendages. In the adult of most insects no appendages are present and the number of segments is usually reduced. In the grasshopper segment I is fused with the metathorax; it consists of a tergum only, on either side of which is an oval *tympanic membrane* covering an *auditory sac.* Segments IX and X have their terga partly fused together and their sterna completely fused. Segment XI consists of a tergum only; this forms the so-called *genitalia* consisting in the male of a *subgenital plate,* two *podical plates,* and two *cerci,* and in the female of two podical plates, two cerci, and three pairs of movable plates, which form the *ovipositor* or egg-laying apparatus.

Internal Anatomy

The systems of organs within the body of the insect (Fig. 11-8) lie in the body cavity, which is filled with blood and is a *hemocoel* and not a coelom. All of the systems characteristic of higher animals are represented.

Muscular System (Fig. 11-9). The muscles are of the *striated* type, very soft and delicate but strong. The number present is very large. They are segmentally arranged in the abdomen but not in the head and thorax. The most conspicuous muscles are those that move the mandibles, the wings, the metathoracic legs, and the ovipositor.

Digestive System (Fig. 11-8). The principal parts of the alimentary canal are the *fore gut, mid gut,* and *hind gut.* The fore gut consists of the *pharynx* in the head into which the *mouth* opens and on either side of which opens a *salivary gland;* next a tubular *esophagus,* which enlarges into a *crop* in the mesothoracic and metathoracic segments; this leads

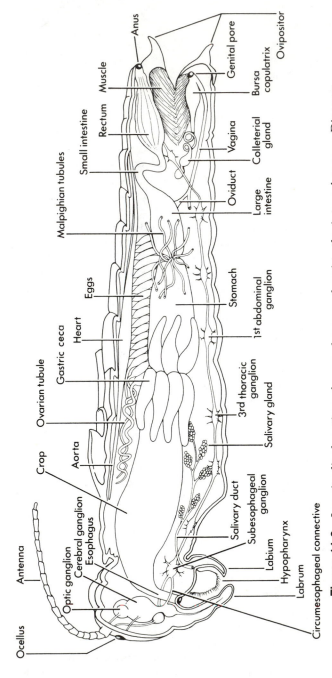

Figure 11-8 Longitudinal section of a grasshopper, showing the internal organs. Diagrammatic. (*After Root.*)

Ocellus

Antenna

Optic ganglion

Cerebral ganglion

Esophagus

Crop

Aorta

Ovarian tubule

Gastric ceca

Heart

Eggs

Malpighian tubules

Small intestine

Rectum

Muscle

Anus

Genital pore

Ovipositor

Bursa copulatrix

Colleterial gland

Vagina

Oviduct

Large intestine

Stomach

1st abdominal ganglion

3rd thoracic ganglion

Salivary gland

Salivary duct

Subesophageal ganglion

Labium

Hypopharynx

Labrum

Circumesophageal connective

580

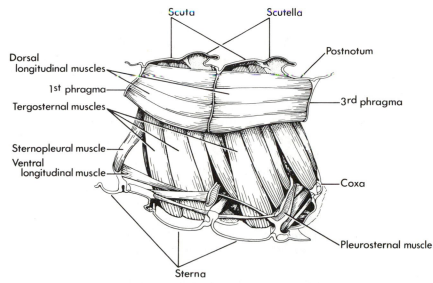

Scuta   Scutella

Dorsal longitudinal muscles

1st phragma

Tergosternal muscles

Sternopleural muscle

Ventral longitudinal muscle

Postnotum

3rd phragma

Coxa

Pleurosternal muscle

Sterna

Figure 11-9   Carolina locust, *Dissosteira carolina*. General view of the musculature in the right half of the mesothorax and metathorax. The dorsal longitudinal muscles, by contraction, arch the tergal plates upward and thereby depress the wings; the tergosternal muscles depress the tergal plates, which in-
directly elevate the wings. The other muscles are muscles of the legs and of the
sterna. (*After Snodgrass.*)

into the *proventriculus*, which is a grinding organ or *gizzard*. The mid
gut, which is the *ventriculus* or stomach, reaches posteriorly into the
abdomen, and into it eight double, cone-shaped pouches, the *gastric
ceca*, pour the digestive juices they secrete. The hind gut is made up of
the *large intestine*, into the anterior end of which the delicate *Malpighian tubules* open; and the *small intestine*, which expands into the
rectum and opens through the *anus*.

Circulatory System (Fig. 11-8). This is not a closed system of blood
vessels as in vertebrates and some invertebrates but consists of a single tube located in the abdomen just under the body wall in the middorsal line. The *heart* is divided into a row of chambers into the base
of each of which opens a pair of *ostia*. These ostia are closed by *valves*
when the heart contracts. The *pericardial sinus* in which the heart lies is
formed by a horizontal partition beneath it. At its anterior end the
heart forms an *aorta* which opens into the body cavity or hemocoel in
the head region. Blood is forced anteriorly through the heart and aorta
into the hemocoel where it bathes all the organs. The blood carries nutriment but does not play a large part in respiration as it does in most
of the Metazoa because of the highly developed tracheal system. It
consists of a *plasma* in which are suspended white blood cells, the
leucocytes.

Respiratory System (Fig. 11-10). The respiratory system consists of a network of ectodermal tubes, the *tracheae,* that communicate with every part of the body. The tracheae consist of a single layer of cells enclosing a chitinous lining, which, in the larger tubes, forms a *spiral thread* that prevents the trachea from collapsing. The tracheae extend from the *spiracles* (Fig. 11-11) to a longitudinal trunk on either side of the body. The finest tracheae, the *tracheoles,* are connected directly with the tissue to which they supply oxygen and from which they carry away carbon dioxide, thus assuming one function performed in other animals by the circulatory system. The small blind endings of the tracheoles, on the muscles and other organs, are filled with fluid. During activity of the muscle the concentration of substances in the body fluid around the tracheoles increases. This causes diffusion of water from the tracheole into the surrounding area, thus bringing oxygen into closer proximity to the site where it is being used as the air moves farther down into the blind tip of the tracheole. After activity

Figure 11-10   Respiratory system of insects. (A) Diagram of tracheae in the body of a beetle. (B) Part of a trachea of a caterpillar. (C) Tracheae, air sacs, and spiracles of a grasshopper. (*A, after Kolbe; B, after Leydig; C, after Vinal.*)

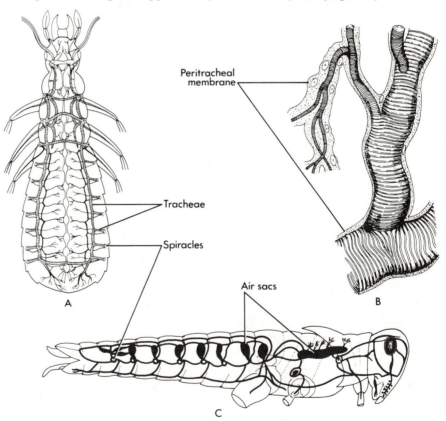

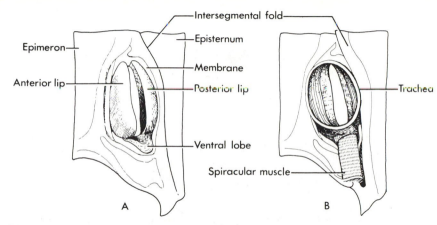

Epimeron

Anterior lip

Intersegmental fold

Episternum

Membrane

Posterior lip

Trachea

Ventral lobe

Spiracular muscle

A   B

Figure 11-11 Carolina locust, *Dissosteira carolina*. (A) Left spiracle of mesothoracic segment, outer view. (B) Right spiracle of mesothoracic segment, inner view. (*After Snodgrass.*)

stops, the metabolic products that changed the osmotic pressure are disposed of and the water returns to the tracheole. In the grasshopper and certain other insects some of the tracheae become expanded into thin-walled air sacs, which are easily compressed and thus aid in the movement of air. Contraction and expansion of the abdomen draw air into and expel it from the tracheal system. In the grasshopper the first four pairs of spiracles are open at inspiration and closed at expiration, whereas the other six pairs are closed at inspiration and open at expiration.

Excretory System. The organs of excretion are the *Malpighian tubules* that are coiled about in the hemocoel and open into the anterior end of the hind gut. Uric acid is the principal nitrogenous end product of protein metabolism in insects. It is of very low solubility. The ability of the Malpighian tubule to excrete uric acid with very little water loss in the process is an important factor in the success of insects in terrestrial environments. Potassium ions from potassium carbonate in the body fluid combine with uric acid under alkaline conditions to form the soluble potassium urate. Potassium urate diffuses into the Malpighian tubule and moves with the fluid in it toward the lumen of the intestine. Near the intestine the pH in the tubule shifts to acid and insoluble uric acid and soluble potassium carbonate are reformed. The latter and water are returned to the body fluid while the uric acid is carried on into the gut with a portion of the fluid. Further uptake of water may occur in the gut leaving the feces quite dry with crystals or uric acid included.

Nervous System (Figs. 11-8, 11-12, and 11-13). There is a *brain*, dorsally located in the head, consisting of three pairs of ganglia fused

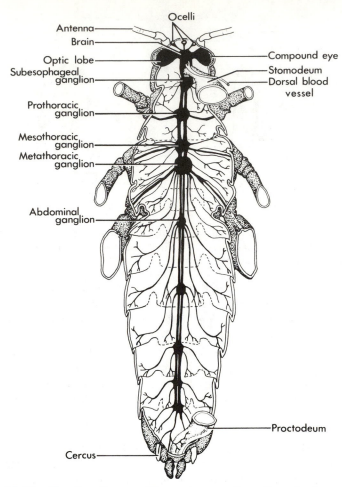

Figure 11-12 Carolina locust, *Dissosteira carolina*. Dorsal view, showing nervous system. (*After Snodgrass.*)

together (Fig. 11-13). These ganglia supply the eyes, antennae, and labrum. The brain is connected by a pair of *circumesophageal connectives* with a *subesophageal ganglion*. This ganglion consists of the three anterior pairs of ganglia of the ventral nerve chain fused together and supplies the mouth parts. The *ventral nerve chain* continues with a pair of large ganglia in each thoracic segment. The ganglia in the metathoracic segment are particularly large and represent the ganglia of this segment and of the first abdominal segment fused. Five pairs of ganglia are present in the abdomen. The pair in the second abdominal segment comprise the pairs from the second and third abdominal somites fused together and the pair in the seventh segment represent the ganglia of the seventh to the eleventh somites combined. Connected

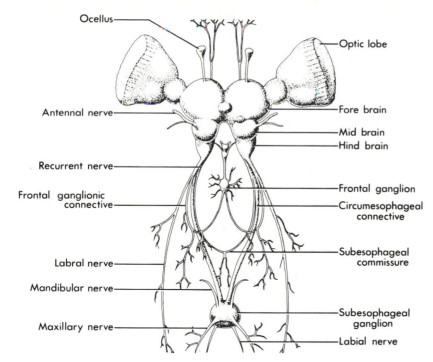

Ocellus
Optic lobe
Antennal nerve
Fore brain
Mid brain
Hind brain
Recurrent nerve
Frontal ganglion
Frontal ganglionic connective
Circumesophageal connective
Subesophageal commissure
Labral nerve
Mandibular nerve
Subesophageal ganglion
Maxillary nerve
Labial nerve

Figure 11-13 Carolina locust, *Dissosteira carolina*. The brain and subesophageal ganglion and their nerves, as seen after removal of the facial wall of the head. (*After Snodgrass.*)

with the brain are ganglia of the *sympathetic nervous system,* which supplies the muscles of the alimentary canal and spiracles.

Sense Organs. Grasshoppers possess organs of sight, hearing, touch, taste, and smell. The *compound eye* and *ocellus* have already been noted. Vision by means of the compound eyes has been described in the crayfish. The ocelli probably do not perceive objects but are merely organs of light perception. The pair of *auditory organs* (Fig. 11-14) are located on the sides of the tergite of the first abdominal segment. They consist of a *tympanum* stretched within an almost circular chitinous ring. The *antennae* (Figs. 11-15 and 11-16) are supplied with the principal organs of smell. Organs of *taste* are located on the mouth parts. The hairlike organs of *touch* are present on various parts of the body but particularly on the antennae.

Reproductive System. Female grasshoppers can easily be distinguished from males because of the presence of the *ovipositor* (Fig. 11-8). The female has two *ovaries.* Each consists of a number of filaments usually called *ovarian tubules* which, however, do not possess a lumen

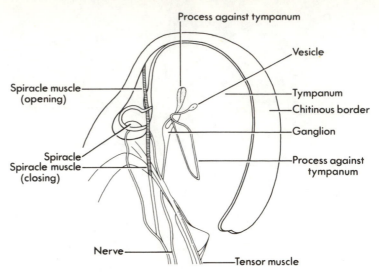

Figure 11-14 Auditory organ of a grasshopper. (*After Graber.*)

(Fig. 11-17). The ovarian filaments contain *oogonia* and *oocytes* arranged in a linear series, *nurse cells,* and other tissue cells. The oocytes grow as they proceed posteriorly down the filament, hence the filament becomes gradually larger toward the posterior end. The filaments of each ovary are attached posteriorly to an *oviduct* into which the eggs are discharged. The two oviducts unite to form a short *vagina,* which leads to the *genital opening* between the plates of the ovipositor.

Figure 11-15 Sense organs of insects. Mexican bean weevil. (A) Antenna, dorsal surface. (B) Antenna, section through tenth segment. (*After McIndoo.*)

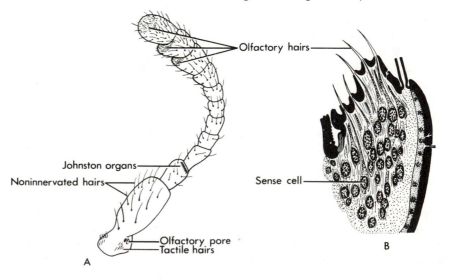

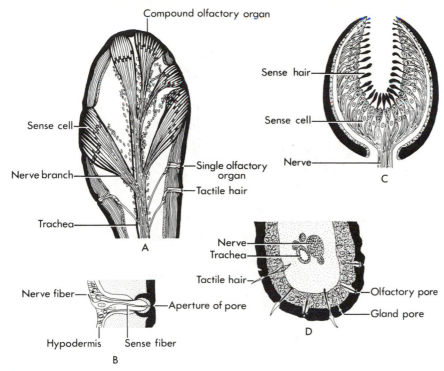

Figure 11-16 Sense organs of insects. (A) Antenna of larva of a beetle, *Cotinis nitida*, longitudinal section of tip. (B) Single olfactory organ from antenna of *Cotinis*, cross section. (C) Sense organ, probably static or balancing, in labial palp of codling moth. (D) Antenna of Mexican bean weevil, cross section of first segment. (*After McIndoo.*)

A tubular *seminal receptacle*, or spermatheca, which opens dorsal to the vaginal pore, receives the *spermatozoa* during copulation and releases them when the eggs are fertilized. In the male are two *testes* in which spermatozoa develop (Fig. 11-18). These are discharged into a *vas deferens.* The two vasa deferentia unite to form an *ejaculatory duct,* which opens on the dorsal surface of the *subgenital plate. Accessory glands* are present at the anterior end of the ejaculatory duct, which apparently secrete a fluid that aids in the transfer of spermatozoa to the female.

Embryonic Development and Growth

The eggs are fertilized at the time they are deposited, by the entrance of spermatozoa through an opening in one end of the egg shell called the *micropyle.* One sperm nucleus unites with the nucleus of the mature egg; a *blastoderm* is formed around the periphery of the egg from which an *embryo* develops. The young grasshopper that hatches from the egg is called a *nymph* (Fig. 11-19). It resembles its parent but has a

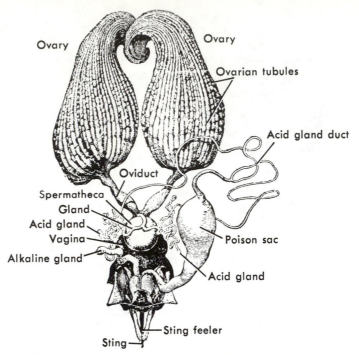

Figure 11-17 Reproductive organs, sting, and poison gland of queen honey bee. (*From Snodgrass.*)

Figure 11-18 Reproductive organs of drone honey bee, dorsal view, natural position. (*From Snodgrass.*)

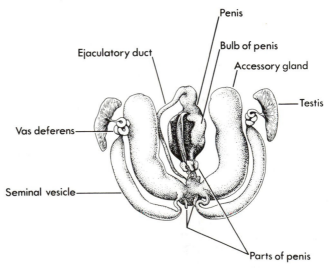

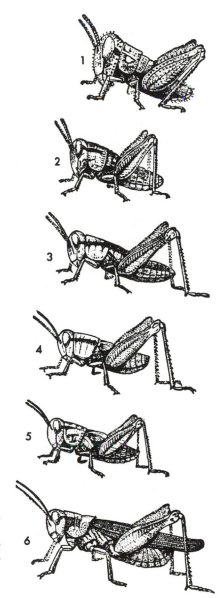

Figure 11-19 Metamorphosis of a grasshopper, *Melanoplus femur-rubrum,* showing the five nymph stages and the gradual growth of the wings. Earlier stages drawn greatly enlarged. (*From Packard, after Emerton.*)

large head compared with the rest of the body and lacks wings. As it grows its body becomes too large for the inflexible chitinous exoskeleton and the latter is shed periodically. Wings are gradually developed from wing buds and the adult condition is finally assumed. This type of development is called *simple metamorphosis* in contrast to the *complete metamorphosis* of many insects which involves larval and pupal stages.

The Insecta are air-breathing Arthropoda with bodies divided into head, thorax, and abdomen. The head bears one pair of antennae and the thorax three pairs of legs and usually one or two pairs of wings in the adult stage. Insects are more numerous in species than all other animals taken together; over 700,000 species have been described and no doubt a few hundred thousand remain to be discovered. They live in almost every conceivable type of environment on land and in the water and their structure, habits, and life cycles are correspondingly modified. Nevertheless, it is possible to separate this vast assemblage into orders, families, etc., although there is no unanimity of opinion with respect to the number of these that should be recognized and to the names that should be applied to them. The following classification into subclasses and orders is one commonly used by entomologists in this country.

The order names have only been partially modified to conform with the -ida endings used for orders. The -ida ending is used here for wingless groups. The -ptera ending for winged orders is so well established they were not modified. The other names left unchanged memorialize the contributions of many early zoologists, including Aristotle, in at least a token form.

Class Insecta

SUBCLASS 1. APTERYGOTA.   Ametabolous: no metamorphosis. Usually included in this subclass is an order of small eyeless insectlike invertebrates that lack antennae, the Proturida. They are whitish, up to 1.5 mm long, and live in soil and decaying vegetation. Wingless.

    ORDER 1. THYSANURIDA.   Bristletails, silverfish, etc.

    ORDER 2. COLLEMBOLIDA.   Springtails.

SUBCLASS 2. PALEOPTERYGOTA.   Wings of adults do not fold flat over or against abdomen.

    ORDER 1. EPHEMEROPTERA.   Mayflies.

    ORDER 2. ODONATA.   Dragonflies.

SUBCLASS 3. EXOPTERYGOTA (NEOPTERA, in part).   Hemimetabolous: simple metamorphosis.

SUPERORDER 1. ORTHOPTEROIDEA.   With cerci.

    ORDER 1. ORTHOPTERA.   Cockroaches, crickets, grasshoppers, locusts, etc.

    ORDER 2. ISOPTERA.   Termites or white ants.

    ORDER 3. EMBIOPTERA.   Web-spinners.

    ORDER 4. PLECOPTERA.   Stoneflies.

    ORDER 5. DERMAPTERA.   Earwigs.

    ORDER 6. ZORAPTERA.   Zorapterans.

SUPERORDER 2. HEMIPTEROIDEA.   Without cerci.

    ORDER 1. PSOCOPTERA.   Book lice, etc.

ORDER 2. THYSANOPTERA. Thrips.
ORDER 3. HOMOPTERA. Plant lice, cicadas, etc.
ORDER 4. HEMIPTERA. Bugs.
ORDER 5. MALLOPHAGIDA. Bird lice.
ORDER 6. ANOPLURIDA. Sucking lice.
SUBCLASS 4. ENDOPTERYGOTA (NEOPTERA, in part). Holometabolous: complete metamorphosis.
ORDER 1. NEUROPTERA. The dobson, aphis-lions, etc.
ORDER 2. COLEOPTERA. Beetles.
ORDER 3. STREPSIPTERA. Stylopids.
ORDER 4. MECOPTERA. Scorpionflies, etc.
ORDER 5. TRICHOPTERA. Caddisflies.
ORDER 6. LEPIDOPTERA. Moths, skippers, and butterflies.
ORDER 7. DIPTERA. Flies.
ORDER 8. SIPHONAPTERA. Fleas.
ORDER 9. HYMENOPTERA. Bees, wasps, ants, etc.

Structural Modifications

Before discussing the characteristics of these orders it seems desirable to point out certain structural modifications that may be encountered in the various groups. The antennae, mouth parts, legs, and wings are among the most interesting external feature of insects.

Wings. The mesothorax and metathorax bear each a pair of wings in most insects. Certain simple species (Thysanurida, Fig. 11-24) do not possess wings; others (lice and fleas, Figs. 11-45 and 11-69) have no wings, but this is because they are degenerate. The flies (Diptera, Fig. 11-61, B) have a pair of clubbed threads, called balancers of halteres, in place of the metathoracic wings. Attached to each thoracic segment is a pair of legs. Wings enable their owners to fly rapidly from place to place and thus to escape from enemies and to find a bountiful food supply. The success of insects in the struggle for existence is in part attributed to the presence of wings. Wings are outgrowths of the skin strengthened by a framework of chitinous tubes, called veins or nerves, which divide the wing into cells. The veins vary in distribution in different species, but are quite constant in individuals of any given species; they are consequently used to a considerable extent for purposes of classification. From comparisons of various wings a hypothetical ancestral wing was derived with the following principal longitudinal veins in sequence from anterior to posterior. (1) *Costa;* unbranched and at, or near, the leading edge of the wing. (2) *Subcosta:* with a terminal fork, near the basal end it may be connected to the costa by the *humeral* crossvein. (3) *Radius:* with two branches, the first branch unbranched, the second is called the *radial sector* and forks twice (Fig. 11-6). (4) *Media: forks twice to produce four branches.* (5) *Cubitus:* forks to produce two or three branches. (6) *Anal:* unbranched veins numbered 1st, 2nd, etc. from anterior to posterior. Crossveins occur as indicated in (Fig. 11-6, C). Modifications come about by reduction or by addition. In the beetles (Coleoptera) the fore wings are sheathlike and are called elytra. The fore wings of Orthoptera (grasshoppers, etc.) are leathery and are known as tegmina.

Legs. The legs of insects are used for various purposes and are highly modified for special functions, for example, those of the honey bee (Fig. 11-77). A typical leg consists of five parts—coxa, trochanter, femur, tibia, and tarsus. The tarsus (Fig. 11-5) is usually composed of five segments and bears at the end a pair of claws, between which is a fleshy lobule, the pulvillus. Running insects possess long, slender legs (Fig. 11-47); the mantis has its fore legs fitted for grasping (Fig. 11-29); the hind legs of the grasshopper are used in leaping (Fig. 11-5); the fore legs of the mole cricket are modified for digging (Fig. 11-32, C); and the hind legs of the water beetle are fitted for swimming (Fig. 11-48, A). Many other types could be mentioned.

Mouth Parts. The mouth parts of insects are in most cases fitted either for biting (mandibulate) or sucking (suctorial). The grasshopper possesses typical mandibulate mouth parts (Fig. 11-3). The mandibles of insects that live on vegetation are adapted for crushing; those of carnivorous species are usually sharp and pointed, being fitted for bitting and piercing. Suctorial mouth parts are adapted for piercing the tissues of plants or animals and sucking juices. The mouth parts of the honey bee (Fig. 11-76) are suctorial, but highly modified. In the female mosquito (Fig. 11-20) the labrum and epipharynx combined form a sucking tube; the mandibles and maxillae are piercing organs; the hypopharynx carries saliva; and the labium constitutes a sheath in which the other mouth parts lie when not in use. The proboscis of the butterflies and moths is a sucking tube formed by the maxillae.

The mouth parts of insects are of considerable importance from an economic standpoint, because insects that eat solid food can be destroyed by spraying the food with poisonous mixtures, whereas those that suck juices must be smothered with gases or have their spiracles closed with emulsion.

Antennae. The antennae of insects are usually tactile, olfactory, or auditory in function. They differ widely in form and structure (Fig. 11-21) Often the antennae of the male differ from those of the female.

Alimentary Canal. Of the internal organs of insects the alimentary canal and respiratory systems are of particular interest. The alimentary canal is modified according to the character of the food. An insect with mandibulate mouth parts (Fig. 11-8; 11-22, A–D) usually possesses (1) an esophagus, which is dilated to form a crop in which food is stored,

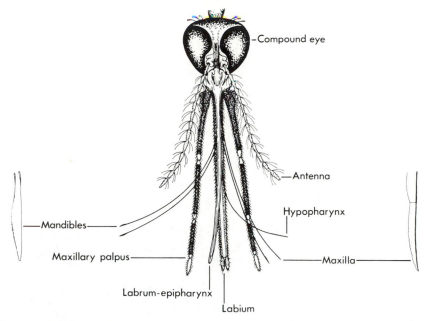

Figure 11-20 Sucking mouthparts of an anopheline mosquito (*Anopheles costalis*), female. (*After Carter.*)

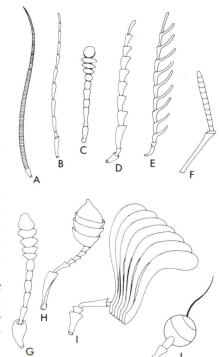

Figure 11-21 Different forms of antennae of insects. (A) Bristlelike antenna of a grasshopper, *Locusta*. (B) Filiform, of a beetle, *Carabus*. (C) Moniliform, of a beetle, *Tenebrio*. (D) Dentate, of a beetle, *Elater*. (E) Pectinate, of *Ctenicera*. (F) Crooked, of honey bee. (G) Club-shaped, of beetle *Silpha*. (H) Knobbed, of beetle, *Necrophorus*. (I) Lamellated, of beetle, *Melolontha*. (J) With bristle, from fly, *Sargus*. (*From Sedgwick, after Burmeister.*)

593

(2) a muscular gizzard or proventriculus, which strains the food and may aid in crushing it, (3) a stomach or ventriculus into which a number of glandular tubes (gastric ceca) pour digestive fluids, and (4) an intestine with urinary or Malpighian tubules at the anterior end. Suctorial insects, like the butterflies and moths (Fig. 11-22, E), are

Figure 11-22 Alimentary canals of insects. (A-D, scanning electron micrographs of the cricket, a chewing insect, alimentary canal.) (A) Crop, freeze-fractured; 3,000×. (B) Inner surface of proventriculus; 75×. (C) Inner surface of midgut; 300×. (D) Inner surface of hindgut; 275×. (E) Diagram of the alimentary canal of a sucking insect, the moth, *Sphinx*. (*A–D, courtesy of R. Ulrich and D. Buthala, Western Michigan University; E, after Wagner, modified.*)

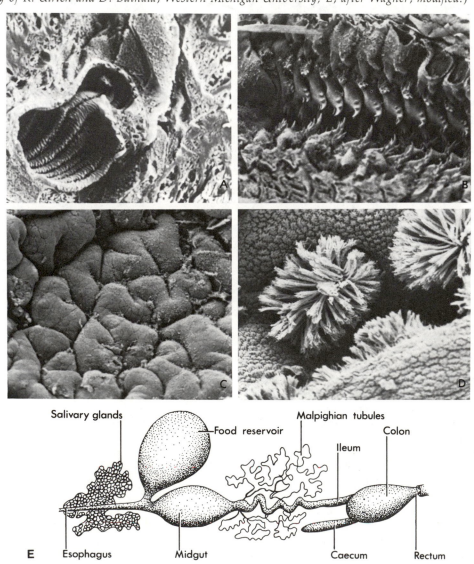

provided with a muscular pharynx which acts as a pumping organ and a sac for the storage of juices.

Respiratory System. The respiratory system of insects is in general like that of the grasshopper (Fig. 11-10), but modifications occur in many species, especially in the larvae of those that live in water. Aquatic larvae, in many cases, do not have spiracles, but get oxygen by means of threadlike or leaflike cuticular outgrowths at the sides or posterior end of the body, termed tracheal gills (Fig. 11-27, B). Damselfly larvae possess caudal tracheal gills, and the larvae of dragonflies take water into the rectum which is lined with papillae abundantly supplied with tracheae.

Other structural modifications, peculiar habits, and life cycles are noted in the discussion of the orders.

Embryonic Development

The embryonic development (Fig. 11-23) of insects is a fascinating study. Eggs of the potato beetle, for example, consist largely of yolk globules with a superficial layer of cytoplasm and thin cytoplasmic strands among the yolk (Fig. 11-23, A). The single fusion nucleus lies in a small island of cytoplasm. Nuclear division is not followed immediately by cell division but hundreds of nuclei are produced before cells are formed (C). Most of these nuclei migrate to the periphery and fuse with the superficial layer of cytoplasm; cell membranes appear and a blastoderm of a single layer of cells (B) covers the entire egg. The rest of the nuclei remain behind among the yolk globules which it is their duty to dissolve. The blastoderm thickens on one side of the egg forming the germ band in which appears a ventral groove (D). This groove is grown over by the amnioserosal fold and the germ band then becomes divided into segments. Appendages that are destined to become mouth parts and legs grow out from the cephalic and thoracic segments (E, F). The insect egg is very highly organized; certain parts of the potato-beetle egg are destined to produce certain definite parts of the embryo even before the blastoderm is formed as can be demonstrated by killing parts of the fresh egg with a hot needle and allowing the rest of the egg to develop. At the posterior end of the eggs of certain insects, especially Diptera, Coleoptera, and Hymenoptera, is a mass of granules that stain deeply with haematoxylin and appear to play an important part in the origin of the primordial germ cells; for this reason they are known as germ-cell determinants (A). These granules have been traced into the germ cells (B) and the latter have been followed from one generation to the next, illustrating very clearly the theory of the continuity of the germ plasm.

Growth and Metamorphosis

The eggs of insects are of various shapes and sizes. Some of them are colored; others are marked with polygonal areas or ridges. Three types

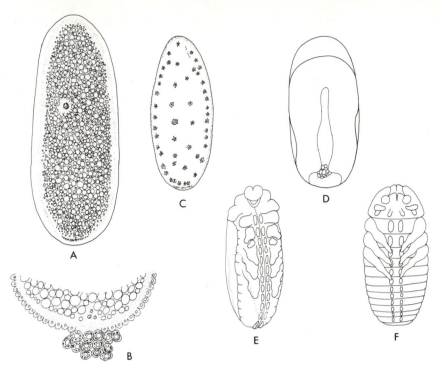

Figure 11-23 Embryonic development of a beetle. (A) Longitudinal section of freshly laid egg showing fusion nucleus, yolk globules, peripheral layer of cytoplasm, and, at the lower, posterior end, the pole disk of granular "germ-cell determinants." (B) Blastoderm stage at end of 24 hours. The primitive germ cells have become extruded at the posterior end. (C) Preblastoderm stage at end of 20 hours, showing cleavage nuclei at periphery. (D) Germ band with ventral groove; primitive germ cells near the posterior end. (E) Embryo with segmentation in progress and appendages of head and thorax growing out. (F) Older embryo with mouthparts and legs further developed. (*From Hegner.*)

of insects may be distinguished with respect to the method of their development: (1) ametabola, (2) hemimetabola, and (3) holometabola.

The ametabolic insects are essentially like the adult, except in size, when they hatch from the egg; they develop to maturity without a metamorphosis. The Thysanurida (Fig. 11-24) and other Apterygota are ametabolic.

The hemimetabolic insects hatch from the egg and develop into adults as in the grasshopper without passing through a true pupal period. Many of the species belonging to the hemimetabolic orders change considerably during the growth period, but are all more or less active throughout their development and are said to undergo simple metamorphosis.

Holometabolic insects, such as the butterflies (Fig. 11-60), pass through both a larval and a pupal stage in their development and are said to undergo complete metamorphosis. The majority of insects belong to this type.

Subclass 1. Apterygota

These wingless insects have no metamorphosis. Immature individuals in the early intermolt stages (instars) resemble the adults.

Some entomologists think the Collembolida should be lumped with the proturans and diplurans as the entognathous hexapods outside the insects proper (Atkins, 1978). Their enclosed mouthparts and simple structure may be indication of adaptation for litter and soil dwelling. Reduction from more complex insects is considered more likely here, hence they are placed after the Thysanurida.

Order 1. Thysanurida

Bristletails (Fig. 11-24, A and B). Primitive wingless insects; ametabolic; chewing mouth parts; 11 abdominal segments; usually two or three long, filiform, segmented, caudal appendages; less than 20 species known from the United States. The pair of caudal appendages, or the outer two, if three are present, are cerci; the antennae are long, filiform, and contain many segments; in some the compound eyes are degenerate or absent; in many the jaws are sunk into a cavity in the head; a few possess appendages of two segments, the styli, on the ventral abdominal segments, which may be remnants of legs of the many-legged ancestors of insects.

Figure 11-24 (A) Thysanurida. *Lepisma saccharina*, the silverfish. (B) Thysanurida. *Campodea staphylinus*. (C) Collembolida. Springtail. (*A, from Wellhouse, after Kellogg; B, after Lubbock; C, after Miall.*)

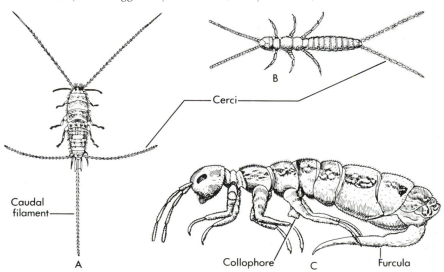

The silverfish or fishmoth, *Lepisma saccharina* (Fig. 11-24, A), is a common species in houses, especially in the warmer parts of the country. It is covered with beautifully marked, shining scales, and often damages books, clothing, etc., because of its fondness for starch. *Campodea staphylinus* (Fig. 11-24, B) is a delicate, whitish species that lives under stones and leaves, and in rotten wood and other damp places. The firebrat, *Thermobia domestica,* prefers a very warm environment such as the vicinity of stoves and fireplaces. *Japyx* has pincerlike caudal appendages and *Machilis maratima* is a type that bears abdominal appendages.

Order 2. Collembolida

Springtails (Fig. 11-24, C). Primitive wingless insects; ametabolic; chewing or sucking mouth parts; four to six segments in the antennae; usually no tracheae; compound eyes absent; Malpighian tubules absent; tarsi absent; six abdominal segments; a springing organ present in most species on the ventral side of the fourth abdominal segment. The springing organ, or furcula, propels the insect into the air when it is released from a catch, the hamula, on the third abdominal segment; mandibles are often overgrown by the cheeks; in some mandibles and maxillae are modified into piercing organs; on the ventral surface of the first abdominal segment is a tubelike projection, or collophore, which holds the insects to the undersurface of an object with a sticky secretion from the labium.

Collembolida are very small insects, often microscopic in size. They are to be found in the crevices of bark, in moss, under stones, leaves, and wood where they feed on decaying matter. The snowflea, *Archorutes nivicolus,* sometimes occurs in winter on the snow in large numbers and is often a nuisance in maple sugar camps. The garden flea, *Bourletiella hortensis,* may become a pest because of its fondness for young vegetables, such as cabbage, cucumbers, turnips, and squashes.

Subclass 2. Paleopterygota

Insects with winged adults incapable of folding the wings in a layered position over the abdomen. Aquatic nymphs with simple metamorphosis. Wing buds evident on older nymphs.

Order 1. Ephemeroptera (*Ephemerida*)

Mayflies (Fig. 11-25). Hemimetabolic; mouth parts of adult, vestigial; two pairs of membranous, triangular wings, the fore wings larger than the hind wings; caudal filament and cerci very long. The family, Ephemeridae, belonging to this order contains a number of species of delicate insects known as mayflies. The larvae live in the water and breathe by means of tracheal gills (Fig. 11-25, B), usually located in a

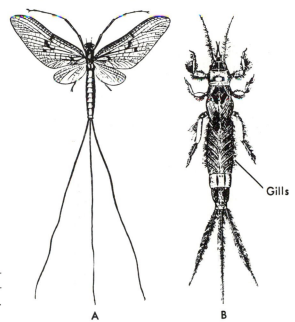

Figure 11-25 Ephemeroptera. *Ephemera*, a mayfly. (A) Adult. (B) Nymph. (*From Kennedy.*)

Gills

A   B

row on either side of the abdomen; these are outgrowths of the integument containing tracheal branches. Oxygen is obtained by them from the water. The adults live only a short time, hence the derivation of the name Ephemeroptera from the Greek word *ephemeros,* meaning lasting but a day. Frequently large numbers of adults emerge from the water at about the same time and huge swarms of them suddenly appear; for example, *Ephemera simulans,* around the lakes of the northern United States. The adults when they leave the water are called subimagos, because they molt within a few minutes to 24 hours after they acquire functional wings; they are the only insects that molt in the winged stage.

Order 2. Odonata

Dragonflies and dameselflies (Figs. 11-26 and 11-27). Hemimetabolic; chewing mouth parts; two pairs of membranous wings, the hind wings as large as or larger than the fore wings; large compound eyes; small antennae; nymphs aquatic; both nymphs and adults, predacious. The dragonflies (suborder Anisoptera, Fig. 11-26) are large insects. The wings are usually held in a horizontal position when at rest and the eyes may be made up of larger facets above and smaller ones below. The nymphs breathe by means of rectal gills, which line the enlarged posterior end of the alimentary canal, and extract oxygen from the water that is drawn in and expelled from this cavity. The labium of the larva is much elongated and can be extended rapidly from its folded resting position beneath the head so as to impale its

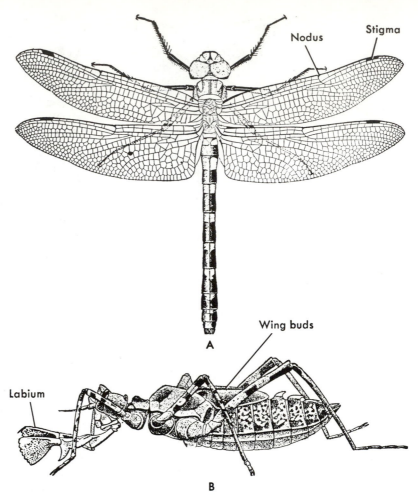

Figure 11-26 Odonata. *Macromia magnifica*, a dragonfly. (A) Adult. (B) Nymph. (*From Kennedy.*)

prey on the hooks at the end. *Anax junius* is a common, widely spread species of the family Aeschnidae. The damselflies (suborder Zygoptera, Fig. 11-27) are smaller. The wings are parallel with the abdomen when at rest, and are delicate compared with those of the dragonflies. *Agrion maculatum* is a common, dark-colored species partial to streams in woods.

Subclass 3. Exopterygota

These winged insects have a simple metamorphosis. Late nymphal instars of winged species show wing buds developing externally.

The exopterygote insects form two clusters: (1) the *paleopterous* insects, including only the Ephemeroptera and Odonata, are thought to

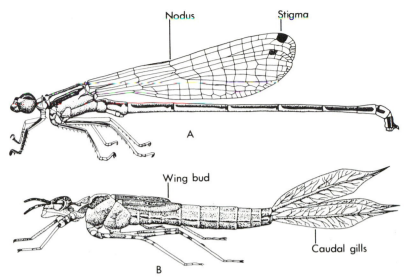

Figure 11-27 *Ischnura cervula*, a damselfly. (A) Adult, male. (B) Nymph. *(From Kennedy.)*

be remnants of the primitive type of winged insects, now mostly extinct; (2) the remaining *neopterous* exopterygote insects have more advanced wing musculature and articulations allowing wing folding over the abdomen. They fall into two groups: (1) the *orthopteroids* include those with many cross-veins, cerci, usually chewing mouthparts, and enlarged hindwings; (2) the *hemipteroids* include those with few cross-veins, no cerci, mouthparts often modified from the chewing type, hindwings equal to or smaller than forewings, and nymphs lack ocelli.

Superorder 1. Orthopteroidea. With cerci.

Order 1. Orthoptera

Grasshoppers, crickets, cockroaches, katydids, locusts, walking sticks, and mantids. Hemimetabolic; chewing mouth parts; typically two pairs of wings, the fore wings often thickened and parchmentlike and called tegmina (singular, tegmen), the hind wings folded like a fan beneath the fore wings; in some, wings are vestigial or absent. Six families in this order contain most of the familiar orthopterans.

Four families are sometimes separated as separate orders because of their distinctive appearances, the Phasmidae as Phasmida or Phasmatodea, the Mantidae as Mantida or Mantodea, the Blattidae as Blattida or Blattodea, and the Grylloblattidae as Grylloblattida or Grylloblattodea. The grylloblattids are wingless, high-altitude, and cold-dwelling insects with similarities to crickets and roaches.

FAMILY 1. PHASMIDAE (Fig. 11-28). Walking sticks and leaf insects. These are herbivorous insects. Some of them are wingless and have a

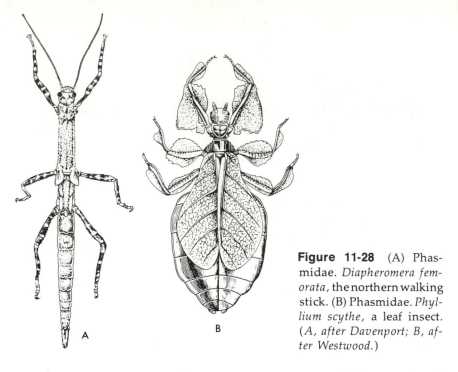

Figure 11-28 (A) Phasmidae. *Diapheromera femorata*, the northern walking stick. (B) Phasmidae. *Phyllium scythe*, a leaf insect. (*A, after Davenport; B, after Westwood.*)

slender body, and long legs and antennae, which give them a sticklike appearance. They are not active and move slowly. Their resemblance to the twigs on which they rest and their habit of feigning death, furnish an example of mimicry. *Diapheromera femorata* (Fig. 11-28, A) is the common walking stick of the northern United States. *Aplopus mayeri* is our only winged phasmid; it occurs in Florida. In the tropics are many species with expansions of the abdomen and leg segments and with wings that give them a close resemblance to leaves.

FAMILY 2. MANTIDAE (Fig. 11-29). Mantids. The mantids are carnivorous, feeding on other insects. The prothorax is much elongated and the prothoracic legs are modified as grasping organs. Spines are present on the tibiae and femora and the former can be bent back so as

Figure 11-29 Mantidae. *Stagmomantis carolina*, the praying mantis. (*After Packard.*)

to hold any insect grasped by them. As they lie in wait for their prey, the fore legs are raised in an attitude of prayer, hence the common name, praying mantis. *Stagmomantis carolina* (Fig. 11-29) is a species that occurs as far north as Maryland and southern Indiana. Other species live in the southern United States; many of those that live in the tropics have wings that resemble the leaves of plants in shape and color.

FAMILY 3. BLATTIDAE (Fig. 11-30). Cockroaches. These are omnivorous insects with dorso-ventrally flattened body, head bent down beneath the thorax, long antennae, and legs adapted for running. They are often a nuisance in houses. In nature, they live under stones, sticks, and other objects. The eggs are enclosed in capsules, the oothecae. Some forms are wingless. Three species that occur in the United States are *Periplaneta americana* (Fig. 11-30, A), the American cockroach; *Blattella germanica* (C), the croton-bug, which was introduced from Europe; and *Blatta orientalis* (B), the oriental cockroach, which probably reached America from Asia.

FAMILY 4. TETTIGONIIDAE (Fig. 11-31). Long-horned grasshoppers. The antennae of this group are very long and slender; the tarsi have four segments; the ovipositor is long and sword-shaped; the body is often green and is hence inconspicuous; the auditory organs are located in the base of the tibiae of the prothoracic legs. In many species the males have certain veins and cells of each wing-cover near the base so modified that when rubbed together they vibrate and produce a sound. The katydid, *Pterophylla camellifolia*, owes its name to its rasping call notes. Meadow grasshoppers, such as *Conocephalus*, are light-green forms that live among the grass in meadows. In a related family, Gryllacrididae, are certain cricketlike species, such as *Ceuthophilus*, known as cave crickets or camel crickets; they live in dark moist places in caves, cellars, under stones and wood, etc. The shield-backed grass-

Figure 11-30 Blattidae. (A) *Periplaneta americana*, the American cockroach. (B) *Blatta orientalis*, the "black beetle"; nymph. (C) *Blattella germanica*, the "croton bug." (*A and B, from Herrick; C, from Essig.*)

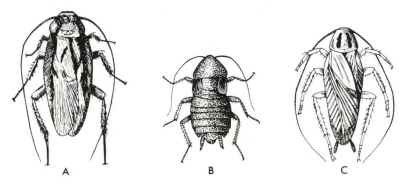

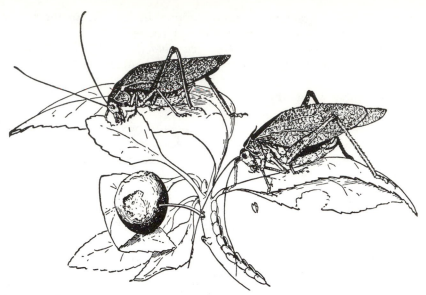

Figure 11-31 Tettigoniidae. *Microcentrum rhombifolium,* a katydid; male and female are shown feeding on apple leaves; eggs are deposited along stem. (*From Essig.*)

hoppers are mostly wingless or nearly wingless species, such as the genus *Atlanticus,* and also resemble crickets somewhat in appearance. They live in grassy fields or open woods.

Family 5. Gryllidae (Fig. 11-32). Crickets. Most crickets have long antennae, metathoracic legs for leaping, and a spear-shaped ovipositor. The males have a highly differentiated stridulating apparatus consisting of a file on the base of one tegmen and a scraper on the other. When the wings are held up over the body, the file is rubbed over the scraper as the wings vibrate and produce the characteristic chirp. Many of the tree-crickets are light green and live in trees. *Oecanthus niveus,* the snowy tree-cricket, is a common species. When a number of individuals are close together they chirp in unison. Field crickets are brown and are abundant in fields, where they usually hide in the daytime under stones, etc., or in burrows in the earth but are active at night. Some of them enter houses, including the house cricket of Europe, *Acheta domesticus* (Fig. 11-32, A), which has been introduced into the northeastern United States. Mole-crickets (*Gryllotalpa*) are likewise nocturnal. They burrow in the ground and have the tibiae of the prothoracic legs broadened and provided with stout spines well adapted for digging (Fig. 11-32, C).

Family 6. Acrididae (Locustidae) (Fig. 11-1). Locusts or short-horned grasshoppers. Here the antennae are not as long as the body and consist of less than 26 segments; the ovipositor is composed of several short plates; and the first abdominal segment bears a tym-

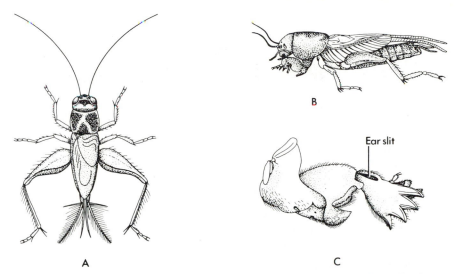

Figure 11-32 Gryllidae. (A) *Acheta domesticus*, the house cricket. (B) *Scapteriscus didactylus*, a mole cricket. (C) Leg of a mole cricket, showing ear slit. (*A, from Herrick; B, after Barrett; C, from Sedgwick.*)

panum on either side. Locusts are vegetarians and often cause great damage to crops. The males produce sounds either by rubbing the row of spines on the inner surface of the femur of the metathoracic leg against the outer surface of the tegmina, or while flying, by rubbing the upper surface of the anterior edge of the hind wings against the undersurface of the tegmina. Common species are the red-legged locust, *Melanoplus femur-rubrum* (Fig. 11-19), the Carolina locust, *Dissosteira carolina*, the American locust, *Schistocera americana*, the lubber grasshopper, *Romalea* sp., and the Rocky Mountain locust, *Melanoplus spretus*, which has been so destructive to crops in Kansas, Iowa, and Nebraska during its migrations. The locusts that were responsible for the plagues of the Pharaohs belonged to this family. In one of the closely related families are certain small active pigmy locusts, such as *Acrydium*, characterized by the presence of a pronotum that extends back over the dorsal surface of the abdomen, sometimes beyond the posterior end of the body.

Order 2. Isoptera

Termites or white ants (Fig. 11-33). Hemimetabolic; chewing mouth parts; two pairs of long, narrow wings laid flat on the back when at rest, or wingless; abdomen joined directly to thorax. These social insects live in colonies much like ants, and are especially abundant in the tropics. They live mostly under cover and the cuticle is light in color and delicate. The adult sexual males and females after their nuptial flight shed their wings near the base where there is a transverse,

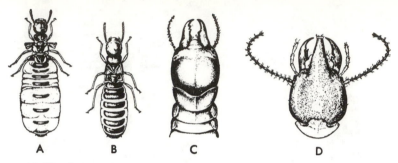

Figure 11-33 Isoptera; termites. (A) *Reticulitermes flavipes*. Mature egg-laying female or queen with abdomen distended. (B) *R. flavipes*. Mature male. (C) *Kalotermes occidentis*. Anterior end of soldier, showing large mandibles and wing pads. (D) *Armitermes intermedius*. Anterior end of soldier, showing large biting mandibles and nasus. (*After Snyder.*)

humeral suture. The castes of termites are more numerous than those of other social insects, such as the bees, wasps, and ants. The species differ with respect to the number of castes present, but each caste contains both males and females. Four castes are usually present: (1) the first reproductive caste, with wings that are shed after the nuptial flight; these are known as king and queen or the primary royal pair; (2) the second reproductive caste, sexually mature but nymphal in form, and known as substitute or supplemental king and queen; (3) sterile workers, which carry on the various activities necessary for the maintenance of the colony, are wingless and usually blind; and (4) sterile soldiers, which are supposed to protect the colony, are wingless, and possess very large heads and mandibles. *Reticulitermes flavipes* is the common species in the northeastern United States. As in many other species, its food consists of dead wood, and it works in the dark, hence, the inside of timbers in buildings are often eaten away without evidence of damage until nothing but a shell remains. Huge nests in the form of mounds over 3 meters in height are built by certain species that live in the tropics. Termite nests are often inhabited by other species of insects; these are called termitophiles. Over 100 species of termitophiles have been recorded. The relation between termites and their intestinal protozoa has already been described.

Order 3. Embioptera

Web-spinners (Fig. 11-34, A). Hemimetabolic; chewing mouth parts; wingless or with two pairs of delicate, membranous wings, containing few veins, and folded on the back when at rest; and cerci of two segments present. The males are usually winged and the females wingless. Embiids live in warm countries and have been reported from the southern states. They live under stones and other objects in tunnels

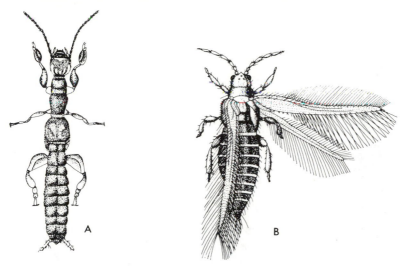

Figure 11-34 (A) Embioptera. *Oligotoma california*, the California embiid. (B) Thysanoptera. A thrips. (*A, from Essig; B, from Moulton.*)

formed of silk produced by tarsal glands, and some of the 60 or more species known are gregarious. *Anisembia texana* is a species discovered in Texas, and *Oligotoma california* (Fig. 11-34, A) a species from California.

Order 4. Plecoptera

Stoneflies (Fig. 11-35). Hemimetabolic; chewing mouth parts, often undeveloped in adults; two pairs of wings, the hind wings usually larger and folded beneath the fore wing; tarsus with three segments; nymphs aquatic, usually in flowing water under stones, often with tufts of thoracic tracheal gills. Ten families have representatives in this

Figure 11-35 Plecoptera. Stonefly. (A) Adult. (B) Nymph showing tracheal gills. (*A, after Comstock; B, after Sharp.*)

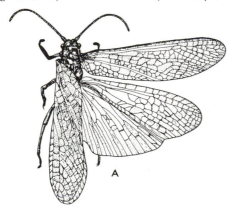

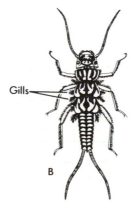

Gills

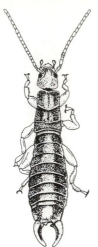

Figure 11-36 Dermaptera. *Anisolabis maritima*, the seaside earwig. (*From Davenport.*)

country. *Allocapnia pygmae* is a common species that may appear on the snow in winter and hence is known as the snowfly. *Taeniopteryx pacifica*, the salmon-fly, is a pest in parts of the state of Washington because it destroys the buds of fruit trees.

Order 5. Dermaptera

Earwigs (Fig. 11-36). Hemimetabolic; chewing mouth parts; wingless or with one or two pairs of wings, the fore wings small, leathery, and meeting in a straight line along the back, the hind wings, large, membranous, and folded lengthwise and crosswise under the fore wings; forcepslike cerci at posterior end of abdomen. The earwigs are nocturnal insects that feed principally on vegetation. The European earwig, *Forficula auricularia*, has been introduced into this country; the seaside earwig, *Anisolabis maritima* (Fig. 11-36), occurs along the Atlantic and Pacific coasts; and little earwig, *Labia minor,* introduced from Europe, has become widely distributed in the United States.

Order 6. Zoraptera

Zorapterans. A small order of less than two dozen species of tiny, termitelike, colonial insects. Probably feed on animal material. Ex. *Zorotypus hubbardi;* southern United States.

Superorder 2. Hemipteroidea. Without cerci.

Order 1. Psocoptera

Psocids and book lice (Fig. 11-37, A). Hemimetabolic; chewing mouth parts; wingless or with two pairs of membranous wings that have few, prominent veins, the fore wings larger than the hind wings; wings, when at rest, held over body like the sides of a roof. Two families

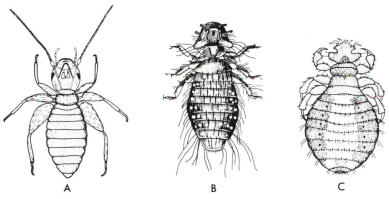

Figure 11-37 (A) Psocoptera. *Liposcelis divinatorius*, a book louse. (B) Mallophagida. *Menopon pallidum*, a chicken louse. (C) Mallophagida. *Gyropus ovalis*, a guinea-pig louse. (*A, from Herrick; B, after Piaget; C, after Osborn.*)

occur in the United States. The family Psocidae contains winged species that live on trees, feeding on dry vegetable matter. They are known as psocids or bark-lice. The Liposcelidae are wingless and called book-lice because they are often observed in old books. *Liposcelis divinatorius* is a species of this type, which feeds on the starch in bookbindings (Figs. 11-37, A).

Order 2. Thysanoptera

Thrips (Fig. 11-34, B). Hemimetabolic; piercing mouth parts; wingless or with two pairs of similar long, narrow, membranous wings with few or no veins, and fringed with long hairs; prothorax large and free; tarsi with two or three segments terminating in a bladder-like, protrusible vesicle. Some species of thrips are injurious to cultivated plants; others are carnivorous and feed on aphids, red spiders, etc. Two suborders are recognized. The Terebrantia contains many injurious species, the females of which lay their eggs in the tissues of plants with their sawlike ovipositors. Here belong the onion thrips, *Thrips tabaci*, the greenhouse thrips, *Heliothrips haemorrhoidalis*, pear thrips, *Taeniothrips inconsequens*, strawberry thrips, *Frankliniella tritici*, and many others. The members of the suborder Tubulifera are not so injurious, and the females do not possess a sawlike ovipositor. Common species are the mullein thrips, *Haplothrips verbasci*, and the camphor thrips, *Liothrips floridensis*.

Order 3. Homoptera

Cicadas, leafhoppers, aphids, scales (Figs. 11-38 to 11-41). Hemimetabolic; mouth parts for piercing and sucking; usually two pairs of wings of uniform thickness held over the back like the sides of a roof. Many economically important insects belong to this order. Several of the families are as follows. The cicadas, and harvestflies, Cicadidae

Figure 11-38 Cicadidae. Six stages in the emergence of an adult harvest fly, or cicada, from the nymph. (*After Snodgrass.*)

(Fig. 11-38), are large, noisy insects; the best known species is the periodical cicada, or 17-year locust, *Magicicada septendecim,* the nymphal stages of which live in the ground for 17 years before the imago emerges. The spittle insects, or frog hoppers, Cercopidae (Fig. 11-39, A), live in a mass of froth elaborated from matter voided from the anus and mixed with a gluelike secretion from abdominal glands. The tree hoppers, Membracidae (Fig. 11-39, B), have a prothorax that extends back over the body and sometimes forms hornlike projections, as in

Figure 11-39 (A) Cercopidae. *Aphrophora,* a spittle insect. (B) Membracidae. *Platycotis vittata,* an oak-tree hopper. (C) Cicadellidae. *Empoasca mali,* the apple-leaf hopper. (*A, after Morse; B, after Woodworth; C, from Essig.*)

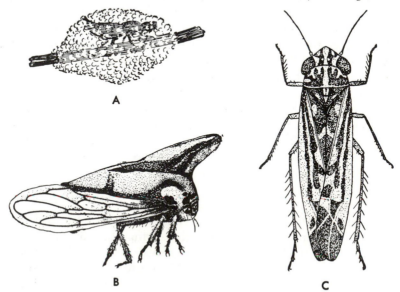

the buffalo tree hopper, *Stictocephala bubalus,* or a hump-back, as in *Telamona ampelopsidis.* The leaf hoppers, Cicadellidae (Fig. 11-39, C), are small and abundant on grass; some are destructive, such as *Euscelis exitiosus* on grain, *Erythroneura comes* on grape leaves and *Empoasca fabae,* on potatoes. The plant lice or aphids, Aphididae (Fig. 11-40, A), are mostly small, green insects with or without wings, that suck juices from plants; they pass from the anus a sweet substance, known as honeydew, which is attractive to ants, bees, and wasps, sometimes bringing about a sort of symbiotic relationship between the aphids and other insects. The life cycle of aphids is of great interest. Females that are parthenogenetic and viviparous hatch from eggs that have lived through the winter. Their offspring are wingless, parthenogenetic females. These give rise to wingless parthenogenetic females, but after a time produce winged females, which migrate to other plants and produce several wingless female generations. As fall approaches, winged males and winged viviparous females appear, they return to the original plant host, mating occurs, and fertilized eggs are laid that remain dormant over winter and give rise to parthenogenetic females in the spring. Among the aphids are the green bug, *Schizaphis graminum,* a pest on wheat and oats, the corn-root aphis, *Aphis maidiradicis.* In the related family, Eriosomatidae, are the woolly apple aphid, *Eriosoma lanigera,* and a number of species that produce plant galls, such as the witch-hazel cone gall aphid, *Hormaphis hamamelidis.* The phylloxerids, Phylloxeridae, are also destructive to vegetation, especially the grape phylloxera, *Phylloxera vitifoliae* (Fig. 11-40, B). The white flies, Aleyrodidae, are covered with a whitish powder, hence their name; the greenhouse white fly, *Asterochiton vaporariorum,* is an important pest. The scale insects and mealy bugs, superfamily Coccoidea (Fig. 11-41), are the most serious pests of horticulturists; they include the San Jose scale, *Aspidiotus perniciosus* (Fig. 11-41, B), the cottony maple scale, *Pulvinaria innumerabilis* (C), the cottony cushion scale,

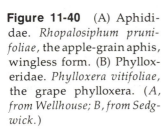

Figure 11-40 (A) Aphididae. *Rhopalosiphum prunifoliae,* the apple-grain aphis, wingless form. (B) Phylloxeridae. *Phylloxera vitifoliae,* the grape phylloxera. (*A, from Wellhouse; B, from Sedgwick.*)

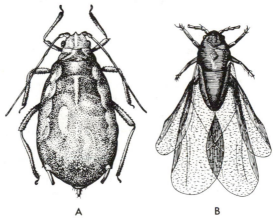

A   B

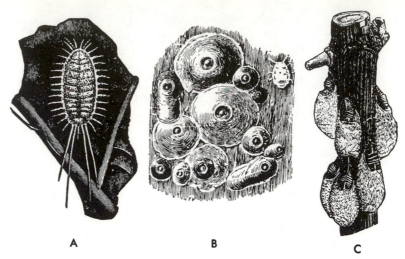

Figure 11-41   Coccoidea. (A) *Pseudococcus longispinus*, the long-tailed mealy bug. (B) *Aspidiotus perniciosus*, the San Jose scale. (C) *Pulvinaria innumerabilis*, the cottony maple scale, females with egg sacs. (*After Comstock.*)

Icerya purchasi, and the mealy bug, *Pseudococcus longispinus* (A); several species are useful to man, such as the lac insect, *Laccifer lacca,* the cochineal insect, *Dactylopius coccus,* and the China wax insect, *Ericerus pe-la.*

Order 4. Hemiptera

True bugs (Figs. 11-42 to 11-44). Hemimetabolic; piercing and sucking mouth parts; wingless or with two pairs of wings, the fore wings thickened at the base (hemelytra). The labium forms a jointed beak into which the slender, piercing maxillae and mandibles move. This order contains many families of interesting and economically important species, a few of which are as follows. The water boatmen, Corixidae (Fig. 11-42, A), have long, flat, fringed metathoracic legs adapted for swimming; they carry a film of air about the body while under the water. The back swimmers, Notonectidae, likewise have oarlike hind legs, but swim on their backs. The water scorpions, Nepidae (Fig. 11-42, B), obtain air through a long caudal tube that is thrust through the surface; a common long, slender species is *Ranatra americana.* The giant water bugs, Belostomatidae (Fig. 11-42, C), are sometimes called electric-light bugs; their hind legs are also adapted for swimming; some tropical species reach a length of 10 cm; common species are *Lethocerus americanus,* which has a femoral groove on each front leg into which the tibia fits, and *Belostoma flumineum* the female of which fastens her eggs on the back of the male. The water striders, Gerridae (Fig. 11-42, D) have long, slender middle and hind legs, which do not break through the surface film as they skim about over the water. The

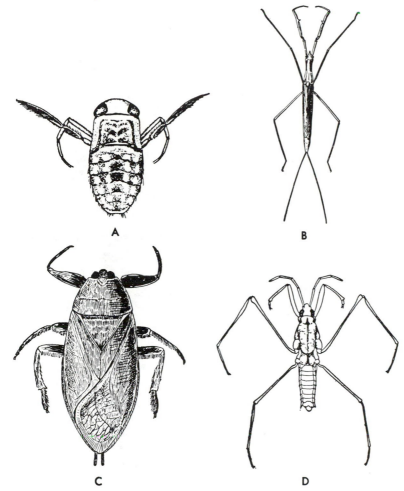

Figure 11-42 Hemiptera. (A) Corixidae. *Artocorixa alternata*, a water boat-man. (B) Nepidae. *Ranatra linearis*, a water scorpion. (C) Belostomatidae. *Lethocerus*, an electric-light bug. (D) Gerridae. *Gerris remigis*, a water strider. (*A, after Hungerford; B, after Sharp; C, from Herrick; D, after Woodworth.*)

leaf bugs, Miridae, are very numerous and often injurious to plants, and include such common species as the tarnished plant bug, *Lygus lineolaris,* and the apple red bug, *Lygidea mendax.* The bed bugs, Cimi-cidae, are nocturnal insects that live on warm-blooded animals; the human species, *Cimex lectularius* (Fig. 11-43, A), has been accused of transmitting various diseases but has not been definitely incrimi-nated. The assassin bugs, Reduviidae, form a large family that con-tains among others, certain species known as kissing bugs; several of these, especially *Panstrongylus megistus* (Fig. 11-43, B) and species of *Triatoma,* are responsible for the transmission of the Protozoa (trypanosomes) responsible for the human disease of South America

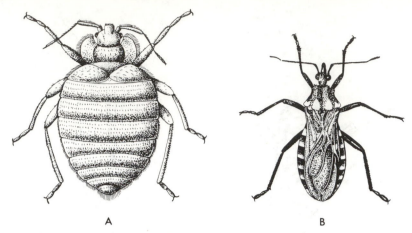

Figure 11-43 (A) Cimicidae. *Cimex lectularius*, the bed bug. (B) Reduviidae. *Panstrongylus megista*, a kissing bug. (*From Brumpt*.)

known as Chagas disease. The lace bugs, Tingidae (Fig. 11-44, B), have front wings, or hemelytra, as well as expansions of the prothorax, of lacelike structure, and are hence easily recognized. The chinch bugs, Lygaeidae, belong to a large family, the most notorious species being the common chinch bug, *Blissus leucopterus* (Fig. 11-44, C), that has been so destructive to grain, especially in the Mississippi Valley. The squash bug family, Coreidae, also contains many species, a good example of which is the common squash bug, *Anasa tristis*. The stink bugs, Pentatomidae (Fig. 11-44, A), secrete a fluid with a nauseous odor, hence their name; the harlequin bug, *Murgantia histrionica*, is destructive to garden vegetables, but another species, *Perillus biocula-tus* is beneficial because it destroys potato beetles.

Figure 11-44 (A) Pentatomidae. *Podisus spinosus*, a stink bug or soldier bug. (B) Tingidae. *Corythuca arcuata*, a lace bug. (C) Lygaeidae. *Blissus leucopterus*, a chinch bug. (*A, after Lugger; B, after Comstock; C, after Webster and Riley*.)

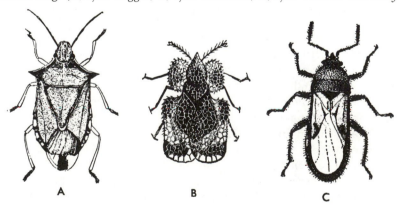

Bird lice (Fig. 11-37, B and C). Hemimetabolic; chewing mouth parts; wings absent; eyes degenerate. The Mallophagida are ectoparasites of birds and, less frequently, of mammals. They feed on hair, feathers, and dermal scales. Those that live on mammals usually possess tarsi with one claw adapted for grasping hair; whereas those that live on birds possess tarsi with two claws which aid them in moving among feathers. Representative species are *Menopon pallidum* (Fig. 11-37), the common chicken louse, *Chelopistes meleagridis,* the large turkey louse, and *Bovicola bovis* that lives on cattle.

Order 6. Anoplurida

Sucking lice (Fig. 11-45). Hemimetabolic; piercing and sucking mouth parts; wingless; ectoparasites on mammals; eyes poorly developed or absent; tarsus with one segment bearing a single, large, curved claw adapted for clinging to the hair of the host. The eggs of Anoplurida are fastened with a gluelike substance to the hairs of the host and are known as nits. Two species occur on humans, the head and body louse, *Pediculus humanus* (Fig. 11-45, A and B), and the crab louse, *Phthirius pubis* (Fig. 11-45, C). Several important diseases are transmitted from person to person by the body louse, including typhus, relapsing, and trench fevers. Among the Anoplurida of domestic animals are *Linognathus setosus,* the dog louse; *Haematopinus suis,* the hog louse; and *L. vituli,* the ox louse (Fig. 11-45, D).

Subclass 4. Endopterygota

These winged insects have a complex metamorphosis. The larval stage is followed by an inactive pupal stage from which the adult eventually emerges. No wing buds develop externally in the larvae of winged species.

Order 1. Neuroptera

The Dobson, aphis lions, ant lions, and others (Fig. 11-46). Holometabolic; chewing mouth parts; four similar membranous wings, usually with many veins and cross veins; no abdominal cerci; larvae carnivorous, some with suctorial mouth parts; tracheal gills usually present on larvae that are aquatic. Thirteen of the 20 families have representatives in North America. Among the interesting or common species are the horned corydalus, *Corydalus cornutus,* of the family Corydalidae, whose larvae (used for bait by fishermen) are known as dobsons or hellgramites; the lacewings of the family Chrysopidae (Fig. 11-46), such as *Chrysopa oculata,* whose larvae are called aphis lions because they feed on aphids; and the ant lions, larvae of members of the family Myrmeleontidae, such as *Myrmeleon immacula-*

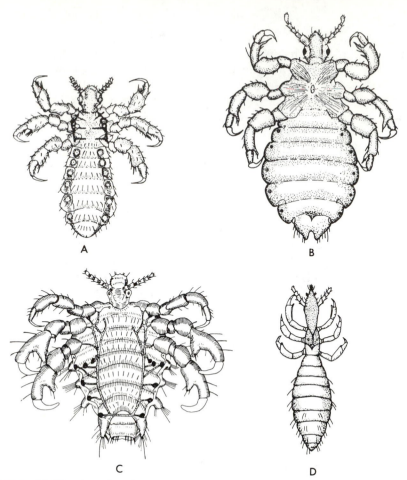

Figure 11-45 Anoplurida. (A) *Pediculus humanus capitis*, the human head louse. (B) *Pediculus humanus corporis*, the human body louse. (C) *Phthirius pubis*, the human crab louse. (D) *Linognathus vituli*, the long-nosed ox louse. (*A–C, from Herrick; D, after Osborn.*)

tus, which live at the bottom of a small pit in the sand and grasp with their strong suctorial jaws any insects that may chance to slide down the precipitous sides.

Order 2. Coleoptera

Beetles (Figs. 11-47 to 11-53). Holometabolic; chewing mouth parts; wingless or with two pairs of wings, the fore wings being hard and sheathlike (elytra) and the hind wings being membranous and folded under the elytra; the prothorax large and movable. The Coleoptera are so numerous and separated into so many families that they are difficult to discuss in a short space. The plan adopted here is to list some

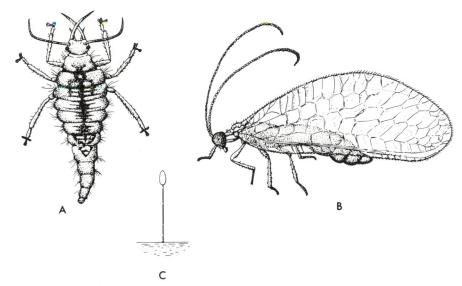

A

B

C

Figure 11-46 Neuroptera. *Chrysopa californica,* a lacewing. (A) Larva. (B) Adult. (C) Stalked egg. (*After Quayle.*)

of the more important groups and point out their more interesting and important characteristics. The species of this largest order are thought to exceed 250,000 kinds.

Suborder 1. Adephaga. Carnivorous beetles; first abdominal sternite divided by hind coxae.

FAMILY CICINDELIDAE (Fig. 11-47, A). Tiger beetles. Predacious; diurnal; usually metallic green or bronze, banded or spotted with yellow; legs adapted for running; larvae in vertical burrows in ground Ex. *Cicindela dorsalis, C. punctulata, Tetracha carolina.*

FAMILY CARABIDAE (Fig. 11-47, B). Ground beetles. Predacious, nocturnal; usually black and shiny; legs adapted for running. Ex. *Calosoma scrutator, Poecilus lucublandus, Harpalus caliginosus.*

Figure 11-47 Coleoptera. (A) Cicindelidae. A tiger beetle. (B) Carabidae. A ground beetle. (*After Bruner and Howard.*)

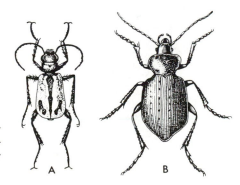

A

B

Phylum Arthropoda / Class II. Insecta **617**

FAMILY DYTISCIDAE. Predacious diving beetles. Aquatic; antennae threadlike; hind legs adapted for swimming; tarsi of fore legs of some species with suckers; air held under elytra while under water; larvae are "water tigers." Ex. *Dytiscus fasciventris, Colymbetes sculptilis, Cybister fimbriolatus.*

FAMILY GYRINIDAE. Whirligig beetles. Aquatic; social in habit; eyes divided into lower and upper halves; middle and hind legs adapted for swimming; secrete ill-smelling whitish liquid. Ex. *Dineutus vittatus, Gyrinus borealis.*

Suborder 2. Polyphaga. Herbivorous beetles; first abdominal sternite not completely divided by hind coxae.

FAMILY HYDROPHILIDAE. Water scavenger beetles. Aquatic; antennae club-shaped; carry film of air underneath the body. Ex. *Hydrophilus obtusatus, H. triangularis* (Fig. 11-48, A).

FAMILY SILPHIDAE. Carrion beetles. Antennae club-shaped; legs adapted for running; feed mostly on decaying animals. Ex. *Necrophorus marginatus* (burying beetle), *Silpha noveboracensis.*

FAMILY STAPHYLINIDAE. Rove beetles. Small, slender; elytra short; abdominal segments movable; legs adapted for running. Ex. *Staphylinus maculosus, Creophilus maxillosus* (Fig. 11-48, B).

FAMILY LAMPYRIDAE. Fireflies. Antennae sawlike; prothorax expanded over head; noctural; emit light; larvae and wingless females are "glow worms." Ex. *Photinus scintillans, Photuris pennsylvanica.*

Figure 11-48 (A) Hydrophilidae. *Hydrophilus triangularis,* the giant water-scavenger beetle. (B) Staphylinidae. *Creophilus maxillosus,* the hairy rove beetle. (*From Essig.*)

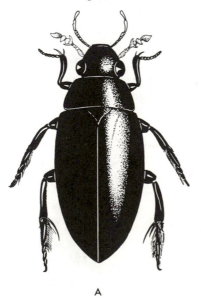

A

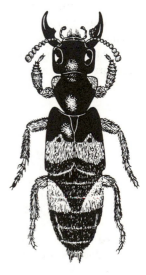

B

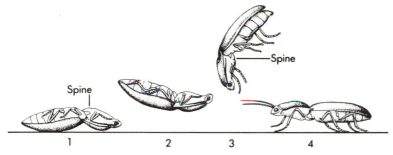

Figure 11-49 Elateridae. A click beetle turning over in the air to its normal position. (*From Schmeil.*)

FAMILY MELOIDAE. Blister beetles and oil beetles. When dried and pulverized produce blister on skin; European species called "Spanish fly"; undergo hypermetamorphosis. Ex. *Epicauta vittata, Meloe angusticollis.*

FAMILY ELATERIDAE (Fig. 11-49). Click beetles. Antennae serrate; body flattened; larvae are wireworms; leaps (clicks) by means of action of prosternal process in metasternal groove. Ex. *Alaus oculatus, Elater nigricollis.*

FAMILY BUPRESTIDAE (Fig. 11-50, A). Metallic wood borers. Antennae serrate; body hard, inflexible, flattened; bronze or metallic in color; larvae mostly "flat-headed" borers. Ex. *Chalcophora virginica, Chrysobothris femorata, Agrilus ruficollis* (causes raspberry gouty gall).

FAMILY DERMESTIDAE. Dermestids. Small; elytra cover abdomen; adults feign death; feed on furs, wool, household goods. Ex. *Dermestes lardarius* (larder beetle), *Anthrenus scrophulariae* (carpet beetle, Fig. 11-50, B), *Anthrenus museorum* (museum pest).

FAMILY TENEBRIONIDAE. Darkling beetles. Black; fore and middle tarsi with five segments, hind tarsi with four segments; food, dry vegetable matter. Ex. *Tenebrio molitor* (meal worm, Fig. 11-50, C), *Bolitotherus cornutus* (fungus beetle).

Figure 11-50 (A) Buprestidae. A metallic wood borer. (B) Dermestidae. *Anthrenus scrophulariae*, the "buffalo bug" or "buffalo moth." (C) Tenebrionidae. *Tenebrio molitor*, the meal worm; larva, pupa and adult. (*From various authors.*)

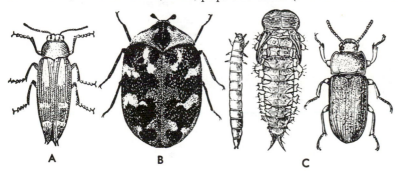

A   B   C

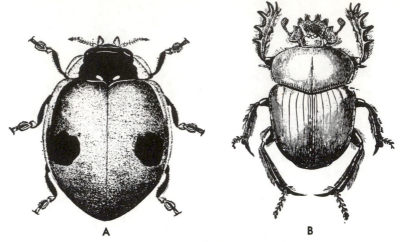

Figure 11-51 (A) Coccinellidae. *Adalia bipunctata,*the two-spotted ladybird beetle. (B) Scarabaeidae. *Scarabaeus sacer,* the sacred beetle of the Egyptians. (*A, from Essig; B, from Sharp.*)

FAMILY COCCINELLIDAE. Ladybird beetle. Mostly predacious; some feed on scale insects; hemispherical in shape; red, yellow, or black with black, white, red, or yellow spots. Ex. *Rodolia cardinalis, Adalia bipunctata* (Fig. 11-51, A).

FAMILY SCARABAEIDAE. Scarabaeids or lamellicorn beetles. Club of antennae lamellate. Ex. *Scarabaeus sacer* (Egyptian scarab, Fig. 11-51, B), *Canthon laevis* (tumblebug), *Phyllophaga fusca* (June beetle), *Macrodactylus subspinosus* (rosebug), *Dynastes tityrus* (rhinoceros beetle), *Euphoria inda* (bumble flower beetle), *Popillia japonica* (Japanese beetle).

FAMILY LUCANIDAE. Stag beetles. Large, sometimes branched, mandibles; club of antennae of flattened plates. Ex. *Lucanus dama, Dorcus parallelus.*

FAMILY PASSALIDAE. One species, *Passalus cornutus,* is common in the United States.

FAMILY CERAMBYCIDAE (Fig. 11-52, A). Long-horned beetles. Antennae often longer than body, of 11 segments; tarsi with five segments, third bilobed, fourth very small; larvae, wood-boring grubs. Ex. *Prionus laticollis, Megacyllene robiniae* (locust borer), *Tetraopes tetraophthalmus* (milkweed beetle).

FAMILY CHRYSOMELIDAE (Fig. 11-52, B and C). Leaf beetles. Body, oval; antennae and legs short; feed on leaves. Ex. *Leptinotarsa decemlineata* (potato beetle, Fig. 11-52, B), *Altica chalybea* (flea beetle), *Cassida nigripes* (tortoise beetle).

FAMILY CURCULIONIDAE. Snout beetles. Head prolonged into snout. Ex. *Anthonomus grandis* (cotton-boll weevil, Fig. 11-53, A), *Curculio rectus* (acorn weevil).

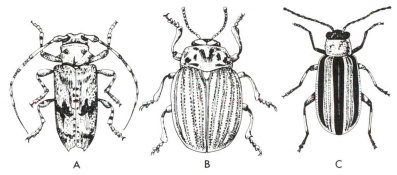

Figure 11-52 (A) Cerambycidae. *Acanthoderes decipiens,* a long-horned beetle. (B) Chrysomelidae. *Leptinotarsa decemlineata,* the potato beetle. (C) Chrysomelidae. *Acalymma vittata,* the striped cucumber beetle. (*After various authors.*)

FAMILY SCOLYTIDAE (Fig. 11-53, B, C). Engraver and ambrosia beetles. Some live on fungus (ambrosia); others are called timber beetles. Ex. *Scolytus multistriatus* (European elm bark beetle, transmits fungus causing Dutch elm disease), *Dendroctonus punctatus* (spruce beetle).

Order 3. Strepsiptera

Stylopids (Fig. 11-54, A). Hypermetamorphosis; mouth parts vestigial or absent; endoparasitic in other insects; male with club-shaped fore wings and large membranous hind wings; female, wingless and legless, nutrition by absorption; life cycle complex. The stylopids live principally in bees, wasps, and homopterous bugs. Ex. *Xenos wheeleri,* a parasite of the wasp, *Polistes metricus.*

Figure 11-53 (A) Curculionidae. *Anthonomus grandis,* the cotton-boll weevil. (B) Scolytidae. Galleries of an engraver beetle under the bark of a tree. (C) Scolytidae. *Dendroctonus frontalis,* the southern pine beetle. (*After Felt and Hopkins.*)

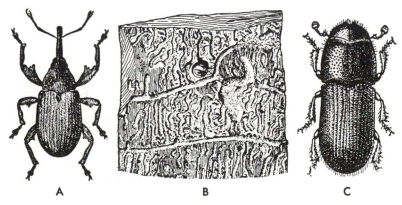

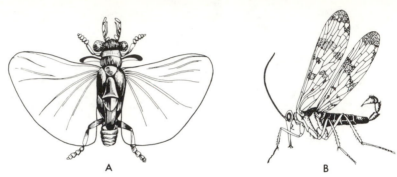

Figure 11-54 (A) Strepsiptera. A twisted-winged insect. (B) Mecoptera. *Panorpa*, a scorpionfly. (*A, after Packard; B, after Sharp.*)

Order 4. Mecoptera

Scorpionflies (Fig. 11-54, B). Holometabolic; chewing mouth parts; antennae long and slender; head prolonged into beak; wingless or with two pairs of long, narrow, membranous wings; males with clasping organ at caudal end resembling the sting of a scorpion; food, fruit and dead insects. Ex. *Panorpa rufescens, Bittacus strigosus.*

Thornhill (1979) observed the first case of "transvestism" in an invertebrate. Some *Hylobittacus apicalis* males mimicked the female to enable them to steal nuptial prey offered by other males. The thefts enabled them to mate with higher frequency with females since the other males had already completed the process of selecting a suitable enticement for the female. The female feeds on the gift during mating.

Order 5. Trichoptera

Caddisflies (Fig. 11-55). Holometabolic; vestigial mouth parts in adult; two pairs of membranous wings clothed with long, silky hairs; the aquatic larvae build portable cases of sand grains or vegetable matter fastened together with silk secreted by modified salivary glands. Members of the genus *Hydropsyche* construct nonportable nets in

Figure 11-55 Trichoptera. An adult caddisfly, with wings spread, and three types of larval cases. (*After several authors.*)

moving water. Ex. *Phryganea interrupta* (case of pieces of leaves), *Molanna cinerea* (case of sand).

Order 6. Lepidoptera

Butterflies, skippers, and moths (Figs. 11-56 to 11-60). Holometabolic; sucking mouth parts; wingless or with two pairs of membranous wings covered with overlapping scales. The sucking apparatus, which is coiled underneath the head, consists of the two half-round maxillae fastened together so as to form a tube. The larvae are called caterpillars. They sometimes spin a cocoon in which they pupate. The pupa of a butterfly is often called a chrysalis. It is customary to divide the Lepidoptera into two groups, the moths (Heterocera) and butterflies (Rhopalocera). One family of butterflies, the Hesperiidae, are often set apart and called skippers. Moths are mostly nocturnal and their antennae are usually threadlike or featherlike and without a terminal knob. Butterflies are diurnal; the antennae are threadlike and have a knob at the end; and the wings are held vertically over the back when at rest. Skippers are diurnal; the antennae are threadlike and have a subterminal knob and a terminal recurved hook; and they have a skipping mode of flight. Space allows listing of only a few families.

FAMILY PRODOXIDAE. Ex. Yucca moth, *Tegeticula alba*. The flowers of the genus *Yucca* depend upon this species for their cross-pollination. The moth visits the flowers in the evening, scrapes some pollen from a stamen, holds it underneath its head, and carries it to another flower. It clings to the pistil of this, and, thrusting its ovipositor through the wall of the ovary, lays an egg. It then mounts the pistil, and forces the pollen it has brought down into the stigmatic tube. Another egg is laid in another part of the ovary, and more pollen is inserted into the stigmatic tube. These processes may be repeated half a dozen times in a single flower. The advantage to the flower is, of course, the certainty of being cross-pollinated and of producing seeds. These seeds provide a supply of food required by the larvae that hatch from the eggs laid by the moth in the ovary. The seeds are so numerous that the few eaten by the larvae may well be spared. Plant and insect are so interdependent that natural reproduction of the plant occurs only where the insect occurs and the insect only lives where the yucca is found.

FAMILY EUCLEIDAE. Slug caterpillars (Fig. 11-57). Medium-sized moths; larvae often unusually shaped and with stinging hairs which are skin irritants. Ex. Saddleback caterpillar, *Sibine stimulea*.

FAMILY TINEIDAE. Clothes moths. Ex. *Tinea pellionella* (Fig. 11-56, A).

FAMILY GELECHIIDAE. Ex. Angoumois grain moth, *Sitotroga cerealella*; Pink boll worm, *Pectinophora gossypiella*.

FAMILY TORTRICIDAE. Ex. Spruce budworm, *Choristoneura fumiferana*.

FAMILY OLETHREUTIDAE. Ex. Codling moth, *Carpocapsa pomonella*.

 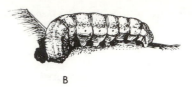

Figure 11-56 Lepidoptera. (A) Tineidae. *Tinea pellionella*, a clothes moth. (B) Noctuidae. *Heliothis*, the cotton-boll worm or corn earworm. (*From Herrick.*)

FAMILY PYRALIDIDAE. Ex. European corn borer, *Pyrausta nubilalis;* bee moth, *Galleria mellonella; Mediterranean flour mouth, Anagasta kuhniella.*

FAMILY SPHINGIDAE (Fig. 11-58). Hawk moths. Ex. White-lined sphinx, *Celerio lineata;* tomato hornworm, *Manduca quinquemaculata.*

FAMILY GEOMETRIDAE. Measuring worms. Ex. Fall canker worm, *Alsophila pometaria.*

FAMILY LIPARIDAE. Tussock moths. Ex. White-marked tussock

Figure 11-57 Caterpillar modifications. (A) Eucleidae larva, the slug caterpillar. (B) Geometridae, the "inch worm" caterpillar. (*A, photo courtesy the Kalamazoo Gazette.*)

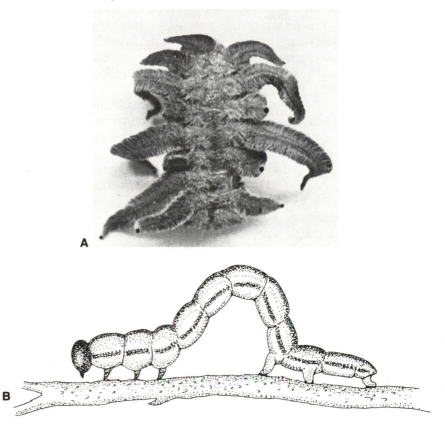

Figure 11-58 Sphingidae. A hawk moth. (A) Adult. (B) Larva. (C) Pupa. (*From U.S. Dept. Agric.*)

moth, *Hemerocampa leucostigma;* gipsy moth, *Porthetria dispar;* brown-tail moth, *Nygmia phareorrhoea.*

FAMILY NOCTUIDAE. Owlet moths. Ex. Cotton leaf-worm, *Alabama argillacea;* army worm, *Pseudaletia unipuncta;* cotton-boll worm, *Heliothis obsoleta* (Fig. 11-56, B).

FAMILY ARCTIIDAE. Tiger moths. Ex. Fall webworm, *Hyphantria cunea;* banded woollybear caterpillar, *Isia isabella.*

FAMILY CITHERONIIDAE. Ex. Regal moth, *Citheronia regalis* (larva in hickory trees); Imperial moth, *Eacles imperialis.*

FAMILY SATURNIIDAE. Giant silkworm moths. Ex. Polyphemus, *Antheraea polyphemus;* luna, *Actias luna;* promethea, *Callosamia promethea;* cecropia, *Hyalophora cecropia.*

FAMILY BOMBYCIDAE. Silkworm moths. Ex. Silkworm, *Bombyx mori.*

Figure 11-59 Hesperiidae. *Calpodes ethlius,* a skipper; larva, pupae, and adult. (*After Chittenden.*)

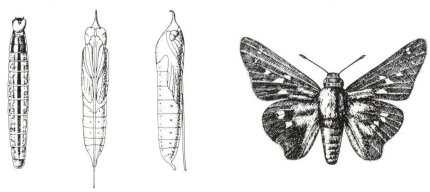

Family Hesperiidae (Fig. 11-59). Skippers. Silver-spotted skipper, *Epargyreus clarus*; cloudy wing, *Thorybes pylades*.

Family Papilionidae. Swallow-tailed butterflies. Ex. Tiger swallowtail, *Papilio glaucus*. The adults occur in two forms, *Papilio glaucus turnus* and *Papilio glaucus glaucus*. *Papilio polyxenes* is known as the black swallow-tail (Fig. 11-60).

Family Pieridae. Ex. Cabbage butterfly, *Pieris rapae*; common sulphur, *Colias philodice*.

Family Danaidae. Milkweed butterflies. Ex. *Danaus plexippus*, the monarch (Fig. 11-60, D), adult distasteful to birds; "migratory."

Family Nymphalidae. Ex. Spangled fritillary, *Speyeria cybele*; mourning cloak, *Nymphalis antiopa*; viceroy, *Limenitis archippus* (mimics the monarch, Fig. 11-60, D, E).

Family Lycaenidae. Gossamer-winged butterflies. Ex. Banded hair streak, *Thecla calanus*; American copper, *Lycaena phleas*; spring azure, *Celastrina argiolus* (very polymorphic, over a dozen forms named from North America).

Figure 11-60 (A–C) Papilionidae; *Papilio polyxenes*, the black swallowtail. (A) Larva. (B) Pupa. (C) Adult. (D) Danaidae; *Danaus plexippus*, the monarch. (E) Nymphalidae; *Limenitis archippus*, the viceroy. (*A–C, after Webster.*)

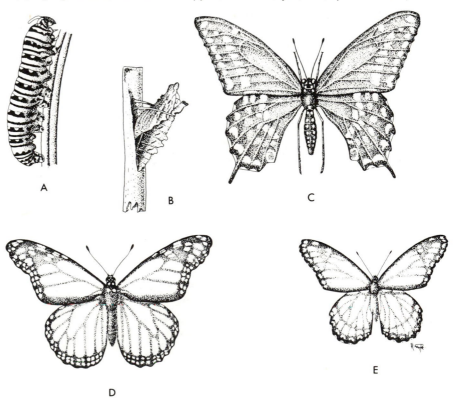

Flies (Figs. 11-61 to 11-68). Holometabolic; piercing and sucking or sponging mouth parts forming a proboscis; wingless or with one pair of membranous fore wings, the hind wings being represented by knobbed threads called *halters;* larvae known as maggots; larval skin sometimes serves as a cocoon and called a puparium. The labium usually serves as a guide for the slender, piercing mandibles and maxillae, and for the sucking tube, consisting of an elongated epipharynx from the roof of the mouth and an elongated hypopharynx from the floor of the mouth, which is thrust into the wound and through which juices are sucked up by means of a pharyngeal pump in the head. Many variations exist in the mouth parts of Diptera.

Suborder 1. Nematocera. Larva with well-chitinized head capsule; pupa enclosed in pupal skin; adult escapes from pupa by longitudinal slit in thoracic region; flies with antennae of more than five segments and usually long and slender.

FAMILY TIPULIDAE (Fig. 11-61, A). Crane flies. Slender body; narrow wings; long legs; transverse V-shaped suture on dorsal surface of mesothorax. Ex. *Tipula abdominalis.*

Figure 11-61 Diptera. (A) Tipulidae. *Tipula,* a crane fly. (B) Ceratopogonidae. *Culicoides guttipennis,* a punkie or "no-see-um." (C) Psychodidae. *Phlebotomus papatasi,* a sand fly. (D) Tendipedidae. A midge larva. (*A, after Weed; B, after Pratt; C, after Newstead.*)

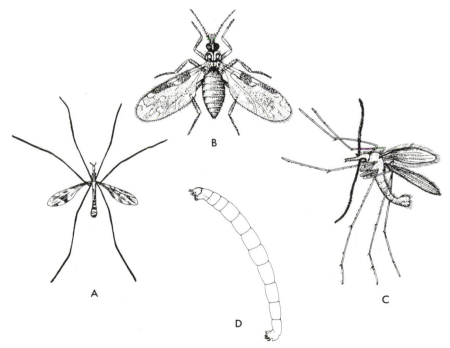

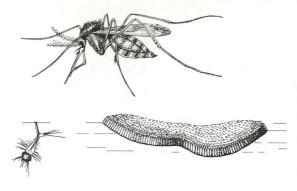

Figure 11-62 Culicidae. *Culex pipiens,* a common mosquito. Adult female, egg mass on surface of water, and young hanging from surface of water. *(From Howard.)*

FAMILY CHIRONOMIDAE (TENDIPEDIDAE). Midges. Larvae (Fig. 11-61, D) important members of aquatic bottom communities. Some are tolerant of organic pollution. Adults do not bite. Ex. *Chironomus riparius* in shallow water with low oxygen content. *C. plumosus* in deep lakes.

FAMILY CERATOPOGONIDAE. Biting midges. Ex. *Culicoides guttipennis,* a punkie or "no-see-um" (Fig. 11-61, B).

FAMILY PSYCHODIDAE. Moth flies. Ex. *Psychoda alternata.* Sand flies of the genus *Phlebotomus* (Fig. 11-61, C) transmit pappataci fever, oriental sore, and oroya fever from person to person.

FAMILY CULICIDAE. Mosquitoes. Ex. *Culex pipiens,* house mosquito (Fig. 11-62); *Anopheles quadrimaculatus* (transmits malaria, Fig. 11-63, A); *Aëdes aegypti* (transmits yellow fever, Fig. 11-63, B). Mosquitoes also transmit dengue and filariasis.

FAMILY CECIDOMYIDAE. Gall gnats. Ex. *Rhabdophaga strobiloides*

Figure 11-63 Culicidae. (A) *Anopheles quadrimaculatus,* a malaria mosquito. (B) *Aëdes aegypti,* the yellow-fever mosquito. *(After Herrick.)*

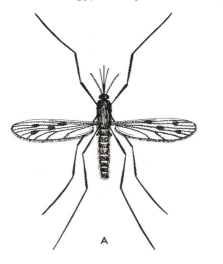
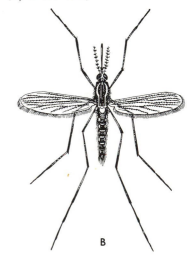

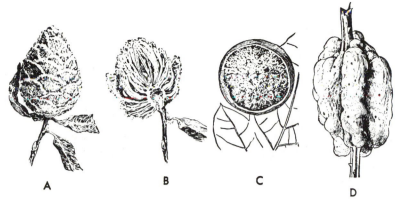

Figure 11-64 Plant galls due to the presence of insects. (A) Willow-cone gall caused by a fly of the family Cecidomyidae. (B) Same, cut open to show the maggots within. (C) Oak-apple gall caused by a hymenopterous insect, *Amphibolips confluens,* of the family Cynipidae. (D) Blackberry-knot gall due to member of the family Cynipidae. (*A and B, after Washburn; C and D, after Beutenmuller.*)

(causes willow-cone gall, Fig. 11-64); Hessian fly, *Phytophaga destructor.*

FAMILY SIMULIIDAE (Fig. 11-65, A). Black flies. Ex. Adirondack black fly, *Prosimulium hirtipes;* buffalo gnat, *Cnephia pecuarum.* These are blood-sucking pests of humans and animals. *Simulium damnosum* transmits the human filarial worm, *Onchocerca volvulus* of Africa.

Suborder 2. Brachycera Flies with short, thick antennae containing five or less segments; larvae with well-chitinized head capsule.

FAMILY TABANIDAE. Horse flies, deer flies. Ex. *Tabanus atratus; Chrysops silacea* (transmits human filarial worm, *Loa loa*); *Chrysops discalis* (transmits tularaemia from jack rabbits to humans, Fig. 11-65, B).

FAMILY BOMBYLIIDAE. Bee flies. Ex. *Bombylius major* (Fig. 11-65, C). Adults resemble bees; larvae feed on egg sacs of Orthoptera and on hymenopterous and lepidopterous larvae.

FAMILY ASILIDAE (Fig. 11-65, D). Robber flies. Ex. *Erax aestuans; Asilus notatus.*

Suborder 3. Cyclorrhapha. Pupa enclosed in last larval skin strengthened with chitin (puparium); head of larva not sclerotized; adult with five or fewer antennal segments.

FAMILY SYRPHIDAE. Flower flies. Ex. Drone fly, *Tubifera tenax* (larva known as rat-tailed maggot); *Syrphus perplexus.*

FAMILY TEPHRITIDAE. Ex. Mediterranean fruit fly, *Ceratitis capitata* (Fig. 11-66, A); goldenrod gall fly. *Eurosta solidaginis.*

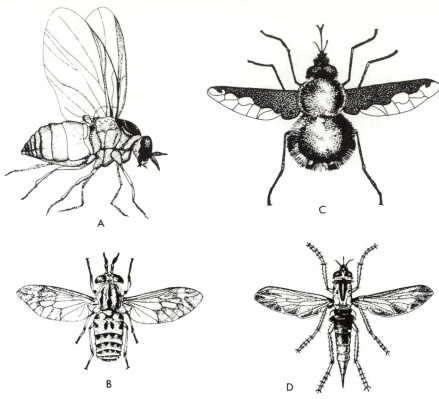

Figure 11-65 (A) Simuliidae. *Simulium pictipes*, a black fly. (B) Tabanidae. *Chrysops discalis*, the vector of tularemia. (C) Bombyliidae. *Bombylius major*, a bee fly. (D) Asilidae. *Asilus cabroniformis*, a robber fly. (*A, from Herrick; B and D, from Brumpt; C, from Essig.*)

FAMILY PIOPHILIDAE. Ex. Cheese skipper, *Piophila casei* (Fig. 11-66, B, C).

FAMILY EPHYDRIDAE. Ex. Petroleum fly, *Psilopa petrolei* (larva lives in crude petroleum).

FAMILY DROSOPHILIDAE. Pomace flies. Fruit fly, *Drosophila melanogaster* (Fig. 11-66, D), employed more extensively than any other animal for the study of genetics.

FAMILY GASTEROPHILIDAE. Bot flies. Ex. Horse bot fly. *Gasterophilus intestinalis*.

FAMILY OESTRIDAE. Bot and warble flies. Ex. Sheep bot fly, *Oestrus ovis*; heel fly, *Hypoderma lineatum*.

FAMILY CALLIPHORIDAE. Ex. Blow fly, *Calliphora vomitoria*; greenbottle fly, *Lucilia caesar*; screwworm fly, *Cochiliomya hominivorax*, larvae parasitic in cattle, ruins hides, eradicated in Florida by release of many sterile males. Stubborn cases of osteomyelitis have been treated successfully with the aid of blow fly maggots.

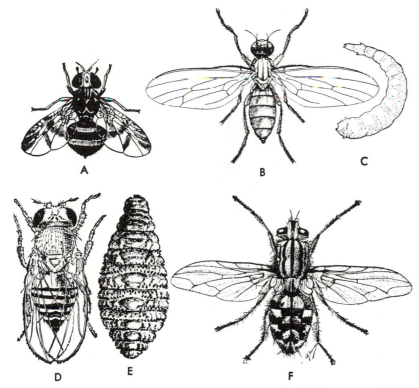

Figure 11-66 (A) Tephritidae; *Ceratitis capitata*, the Mediterranean fruit fly. (B) Piophilidae; *Piophila casei*, the cheese skipper. (C) Cheese skipper, the maggot of *P. casei*. (D) Drosophilidae; *Drosophila melanogaster*, a fruit fly. (E) Cuterebridae; *Dermatobia cyaniventris*, larva of a bot fly. (F) Sarcophagidae; *Sarcophaga carnaria*, a flesh fly. (*A, after Fuller; B and C, after Herrick; D, after Bridges; E, after Blanchard; F, after Brumpt.*)

FAMILY SARCOPHAGIDAE (Fig. 11-66, F). Ex. *Sarcophaga haemor-rhoidalis* (may cause intestinal myiasis in people); *Wohlfahrtia vigil* (may cause cutaneous myiasis in humans). Several species live on dead insects in the cups of pitcher plants.

FAMILY TACHINIDAE. Ex. *Trichopoda pennipes*. This and other species are important parasites of injurious insects.

FAMILY MUSCIDAE. Ex. House fly, *Musca domestica* (Fig. 11-67), has an extremely valuable research literature bibliographed by West and Peters (1973); stable fly, *Stomoxys calcitrans;* tsetse fly, *Glossina palpalis* (Fig. 11-68, A, B, transmit the protozoan parasite of sleeping sickness in Africa, see *Trypanosoma*).

FAMILY HIPPOBOSCIDAE. Louse flies. Ex. Sheep tick, *Melophagus ovinus* (Fig. 11-68, D).

FAMILY BRAULIDAE. Bee lice. Ex. *Braula caeca*, a parasite of the honey bee (Fig. 11-68, C).

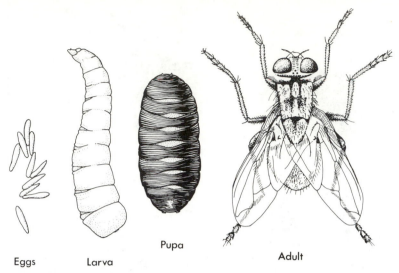

Eggs   Larva   Pupa   Adult

Figure 11-67   Muscidae. *Musca domestica*, the house fly. (*From Woodruff.*)

Order 8. Siphonapterida

Fleas (Fig. 11-69). Holometabolic; piercing and sucking mouth parts; wingless; body laterally compressed; head small; no compound eyes; legs adapted for leaping; ectoparasites of mammals, a few of birds. Ex. Cat flea, *Ctenocephalus felis* (also attacks dog and human); dog flea, *C. canis* (also attacks cat and human); human flea, *Pulex irritans;* rat flea, *Xenopsylla cheopis*, (transmits bubonic plague); sticktight flea, *Echidnophaga gallinacea* (serious pest of poultry); chigoe or jigger, *Tunga penetrans.* (See also Acarinida, Family Trombidiidae.)

Order 9. Hymenoptera

Social insects; bees, wasps, hornets, ants, and sawflies (Figs. 11-70 to 11-78). Holometabolic; chewing or sucking mouth parts; wingless or with two pairs of membranous wings, fore wings larger, venation reduced; wings on each side held together by hooks (hamuli); females usually with sting, piercer, or saw; some parasitic on other insects.

FAMILY TENTHREDINIDAE. Sawflies. Ex. Currant worm, *Nematus ribesii* (Fig. 11-70, A. B); rose slug, *Cladius isomerus.*

FAMILY BRACONIDAE (Fig. 11-70, C). Ex. *Microgaster facetosa* (parasitic on moths); *Aphidius rosae* (parasitic on aphids).

FAMILY ICHNEUMONIDAE (Fig. 11-70, D.). Ichneumon wasps. Ex. *Megarhyssa lunator* (parasitic on larva of pigeon horn-tail, *Tremex columba*); *Ophion bilineatum* (parasitic on skippers).

FAMILY CYNIPIDAE. Gall wasps and others. Ex. Oak hedgehog gall, *Andricus erinacei;* oak apple, *Amphibolips confluens* (Fig. 11-64, C); mossy rose gall, *Diplolepis rosae* (Fig. 11-71, A, B).

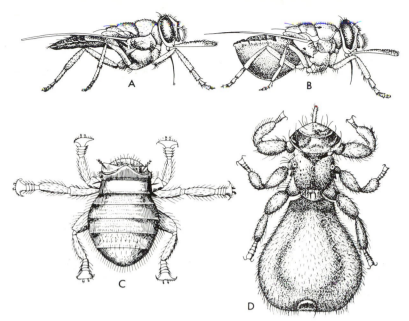

Figure 11-68 (A, B) Muscidae; *Glossina palpalis,* the vector of African sleeping sickness, (A) before and (B) after a meal. (C) Braulidae; *Braula caeca,* a bee louse parasitic on the honey bee. (D) Hippoboscidae; *Melophagus ovinus,* a sheep tick. (*A, B, and D, from Brumpt; after Meinert.*)

Figure 11-69 Siphonapterida. Diagram showing principal external features of fleas. *Pulex irritans,* the human flea, and *Xenopsylla cheopis,* the oriental rat flea and vector of bubonic plague; both lack pronotal and genal ctendia. (*From Herms.*)

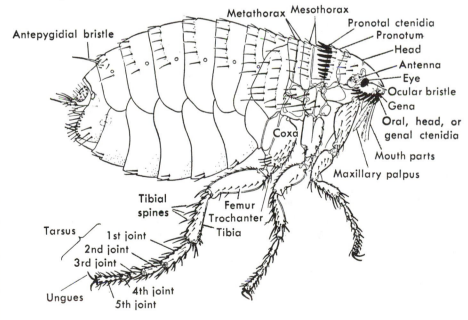

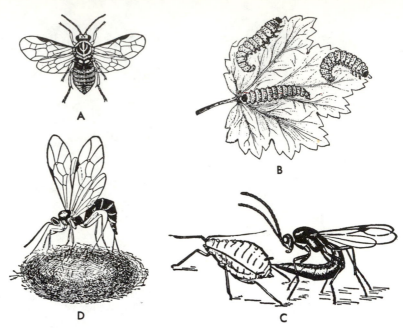

Figure 11-70 Hymenoptera. (A and B) Tenthredinidae; *Nematus*. (A) Adult sawfly. (B) Larva or currant worm. (C) Braconidae; *Lysiphlebus,* a parasite on aphids. (D) Ichneumonidae; *Itoplectis conquisitor;* female laying eggs in cocoon of tent caterpillar. (*After various authors.*)

FAMILY EULOPHIDAE. Ex. *Aphelinus jucundus*, parasitic on aphids. This family and a number of closely related families in the superfamily Chalcidoidea contain many tiny chalcid wasp species. Most are parasites of other insects, a few are plant pests or seed eaters. *Blastophaga psenes* (Fig. 11-71, C) develops in a gall in wild fig flowers in California. When it emerges covered with pollen it visits other flowers and in the process insures pollination of the cultivated Smyrna fig.

FAMILY FORMICIDAE. Ants. Ants are all social insects and each colony contains several castes as in termites. Ex. Red ant, *Monomorium pharaonis* (Fig. 11-72, B, a household nuisance); agricultural ant, *Pogonomyrmex barbatus;* fire ant, *Solenopsis saevissima richteri;* shed builder, *Crematogaster lineolata* (builds sheds for aphids); fungus-growing ant, *Atta texana;* Argentine ant, *Iridomyrmex humilis* (Fig. 11-72, C, a household and orchard pest); carpenter ant, *Camponotus herculeanus pennsylvanicus* (Fig. 11-72, A); mound-building ant, *Formica exsectoides;* slave maker, *Formica sanguinea;* honey ant, *Myrmecocystus melliger* (certain individuals repletes, serve as storage reservoirs).

FAMILY VESPIDAE. Wasps. Ex. Solitary jug builder, *Eumenes fraternus;* social polistes, *Polistes fuscatus pallipes;* white-faced hornet, *Vespa maculata* (Fig. 11-73, B); yellow jacket, *Vespa maculifrons* (Fig. 11-73, A).

634 *Class Insecta*

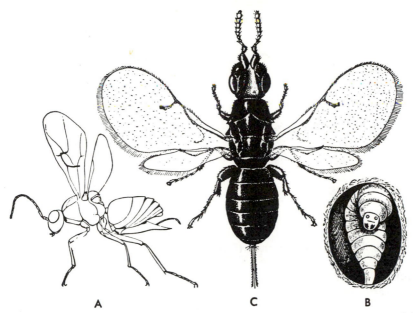

Figure 11-71 (A and B) Cynipidae. (A) *Diplolepis rosae,* causative agent of the mossy rose gall. (B) Larva of same in its cell. (C) Chalcidoidea. *Blastophaga psenes,* adult female fig insect. (*A and B, after Sharp; C, from Essig.*)

FAMILY SPHECIDAE. Digger wasps. Ex. Mud dauber, *Trypoxylon albitarsis;* tool-using wasp, *Sphex urnaria* (Fig. 11-74, A).

FAMILY ANDRENIDAE. Ex. Mining bee, *Andrena vicina* (Fig. 11-74, B).

FAMILY XYLOCOPIDAE. Carpenter bees. Ex. *Ceratina dupla.*

Figure 11-72 Formicidae. (A) *Camponotus herculeanus pennsylvanicus,* the large black carpenter ant. (B) *Monomorium pharaonis,* the red ant. (C) *Iridomyrmex humilis,* the Argentine ant queen. (*From Herrick.*)

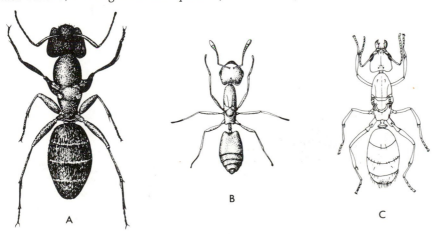

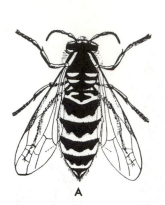

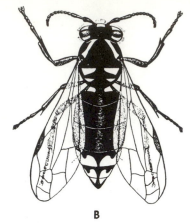

Figure 11-73 Vespidae. (A) *Vespa maculifrons*, the yellow jacket. (B) *Vespa maculata*, the bald-faced hornet. (*From Herrick.*)

Figure 11-74 (A) Sphecidae. *Sphex urnaria;* a digger wasp using a stone to pack earth over its nest. (B) Andrenidae. *Andrena*, a mining bee. (C) Bombidae. *Bombus,* a bumble bee pollinating a flower of the orchid *Cypripedium.* (*After various authors.*)

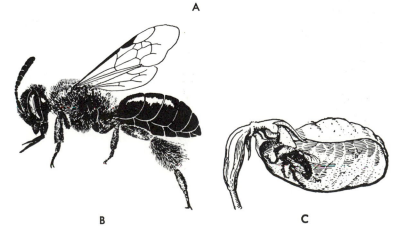

636

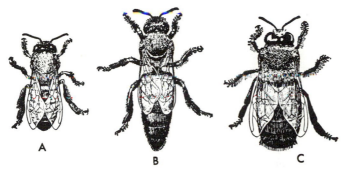

Figure 11-75 Honey bees. (A) Worker. (B) Queen. (C) Drone. (*From Phillips.*)

FAMILY MEGACHILIDAE. Leaf-cutter bees. Ex. *Megachile brevis.*

FAMILY BOMBIDAE (Fig. 11-74, C). Bumble bees. Nest-building bumble bee, *Bombus vagans;* usurper bumble bee, *Psithyrus laboriosus* (social parasite of other bumble bees).

FAMILY APIDAE. Honey bees. One introduced species in the United States, *Apis mellifera* (Fig. 11-75).

The Honey Bee

The mouth parts, legs, and sting of the worker honey bee are so marvelously adapted for various functions that a description of them is presented here and the student is advised to study these structures in detail.

Mouth Parts (Fig. 11-76). The mouth parts consist of a labrum, or upper lip, the epipharynx, a pair of mandibles, two maxillae, and a labium, or under lip. The *labrum* is joined to the *clypeus,* which lies just above it. From beneath the labrum projects the fleshy *epipharynx;* this is probably an organ of taste. The *mandibles,* or jaws, are situated one on either side of the labrum; they are notched in the queen and drone, but smooth in the worker. The latter makes use of them in building honeycomb. The *labium* is a complicated median structure extending downward from beneath the labrum. It is joined to the back of the head by a triangular piece, the *submentum.* Next to this is a chitinous, muscle-filled piece, the *mentum,* beyond which is the *ligula* (*glossa* or tongue) with one *labial palpus* on each side. The ligula may be drawn in or extended. It is long and flexible, with a spoon or *bouton* at the end. Hairs of various kinds are arranged upon in it regular rows; these are used for gathering nectar, and as organs of touch and taste. The *maxillae,* or lower jaws, fit over the mentum on either side. Along their front edges are rows of stiff hairs. *Maxillary palpi* are also present.

Nectar is collected in the following manner. The maxillae and the labial palpi form a tube, in the center of which the tongue moves backward and forward. When the epipharynx is lowered, a passage is com-

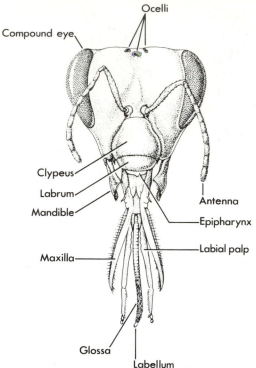

Ocelli

Compound eye

Clypeus
Labrum
Mandible

Maxilla

Antenna
Epipharynx
Labial palp

Glossa
Labellum

Figure 11-76 Head of honey
bee. (*From Woodruff.*)

pleted into the esophagus. The nectar is first collected by the hairs on
the ligula; it is then forced upward by the pressing together of the
maxillae and labial palpi.

Legs (Fig. 11-77). The *prothoracic legs* possess the following useful
structures. The femur and the tibia are clothed with *branched hairs* for
gathering pollen. Extending on one side from the distal end of the
tibia are a number of curved bristles, the *pollen brush,* which are used
to brush up the pollen loosened by the coarser spines; on the other
side is a flattened movable spine, the *velum,* which fits over a curved
indentation in the first tarsal joint or metatarsus. This entire structure
is called the *antenna cleaner* and the row of teeth which lines the iden-
tation is known as the *antenna comb.* On the front of the metatarsus is
a row of spines called the *eye brush,* which is used to brush out any
pollen or foreign particles lodged among the hairs on the compound
eyes. The last tarsal joint of every leg bears a pair of notched claws,
which enable the bee to obtain a foothold on rough surffaces. Between
the claws is a fleshy glandular lobule, the *pulvillis,* whose sticky secretion makes it possible for the bee to cling to smooth objects. Tactile
hairs are also present.

The middle, or *mesothoracic legs,* are provided with a *pollen brush,*
but, instead of an antenna cleaner, a *spur* is present at the distal end of

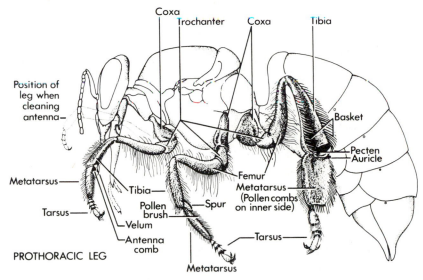

Figure 11-77 Details of the legs of the honey bee. (*From Woodruff.*)

the tibia. This spur is used to pry the pollen out of the pollen baskets on the third pair of legs, and to clean the wings.

The *metathoracic legs* possess three very remarkable structures, the pollen basket, the wax pinchers, and the pollen crumbs. The *pollen basket* consists of a concavity in the outer surface of the tibia with rows of curved bristles along the edges. By storing pollen in this basket-like structure, it is possible for the bee to spend more time in the field, and to carry a larger load at each trip. The pollen combs serve to fill the basket by combing out the pollen, which has become entangled in the hairs on the thorax, and transferring it to the concavity in the tibia of the opposite leg. At the distal end of the tibia is a row of wide spines; these are opposed by a smooth plate on the proximal end of the meta-tarsus. The term *wax pinchers* has been applied to these structures, since they are used to remove the wax plates from the abdomen of the worker.

Sting (Fig. 11-78). The *sting* is a very complicated structure. Before the bee stings, a suitable place is usually selected with the help of the *sting feelers;* then the two *barbed darts* are thrust forward. The *sheath* serves to guide the darts, to open up the wound, and to aid in conducting the poison. The *poison* is secreted in a pair of *glands,* one acid, the other alkaline, and is stored in a *reservoir.* Generally the sting, poison glands, and part of the intestine are pulled out when a bee stings, so that death ensues after several hours, but if only the sting is lost, the bee is not fatally injured. The queen seldom uses her sting except in combat with other queens.

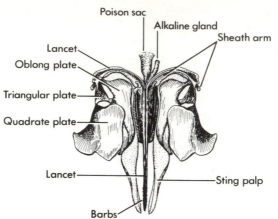

Poison sac
Alkaline gland
Sheath arm
Lancet
Oblong plate
Triangular plate
Quadrate plate
Lancet
Sting palp
Barbs

Figure 11-78 Sting of the worker honey bee. (*After Snodgrass.*)

Sex Determination. As with typical hymenopterous insects, the bees exhibit *haplodiploidy;* the male is haploid, the female diploid. When males are needed, unfertilized eggs are laid. Other orders usually have heterogametic males, either XY or XO for the sex chromosome condition.

Killer Bees. An African strain of the honey bee is especially aggressive and vigorous. Cross-breeding with docile European strains normally used by beekeepers was attempted to improve honey production in Brazil. Unfortunately, a few African strain queens escaped and rapidly established wild colonies.

Escaped bees were more successful than native bees and colonized the Amazon Basin, progressing across it at speeds sometimes exceeding 200 miles per year.

The name "killer bee" comes from the fact that they tend to pursue tormentors farther, attack with less provocation, and attack in greater numbers than the usual domesticated varieties. They can be used domestically, although beekeepers do need protective clothing more urgently than many are accustomed to in normal activities with European strains.

Competition with native bees, interbreeding and dilution of aggressive tendencies, and natural barriers slowed their northward spread less than expected at first and fueled entertainment "scare" features, including at least two motion pictures.

Invasion of temperate areas has proceeded at a much slower rate. Some hybridization does seem to be occurring and the threat, sensationalized out of proportion, will not be much greater in the southern United States, when and if they reach there, than the danger from other bees.

Multiple bee stings of any strain are dangerous to anyone. Near a hive, pheromones released with one bee sting tend to attract additional defenders. Bee and wasp stings annually cause dozens of deaths in the United States, in most cases due to hyperallergic responses to one or a few stings.

Communication Among Bees. Scout bees are able to communicate the direction and distance of a food source to other honey bees by a dance. The dance within the hive can indicate the direction of a food source some distance from the hive, a figure-8 dance has a straight run on the face of the honeycomb, the angle of the run from vertical corresponds to the angle of the food from the direction of the sun. Vigor of the dance, odor, and sound give added information. There has been some controversy over the mechanism of communication. For additional details see Wenner, Wells, and Johnson (1969), von Frisch (1974), and Gould (1975). Jaycox (1976) provides a good guide to basic beekeeping.

INSECTS IN GENERAL

The insects are the largest class of invertebrates. They are also the most important group economically. Enumeration of the general ways they are important will emphasize this point.

A. Positive values and examples.
   1. Pollination of plants. Fruit trees, by honey bee.
   2. Control of insect pests. Cottony cushion scale by ladybird beetle.
   3. Source of food. Honey from bee, roast grasshoppers for the gourmet.
   4. Source of materials. Beeswax, lac, silk.
   5. Food chain intermediates. Aquatic insects.
   6. Scavengers. Blow flies, ants, dermestid beetles.
   7. Medical use. Fly larvae for burns.
   8. Research tools. *Drosophila* for genetics.
   9. Aesthetic value. Butterfly collections, inspiration from their sounds.
B. Negative values.
   1. Parasites of humans, animals, and plants.
   2. Vectors for diseases of people, animals, and plants.
   3. Destroyers of crops and stored food.
   4. Destroyers of fibers, structures.
   5. Nuisance value of bites, stings, carcasses on windshields.

The positive value of crops which require insect pollination outweighs the combined economic losses caused by insects in this country.

Biological Control

The most toxic insecticides, such as the chlorinated hydrocarbons and organophosphates, attack the insect synapses as their mode of action. Unfortunately, they do the same to cholinergic synapses of other animals and people. Some give unwanted bonuses like the DDT-caused egg-shell thinning of birds that has caused near extinction of some fish-eating species due to food-chain magnification of concentration.

These dangers have been a major impetus for the use of biological control methods. *Biological control,* usually means the use of organisms to eliminate or suppress other organisms.

The use of living organisms for the elimination or reduction of pest organisms frequently involves the use of insects. The earliest instance of effective biological control was the introduction of an Australian ladybird beetle, *Rodolia cardinalis,* which effectively controlled a California citrus pest, *Icerya purchasi,* the cottony cushion scale, by feeding upon it.

Perhaps the best method for controlling the Japanese beetle has been the introduction of milky spore disease into soil containing the beetles.

A unique approach to biological control by using sexual competition was successful in eliminating the screwworm fly from Florida. The money invested in the research program and control measures was repaid a hundred-fold or more by the value of the resulting cattle industry in Florida. Control was achieved by the following fortunate circumstances. The female screwworm fly *Cochliomya hominivorax,* mates once for life. Flies can be sterilized by irradiation during the pupal stage without eliminating the ability of the male to mate. By releasing large numbers of sterile flies among the natural population the wild females often layed sterile eggs because of the large percentage that had mated with sterile males. The process was continued until the natural population was eliminated.

Attempts to find biological means of controlling other pests are desirable because use of chemical control measures may pollute food and water with compounds of possible danger to humans. One encouraging line of research in chemical controls is the effort to develop compounds which mimic hormones, enzymes, etc. in their uptake, effectiveness in very low concentrations, and localization in the cells where they will interfere with vital processes. If these can be tailored to species-specific control compounds, with low or no activity in other organisms, concern about pollution will be less than with compounds of broad spectrum activity and low breakdown rates.

Integrated Control

Much rhetoric has been wasted advocating either chemical controls or biological controls. Those involved have tried to inject "common sense" into the controversy by advocating *integrated control*—the judicious use of any and all techniques in proper blend when all factors have been considered. The following discussion will include some of the types of activities that can supplement conventional chemical and biological methods.

Sex pheromones have been used successfully in disrupting successful mating of gypsy moths (Cameron et al., 1974) and pink bollworm (Gaston et al., 1977). They are also useful in baited traps for censusing pest populations. The potential exists for saturating an area with traps as a direct control method. Knowledge of pest distribution can enable selective treatment with less waste in areas of low pest populations.

Use of *resistant varieties* of plants is of value in some cases. As an illustration of a possible problem, Campbell and Duffey (1979) found an instance where a plant toxin was passed through a lepidopteran pest without seriously damaging it, but the toxin did have a severely toxic effect on the parasitic wasp that was intended to help control the pest. The parasite got the poison from the pest.

A more attractive variety can be used to lure pests to a localized area for chemical eradication.

Crop rotation and destruction of infested plant residues are old, but important, techniques that may be incorporated in modern programs. Quarantines, legal, and group action are sometimes helpful in keeping a problem localized for easier treatment. Such procedures are blended with other control procedures to maximize effectiveness while minimizing costs, both direct short-term economic costs and indirect long-term economic costs. The latter are the ecological costs that are ultimately economic to someone. Quantifying these costs provide the best evidence for success in getting action from the political and business communities.

Behavior

In the insects, as in other areas of zoology, behavioral studies are increasing. Among insects we find more highly developed social behavior than in any other invertebrate group.

Construction by Insects. Among the varied things insects make and do, one of the more interesting is the construction of cases by immature caddisflies. The legs and mouthparts of some larvae assemble a tube of materials held together with silk or secretions. The tube is approximately the length of the larva. When certain sensory receptors at the larval posterior are removed a much longer than normal case is produced which extends far beyond the larval posterior. Some bee, wasp, and termite nests are engineering achievements. Some termite mounds are nearly 10 meters tall.

Circadian Rhythms. Many insects show peaks of activity at characteristic times of the day. Thus, the cockroach is most active in the early hours of darkness. The compound eyes, not the ocelli, are essential to keep the activity peak entrained on a 24-hour cycle (Roberts, 1965).

Pheromones. Pheromones of insects are secretions that effect other insects of the same species. They are effective in very low concentrations. Thus, pheromones of some female moths may attract males from a mile or more away. The male cockroach responds with changed activity to air-borne pheromones from females. Ants use them for marking trails.

Chemical identification of many pheromones has been achieved. A sampling of some is given in the table, Some Insect Pheromones. Blum and Brand (1972) have reviewed social insect pheromones.

In some cases an isomer of a pheromone may suppress pheromone activity. Sometimes, as in the *Anthonomous grandis* male-produced sex stimulant, one of the ingredients may be either the *cis* or *trans* form or their combination (Tumlinson et al., 1969).

Regnier and Wilson (1971) have found that the decyl acetate, dodecyl acetate, and tetradecyl acetate from the slave-maker ant, *Formica subintegra,* serves several purposes. They can function as weapons, as alarm pheromones, as trail-marking substances, and as "propaganda" substances for confusing or subduing *F. subsericea* being raided for slavery recruits.

Dietary substances are sometimes needed and modified slightly to form a pheromone. Some cases have been noted where diet has little effect.

Some related species use the same pheromone. Some related species use different percentage mixtures of the same pheromones and respond only to the proportions of their species. Many different types of small molecules of volatile substances are used in addition to those listed in the table. For example, Calman (1971) notes that *Bombus terrestris* and *B. lucorum* are bumble bees using fatty acid ethyl esters for their territory marking scents.

Gilbert (1976) has found indications that *Heliconius erato* butterfly males transfer a pheromone when mating that prevents further matings by repelling males from the female.

Pheromones are thought to regulate development of castes in social insects. Undoubtedly, more behavioral features under pheromonal control remain to be discovered.

Ecology

Insects have radiated to fill most types of niches in terrestrial and fresh-water environments. Most species have certain requirements that determine their distribution. Thus, in aquatic environments low in oxygen, only those with certain respiratory modifications such as

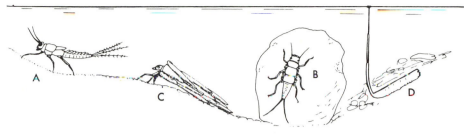

Figure 11-79 Semiaquatic insects that spend only part of their life cycle in the water. Aquatic larvae. (A) Mayfly. (B) Stonefly. (C) Caddisfly. (D) Rattailed maggot. Larvae usually show a strong preference for a particular bottom type.

the hemoglobin of some midge larvae or respiratory tubes that reach the air in other dipteran larvae will be found. By using faunal lists of insects from fresh water, it is possible to deduce certain facts about the water quality. In interpreting such information it is necessary to consider many things such as the season at which such information is gathered because some aquatic insects spend only a portion of their life cycle in the water (Fig. 11-79).

The abundance of insects, their prevalence most places, and small size have made them an important food source for many animals and even some plants. Their importance in the food chain or energy cycle is increased by the variety of nutrients they utilize.

The profound effect insects can have on man's use of the earth is illustrated by the slow development of regions where the mosquito and tsetse fly have actively transmitted severe diseases. A more obvious change in this country is the loss of millions of elm trees, which has already occurred in many places due to an insect-borne disease (see Scolytidae).

Physiological Ecology. Many events in insect life cycles are under partial control of the environment. Thus, food, temperature, and photoperiod are some of the things which may regulate events in development. Selection of the proper responses correlated with seasonal events has contributed greatly to the success of insects in terrestrial environments.

Diapause. Arrested development during certain, usually unfavorable, periods is called diapause. Short photoperiods are often the environmental cue. Tauber and Tauber (1970) have demonstrated that not only absolute length of photoperiod but direction of change can be the cue. They found that *Chrysopa carnea* entered diapause with shortened daylength and began to reproduce with increased daylength even though absolute photoperiods were in the range normally causing the opposite response. Saunders (1973) demonstrated that temperature can control diapause of the parasitic wasp, *Nasonia vitripennis,* in the ab-

Some Insect Pheromones

Insect	Active Substance	Reference
	Female-produced sex attractants	
Pink bollworm (Lepid., Gelechiidae)	gossyplure (mixed isomers of hexa-decadienyl acetate)	Gaston et al., 1977
Gypsy moth (Lepid., Liparidae)	disparlure (cis-7,8-epoxy-2-methyl-octadecane)	Bierl, Beroza, and Collier, 1970 Cameron et al., 1974
Archips semiferanus*	67:33 ratio of (trans-11- and cis-11-tetradecenyl acetate)	Miller et al., 1976
Archips argyrospilus*	30:70 ratio of above plus dodecyl acetate	Miller et al., 1976
Argyrotaenia velutinana*	7:93 ratio of above plus dodecyl acetate	Miller et al., 1976
Archips podana†	50:50 ratio of (trans-11- and cis-11-tetradecenyl acetate)	Miller et al., 1976
Japanese beetle Popillia japonica	(Z)-5-(1-decenyl) dihydro-2(3H)-furanone	Tumlinson et al., 1977
House fly	(Z)-9-tricosene	Carlson et al., 1971

Insect	Active Substance	Reference
	Male-produced sex stimulants	
Queen butterfly	2,3-dihydro-7-methyl-1H-pyrrolizin-1-one; this ketone is adherant to a terpinoid alcohol	Pliske and Eisner, 1969 Meinwald, Meinwald, and Mazzocchi, 1969
Boll weevil *Anthonomus grandis*	mixed I,II, and III or IV or III + IV I = (+)-*cis*-2-isopropenyl-1-methylcyclobutaneethanol II = *cis*-3,3-dimethyl-$\Delta^{1,\beta}$-cyclohexaneethanol III = *cis* $\Big\}$ -3,3-*dimethyl*-$\Delta^{1,\alpha}$- IV = *trans* cyclohexaneacetaldehyde	Tumlinson et al., 1969
	Alarm pheromones	
Homoptera, Aphididae	*trans*-β-farnesene	Bowers et al., 1972
Hymenoptera Formicidae *Formica*	methylheptanone formic acid	Blum and Brand, 1972

*One of three sympatric Tortricidae.   †A European species of Tortricidae.

sence of light. Sonobe and Ohnishi (1971) demonstrated that the hormone mediating diapause is a protein originating in the subesophageal ganglion in the case of the silkworm, *Bombyx mori* L.

Bradshaw (1972) demonstrated that light of 540 nm wavelength is most effective in terminating diapause, especially if it immediately precedes the white light period for *Chaoborus americanus* larvae.

The survival during diapause, whether for lack of food, extreme temperature, or low humidity, presumably requires physiological changes. Many adult insects are killed by freezing, but Miller (1969) demonstrated that winter beetles of the carabid beetle *Pterostichus brevicornis* of Alaska can tolerate freezing temperatures far below those tolerated by the same beetle in the summer. Slow cooling was less damaging than rapid rates of cooling. Glycerol was about 25 percent of body weight in winter beetles compared to negligible amounts in summer.

I have observed large numbers of caliphorid flys forming a nearly solid layer under sheets of tarpaper covering an unheated shed in wintertime Germany. During a sunny period of thaw, many that were uncovered showed sluggish activity. The behavioral massing in protected locations may not prevent low temperatures but will still allow survival by slowing the rate of cooling.

Some insects thermoregulate to limited extent (Heinrich, 1974). Thermoregulation is most readily achieved by large-bodied insects with (1) active flight to achieve heat production, (2) adequate food reserves or nectar supplies to sustain activity, and (3) hairy bodies to conserve heat. These conditions are met by *Bombus* and many Sphingidae that can thus be active in cooler weather or at night. Other insects may maintain appropriate temperatures more effectively by selecting orientation to the sun, degree of exposure to the sun, and times of appropriate ambient temperatures. Hamilton (1973) presents a convincing case for behavior and coloration being selected for adaptation to temperature in a group of desert beetles.

This is but a minute sample of the many interesting and significant physiological–ecological adaptations encountered in insects. Much research in neuronal mechanisms of behavioral, sensory, and motor adaptations is in progress. Olfaction involves receptors in various cuticular locations, such as antennae and tarsi. Sound production, especially in the crickets and cicadas, and its reception involve assorted mechanisms. The light-production mechanism of lamprid beetles for their sexual communication has been commercialized in test tubes for assay of ATP.

Physiology

Among the areas of physiological research on insects where new developments are occurring most rapidly is the field of hormones and neurosecretions. Related to this is one of the most interesting developmental and genetic stories dealing with chromosome puffs (Beerman and Clever, 1964). In midge larvae of the family Chironomidae (Tendipedidae) some morphological events of development have been correlated with production of some hormones and enzymes. Interrelated with this is a sequential puffing of certain chromosomal regions. The puffs are thought to represent an opening up of the DNA helix where messenger RNA is being formed.

The presence of respiratory and excretory tubules, which reach to most parts of the body, have allowed a reduction of the circulatory system in insects. They typically have only a single dorsal vessel, part of which is the heart. The rest of the system is one of spaces.

The localization of mitochondrial enzymes in relation to structure (Candy and Kilby, 1975, Table 1.4) is very similar to those in mitochondria of other higher organisms. However, in insects they are adapted to much greater rates of respiratory exchange rate changes than are vertebrates. The 100-fold change in rate of respiratory exchange possible for nondiapausing insects is made possible by a phosphoralase-b-kinase activated by divalent calcium and stimulated by high inorganic phosphate levels. The kinase enables appropriate transitions back and forth for phosphorylase b, which is inhibited by ATP, to or from phosphorylase a, which is not so inhibited by ATP. The phosphorylases act on glycogen to make glucose-1-P available; phosphorylase a does it at extremely high rates for sustained flight activity. A variety of fuels besides glycogen can be stored in insect flight muscles. They include trehalose, glucose, amino acids, and lipids.

Molting (Fig. 11-80). This important part of development is under the control of a hormonal mechanism. Unknown cues related to development trigger production of a *brain hormone* in neurosecretory cells of the *pars intercerebralis*. From there the hormone is transported along the neuronal axons to the *corpora cardiaca,* a small body attached to the posterior of the brain, where it is stored until released into the hemolymph. The brain hormone stimulates the *prothoracic gland* to produce and release its hormone, *ecdysone* (Fig. 1-1). The ecdysone stimulates the production of cuticle by the hypodermis.

The *corpus allatum* produces *juvenile hormone* that interacts with the effect of ecdysone. When high levels of juvenile hormone are produced, the molt produces a larva, lower levels result in a pupa, and no juvenile hormone release results in an adult produced at that molt.

A small related variety of the hormones involved are naturally occurring. A number of noninsectan compounds have juvenile hormone activity. Carlson and Bentley (1977) have recorded the neural aspect of the movements for final shedding of old exoskeleton of the cricket. Trueman and Riddiford (1970) have shown that coordination of time of ecdysis (molting) is correlated with photoperiod time by action on the nervous system by a hormone released from the brain. Cross-species

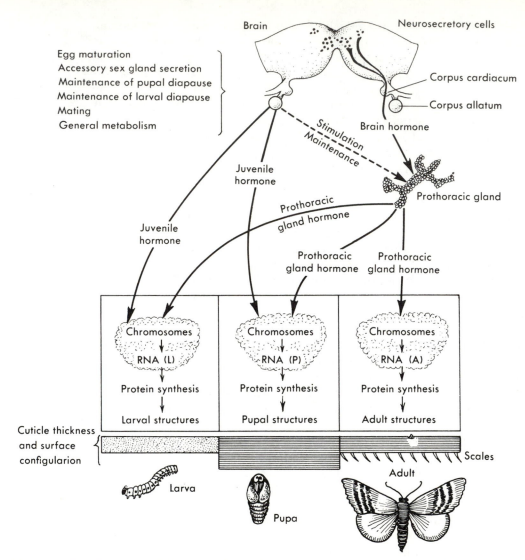

Figure 11-80 Diagram of the interrelationship between hormones and development in the giant silkworm moth. (*From Atkins, 1978, as modified from Gilbert, 1964. Reprinted with permission of Macmillan Publishing Co., Inc., from* Insects in Perpective *by M. D. Atkins. Copyright 1978 by M. D. Atkins. Published by Macmillan Publishing Co., Inc., New York.*)

implantation of replacement brains caused molting times characteristic of the brain's species.

Hormones and analogs for the molting process have been used experimentally with some success for insect control by causing fatal interference with the development process.

Insects are very important in the lives of other organisms. Other organisms are frequently important in the lives of insects. In some cases they have evolved along together as cases of *coevolution.*

Flowers and insects provide some of the best documented cases. Figs and fig wasps are a classic example studied more recently by Ramirez (1969). The yucca and yucca moth were discussed earlier. Some orchids have flower parts that mimic female bees; the copulatory movements of the bees visiting for food are effective in transmitting orchid pollen. Ehrlich and Raven (1967) have investigated the coevolutionary race between plants and insects. Eisner et al. (1969) have indicated an unseen facet in the ultraviolet range with both insects and plants presenting patterns in UV that are not visible in the visible range.

Insects show preferences for both color in the visible range and pattern (Fig. 11-81) of artificial test flowers; the preferences vary with the insect group.

Toxic substances that plants produce for their own protection are sometimes detoxified (Rosenthal, Janzen, and Dahlman, 1977) or they may be retained to produce unpalatability, as in the case of the monarch butterfly and numerous other insects.

The toxic or unpalatable nature of many insects has enabled some to develop bright warning colors. Examples are the yellow and black banding of many stinging hymenopterans and their harmless mimics from the Diptera and Lepidoptera.

Mimicry Rings. An outstanding study of mimicry by Papageorgis (1975) demonstrates five separate mimicry rings in neotropical butterflies. For five rings each set of models and mimics dominates at different levels. The transparent complex flys about 1 meter (modal value) above the forest floor; the tiger complex, slightly higher; the red com-

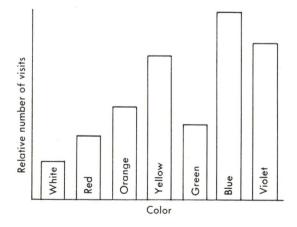

Figure 11-81 Relative frequency of visits to artificial flowers by butterflies, bees, and wasps according to color. Equal numbers of each color of flower were presented for each construction tested. (*From data courtesy of Rex Hollinger.*)

plex, nearly 10 meters; whereas the blue complex is at 10 to 15 meters, near the top of the canopy; and the orange complex flys above the canopy. A few dozen species drawn from as many as six families were involved in the largest rings at the lower levels.

The *coevolution of insects and birds* has been less species-specific than some of the plant–insect coevolutionary examples. Most birds are partially or completely insectivorous. As such they have had a dominant role on the form, coloration, and behavior of many insects, as has been demonstrated by Kettlewell (1965) and others. Treehoppers that resemble thorns, stick insects, insects with cryptic colors, insects with warning colors, and industrial melanism in moths are examples. It is less generally appreciated that insects have had an influence on the direction of selection in birds.

The insectivorous bird bill has been selected for more than mechanical efficiency in grasping insects. Insectivorous birds have bills and skulls that form an angle at the tip of the bill of within a few degrees of 30° (Engemann, unpublished). Raptors that catch prey with their talons have more obtuse angles, as do seed eaters. Birds preying on vertebrate prey possessing more acute vision than insects have bills with more acute angles if they catch prey with their bills. Insects may detect an angle of about 1°, which is much less acute than angles many vertebrates can detect (Chapman, 1971).

A sampling of bird species in the family Parulidae had a significant correlation of width of bill and development of eyeline. Wide bills were associated with prominent eyelines that presumably helped disguise the eye and improve abilities as an insect predator. A geometrical analysis shows that eyelines may even provide an illusion that an object is moving laterally rather than approaching as it really is. Feathers and sharp vision are also assets to an insectivore; insects may be the selective force by feeding those possessing them. The feathers would be expected to muffle airborne shock waves that might be perceived by insect vibration receptors. The value of vision in feeding on insects is obvious.

Insect Habitats. Insects are predominately terrestrial. A water strider is reported from ocean surfaces. Many are important in fresh water but few have evolved to complete the life cycle without emerging. One example of extreme adaptation is a group of stoneflies in subterranean waters reported by Stanford and Gaufin (1974) that apparently complete their life cycles underground and mostly in the water.

Reproductive isolation is reflected in increasing genetic differences through time. A very convenient way of making a determination of this is by comparing isozymes of different populations. There is usually considerable within-population variation, due to such factors as new mutations, hybrid vigor, sexual selection of rare males, and shifting environmental selection of heterozygous populations. By ap-

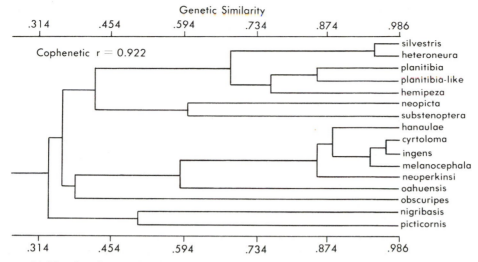

Figure 11-82 Similarity dendrogram among species of the *D. planitibia* subgroup derived from a matrix of Rogers' coefficients by the unweighted pair-group method, using arithmetic means (Sneath and Sokal, 1973). (*Reprinted with permission of Academic Press from W. E. Johnson, H. L. Carson, K. Y. Kaneshiro, W. W. M. Steiner, and M. M. Cooper. Genetic variation in Hawaiian* Drosophila *II. Allozymic differentiation in the* D. planitibia *subgroup.* Isozymes, *vol. IV, Genetics and Evolution. Copyright 1975 by Academic Press, Inc., New York.*)

propriate statistical tools such as indicated in Sneath and Sokal (1973) it is possible to draw dendrograms of relationships (Fig. 11-82), such as was done by Johnson et al. (1975) for some Hawaiian *Drosophila*. Comparable procedures can be used for distantly related groups or higher taxonomic levels if characteristics more conservative than enzyme polymorphism are used.

Phylogeny

The present state of knowledge of relationships of insect orders to each other and to other groups is based predominantly on comparative morphology. Knowledge of comparative biochemistry and the fossil record is not yet complete enough to be of much help. The magnitude of ecological selection and adaptation has been overlooked by those who would postulate a uniramous ancestor rather than one with crustacean affinities for the ancestor of primitive insects. The ecological principle has been pointed out earlier as a probable explanation indicating that the entognathous arthropods are derived from other insects. In general, the insects represent the pinnacle of arthropodization and are not regarded as anywhere near ancestral to other groups.

History of Entomology

Insects have been observed and studied through historical times and probably prehistorically. North American Indians had an extensive vocabulary of insect names in their languages. The golden scarab beetle was symbolic of immense forces for the ancient Egyptians. Aristotle (third century B.C., Greece) provided some order names of insects still in use today.

Fabre contributed many observations on the natural history of insects in nineteenth-century France. Thomas Say (1787–1834) is called the father of American entomology (Atkins, 1978).

Many have made notable contributions to the past and present of entomology: Wigglesworth to physiology; Snodgrass to anatomy; Borror and DeLong meant entomology to recent generations as Essig did to earlier ones. Currently, some leaders associated with various fields include: Alexander, insect sounds; Eisner, chemical secretions; Ehrlich, coevolution; Wilson, social insect behavior; and Williams, hormones.

Numerous excellent entomology texts are in print, a few are listed. Borror, DeLong, and Triplehorn (1976) remains excellent for taxonomic emphasis; Chapman (1971) covers anatomy and physiology in depth; Romoser (1973) is an excellent balanced treatment with an emphasis on modern biology; Little (1972) has abundant information on individual pest species; Atkins (1978) provides a modern treatment with some emphasis on applied entomology; Daly, Doyen, and Ehrlich (1978) have a briefer modern treatment but more on taxonomy than most other texts; Eisner and Wilson (1953–1977) have selected an informative set of *Scientific American* articles; and Barbosa and Peters (1972) provide useful readings from a wider range of journals.

A Possible Effect of Insect Study

The number of entomologists that have achieved prominence outside their specialty seems to be related to the study of insects. Other invertebrate zoologists have had similar career (benefits?) effects. The capacity for understanding complex interactions seems to be developed by practice with situations where such problems can be mastered. Insect populations, their fluctuations, and interactions provide such a training ground. Two outstanding current examples are Edward O. Wilson and Paul Ehrlich.

Wilson's studies of insect societies was probably the catalyst for his broader theories of sociobiology. Ehrlich's knowledge of insect populations undoubtedly was influential in developing his ability to provoke popular concern for population problems. The small size, short life cycles in most cases, and numerous populations in more restricted areas make it possible to develop these skills studying insects better than with any other group.

ARNETT, R. H., Jr. 1967. Present and future systematics of the Coleoptera in North America. *Ann. Ent. Soc. Am.,* **60:**67–72.

ATKINS, M. D. 1978. *Insects in Perspective.* Macmillan, New York. 513 pp.

BAKER, J. M. 1963. Ambrosia beetles and their fungi, with particular reference to *Platypus cylindrus* Fab. Pp. 232–265 + 2 Pl. in *Symbiotic Associations.* 13th Symp. Soc. Gen. Microbiol. Cambridge Univ. Press, New York. 356 pp.

BARBOSA, P., and T. M. PETERS. 1972. *Readings in Entomology.* Saunders, Philadelphia. 450 pp.

BATES, M. 1949. *The Natural History of Mosquitoes.* Macmillan, New York. 379 pp.

BEERMAN, W., and U. CLEVER. 1964. Chromosome puffs. *Sci. Am.,* **210:**50–58.

BELL, R. A., and P. L. ADKISSON. 1964. Photoperiodic reversibility of diapause induction in an insect. *Science,* **144:**1149–1151.

BIERL, B. A., M. BEROZA, and C. W. COLLIER. 1970. Potent sex attractant of the gypsy moth: its isolation, identification, and synthesis. *Science,* **170:**87–89.

BLUM, M. S., and J. M. BRAND. 1972. Social insect pheromones: their chemistry and function. *Am. Zoologist,* **12:**553–576.

BORROR, D. J., D. M. DeLONG, and C. A. TRIPLEHORN. 1976. *An Introduction to the Study of Insects.* 4th Ed. Holt, Rinehart and Winston, New York. 852 pp.

BOWERS, W. S., L. R. NAULT, R. E. WEBB, and S. R. DUTKY. 1972. Aphid alarm pheromone: isolation, identification, synthesis. *Science,* **177:**1121–1122.

BRADSHAW, W. E. 1972. Action spectra for photoperiodic response in a dispausing mosquito. *Science,* **175:**1361–1362.

CALAM, D. H. 1971. Natural occurrence of fatty acid ethyl esters. *Science,* **174:**78.

CAMERON, E. A., C. P. SCHWALBE, M. BEROZA, and E. F. KNIPLING. 1974. Disruption of gypsy moth mating with microencapsulated disparlure. *Science,* **183:**972–973.

CAMPBELL, B. C., and S. S. DUFFEY. 1979. Tomatine and parasitic wasps: potential incompatability of plant antibiosis with biological control. *Science,* **205:**700–702.

CANDY, D. J., and B. A. KILBY (Eds.). 1975. *Insect Biochemistry and Function.* Chapman & Hall, London. 314 pp.

CARLSON, D. A., M. S. MAYER, D. L. SILHACEK, J. D. JAMES, M. BEROZA, and B. A. BIERL. 1971. Sex attractant pheromone of the house fly: isolation, identification and synthesis. *Science,* **174:**76–78.

CARLSON, J. R., and D. BENTLEY. 1977. Ecdysis: neural orchestration of a complex behavioral performance. *Science,* **195:**1006–1008.

CAVILL, G. W. K., and P. L. ROBERTSON. 1965. Ant venoms, attractants, and repellents. *Science,* **149:**1337–1345.

CHAMBERS, D.. L., and V. J. BROOKES. 1967. Hormonal control of reproduction—I. Initiation of oocyte development in the isolated abdomen of *Leucophaea maderae. J. Insect Physiol.,* **13:**99–111.

CHAPMAN, R. F. 1971. *The Insects.* 2nd Ed. Elsevier, New York. 819 pp.

CHU, H. F. 1949. *How to Know the Immature Insects*. Brown, Dubuque. 234 pp.

CLEMENTS, A. N. 1963. *The Physiology of Mosquitoes*. Macmillan, New York. 393 pp.

COYNE, J. A., W. F. EANES, J. A. M. RAMSHAW, and R. K. KOEHN. 1979. Electrophoretic heterogeneity of α-glycerophosphate dehydrogenase among many species of *Drosophila*. *Syst. Zool.*, **28**:164–175.

DALY, H. V., J. T. DOYEN, and P. R. EHRLICH. 1978. *Introduction to Insect Biology and Diversity*. McGraw-Hill, New York. 564 pp.

DEMEREC, M. 1950. *Biology of Drosophila*. Wiley, New York. 632 pp.

DERR, R. F., and D. D. RANDALL. 1966. Trehalase of the differential grasshopper, *Melanoplus differentialis*. *J. Insect Physiol.*, **12**:1105–1114.

EHRLICH, P. R., and P. H. RAVEN. 1967. Butterflies and plants. *Sci. Am.*, **216**:104–113. Reprinted in Eisner and Wilson (1953–1977).

EISNER, T., and E. O. WILSON. (1953–1977). *The Insects. Scientific American* reprints. Freeman, San Francisco. 334 pp.

——, R. E. SILBERGLIED, D. ANESHANSLEY, J. CARREL, and H. C. HOWLAND. 1969. Ultraviolet viedo-viewing: the television camera as an insect eye. *Science*, **166**:1172–1174.

ESAU, K. 1961. *Plants, Viruses, and Insects*. Harvard Univ. Press, Cambridge. 110 pp.

EVANS, H. E. 1963. Predatory wasps. *Sci. Am.*, **208(4)**:144–154.

GASTON, L. K., R. S. KAAE, H. H. SHOREY, and D. SELLERS. 1977. Controlling the pink bollworm by disrupting sex pheromone communication between adult moths. *Science*, **196**:904–905.

GILBERT, L. E. 1976. Postmating female odor in *Heliconius* butterflies: a male-contributed antiaphrodisiac? *Science*, **193**:419–420.

GILLETT, J. D. 1973. The mosquito: still man's worst enemy. *Am. Scientist*, **61**:430–436.

GOULD, J. L. 1975. Honey bee recruitment: the dance-language controversy. *Science*, **189**:685–693.

HAMILTON, W. J., III. 1973. *Life's Color Code*. McGraw-Hill, New York. 238 pp.

HARTMAN, H., W. W. WALTHALL, L. P. BENNETT, and R. R. STEWART. 1979. Giant interneurons mediating equilibrium reception in an insect. *Science*, **205**:503–505.

HEINRICH, B. 1974. Thermoregulation in endothermic insects. *Science*, **185**:747–756.

HENDRY, L. B. 1976. Insect pheromones: diet related? *Science*, **192**:143–145.

HERMS, W. B., and M. T. JAMES. 1961. *Medical Entomology*. Macmillan, New York. 616 pp.

HOFFMAN, R. J. 1972. Environmental control of seasonal variation in *Colias* butterflies. *Am. Zoologist*, **12**:711.

HOWE, W. H. (Ed.). 1975. *The Butterflies of North America*. Doubleday, Garden City. 644 pp.

JACOBSON, M., and M. BEROZA. 1963. Chemical insect attractants. *Science*, **140**:1367–1373.

JACQUES, H. E. 1947. *How to Know the Insects*. Brown, Dubuque. 205 pp.

JAYCOX, E. R. 1976. *Beekeeping in the Midwest*. Circular 1125. Coop. Ext. Serv., Univ. of Illinois, Urbana-Champaign. 169 pp.

Johnson, C. G. 1963. The aerial migrations of insects. *Sci. Am.*, **209**(6):132–138.

Johnson, W. E., H. L. Carson, K. Y. Kaneshiro, W. W. M. Steiner, and M. M. Cooper. 1975. Genetic variation in Hawaiian *Drosophila* II. Allozymic differentiation in the *D. plantibia* subgroup. Pp. 563–584 in *Isozymes*. Vol. IV. Academic Press, New York.

Kafatos, F. C., and C. M. Williams. 1964. Enzymatic mechanisms for the escape of certain moths from their cocoons. *Science*, **146**:538–540.

Kettlewell, H. B. D. 1965. Insect survival and selection for pattern. *Science*, **148**:1290–1296.

Knipling, E. F. 1960. The eradication of the screw-worm fly. *Sci. Am.* **203**(4):54–61.

Levins, R., M. L. Pressick, and J. Heatwole. 1973. Coexistence patterns in insular ants. *Am. Scientist*, **61**:463–472.

Lindauer, M. 1967. Recent advances in bee communication and orientation. *Ann. Rev. Ent.*, **1967**:439–470.

Linsley, E. G. 1963. *The Cerambycidae of North America*. Part IV. Univ. Calif. Publ. Ent., **21**.

Little, V. A. 1972. *General and Applied Entomology*. 3rd Ed. Harper & Row, New York. 527 pp.

Mani, M. S. 1962. *Introduction to High Altitude Entomology*. Methuen, London. 302 pp.

Marx, J. L. 1973. Insect control (I): use of pheromones. *Science*, **181**:736–737.

Meinwald, J., Y. C. Meinwald, and P. H. Mazzocchi. 1969. Sex pheromone of the queen butterfly: chemistry. *Science*, **164**:1174–1175.

Miller, J. R., T. C. Baker, R. T. Carde, and W. L. Roelofs. 1976. Reinvestigation of oak leaf roller sex pheromone components and the hypothesis that they vary with diet. *Science*, **192**:140–143.

Miller, L. K. 1969. Freezing tolerance in an adult insect. *Science*, **166**:105–106.

Miller, T. 1979. Nervous *versus* neurohormonal control of insect heartbeat. *Am. Zoologist*, **19**:77–86.

Mote, M. I., and T. H. Goldsmith. 1971. Compound eyes: localization of two color receptors in the same ommatidium. *Science*, **171**:1254–1255.

Mound, L. A., and N. Waloff (Eds.). 1978. *Diversity of Insect Faunas*. Symp. Roy. Ent. Soc. Lond. Blackwell, Oxford. 204 pp.

O'Brien, R. D., and L. S. Wolfe. 1964. *Radiation, Radioactivity, and Insects*. Academic Press, New York. 211 pp.

Papageorgis, C. 1975. Mimicry in neotropical butterflies. *Am. Scientist*, **63**:522–532.

Patterson, J. T., and W. S. Stone. 1952. *Evolution in the Genus Drosophila*. Macmillan, New York. 610 pp.

Patton, R. L. 1963. *Introductory Insect Physiology*. Saunders, Philadelphia. 245 pp.

Pfadt, R. E. 1962. *Fundamentals of Applied Entomology*. Macmillan, New York. 668 pp.

Pipkin, S., R. Rodriguez, and J. Leon. 1966. Plant host specificity among flower-feeding neotropical *Drosophila* (Diptera: Drosophilidae). *Am. Nat.*, **100**:135–156.

PLISKE, T. E., and T. EISNER. 1969. Sex pheromone of the queen butterfly: biology. *Science,* **164:**1170–1172.

RAMIREZ, B. W. 1969. Fig wasps: mechanism of pollen transfer. *Science,* **163:**580–581.

REGNIER, F. E., and E. O. WILSON. 1971. Chemical communication and "propaganda" in slave-maker ants. *Science,* **172:**267–269.

RICHARDS, A. J. (Ed.). 1978. *The Pollination of Flowers by Insects.* Academic Press, London. 213 pp.

RIEMAN, J. G. 1965. The development of eggs of the screw-worm fly *Cochliomya hominivorax* (Coquerel) (Diptera: Calliphoridae) to the blastoderm stage as seen in whole-mount preparations. *Biol. Bull.,* **129:**329–339.

ROBERTS, S. K. 1965. Photoreception and entrainment of cockroach activity rhythms. *Science,* **148:**958–959.

ROCKSTEIN, M. 1964–1965. *The Physiology of Insecta.* Vols. I and II. Academic Press, New York.

ROEDER, K. D. 1965. Moths and ultrasound. *Sci. Am.,* **212**(4):94–102.

ROMOSER, W. S. 1973. *The Science of Entomology.* Macmillan, New York. 449 pp.

ROSENTHAL, G. A., D. H. JANZEN, and D. L. DAHLMAN. 1977. Degradation and detoxification of canavanine by a specialized seed predator. *Science,* **196:**658–660.

SAUNDERS, D. S. 1973. Thermoperiodic control of diapause in an insect: theory of internal coincidence. *Science,* **181:**358–360.

SCHNEIDERMAN, H. A., and L. I. GILBERT. 1964. Control of growth and development in insects. *Science,* **143:**325–333.

SLATER, J. A., and R. M. BARANOWSKI. 1978. *How to Know the True Bugs (Hemiptera-Heteroptera).* Brown, Dubuque. 256 pp.

SMITH, C. F., and J. GRAHAM. 1967. Life history, synonymy, and description of *Neoprociphilus aceris* (Homoptera: Aphididae). *Ann. Ent. Soc. Am.,* **60:**67–72.

SNEATH, P., and R. SOKAL. 1973. *Numerical Taxonomy.* Freeman, San Francisco. 573 pp.

SONOBE, H., and E. OHNISHI. 1971. Silkworm *Bombyx mori* L.: nature of diapause factor. *Science,* **174:**835–838.

SOUTHWOOD, T. R. E. 1978. *Ecological Methods With Particular Reference to the Study of Insect Populations.* Chapman & Hall, London. 524 pp.

STANFORD, J. A., and A. R. GAUFIN. 1974. Hyporheic communities of two Montana rivers. *Science,* **185:**700–702.

TAUBER, M. J., and C. A. TAUBER. 1970. Photoperiodic induction and termination of diapause in an insect: response to changing day lengths. *Science,* **167:**170.

THORNHILL, R. 1979. Adaptive female-mimicking behavior in a scorpionfly. *Science,* **205:**412–414.

TRIVERS, R. L., and H. HARE. 1976. Haplodiploidy and the evolution of the social insects. *Science,* **191:**249–263.

TRUEMAN, J. W., and L. M. RIDDIFORD. 1970. Neuroendocrine control of ecdysis in silkworms. *Science,* **167:**1624–1626.

TUMLINSON, J. H., D. D. HARDEE, R. C. GUELDNER, A. C. THOMPSON, P. A. HEDIN, and J. P. MINYARD. 1969. Sex pheromones produced by male boll weevil: isolation, identification, and synthesis. *Science,* **166:**1010–1012.

————, M. G. KLEIN, R. E. DOOLITTLE, T. L. LADD, and A. T. PRO-VEAUX. 1977. Identification of the female Japanese beetle sex pheromone: inhibition of male response by an enantiomer. *Science*, **197**:789–792.

TURNER, J. R. G. 1977. Butterfly mimicry: the genetical evolution of an adaptation. Pp. 163–206 in *Evolutionary Biology*. Vol. 10. Plenum Press, New York.

VON FRISCH, K. 1974. Decoding the language of the bee. *Science*, **185**:663–668.

WAKELAND, C. 1958. *The High Plains Grasshopper*. U.S.D.A. Tech. Bull. No. 1167. 168 pp.

WENNER, A. M., P. H. WELLS, and D. L. JOHNSON. 1969. Honey bee recruitment to food sources: olfaction or language? *Science*, **164**:84–86.

WEST, L. S. 1951. *The Housefly*. Comstock, Ithaca. 584 pp.

————, and O. B. PETERS. 1973. *An Annotated Bibliography of* Musca domestica *Linnaeus*. Dawsons of Paul Mall, Folkestone, England. 743 pp.

WIGGLESWORTH, V. B. 1956. *Insect Physiology*. Methuen, London. 130 pp.

WILSON, E. O. 1965. Chemical communication in the social insects. *Science*, **149**:1064–1071.

————. 1971. *The Insect Societies*. Harvard Univ. Press, Cambridge. 548 pp.

————. 1975. *Sociobiology*. Harvard Univ. Press, Cambridge. 697 pp.

12

Phylum Echinodermata

The echinoderms are the starfishes, brittle stars, sea urchins, sea cucumbers, and sea lilies. They are coelomate animals, bilaterally symmetrical in the larval stage but radically symmetrical as adults; usually five antimeres are present. A water-vascular system including organs known as tube feet is characteristic; also a spiny skeleton of calcareous plates. There are five classes of living echinoderms. All living echinoderms are marine species and so, presumably, were the fossil species. Four of the five classes were formerly grouped together in the subphylum Eleutherozoa, because they are able to move about freely; the class Crinoidea together with four extinct classes then comprised the subphylum Pelmatozoa, which are stalked and attached during all or part of their lives.

The sessile condition was a characteristic of the echinoderms from which unattached forms evolved. Since loss of the sessile habit may have occurred at various times in different groups, it may not indicate a close phylogenetic relationship among the Eleutherozoa. Thus, the concept of Eleutherozoa and Pelmatozoa may be abandoned as there is no strong indication that it is a phylogenetic classification.

Four subphyla are currently recognized. The subphylum HOMALOZOA includes only extinct forms with asymmetry or imperfect bilateral symmetry.

Phylum Echinodermata

Subphylum 1. **Asterozoa**.   One superclass, STELLEROIDEA; with radial symmetry; never stalked; movable arms.

660

CLASS I. ASTEROIDEA. Starfishes. Typically pentamerous; arms usually not sharply marked off from the disk; ambulacral groove present. *Asterias forbesi.*

CLASS II. OPHIUROIDEA. Brittle stars. Typically pentamerous; arms sharply marked off from the disk; no ambulacral groove. *Ophiura sarsi.*

Subphylum 2. **Echinozoa.** Tube feet restricted to the compact body; never stalked; radial symmetry, often with superimposed bilateral appearance.

CLASS I. ECHINOIDEA. Sea urchins. Pentamerous, without arms or free rays; test of calcareous plates bearing movable spines. *Strongylocentrotus, Arbacia.*

CLASS I. HOLOTHUROIDEA. Sea cucumbers. Long ovoid; muscular body wall; tentacles around mouth. *Thyone, Cucumaria.*

Subphylum 3. **Crinozoa.** Several classes of radially symmetrical attached forms; one surviving class.

CLASS I. CRINOIDEA. Sea lilies. Arms generally branched and with pinnules; aboral pole sometimes with cirri but usually with stalk for temporary or permanent attachment. *Antedon.*

Class Asteroidea

ASTERIAS—A STARFISH

External Features

The starfishes are common along many seacoasts, where they may be found usually upon the rocks with the mouth down. The upper surface is therefore *aboral*. On the *aboral surface* are many *spines* (Fig. 12-1) of various sizes, *pedicellariae* (Fig. 12-2) at the base of the spines, a *madreporite*, which is the entrance to the water-vascular system, and the anal opening (*anus*). A glance at the *oral surface* reveals a *mouth* centrally situated in the membranous *peristome*, and five *grooves* (*ambulacral*), one in each arm, from which two or four rows of *tube feet* extend.

The Skeleton

The skeleton is made up of *calcareous plates* or *ossicles* bound together by fibers of connective tissue. The ossicles are regularly arranged about the mouth and in the ambulacral grooves and often along the sides of the arms, but are more or less scattered elsewhere. The *ambulacral* and *adambulacral ossicles* (Fig. 12-1) have muscular attachments and are so situated that when the animal is disturbed they are able to close the groove and thus protect the tube feet. The *spines* of the starfish are short and blunt and covered with *ectoderm* (Fig. 12-1). Around their bases are many whitish modified spines called *pedicellariae* (Fig.

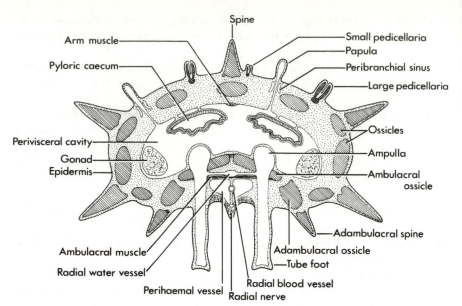

Figure 12-1 Diagrammatic transverse section of the arm of a starfish. (*From Borradaile and Potts, modified.*)

12-2). These are little jaws, which when irritated may be opened and closed by several sets of muscles. Their function is to protect the *papulae* (*dermal branchiae* or *gills*) (Fig. 12-1), to prevent debris and small organisms from collecting on the surface, and to capture food. The skeleton serves to give the animal definite shape, to strengthen the body wall, and as a protection from the action of waves and from other organisms.

The Muscular System

The *arms* of the starfish are not rigid, but may be flexed slowly by a few muscle fibers in the body wall (Fig. 12-1). The *tube feet* are also supplied with muscle fibers.

Coelom

The true body cavity of the starfish is very large and may be separated into several distinct divisions. The *perivisceral part of the coelom* (Figs. 12-1 and 12-4) surrounds the alimentary canal and extends into the arms. It is lined with *peritoneum* and filled with seawater containing some albuminous matter. Oxygen is taken into the coelomic fluid and carbon dioxide given off through outpushings of the body wall known as *papulae* or gills. The coelom also has an *excretory function*, since cells from the peritoneum are budded off into the coelomic fluid, where they move about as amoebocytes gathering waste matters. These cells

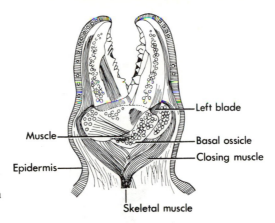

Figure 12-2 Pedicellaria of a starfish. (*After Cuenot.*)

Muscle

Epidermis

Left blade

Basal ossicle

Closing muscle

Skeletal muscle

make their way into the gills, through the walls of which they pass to the outside, where they disintegrate.

The Water-Vascular System

The water-vascular system (Fig. 12-3) is a division of the coelom peculiar to echinoderms. Beginning with the *madreporite* the following structures are encountered: the *stone canal* running downward enters the *ring canal,* which encircles the mouth; from this canal five *radial canals* (Fig. 12-1), one in each arm, pass outward just above the ambulacral grooves. The radial canals give off side branches from which arise the *tube feet* (Fig. 12-1) and *ampullae.* The ampullae are bulblike sacs extending into the coelom; they are connected directly with the

Figure 12-3 Ambulacral system of a starfish. (*After Delage and Herouard.*)

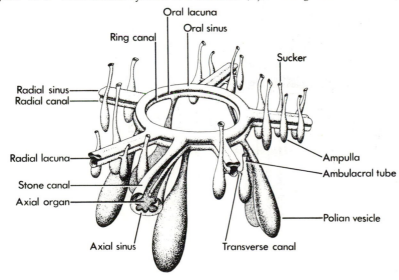

Oral lacuna

Oral sinus

Ring canal

Sucker

Radial sinus
Radial canal

Radial lacuna

Ampulla
Ambulacral tube

Stone canal
Axial organ

Polian vesicle

Axial sinus

Transverse canal

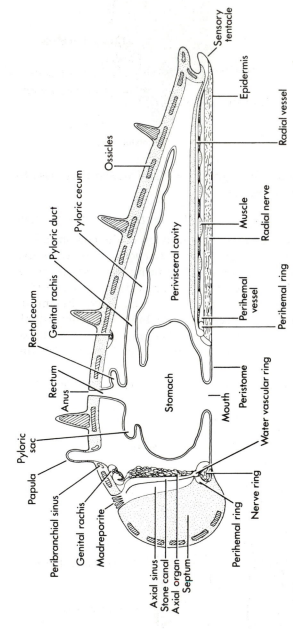

Figure 12-4 Diagrammatic section through the disk and one arm of a starfish. (*From Borradaile and Potts, modified.*)

tube feet, which pass through tiny pores between the ambulacral ossicles. Seawater is forced into this system of canals by cilia, which occur in grooves on the outer surface of the madreporite and in the canals which penetrate it. Arising from the ring canal near the ampullae of the first tube feet are nine vesicles called, after the name of their discoverer, "Tiedemann's bodies." These structures produce amoebocytes, which pass into the fluid of the water-vascular system. *Polian vesicles* (Fig. 12-3) are present in some starfishes, but not in *Asterias.*

The most interesting structures of the water-vascular system are the *tube feet* (Figs. 12-1 and 12-3). They have only longitudinal muscles. A uniform diameter is maintained by circular connective tissue fibers. Fluid forced out of a contracting ampulla causes extension of the tube foot. When the suckered end adheres to an object, contraction of the longitudinal muscles draws the animal forward and returns the fluid to the ampulla. Tube feet are primarily locomotory. They have some sensory and respiratory functions.

Digestion

The *alimentary canal* of the starfish (Figs. 12-4, 12-5) is short and greatly modified. The *mouth* opens into an *esophagus* which leads into

Figure 12-5 Starfish, *Asterias rubens.* Aboral view with two arms and part of skeleton removed; one lobe of stomach cut away and another lobe turned back. (*After Borradaile, modified.*)

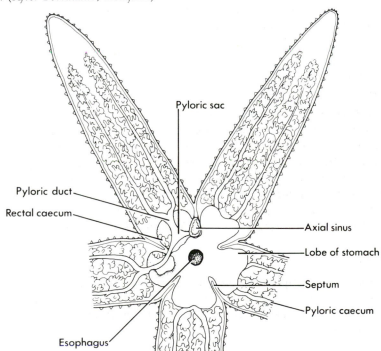

a thin-walled sac, the *stomach*. Following this is the *pyloric sac*. From the pyloric sac a tube passes into each arm, then divides into two branches, each of which possesses a large number of lateral green pouches; these branches are called *pyloric* or *hepatic ceca*. Above the pyloric sac is the slender *rectum*, which may open to the outside through the *anus*. Two branched brown pouches arise from the rectum and are known as *rectal ceca*.

The *food* of the starfish consists of fish, oysters, mussels, barnacles, clams, snails, worms, Crustacea, etc. When a mussel is to be eaten it is held under the mouth and the arms pull on the valves with a force of up to 1,350 grams. The bivalve can withstand a greater pull for a short time, but it eventually yields to sustained pull accompanied by a digestive action of the stomach that has been everted and applied against it. Small bivalves are completely ingested and undigested parts are later expelled through the mouth. Only a very narrow opening is necessary for the starfish to evert its stomach through and begin digestion. Digestive action on the adductor muscles enables the bivalve to be easily opened the rest of the way. Some species of starfish feed on attached microorganisms by everting their stomach over the rock or other materials where they are attached.

The lining of the stomach secrets mucus; that of the pyloric sac and ceca secretes enzymes; these change protein into diffusible peptones, starch into maltose, and fats into fatty acids and glycerine. Thus, digestion is accomplished. Undigested matter is ejected through the mouth, and very little, if any, matter passes out of the anus. The rectal ceca secrete a brownish material of unknown function, probably excretory.

Circulation

The fluid in the coelom is kept in motion by cilia and carries the absorbed food to all parts of the body.

A number of other spaces and canals have been considered parts of the coelom and some consider them to be a "blood"-vascular system. These are the *axial sinuses* lying along the stone-canal and opening to the outside through the madreporite, the *inner, circumoral perihemal canal*, the *outer perihemal canal* beneath the ring canal, the *aboral sinus*, and the *peribranchial spaces*. A lymphatic or phagocytic function is indicated for these structures also (Ferguson, 1966; Millott, 1966).

Excretion

Excretion is accomplished by the *amoebocytes* in the coelomic fluid, probably aided by the rectal ceca. The tube feet and papulae, no doubt, serve as areas where soluble waste products diffuse from the echinoderm.

The papulae and tube feet (Fig. 12-1) function as respiratory organs. Cilia covering the papulae as well as those lining their coelomic side circulate seawater and coelomic fluid thus aiding respiratory efficiency.

The Nervous System

Besides many *nerve cells,* which lie among the epidermal cells, ridges of nervous tissue, the *radial nerve cords* (Fig. 12-4), run along the ambulacral grooves, and unite with a *nerve ring* encircling the mouth. The *apical nervous system* consists of a trunk in each arm which meets the other trunks at the center of the disk; these trunks innervate the dorsal muscles of the arms.

Sense Organs

The *tube feet* are the principal sense organs. They receive nerve fibers from the radial nerve cords. At the end of each radial canal the radial nerve cord ends in a *pigmental mass;* this is called the *eye,* since it is a light-perceiving organ. The papulae are probably sensory, also.

Reproduction

The *sexes* of starfishes are distinct. The *reproductive organs* are dendritic structures, two in the base of each arm; they discharge the eggs or sperm out into the water through pores in the aboral surface at the interspace between two adjacent arms. Most starfishes have an annual breeding season which seems to be induced by rising spring temperatures. Substances affecting release of gametes have been located in the radial nerves. The *eggs* of many starfishes are fertilized in the water; they are *holoblastic,* undergo *equal cleavage,* and form a *blastula* and *gastrula* similar to those shown in Figure 12-28. The opening (*blastopore*) of the gastrula becomes the anus, and a new opening, the mouth, breaks through. Ciliated projections develop on either side of the body, and a larva called a *bipinnaria* (Fig. 12-29) results. This changes (metamorphosis) into the starfish.

Behavior

The starfish moves from place to place by means of its tube feet. During the day it usually remains quiet in a crevice, but at night it is most active.

The *responses* of the starfish to *stimuli* are too complex to be stated definitely. When a starfish is placed on its aboral surface it performs the "righting reaction," i.e., it turns a sort of handspring by means of its arms. Professor Jennings taught individuals to use a certain arm in turning over. One animal was trained in 18 days (180 lessons), and after an interval of seven days apparently "remembered" which arm to use. Old individuals could not be trained as readily as young specimens.

Regeneration

The starfish has remarkable powers of regeneration. A single arm with part of the disk will regenerate an entire body. If an arm is injured, it is usually cast off near the base at the fourth or fifth ambulacral ossicle. This is *autotomy*.

Some species autotomize spontaneously as a method of reproduction and few individuals are found with all five rays of equal size. Because of regenerative ability starfish predators in oyster beds are no longer cut in two and thrown back by the oysterfishermen, but are now completely removed.

OTHER ASTEROIDEA

The starfishes usually possess five arms that are not sharply marked off from the central disk; in some species, however, as many as 40 arms may be present. The body is flattened. They move about with the oral surface down. The viscera extend into the arms and the anus is on the aboral surface. Along the oral surface of each arm is a median ambulacral groove from which project either two or four rows of tube feet. The madreporite is located on the aboral surface between the bases of

Figure 12-6 Asteroidea. The cushion star of the Great Barrier Reef of Australia. (*Photo by author.*)

two of the arms. The outside of the body is covered by a cilated epithelium and characterized by the presence of spines or tubercles, some of which are movable and others modified into pedicellariae. The only special sense organs present are a red eye spot at the end of each arm.

The living species of Asteroidea are in about 20 families in the five orders below. The first three orders are in the Phanerozonia and have in common one or two rows of large marginal plates along the arms; two rows of tube feet in each arm; and pedicellariae, when present, usually sessile or in pits.

Order 1. Platyasterida

Luidia, with spiny fringe on arms; *Platasterias*.

Order 2. Paxillosida

Usually without suckers on tube feet.

FAMILY 1. ASTROPECTINIDAE. Arms long and slender; paxillae (skeletal tubercles bearing many spines) covering membranous aboral wall; conical tube feet without sucking disks; ampullae bifurcate; usually no anus. Ex. *Astropecten articulatus*.

FAMILY 2. GONIOPECTINIDAE. Ampullae simple, anus usually well developed. Ex. *Ctenodiscus crispatus*.

Order 3. Valvatida

Usually with suckers on tube feet and with a relatively large disk.

FAMILY 1. GONIASTERIDAE. (Fig. 12-7, A) Usually flattened, with prominent marginal plates, often colorful. Ex. *Mediaster aequalis*, California; *Hippasteria phrygiana*, East Coast.

FAMILY 2. PORCELLANASTERIDAE. Arms short; body more or less pentagonal; paxillae covering membranous aboral wall; no anus;

Figure 12-7 Asteroidea. (A) *Pentagonaster japonicus*, a type with short arms and pentagonal body. (B) *Odinia*, a type with many long arms and small disk. (*A, after Sladen; B, after Perrier.*)

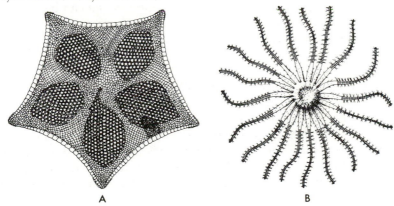

A   B

cribriform bodies between certain of the marginal plates; on deep ocean mud bottoms. Ex. *Porcellanaster coeruleus.*

FAMILY 3. OREASTERIDAE. Body thick; marginal plates small; aboral plates arranged in a reticulum. Ex. *Oreaster reticulatus.*

FAMILY 4. LINCKIIDAE. Disk small, arms cylindroid. Ex. *Linckia*, capable of regenerating from an arm lacking any portion of the central disk.

Order 4. Spinulosida

Marginal plates inconspicuous; tube feet with sucking disks; pedicellariae pedunculate, usually absent (Fig. 12-6).

FAMILY 1. ASTERINIDAE. Body pentagonal; aboral plates granular. Ex. *Patiria miniata.*

FAMILY 2. ECHINASTERIDAE. Arms long and slender; no pedicellariae. Ex. *Echinaster spinulosus.*

FAMILY 3. ACANTHASTERIDAE. Very spiny, multi-rayed, bifurcated ampullae. Ex. *Acanthaster planci* (Fig. 12-30), the Crown-of-Thorns starfish, predaceous on corals of reef slopes, circumtropical.

Order 5. Forcipulatida

Marginal plates inconspicuous; pedicellariae stalked; prominent spines (Fig. 12-7, B).

FAMILY 1. ZOROASTERIDAE. Arms long and slender; disk small; four rows of tube feet; deep sea. Ex. *Zoroaster evermanni.*

FAMILY 2. ASTERIIDAE. Arms tapering and often more than five; medium-sized disk; four rows of tube feet; widely distributed. Ex. *Asterias forbesi, A. vulgaris, Pisaster ochraceus.*

Class Ophiuroidea

The brittle stars and basket stars that belong to this class resemble the starfish in many respects. Although they are not satisfactory for general dissection, species such as *Ophioderma brevispina* can be examined externally for the main features of the class. Only reference to their peculiarities as compared to asteroids is made here.

Skeleton

The disk in the Ophiuroidea (Figs. 12-8 and 12-9) is usually very distinct. The arms are slender and exceedingly flexible. The ambulacral groove is absent, being covered over by skeletal plates and converted into the epineural canal. Each arm is covered by four rows of plates, one aboral, one oral, and two lateral. Spines are restricted to the lateral plates. Within the arm are plates that have fused together and are known as vertebrae. The muscular system of the arm is well developed.

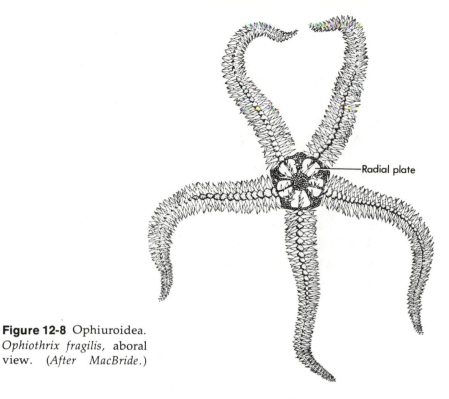

Figure 12-8 Ophiuroidea. *Ophiothrix fragilis,* aboral view. *(After MacBride.)*

Water-Vascular System

This differs in several respects from that of the starfish. The madreporite is on the oral surface. The tube feet have lost their locomotor function and serve as tactile organs; the ampullae have consequently disappeared.

Digestive System

The food of the brittle stars consists of minute organisms and decaying organic matter lying on the mud of the sea bottom. It is scooped into

Figure 12-9 (A) *Ophiura;* oral surface of disk. (B) *Ophioglypha;* aboral surface of disk. *(From Woods.)*

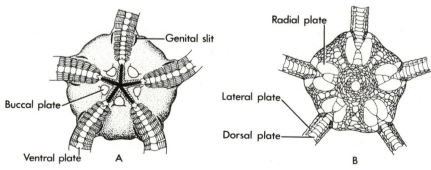

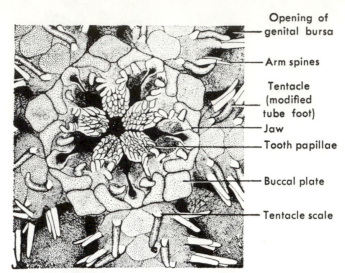

Opening of genital bursa

Arm spines

Tentacle (modified tube foot)

Jaw

Tooth papillae

Buccal plate

Tentacle scale

Figure 12-10   *Ophiopteris papillosa;* mouthparts. *(From Johnson and Snook.)*

the mouth by special tube feet, two pairs to each arm, called the oral tube feet. The rows of spines that extend out over the mouth opening serve as strainers (Fig. 12-10). The stomach is a simple sac without ceca; it cannot be pushed out of the mouth. There is no anus. The basket star is thought to feed on plankton organisms by trapping them in microscopic hooks on the terminal branches of the rays.

Reproductive System

The gonads discharge their products into genital bursae which open, one on either side of the base of each arm in most ophiuroids, but in *O. brevispina* the bursae are divided so that two pair appear beside each arm. One species is viviparous. Some deposit large eggs, which develop directly without a planktonic stage. Many produce small eggs, which undergo holoblastic cleavage, eventually developing into a pluteus (Fig. 12-29) larva before metamorphosis to the adult. In *O. brevispina* the arms of the pluteus do not develop even though the larvae are ciliated and planktonic.

Behavior

The locomotion of brittle stars is comparatively rapid. The arms are bent laterally, and enable animals belonging to certain species to "run," or climb, and probably swim. When pieces of rubber tubing are slipped over the arms the ophiuroids are able to remove them by various methods although they cease trying unsucessful methods.

Regeneration

The term *brittle star* is derived from the fact that these animals break off their arms if they become injured. This autotomy often allows the individual to escape from its enemies, and is of no serious consequence, since new arms are speedily regenerated. In a number of species the aboral covering of the disk is normally cast off, probably for reproductive purposes.

Classification

Two orders contain most of the living species in the class Ophiuroidea.

Order 1. Ophiurida

Five unbranched arms. Disk usually bearing scales or shields.

FAMILY 1. OPHIODERMATIDAE. Disk covered with granules; oral papillae around mouth. Ex. *Ophioderma brevispina.*

FAMILY 2. OPHIOLEPIDIDAE. Disk covered with plates or scales; oral papillae around mouth. Ex. *Ophiura sarsi.*

FAMILY 3. AMPHIURIDAE. Plates on sides of arms with small prominent solid spines; oral papillae around mouth. Ex. *Amphipholis squamata* (a viviparous species).

FAMILY 4. OPHIOTRICHIDAE (Fig. 12-8). Plates small on upper surface of arms; no oral papillae; brachial spines project directly from surface. Ex. *Ophiothrix angulata.*

Order 2. Phrynophiuroida

Disk usually lacking scales and shields.

SUBORDER EURYALAE. Arms may be branched.

FAMILY GORGONOCEPHALIDAE (Fig. 12-11). Arms with many dichotomous branches repeated and giving a dendritic appearance. Ex. *Gorgonocephalus agassiz* (basket star).

Figure 12-11 Ophiuroidea. Euryalae. A basket star.

Class Echinoidea

ARBACIA—A SEA URCHIN

A sea urchin (Fig. 12-12) resembles a starfish whose aboral surface has become exceedingly reduced, being represented by a small area, the *periproct,* and the tips of whose arms have at the same time been bent upward and united near the center of the aboral surface. Both tube feet and spines are used in locomotion. "The spines are pressed against the substratum and keep the animal from rolling over under the pull of the tube feet and also help to push it on."

Skeleton

The skeleton of the sea urchin is known as a shell or *test,* and is shown in detail in Figure 12-13. The *apical system* of plates contains the madreporite, four other genital plates, with genital pores, and five ocular plates, each with a mass of pigmented cells. There are five pairs of col-

Figure 12-12 Echinoidea. (A) *Strongylocentrotus,* aboral view, and (B) oral view. (C) *Arbacia.* (D) A cidaroid sea urchin.

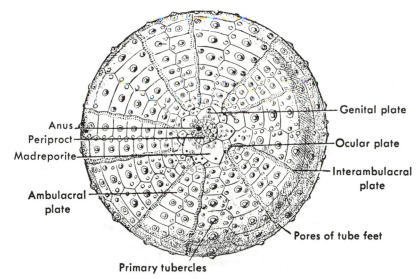

Anus
Periproct
Madreporite
Ambulacral
plate
Genital plate
Ocular plate
Interambulacral
plate
Pores of tube feet
Primary tubercles

Figure 12-13   Aboral view of the test of sea urchin with spines removed.
(*From MacBride.*)

umns of *ambulacral plates,* so called because they are penetrated by
tube feet, and five pairs of columns of *interambulacral plates.* On the
inside of the test around the peristome in many sea urchins are five
arches, often incomplete, called *auricles.* Most of the plates bear
spines, which are attached by muscles and move freely on little knoblike elevations called *tubercles.* The *pedicellariae* are more specialized
than those of the starfish; they commonly have three jaws. The mouth
is provided with five white teeth; these are part of a complicated structure known as "Aristotle's lantern."

Digestive System

The *food* of the sea urchin (Fig. 12-14) consists of marine vegetable
and animal matter which falls to the sea bottom and is ingested by
means of Aristotle's lantern. That sea urchins are capable of capturing
fish may be demonstrated by placing specimens in an aquarium with
a group of *Fundulus heteroclitus.* The head or tail of the fish is secured
by the sea urchin and held against the bottom or side of the aquarium
by the spines and ambulacral feet while the body of the fish is gnawed
away by the jaws of Aristotle's lantern. The *intestine* is very long; it
takes one turn around the inside of the body and then bends upon itself and takes a turn in the opposite direction. A small tube, the *siphon,* accompanies the intestine part way opening into it at either end.
The *anus* of the sea urchin is near the center of the aboral surface.

    Aristotle's lantern (Fig. 12-15) is a complicated apparatus that should
command the admiration of every student of nature. The hollow axis
of the lantern is formed by the *pharynx,* the body of the lantern, which

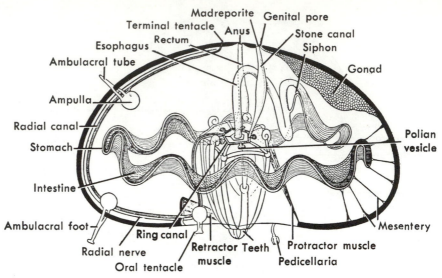

Figure 12-14 Diagram showing the structure of a sea urchin. (*After Petrunkevitch.*)

has the shape of a pyramid with a pentagonal base, is composed of five complicated calcareous parts or *jaws* and as many groups of muscles. When isolated each calcareous jaw appears in the shape of a triangular pyramid. The middle portion of the tooth is enclosed between the two halves of an ossicle called the *alveolus*. The elastic free upper end of the tooth is curved over the base of the pyramid and enclosed in a pouch of the oral sinus. The hornlike processes of the alveoli serving for the attachment of the protractors are termed *epiphyses* and, though fused with the alveoli, are in reality separate ossicles. Radiating from the middle of the lantern at its base are five ossicles, the *rotulae,* articulated to the alveoli. Below the rotulae and also radial in position are five *compasses* or Y-shaped ossicles, so-called because of their two diverging ligaments. These long and thin *ligaments* arise, side by side, from the head or distal enlargement of the compass and are attached to the peristomial edge of the two adambulacral plates on each side adjoining the radius to which the ossicle belongs.

The *muscular apparatus* of the lantern is very complicated. It consists of seven sets of muscles consisting of over 60 individual muscles. Among these are the following: (1) Five short interpyramidal (or interalveolar) muscles are attached to the adjoining radial surfaces of the alveoli; hold the alveoli together; and close the teeth. (2) Five pairs of protractors are attached to the epiphysis and the peristomial edge of the test and run to the inside of, and parallel to, the compass ligaments. (3) Five pairs of retractors are attached to the external surface of the alveoli near the teeth and to the auricles. (4) Five muscles bind together the compasses and form the diaphragm surrounding the eso-

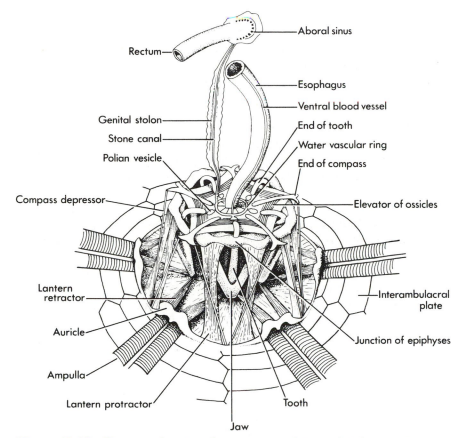

Figure 12-15 Diagram showing the structure of Aristotle's lantern in a sea urchin. (*From MacBride.*)

phagus. All lantern muscles are composed of smooth fibers. The entire lantern with its muscles is enclosed in the *oral sinus* formed by a part of the peritoneum called the *peripharyngeal membrane*.

The antagonistic action of muscles is readily illustrated by the protractor and retractor muscles of the lantern. Muscles work only by contracting and must have either a hydraulic, elastic, or muscular force to return the part moved so further work may be done by the muscle. By examining a side dissection of the lantern (Fig. 12-15) one can see that the protractors must shorten to move the lantern out while the retractors are lengthened. The reverse happens when the lantern is retracted.

Water-Vascular System

One of the genital plates serves also as a sievelike *madreporite* (Fig. 12-14). From the madreporite a *stone canal* leads orally to a *circular canal* surrounding the esophagus aboral to the lantern. From the circular canal arise five *radial canals* and five spongelike bodies called *Tiedemann's bodies or Polian vesicles*. Each radial canal leads orally and outward under the radial plates and ends in a terminal tentacle. *Transverse canals* leading to the *ambulacral feet* and *ampullae* are given off by the radial canals. Two pores for each ambulacral foot are present in the ambulacral plates.

Respiration and the Axial Gland Complex

The tube feet have an important respiratory function in addition to their other functions. Respiration also takes place in most echinoids through ten peristomial gills situated on the area surrounding the mouth, one pair in each angle between the ambulacral plates.

The *axial gland* and associated pulsating vessel complex enveloping the stone canal beneath the madreporite have been shown by Boolootian and Campbell (1964) to have some circulatory function. Farmanfarmaian (1968) found that well over 90 percent of the respiratory function is not dependent on the axial gland/pulsating vessel complex. He had earlier found that tube feet are very important for respiration. Consequently, the axial gland seems predominantly concerned with a lymphatic-type function (Ferguson, 1966; Millott, 1966).

Nervous System

A nerve ring encircles the pharyngeal region. From it five radial nerves extend out under the radial canals. Coordinated movements of locomotion depend on these nerves. Localized spine reactions to mild stimuli are coordinated by a subepidermal nerve net, the spines inclining to the direction of the stimulus.

Reproduction

Spawning can be induced by a factor found in the radial nerves (Cochran and Engelmann, 1972). Sexes are separate and spawning may occur in spring and summer. Fertilization occurs in seawater and development is similar to that described for the starfish in the earliest stages, but the ciliated larva is a pluteus like the larva of a brittle star. The pluteus metamorphoses to become a juvenile urchin. Because eggs and sperm of sea urchins can be taken from urchins, then combined in the laboratory, for successful artificial fertilization, they are important tools of experimental embryology. To apply biochemical techniques to determine sequences of developmental changes in cellular biochemistry, it is much simpler to use quantities of eggs at the same stages of development than to use individual eggs of vertebrates. The earliest experiments involved treatment of eggs with various chemicals to see what aberrations developed. Lithium salts were found to prevent invagination of the endoderm. It would seem that these animals might be useful for preliminary screening of drugs for effects on early development.

The locomotor organs of an echinoid, the sand dollar (*Echinarachnius parma*), have been studied in detail. *Cilia* on the spines seem to perform feeding, cleaning, and respiratory functions. *Ambulacral feet* aid in moving sand up over the anterior edge of the test. *Spines* are of great importance in locomotion. Long and short ones cover both surfaces. They have a power stroke that passes through a plane perpendicular to the body surface. The spine remains close to the body during the recovery stroke. Coordinated spine movements are especially prominent on the anterior oral surface.

OTHER ECHINOIDEA

The Echinoidea are subglobular or disk-shaped and without arms or free rays. The calcareous plates of the skeleton or test are closely fitted together and are usually arranged into five pairs of ambulacral rows and five pairs of interambulacral rows. The surface of the test bears tubercles with which movable spines articulate; these spines are often very long and their tips are in some species poisonous. The pedicellariae are stalked and usually possess three jaws. Over a dozen orders have living species.

Order 1. Cidaroida

Primitive urchins with few, but heavy, primary spines (Fig. 12-12, D). Ex. *Cidaris.*

Order 2. Diadematoida

Spine tubercles perforate, spines long and hollow. Ex. *Diadema,* needle-spined urchins, spines inflict a painful wound, on coral reefs.

Order 3. Arbacioida

Mouth central; anus within apical area; epiphyses separate.

FAMILY 1. ARBACIIDAE. Body subglobular; spines large and solid; aboral tube feet without suckers. Ex. *Arbacia punctulata* (Fig. 12-12, C).

Order 4. Echinoida

Mouth central; anus within apical area; epiphyses united.

FAMILY 1. ECHINIDAE. Body subglobular; spines solid; periproct with many plates; ambulacral plates with three pairs of pores. Ex. *Tripneustes esculentus.*

FAMILY 2. STRONGYLOCENTROTIDAE (Fig. 12-12, A, B). Body subglobular; ambulacral plates with from 4 to 11 pairs of pores. Ex. *Strongylocentrotus dröbachiensis* (Fig. 12-12, A), Atlantic Coast; *S. purpuratus* and *S. franciscanus,* Pacific Coast.

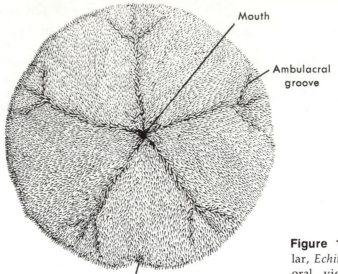

Figure 12-16 Sand dollar, *Echinarachnius parma*, oral view. (*From MacBride.*)

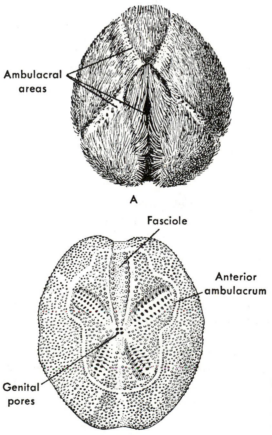

Figure 12-17 (A) Heart urchin, *Echinocardium cordatum*. (B) Test of a heart urchin, *Schizaster*. (*A, from MacBride; B, after Agassiz.*)

680

Order 5. Clypeastroida

Mouth central; anus outside apical area; body usually flattened.

FAMILY 1. CLYPEASTRIDAE. Body bilaterally symmetrical; shell thick; spines short. Ex. *Clypeaster subdepressus.*

FAMILY 2. SCUTELLIDAE. Body flat; spines very small. Ex. *Echinarachnius parama* (sand dollar, Fig. 12-16).

Order 4. Spatangoida

Body cushion-shaped or heart-shaped; anus, and sometimes mouth, eccentric; no Aristotle's lantern.

FAMILY 1. LOVENIIDAE. Heart-shaped. Ex. *Echinocardium cordatum* (Fig. 12-17).

Class Holothuroidea

THYONE—A SEA CUCUMBER

To this class belong the sea cucumbers (Figs. 12-18, 12-20, and 12-21). Their most striking external features are the *muscular body wall* almost devoid of large skeletal plates, the *branching tentacles* surrounding the mouth, and the lateral position when at rest or moving about on the sea bottom. The tube feet, when present, are organs of locomotion. They pull the animal along on its ventral, flattened surface. Waves of muscular contraction which travel from one end of the body to the other are important in locomotion, and the tentacles may also assist.

Body Wall

Instead of an articulated calcareous skeleton, the body wall is soft and *muscular* with irregular plates embedded in the dermal layer. Beneath the epidermis and dermis ia a thick layer of circular muscles overlying five longitudinal muscles, each of which consists of two bundles. Five retractor muscles are connected with the pharynx and a number of muscles extend from the body wall to the *cloaca* which they are able to dilate. The innermost layer of the body wall is a *ciliated peritoneal epithelium.*

Digestive System

The alimentary canal consists of a cylindrical *pharynx,* a short *esophagus,* a small muscular *stomach,* and a long looped *intestine,* the posterior end of which is a muscular enlargement called the cloaca. The pharynx is encircled by a calcareous ring of plates homologous to Aristotle's Lantern of the Echinoidea. The food of most sea cucumbers consists of organic particles extracted from the sand or mud which is taken into the alimentary canal. Some species stretch out their branched tentacles on which many small organisms come to rest. A

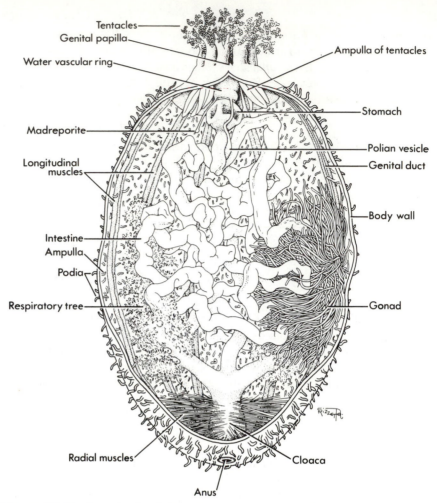

Tentacles
Genital papilla
Water vascular ring
Ampulla of tentacles
Stomach
Madreporite
Polian vesicle
Genital duct
Longitudinal muscles
Intestine
Body wall
Ampulla
Podia
Respiratory tree
Gonad
Radial muscles
Cloaca
Anus

Figure 12-18 Longitudinal dissection of a sea cucumber, showing internal organs. (*Modified after Root.*)

food laden tentacle is inserted into the mouth, then withdrawn so that the food is wiped off.

Hemal System

The greatest development of the echinoderm hemal system occurs in the holothurians. A hemal ring encircles the pharynx and gives off five branches that parallel the radial water canals. The most conspicuous connections are along the intestine, where dorsal and ventral vessels communicate with many finer vessels (Fig. 12-19). The vessels have no special lining. The system is in communication with the water vascular system and coelomic space. The axial complex disappears during development and is apparently not part of this system. The rela-

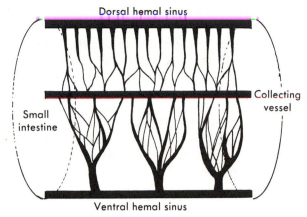

Figure 12-19 Intestine of *Holothuria* showing hemal vessels. (*Adapted from Hyman, after Ludwig.*)

tionship to other systems suggests that the principal function of this system is transport of nutrients.

Water-Vascular System

This is homologous to those of the other classes of echinoderms. There is a *circular canal* around the esophagus, five *radial canals* which end blindly near the anus, and *tube feet.* The circular canal gives off a *Polian vesicle* and one or more *stone-canals* ending in internal *madreporites.* From 10 to 30 of the tube feet surrounding the mouth are modified as *tentacles* for procuring food.

Respiration

Respiration is carried on by the cloaca, *respiratory trees,* tentacles, tube feet, and body wall. The cloaca and respiratory trees also function as excretory organs. Water flows into the cloaca through the anus and passes into two long branching tubes, the respiratory trees; here part of it probably finds its way through the walls into the body cavity.

Behavior

The common sea cucumbers, *Thyone briareus,* are sensitive to contact with solid objects, and many of them burrow in the sand or mud. They are extremely sensitive to a decrease in the light intensity and will contract the body if an object passes between them and the source of light. They are also negative to light since they move away from the source. *Thyone* remains buried in the mud or covered with debris most of the time. This habit probably affords considerable protection from large predators.

Autotomy and Regeneration

Sea cucumbers possess remarkable powers of autotomy and regeneration. When one is irritated it contracts the muscles of the body wall with such great force that the viscera is extruded through the anus (or sometimes the mouth). The respiratory tree is extruded first. It is of some protective value because some branches (Cuvierian organs) of it may swell up into an entangling mass. Parts lost in the process are eventually regenerated.

Autotomy can be induced in *Thyone briareus* in the following manner. When an animal is placed in 7 *N* ammonium hydroxide in the proportion of one part ammonia to 800 parts of seawater, they will rapidly eviscerate. More than 90 percent of specimens eviscerated in this manner live and regenerate lost parts if they are immediately transferred to fresh seawater.

OTHER HOLOTHUROIDEA

Sea cucumbers are elongate cylindrical animals that lie on one flattened side with the mouth at the anterior end and the anus at the posterior end. The muscular body wall contains only small calcareous particles and its external surface is free from cilia, spines, and pedicellariae. There are 10 branched oral tentacles around the mouth, and tube feet and ambulacral tentacles may be present on other parts of the body. Among the South Pacific Islands and on the coasts of Queensland and in southern China, dried holothurians are known as "beche-de-mer" or "trepang" and are used for food. Some may be toxic. Holothurian material is said to be used by some south sea natives to stun fish. Five orders are recognized.

Order 1. Dendrochirotida

Many tube feet; simple or branching oral tentacles. Ex. *Cucumaria frondosa, Thyone briareus.*

Order 2. Aspidochirotida

Many tube feet; peltate oral tentacles. Ex. *Holothuria floridana, Stichopus californicus.*

Order 3. Elasipodida

Tube feet present; no respiratory tree; most are found in deeper regions of the ocean, some are pelagic. Ex. *Kolga hyalina, Pelagothuria natatrix, Scotoplanes.*

Order 4. Molpadonida

Tube feet absent from most of body; respiratory tree present. Ex. *Caudina arenata.*

Order 5. Apodida

Tube feet and respiratory tree absent.

Figure 12-20 *Cucumaria;* a sea cucumber carrying its young. (*After Thomson.*)

Figure 12-21 Holothuroidea. Sea cucumbers; (A) extending from hiding place under coral rock; (B) *Stichopus californicus*. (*B, from Johnson and Snook.*)

A

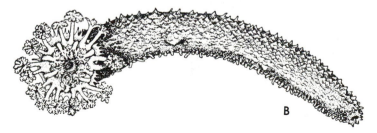

B

685

FAMILY 1. SYNAPTIDAE. Anchor-shaped ossicles in body wall. Ex. *Leptosynapta inhaerens*.

FAMILY 2. CHIRIDOTIDAE. Six-spoked wheel-shaped or sigmoid ossicles in body wall. Ex. *Chiridota laevis*.

FAMILY 3. MYRIOTROCHIDAE. Wheel-shaped ossicles with eight or more spokes. Ex. *Myriotrochus rinkii*.

Class Crinoidea

The crinoids are called sea lilies or feather stars (Figs. 12-22 and 12-26). They are attached by the aboral apex of the body throughout life or during the early stages of their development with the oral surface above. Their arms are usually branched and bear pinnules (Fig. 12-23). The tube feet are like tentacles and without ampullae. No madreporite, spines, nor pedicellariae are present. There are over 600 living representatives of this class belonging to about 150 genera and about 30 families; fossil remains are very abundant in limestone formations.

Figure 12-22 Crinoidea. A feather star, *Antedon adriatica*. The pinnules on three of the arms are omitted. (*After Clark.*)

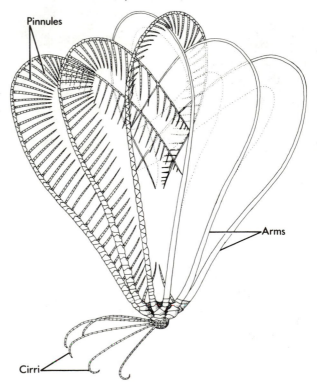

Figure 12-23 Crinoidea. Details of structure of several species. (A) *Comatula pectinata,* grooved anterior arm. (B) *Capillaster multiradiata,* cirrus with dorsal spines. (C) *Leptonemaster venustus,* terminal cone on oral pinnule. (D) *Nanometra bowersi,* large and small cirri. (E) *Pentametrocrinus japonicus,* centrodorsal and articular faces of the radials. This is the only portion of the body that remains of most fossil feather stars. (*After Clark.*)

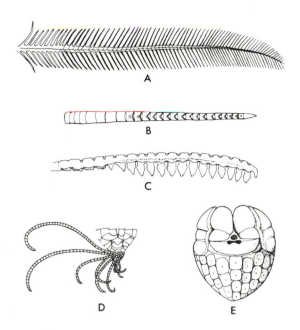

Most of the living crinoids are found at moderate depths, a few are deep-sea forms, and some inhabit shallow water. They are abundant both as individuals and as species where satisfactory living conditions exist. Crinoids are often attached by a jointed stalk. Some species break off from the stalk when they become mature, and swim about by means of muscular contractions of the arms.

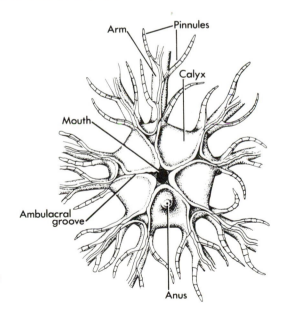

Figure 12-24 Oral view of *Antedon.* (*From MacBride.*)

The arms of crinoids are usually five in number. The apparently greater number is due to branching near the base (Fig. 12-24). The branches may be equal, or one large and the other small; in the latter case the smaller branch is called a pinnule.

Classification

Living crinoids are included in one order, the Articulatida. Three of the more important families are as follows.

FAMILY 1. ISOCRINIDAE. Stalk long, with long cirri; root absent; disk small; arms divided dichtomously up to 10 times; pinnules small; about 80 species. Ex. *Cenocrinus asteria* (West Indies).

FAMILY 2. ANTEDONIDAE (Fig. 12-22). Stalk present only in young; cirri at stalk base in adult; arms long, five to 20 in number; pinnules present; over 130 species; cosmopolitan. Ex. *Antedon tenella* (Atlantic Coast).

FAMILY 3. COMASTERIDAE. Stalk present only in young; cirri few or absent at stalk base in adult; mouth eccentric; many species in tropical seas. Ex. *Neocomatella alata* (West Indies), *Comanthus japonica* (Fig. 12-25).

Figure 12-25   Life cycle of *Comanthus japonica*. (*After Holland and Kubota, 1975*.)

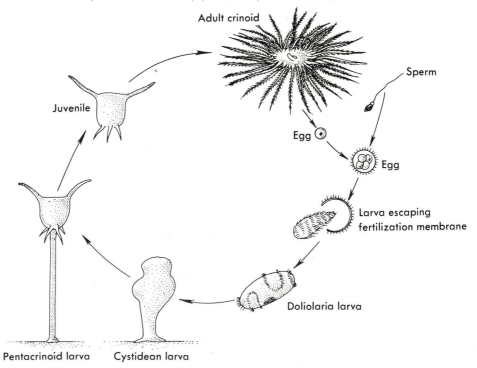

Adult crinoid

Sperm

Juvenile

Egg

Egg

Larva escaping fertilization membrane

Doliolaria larva

Pentacrinoid larva   Cystidean larva

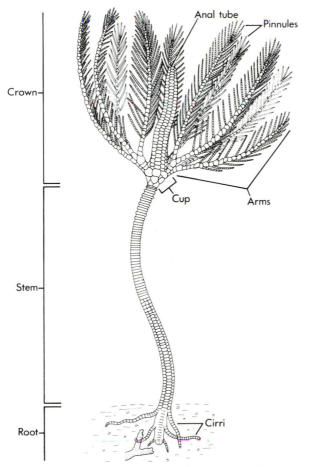

Labels on figure:
Anal tube
Pinnules
Crown
Cup
Arms
Stem
Root
Cirri

Figure 12-26 *Botryocrinus decadactylus;* a fossil crinoid. (*British Museum.*)

Fossil Echinoderms

Every class of echinoderms with living representatives includes also fossil species except the Holothuroidea of which only isolated ossicles have been found. *Eothuria* was perhaps ancestral to the holothuroideans but is generally considered an echinoid because it was covered with plates. The following groups are known from fossil remains only.

MACHAERIDIA. The skeleton in this group consists of two or four rows of plates that resemble those of echinoderms. The animals were elongate, probably bilaterally symmetrical, flexible, and possibly attached. Ex. *Machaeridian lepidocelus.*

HETEROSTELEA. These were stalked forms with bilateral symmetry.

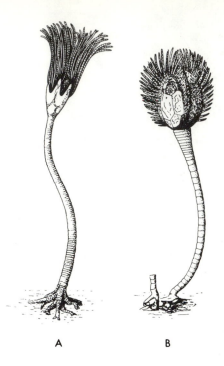

Figure 12-27 Fossil echinoderms. (A) Blastoidea. *Orophocrinus fusiformis.* (B) Cystidea. *Lepadocrinus quadrifasciatus.* (*British Museum.*)

The animal probably lay flat on the bottom. No genital pore or water pore appear to have been present. Ex. *Dendrocystis barrandei.*

CYSTIDEA (Fig. 12-27, B). Here radial symmetry has developed. The body was attached by a stem and food and water were taken in and genital products expelled from the opposite end. Slender, jointed appendages, the brachioles, usually surrounded the distal end. Ex. *Chirocrinus.*

BLASTOIDEA. The main skeleton in these consists of thirteen plates arranged in three circlets of radials, orals, and basals. The body was attached by a jointed stem, sometimes with roots. Many brachioles are present. Ex. *Orophocrinus fusiformis* (Fig. 12-27, A).

EDRIOASTEROIDEA. Here the skeleton forms a radially symmetrical circular test of many irregular plates which was attached directly to the sea bottom. Apparently a well-developed water-vascular system was present including tube feet and ampullae. Brachioles are absent. Cambrian. Ex. *Edrioaster.*

OPHIOCISTIOIDEA. Unattached, discoidal bodies enclosed by plates except in the mouth region. No arms, but five ambulacral areas were present. Ex. *Eucladia.*

CTENOCYSTOIDEA. Very small (3 to 9 mm length) unattached forms with more symmetry than other carpoid Homalozoa. Bilateral symmetry most closely approached by a ctenoid feeding apparatus. From lower Middle Cambrian. Ex. *Ctenocystis.*

HELICOPLACOIDEA. Spindle-shaped and unattached, with a spiraled system of plates and ambulacral areas giving an asymmetry. Adults to about 35 mm long. From Lower Cambrian in California. Ex. *Helicoplacus*.

Echinoderms in General

Anatomy

The echinoderms are unique in the possession of radial symmetry in the adult stage, which is evidently secondary since the larvae are bilaterally symmetrical. The number of radii is usually five.

Body Wall. The epidermis of echinoderms is usually ciliated and contains gland cells and sense cells. The ossicles in the body wall may be few, small, and widely scattered, or large, numerous, and united more or less firmly into a definite skeleton. Certain of the ossicles usually form spines. Pedicellariae are often present.

The ossicles and spines are unusual in that each forms a single crystal of calcite except for polycrystalline structures due to fracturing by the spine base, echinoid teeth, and some holothurian plates (Donnay and Pawson, 1969). The ossicles are *porous crystals* with the crystalline portion forming in an anastomosing network (Nissen, 1969). The collagen of echinoderms may be digested out from these spaces within the network more readily than the collagen found in the more compact minerals of vertebrate bone. Consequently, a conclusion of no collagen (Klein and Currey, 1970) in the ossicles is premature.

Coelom. The coelom is complex, consisting of a number of spaces including a perivisceral cavity, a perihemal system, an aboral sinus system, a water-vascular system, a madreporic vesicle, and an axial sinus.

Systems of Organs. The *alimentary canal* may be axial or coiled and may possess diverticulae. *Respiration* is accomplished in many by the tube feet or by a respiratory tree. The tube feet have various other functions, notably locomotion, but probably originated as sensory or food-collecting organs. No *nephridia* are present, excretory matter being taken up by amoeboid wandering cells that pass through the epithelium. The *circulatory system* consists of strands of peculiar lacunar tissue. Probably circulation is rather slow in species having such a system. The *nervous system* is primitive in type, consisting of nerve rings, radial nerves, and nerves to the tube feet, spines, etc. *Sense organs* are not well developed. The general surface of the body is sensitive to touch and the tube feet and terminal tentacles at the end of each radial vessel are especially sensitive to tactile stimuli. At the base of each terminal tentacle in the Asteroidea is an eye spot. Certain holothurians possess statocysts. The sexes are usually separate. The *reproductive organs* are simple and the ova and spermatozoa are shed directly to the exterior without the aid of accessory glands, penis, seminal vesicle, and seminal receptacle.

Development of Echinoderms

In most of the echinoderms, the eggs pass through a cilated blastula stage (Fig. 12-28), a gastrula stage, and a larval stage, which, in the course of from 2 weeks to 2 months, metamorphoses into an adult. The larvae (Fig. 12-29) of the four principal classes of echinoderms resemble one another, but are nevertheless quite distinct. They are bilaterally symmetrical, and swim about by means of a cilated band, which may be complicated by a number of armlike processes. The alimentary canal consists of a mouth, esophagus, stomach, intestine, and anus. From the digestive tract two coelomic sacs are budded off; these develop into the body cavity, water-vascular system, and other coelomic cavities of the adult.

The larvae of the different classes have been given names as follows: those of the Asteroidea are called *bipinnaria* (Fig. 12-29); Ophiuroidea, *ophiopluteus;* Echinoidea, *echinopluteus;* and Holothuroidea, *auricularia.* The adults which develop from these larvae are, as we have seen, radially symmetrical, although many of them, notably the Holothuroidea, are more or less bilateral in structure. The bilateral condition of the larvae indicates that the ancestors of the echinoderms were either

Figure 12-28 Early stages in the development of the egg of a sea urchin. (*From Woodruff.*)

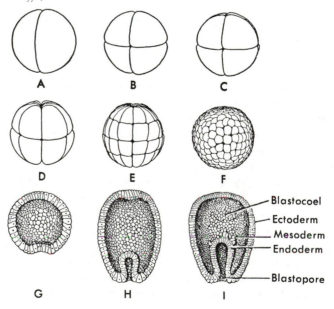

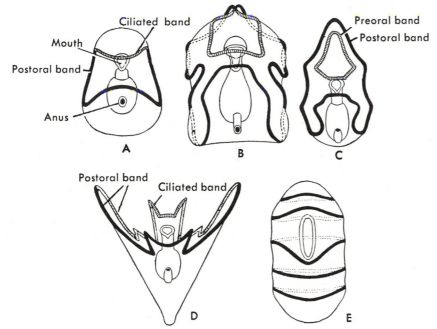

Figure 12-29 Diagrams of larval echinoderms. (A) Early larval stage. (B) Auricularia. (C) Bipinnaria. (D) Pluteus. (E) Crinoid larva. (*From Borradaile and Potts.*)

bilaterally symmetrical or that the larvae have become adapted to an active life in the water.

Artificial Parthenogenesis

The eggs of echinoderms pass through a total and equal cleavage, and are easily fertilized and reared to the larval stage in the laboratory. For these reasons they have become classical material for embryological studies and for experimental purposes.

One of the most interesting phenomena discovered by means of experiments with echinoderm eggs is the development of a larva from an unfertilized egg when subjected to certain environmental conditions. This phenomenon is known as artificial parthenogenesis. The eggs of other animals, for example, annelids, are also capable of developing under certain conditions without fertilization, and those of some species, like plant lice and rotifers are normally parthenogenetic, but echinoderm eggs have been used for experimental purposes more frequently than any others. Large numbers can be fertilized simultaneously in the laboratory. Since all are then in phase in development, periodic sampling and analysis gives more meaningful data on sequential molecular events than with studies of most other animal embryos.

Ecology

The slow-moving, passive appearance of echinoderms does not suggest the considerable impact they have on marine environments. The dominance of brittle stars in abyssobenthic communities may indicate their initial derivation from other stellaroids for that specialization.

Echinoids have a major role in determining the character of marine habitats. Their grazing can eliminate or prevent the development of extensive beds of vegetation. Ogden, Brown, and Salesky (1973) have shown how *Diadema* grazes away vegetation around patch reefs. Estes and Palmisano (1974) show the suppression in sizes and biomass in echinoid populations when sea otters are present. Their observations are consistent with earlier views that sea otter destruction by fur hunters caused decline of California kelp beds due to the explosion of unchecked echinoid populations. A similar phenomenon of decline in New England kelp is thought to be related to overharvesting of lobsters, which are thought to be major predators of echinoids.

Negative growth occurred for some echinoids in tagging experiments by Ebert (1967). Thus, catabolism of their own structure, together with the known abilities of some echinoderms to remove very dilute nutrients from solution in seawater, could prolong the impact of population explosions of echinoderms.

The Crown-of-Thorns Starfish. *Acanthaster planci,* the Crown-of-Thorns starfish (Fig. 12-30), has been associated with some destructive episodes on several coral reefs. Excellent color photos in a popular-type magazine article by Sugar (1970) show the way the stomach can be everted to digest large surface areas of corals.

Chesher (1969) suggests that blasting and dredging are major factors enabling larval starfish to settle and metamorphose without being eaten by the corals that will be fed upon later by the adult starfish. Pollution and predator decline were probably contributing factors.

Others have questioned the causes of the population explosions and the role of *Acanthaster* in producing changes that may be recurring natural phenomena. The awareness of them as a new problem may be a result of lack of observations prior to the development of SCUBA.

The potential importance to the economies of Guam and other regions with coral reefs has prompted much research and publication, ranging from government reports through popular articles. Reviews and many bibliographic references are included in Branham (1973) and Vine (1972).

Branham et al. (1971) found that a large population of *Acanthaster* fed largely on *Montipora* that make up about 5 percent of the coral in Hawaiian waters. The *Montipora* grows very rapidly and seems to maintain its population levels.

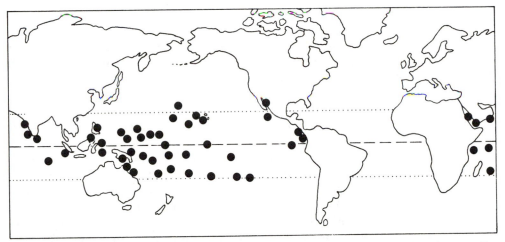

Figure 12-30 *Acanthaster planci,* the Crown-of-Thorns starfish. Map of distribution. Spots indicate known distribution. (*Adapted from Branham, 1973, and Glynn, 1973.*)

Glynn (1973) found that *Pocillopora* reefs off Panama were capable of growing rapidly enough to survive much higher population levels of *Acanthaster.*

Multiple factors are undoubtedly involved in *A. planci* population explosions, perhaps different sets in different locations, including: (1) the destructiveness of *A. planci* may vary geographically; (2) large populations may provide a greater genetic pool, resulting in increased adaptation to more varied coral food; (3) construction or wartime explosions may have improved survival of young starfish by providing coral-free areas in the shallows; (4) destruction of predators by pollution; (5) reduced numbers of large tritons that prey upon starfish, due to collection of tritons for their shells; and (6) natural regulatory mechanisms of reef dynamics may be involved in starfish outbreaks. The precise multiple factors responsible for *Acanthaster* outbreaks will probably remain a subject of controversy since experimental resolution of the problem is impractical.

Physiological Ecology. Lack of effective osmoregulatory capacity seems to limit echinoderms to seawater or saline waters no less than about one-third seawater.

Chemical communication among echinoderms for gamete shedding may be due in at least two classes to substances in the radial nerves. It is probable, but uncertain, that these are the substances inducing natural spawning.

Echinoderms respond behaviorally to substances from their food organisms. *Diadema* give an alarm response to substances released from injured *Diadema* (Snyder and Snyder, 1970).

The axial gland complex of the stelleroids appears to function predominately as a lymphatic mechanism for purifying the internal fluids of particulate contaminants reaching them.

Toxicity is associated with many holothurians. Bakus and Green (1974) have found tropical species most toxic, with 100 percent of the species on some coral reefs toxic.

Interrelations and Phylogeny of the Echinoderms

Echinoderms and coelenterates, because of their radial symmetry, were at one time placed together in a group called Radiata. The anatomy of the adult and the structure of the larvae, however, show that these phyla really occupy widely separated positions in the animal kingdom. The adult echinoderms cannot be compared with any other group of animals, and we must look to the larvae for signs of relationship. The bilateral larva is either a modification for a free-swimming life or an indication of the conditions of its ancestors. The latter view is accepted by most zoologists. The ancestors of echinoderms were doubtless bilateral, wormlike animals which became fixed and were then modified into radially symmetrical adults.

It is interesting to compare the development of echinoderms with that of certain deuterostomes such as *Balanoglossus*. In both, the eggs pass through the indeterminate type of cleavage; the mesoderm arises from the multiplication of cells around the lip of the blastopore; the anus evolves directly from the blastopore; the mouth is formed near the anterior end of the archenteron; the coelom develops from pairs of diverticulae of the archenteron; and the larvae of certain echinoderms are remarkably similar to the *tornaria* larva of *Balanoglossus*. Because of these similarities the echinoderms and chordates appear to be closely related and to have had common ancestors. For this reason these groups are sometimes placed near each other on the phylogenetic tree.

Creatine phosphate is the muscle phosphate of echinoderms as well as chordates. This was once considered strong evidence of a close relationship. However, creatine has since been found in a variety of other invertebrates. There seems to be a high concentration of the enzyme, creatine phosphatase, in forms with low concentrations of creatine phosphate. Thus, it no longer seems to be strong evidence of close phylogenetic relationship. Apparently, it is a case of enzyme– substrate equilibrium, which may be related to food sources of the group. Lack of special relationship to the chordates is supported by the evidence presented by Lovtrup (1975) that chordates and mollusks are much closer to each other than either is to the echinoderms.

Echinoderms are a major phylum of special interest because of three ways in which they differ from other major phyla: (1) they are an old, large phylum with few parasitic members; (2) they are an old, large phylum, which has not invaded freshwater or terrestrial environments; and (3) their radial symmetry seems to be of secondary phylogenetic origin.

Curious Parallels. When relationships seem remote on the conventional grounds used in establishing phylogeny, the presence of two comparable features in unrelated groups is usually ignored or considered to be a case of convergent adaptation. Consequently, the similarities of some isolated echinoderm features to limited features of other phyla has had no attention. The presence of a siphon connecting different regions of the alimentary canal of both echinoids and echiuroids is one such feature. Another is the placement of the retractor muscles for the extrovert of both holothuroids and sipunculids.

The parisimony seen in evolution where major features do not arise *de novo* but often are adapted or modified from existing features should prompt comparison of the excretory systems of most metazoa with the water-vascular system of echinoderms. If the water-vascular system was derived from an excretory system of echinoderm ancestors, the consequent loss of an excretory system, dispensable because the passive diffusion of wastes through thin epithelia is possible, would make reconstitution of such a system difficult. The intimate dependance on seawater related to bathing in isotonic seawater and circulating it in a water-vascular system may explain the freshwater exclusion of echinoderms. As noted for more primitive phyla, excretory organs and organelles have a major water-regulatory function. This lack is undoubtedly why echinoderms are the most stenohaline of the major phyla.

The current view of considering the radial symmetry a secondary feature is based largely on bilateral larval stages and recapitulation theory (biogenetic law). Larval bilaterality could be adaptive to the dispersal phase of the echinoderms. Other similarities with coelenterates, beyond the gross radial symmetry, include ring nerves, the above-noted absence of an excretory system, carbonate deposition for skeletal hard parts, and some gross similarities of life cycles of organisms such as *Aurelia* and *Comanthus.*

This is not a plea for resurrection of the Radiata concept. However, such a possibility should not be totally dismissed until a variety of molecular techniques confirm our present concept of a great divergence for the groups in question.

History of Our Knowledge of the Echinodermata

The name of this phylum was introduced by Klein in 1734 as an appropriate appellation for sea urchins. For many years the echinoderms and coelenterates were included as a class among the Radiata, largely because of the radial symmetry of the adults. The Echinodermata were first recognized as a group distinct from the Radiata by Leuckart in 1847.

REFERENCES TO LITERATURE ON THE ECHINODERMATA

ALTON, M. S. 1966. Bathymetric distribution of sea stars (Asteroidea) off the northern Oregon coast. *J. Fish. Res. Bd. Can.*, **23**:1673–1714.

ANDERSON, J. M. 1965. Studies on visceral regeneration in sea-stars. III. Regeneration of the cardiac stomach in *Asterias forbesi* (Desor). *Biol. Bull.*, **129**:454–470.

BAKUS, G. J., and G. GREEN. 1974. Toxicity in sponges and holothurians: a geographic pattern. *Science*, **185**:951–953.

BOOLOOTIAN, R. A. (Ed.). 1966. *Physiology of Echinodermata*. Interscience, New York. 822 pp.

———, and J. L. CAMPBELL. 1964. A primitive heart in the echinoid *Strongylocentrotus purpuratus*. *Science*, **145**:173–175.

BRANHAM, J. M. 1973. The Crown of Thorns on coral reefs. *BioScience*, **23**:219–226.

———, S. A. REED, J. H. BAILEY, and J. CAPERON. 1971. Coral-eating sea stars *Acanthaster planci* in Hawaii. *Science*, **172**:1155–1157.

CHANGEUX, J. 1961. Contribution à l'étude des animaux associés aux Holothurides. *Act. Sci. Ind.*, No. **1284**. 124 pp.

CHESHER, R. H. 1969. Destruction of Pacific corals by the sea star *Acanthaster planci*. *Science*, **165**:280–283.

COCHRAN, R. C., and F. ENGELMANN. 1972. Echinoid spawning induced by a radial nerve factor. *Science*, **178**:423–424.

COCKBAIN, A. E. 1966. Pentamerism in echinoderms and the calcite skeleton. *Nature*, **212**:740–741.

DONNAY, G., and D. L. PAWSON. 1969. X-ray diffraction studies of echinoderm plates. *Science*, **166**:1147–1150.

DURHAM, J. W. 1966. Evolution among the Echinoidea. *Biol. Rev.*, **41**:368–391.

———, and K. E. CASTER. 1963. Helicoplacoidea: a new class of echinoderms. *Science*, **140**:820–822.

EBERT, T. A. 1967. Negative growth and longevity in the purple sea urchin *Strongylocentrotus purpuratus* (Stimpson). *Science*, **157**:557–558.

ESTES, J. A., and J. F. PALMISANO. 1974. Sea otters: their role in structuring nearshore communities. *Science*, **185**:1058–1060.

FARMANFARMAIAN, A. 1968. The controversial echinoid heart and hemal system–function effectiveness in respiratory exchange. *Comp. Biochem. Physiol.*, 855–863.

FERGUSON, J. C. 1966. Cell production in the Tiedemann bodies and haemal organs of the starfish, *Asterias forbesi*. *Trans. Am. Micr. Soc.*, **85**:200–209.

GLYNN, P. W. 1973. *Acanthaster*: effect on coral reef growth in Panama. *Science*, **180**:504–506.

HILGARD, H. R. and J. H. PHILLIPS. 1968. Sea urchin response to foreign substances. *Science*, **161**:1243–1245.

HOLLAND, N. D., and H. KUBOTA. 1975. Correlated scanning and transmission electron microscopy of larvae of the feather star *Comanthus japonica* (Echinodermata: Crinoidea). *Trans. Am. Micr. Soc.*, **94**:58–70.

HYMAN, L. H. 1955. *The Invertebrates: Echinodermata*. Vol. IV. McGraw-Hill, New York. 763 pp.

KANATANI, H., and M. OHGURI. 1966. Mechanism of starfish spawning. *Biol. Bull.,* **131:**104–114.

KLEIN, L., and J. D. CURREY. 1970. Echinoid skeleton: absence of a collagenous matrix. *Science,* **169:**1209–1210.

LOVTRUP, S. 1975. Validity of the Protostomia-Deuterostomia theory. *Syst. Zool.,* **24:**96–108.

MILLOTT, N. 1966. A possible function for the axial organ of echinoids. *Nature,* **209:**594–596.

NICHOLS, D. 1962. *Echinoderms.* Hillary House, New York. 200 pp.

NISSEN, H.-U. 1969. Crystal orientation and plate structure in echinoid skeletal units. *Science,* **166:**1150–1152.

OGDEN, J. C., R. A. BROWN, and N. SALESKY. 1973. Grazing by the echinoid *Diadema antillarum* Phillipi: formation of halos around West Indian patch reefs. *Science,* **182:**715–717.

ROBISON, R. A., and J. SPRINKLE. 1969. Ctenocystoidea: new class of primitive echinoderms. *Science,* **166:**1512–1514.

SNYDER, N., and H. SNYDER. 1970. Alarm response of *Diadema antillarum. Science,* **168:**276–278.

STEPHENS, G. C., J. F. VAN PILSUM, and D. TAYLOR. 1965. Phylogeny and the distribution of creatine in invertebrates. *Biol. Bull.,* **129:**573–581.

SUGAR, J. A. 1970. Starfish threaten Pacific reefs. *Nat. Geogr.* **137:**340–352.

VINE, P. J. 1972. Recent research on the Crown of Thorns starfish. *Underwater J.,* **4:**64–73.

13

Invertebrates— An Overview

The invertebrates present a tremendous array of diverse types of animals with many specializations and peculiar features. A number of common themes may be seen in various groups. Here, in a few brief pages, some of the general similarities and points of diversity will be offered as a token summary of a subject that would take many lifetimes to present in detail.

Cellular Biology of Invertebrates

Biochemistry. Macromolecules and biosynthetic pathways of invertebrates have much in common within the ranks of invertebrates as well as with other living organisms. Similarities are so great that methods of comparison (Fitch and Margoliash, 1967) are now available that may soon unravel some of the more difficult and intriguing problems of phylogeny. Points of difference, such as formation of unusual compounds, may hold great rewards in searches for new drugs and other materials. An example of their potential value is the use of natural or synthetic pheromones and hormones in pest control with minimal side effects on useful organisms. The invertebrates represent a tremendous resource of varied tissues which may eventually, in tissue culture, biosynthesize needed compounds.

Cytology. Structurally, the cells of protozoans, especially the ciliates, seem to show the most highly evolved forms departing from the "typical" cell.

700

The unusual features were often a point of focus in studying the various phyla, for example sponge choanocytes, coelenterate nematocysts, and flatworm rhabdites and sperm flagella. Not to be overlooked are the more remarkable similarities of chordates and various invertebrate phyla. Intercalated disks are present in human heart muscles as well as in arthropod and mollusk heart muscles. Cartilage of squid and vertebrate are similar. Ciliary-based sensory structures seem to occur in all phyla. Amoeboid cells are nearly universally present in animals. Nervous systems follow similar functional principles: larger fibers are used when rapid transmission is needed in advanced phyla, including chordates, mollusks, annelids, and arthropods; synaptic transmission has been found back to even the sponges; and neurosecretory materials are important in chemical controls, with steroidal materials used widely for regulation.

Lack of distinctive cytology for complex individuals of advanced phyla indicate a recent common origin, or an old one so successfully integrated that little variation is acceptable. Ultrastructure revealed by the electron microscope has shown that the basic structural features include many membranes, fibrils, and microtubules with remarkably similar characteristics. Also revealed are differences that may help to interpret relationships when used in conjunction with other information.

General Morphology

Many aspects of morphology show a close correlation with the physiological function of the structure. Some general correlations with the ecology of the organisms are also present. Thus, larger burrowing forms are typically long and cylindrical in cross section, a shape that allows a larger biomass of organisms to go through small passageways. Because legs or locomotor surfaces from a long linear sequence can all exert their effect on a small cross-sectional area, they are able to force their way through material other organisms could not

Streamlining and organs of attachment are found on many forms adapted to streams and wave-washed areas. Forms lacking these but found in the same habitats are either good swimmers or actually utilizing protected microhabitats with less current.

Sedentary forms often show a radial arrangement. This is most noticeable in feeding structures of those species that do not create a strong feeding current but rely on chance arrival of food organisms. Many sedentary and tube-dwelling forms create a feeding and respiratory current with cilia or specialized appendages. Notable examples of predators that lie in wait for prey to be trapped by devices they have constructed are found in the spiders, some net-building caddisflys, and the ant lion.

Planktonic adaptations often include projecting appendages or parts that slow the rate of sinking, high oil content or gas chambers for buoyancy, or swimming devices such as cilia, nectophores, or swimming appendages.

Morphological modifications for protection of soft parts include covering of spines, plates, or shells. Some shelled forms have developed bivalved coverings as examplified by the brachiopods, ostracods, and pelecypods. Spiraling of a shell (Fig. 13-4) into a compact unit has occurred in the Foraminiferida, Gastropoda, and Cephalopoda. The same effect is achieved in the case of the larva of the caddisfly, *Helicopsyche.*

Smaller-sized species evolved from larger ancestors often have a reduction or loss of circulatory, respiratory, and excretory structures in comparison to their larger relatives.

The spherical or discoidal shape common in dormant stages and cysts presents less surface to what may be an unfavorable environment.

The cuticle of trematodes and cestodes merits special mention, because this rather thick layer is now known to contain many mitochondria and other inclusions. The mitochondria are an indication that the cuticle is performing an active metabolic function. The outer surface of the trematode cuticle is relatively smooth. That of the cestode is covered with microvilli and penetrated by tubules which seem to convey some of the surrounding medium to the interior.

One morphological development may nullify the value of another. Protection in a shell has probably resulted in loss of gross metamerism in mollusks. A tube has probably had a similar effect in obscuring a polychaete origin of the pogonophorans and perhaps other minor phyla. Protected situations of other types often result in loss of distinctive phyletic features as in parasitic reduction, the reduction of the hermit crab abdominal sclerotization, and the interstitial adaptations of numerous micrometazoa.

Feeding and Digestion

Almost any type of natural organic material can be utilized by some invertebrate. Intracellular or extracellular processes may be involved. Food in solution can be utilized by many protozoans and parasites. Particulate food is utilized by most animals. Small particles are acquired by pseudopodial, ciliary, tentacular, mucoid, or setose mechanisms most commonly. Species feeding on larger particles do so in many ways. In addition to simple ingestion of mud or other material by relatively unspecialized mouth parts, many interesting specializations have developed. Fragmentation of larger food masses by the mandibles of many crustaceans and insects, the radula of mollusks, Aristotle's lantern of *Arbacia*, the valves of *Teredo*, chelicerae and gnathobases of arachnids, are types used for solid foods, as is the external digestion by eversion of the starfish stomach. Liquids and soft tissues are utilized by forms with sucking mouth parts or piercing and sucking mouthparts often provided with a muscular pharynx. Among such forms are many leeches, nematodes, suctorians, lepidopterans, and several other arthropod groups.

Internal structural modifications (Fig. 13-1) of the digestive tract include the mastax and gastric mill for grinding, various filtering and sorting devices (Fig. 11-22, A, B), and folds and cecae for increase of digestive surface or storage capacity.

Secretions of salivary and other glands used include anticoagulants, poisons, many enzymes. Remarkably few invertebrates secrete cellulases for digestion of the very abundant material, cellulose.

Figure 13-1 Ultrastructure and function. Scanning electron microscopy reveals some functional aspects of the lining of the cricket alimentary canal (see also Figs. 11-22, A–D). (A) At 1,450× the cell outlines are evident on the convoluted surface of the midgut. (B) Microvilli increase surface area of midgut cells; spacing is much closer than the size of overlying bacteria; 15,000×. Compare these microvilli with those of the bacillary layer of *Ascaris* intestine, Fig. 6-12. (C) The fold of a sphincter separates the midgut from the fibrous appearing hindgut; 275×. (D) The fibrous tufts (shown in Fig. 11-22, D) of the hindgut serve as a haven of safety for many symbiotic bacteria; 6,000×. (*Courtesy of R. Ulrich and D. Buthala, Western Michigan University.*)

Availability of food and frequency of feeding have an influence on growth and rate of development in some species. A most notable example are the ticks.

Saprozoic nutrition and uptake of nutrients from very dilute solutions have been demonstrated for ever more invertebrate groups. For example, uptake of amino acids at ambient concentrations in seawater can be done by some coelenterates, echinoderms, and mollusks. This type of nutrition may be important in survival for dispersal of planktotrophic marine larvae. Direct absorption of nutrients is important for many parasites. It is presumed important for many Protozoa in organically enriched substrates. We can anticipate that saprozoic nutrition will be demonstrated as the feeding method of the Pogonophora, if not already evident, and that many interstitial forms may use it as an important nutritional supplement.

Water Relationships

The osmotic concentration of body fluids of invertebrates as measured by freezing point depression in degrees Celsius ranges from −0.08 for the freshwater mussel to −1.85 for marine invertebrates, which have an osmotic concentration near that of seawater. Since marine invertebrates are nearly isoosmotic with seawater, they have few regulatory problems with water balance although they do regulate the ions present in a different balance than that of seawater.

Body fluids of freshwater invertebrates are hyperosmotic to their environment, so the excretory organs and organelles function to eliminate water which diffuses into the organism in addition to eliminating some metabolic wastes. *Artemia,* the brine shrimp, is peculiar in that it is hypoosmotic to its environment. It absorbs water and salt through the gut and excretes salt from special glands in the gills (Copeland, 1966). The glands consist of mitochondria arranged as nested saucers. It is of interest that the contractile vacuole of protozoans is surrounded by many mitochondria. The mitochondria probably provide energy for active transport.

Terrestrial forms are faced with the problem of retaining water for proper functioning of the cells and tissues while still eliminating metabolic wastes. Great losses occur from respiratory surfaces which are not well protected. Thus, the earthworm loses about a third of its body weight if left in dry air at room temperature for about ½ hour. Many insects have negligible losses under similar conditions. Other terrestrial invertebrates that have intermediate losses under such conditions in nature live in humid environments or live with access to water or fluid. Neurosecretions sometimes have a role in water regulation (Riddiford et al., 1976). Behavioral, metabolic, and structural adaptations for terrestrial existence (Bliss et al., 1968) are many times selected for their value in maintaining water reserves.

Circulation takes place on the cellular level by diffusion and protoplasmic movements. Circulatory systems of larger organisms eliminate most of the restrictive effects of dependence on diffusion over long distances. They also make possible the localization of physiological functions in organs and a great efficiency through specialization. Extracellular transport may depend on one or more of the following.

1. Movement of the surrounding water. Ex. Sponges and some coelenterates.
2. Gastrovascular-type system. Ex. Coelenterates and flatworms.
3. Transport by body fluids of the coelom or pseudocoelom. Two variations of this type occur. (a) Fluid moved by body movement or activities with a noncirculatory function. Ex. Rotifers, nematodes, and annelids (coelomic fluid). (b) Fluid moved by peritoneal cilia, often in a specific circulatory pattern. Ex. Ectoprocta, Brachiopoda, Chaetognatha, Sipunculida, and Echinodermata.
4. Circulatory, or vascular, system. A great number of variations occur ranging from vessels with peristaltic contractions to well developed chambered hearts with valves and complex system of vessels. *Open* types have a heart and arteries with blood returning through sinuses or hemocoelic spaces. *Closed* types have heart, arteries, capillaries, and veins which return the blood to the heart. Heart muscles generally have some spontaneous rhythm of contraction. The intercalated disks of some invertebrate heart muscles (Hill et al., 1979) increase the evidence of close vertebrate–invertebrate relationship.

Blood often contains amoebocytic cells, proteins, and respiratory pigments as well as food, wastes, antibodies, and hormones.

Body fluid or blood is important as a hydrostatic skeleton and in locomotor movements of some groups.

Respiratory Mechanisms

Respiratory surfaces are typically moist, thin walled, and often highly vascularized. There are four general types: the general body surface or surface of appendages, gills, lungs, and tracheae. The Oligochaeta are an example of the body surface type; many mollusks and crustaceans have gills; pulmonate snails have a lunglike mantle cavity, spiders have a lamellated type of lung, holothuroideans a water-lung called the respiratory tree; tracheae of insects and some other arthropods differ from lungs in the extent of penetration of body parts, the tracheal type being less dependent on transport by circulating fluids.

The anthozoan siphonoglyph is probably an aid to circulation of water for respiratory purposes as is the gill bailer, or scaphognathite, of the decapods. The hydrofuge hairs of some aquatic insects are water repellent, thus trapping a plastron of air on the surface of the body for circulation through the trachea. Many other interesting modifications occur in aquatic insects including such things as respiratory tubes in many dipteran larvae and rectal "breathing" in dragonfly naiads.

Respiratory pigments in the blood increase its oxygen carrying capacity. They may facilitate one or more of the following.

1. Proportionally smaller respiratory structures.
2. Greater size.
3. Greater activity.
4. Acquisition of sufficient oxygen at lower oxygen tensions.
5. Reduction of the circulatory system.

Respiratory pigments may be red, blue, green, or brown. Hemoglobin —a red, iron-containing compound—is one of the more common invertebrate respiratory pigments. It is found in many annelids, some mollusks, some echinoderms, and some arthropods. Hemocyanin—a green, copper-containing compound—is found in cephalopods, *Xiphosura,* and the Malacostraca.

Respiration may be aided by activity of the animal. As oxygen in the water decreases, *Tubifex* extends farther from its tube and undulates its posterior end. Leeches may undulate their body while attached by both suckers when oxygen is depleted. Many other aquatic forms increase the rate of respiratory movements as dissolved oxygen decreases until a point near death. This activity enables them to survive under some suboptimal conditions. The opposite approach can have survival value too. Cessation of activity minimizes oxygen requirements under conditions of low oxygen. This reaction if exhibited by *Xiphosura.*

Anaerobic respiration is possible for some invertebrates for short periods of time. A few, such as the flagellates of termites, must live under anaerobic conditions. Species that commonly depend on anaerobic processes (many intestinal parasites are of this type) live in organic rich environments.

Excretion

The elimination of the end products of metabolism is accomplished by a variety of organs. In aquatic forms carbon dioxide and other soluble wastes are partially eliminated by diffusion from respiratory and other surfaces. Water produced by metabolism as well as water taken in by other means usually serves as the medium of transport for nitrogenous and other waste products. Specialized structures for these purposes include contractile vacuoles (Protozoa), flame cells (flatworms), nephridia (annelids), maxillary glands (crustacea), and coxal glands (arachnids). Malpighian tubules, of insects and some other terrestrial arthropods, eliminate uric acid, which can be excreted with very little water loss. Ammonia is the principal nitrogenous waste product of most aquatic species. A few species are known to store some waste products.

Nervous System

Most nervous systems are usually polarized and synaptic although some think a portion of the nerve net of *Hydra* is neither since transmission is in any direction and contact is apparently made between neighboring cells. The system is quite diffuse in coelenterates, flatworms, and echinoderms, but in arthropods and higher mollusks nerve cell bodies are not distributed along the bundles of their processes forming the nerves. Even in forms without an obvious head many have cephalization of the nervous system and cerebral ganglion development. Especially large ganglia may be associated with image forming eyes. Ultrastructural studies have demonstrated the existence of nerve cells and synapses in sponges.

Sensory receptors are known for a wide variety of stimuli. Many groups have eyes. Investigation of color vision has shown that a sample of arthropods have the needed variety of receptors showing at least two different spectral sensitivity curves (Wasserman, 1973). Statocysts responding to gravitational force are common, and some receptors of insects detect sound. Other stimuli responded to include touch, chemical substances in water or air, heat, and humidity. Some snails may respond to magnetism. Response of insects to ionizing radiation is probably not due to a specific receptor.

Neurosecretions are responsible for most of the hormonal activity demonstrated in invertebrates. It seems probable that all invertebrates having nervous systems also produce neurosecretions although only the insects and crustaceans have been intensively studied in this regard. Hormones are also known from the gut of an insect. The most primitive phyla in which neurosecretions have been found are the Porifera, Coelenterata, and Platyhelminthes.

Discovery of neurosecretions was anticipated for the sponges when nerve cells were demonstrated. Demonstration of neurosecretion function will be difficult because of the low level of organization and low level of activity of sponges.

Axonal transport of neurosecretory products has been noted in both arthropods and mollusks. The transport function may have been the original function of elongated cells that preceded nerve impulse transmission.

Pheromones

Pheromones, the chemical substances that modify the behavior of others of the same species, have been found to have a variety of functions in insects (Chapter 11). Such chemical communication is expected in most groups and already noted in several other phyla. The fundamental importance of such substances is shown by their demonstration in the Protozoa (Miyake and Beyer, 1974).

Behavior

Appropriate activities are essential to the survival of organisms. Such activities and the neuromuscular basis are the subject of intense research activity (Witt et al., 1972; Sherman et al., 1973). Redundancy may be built in to important systems. Gardner (1971) notes the bilateral duplication of functional nervous system units in *Aplysia*. Different physiological types of muscles, fast or slow, increase ranges of response in some mollusks and some arthropods. Some increased control options result from having both excitatory and inhibitory innervation of some invertebrate muscles.

Immune Responses and Cell Compatibility

Immune responses may be the cause of failure of some cross-species grafts. Immunity may be the basis of resistance to additional infections seen in some snail–host, trematode–parasite relationships. An inducible antibody to honey bee venom can be produced by the cockroach (Rheins, Karp, and Butz, 1978).

Selective suppression of immune responses seems to have been an important feature in animal evolution. The mitochondria of all organisms and the chloroplasts of plants and phytomastigophorans are thought to be derived from internal symbionts. Utilization of intact organelles from other organisms have been observed in flatworms and mollusks. Opisthobranch mollusks seem to be among the most successful at this activity, some having the capacity to convert coelenterate nematocysts or algal chloroplasts to their own use. We do not know whether the almost exclusive role of gastropods as digenetic trematode first intermediate hosts is due to a special capacity for symbiosis or if it is a historical artifact of an original ancestral parasitic symbiosis.

Reproduction

Asexual reproductive phenomena include fission, budding, and regeneration of fragments. Sponge gemmules and bryozoan statoblasts may be considered types of internal buds.

Reproduction from eggs may occur parthenogenetically in some groups (Oliver et al., 1971). Fertilized sexual eggs are produced by both hermaphroditic and dioecious species. The mechanisms of male production varies. Some produce males from unfertilized eggs, females from fertilized eggs. In many, sex is determined by the chromosomes inherited. Sex ratios are sometimes environmentally regulated. *Protandry*, or reversal of the functional sex of a hermaphroditic individual, occurs in several phyla.

Internal fertilization or other types of transfer of sperm by individuals occurs in most parasitic species and species producing small numbers of large eggs. External fertilization may occur in the seawater for many species of marine invertebrates which (more often than freshwater species) produce great quantities of eggs of small size. In the latter case coordination of release of gametes may be by environmental conditions such as temperature or phase of the moon or by pheromones released by gamete shedding individuals. Thus, coordination in species not transferring sperm directly minimizes waste of gametes.

Planktonic larvae are common in marine species with small eggs. The enormous numbers enable a few to survive predation and slight chances of reaching favorable environments.

Reproductive success is the ultimate criterion of group survival. An amazing array of behavioral, structural, biochemical, and ecological adaptations are used to achieve such success.

Knowledge of chemical coordination of mating activity continues to expand as sex pheromones are one of the pheromones more easily demonstrated behaviorally. Coordinated release of gametes for external fertilization is known in some instances, and we may expect demonstration in many more instances that it often is chemically (pheromonally) coordinated.

Reproductive success is not only determined by success in the physical environment and competition with other species, but by the results of intraspecies competition as well. The competition can be for either resources or mating. Documentation for mating competition adaptations include the previously noted effect of cement glands of Acanthocephala in hampering insemination by competing males, the behavioral tricks of some scorpionflys to rob the efforts of other competing males, and structural modifications of damselfly males that remove sperm from prior matings while inseminating the female (Waage, 1979). Gamete competition may be a factor determining sperm structure as noted for the gnathostomulids.

Large size of males is often associated with aggressive behavior in competition with other males as recently noted in mites as well as in theory. The small size of males is more efficient for the species in energetic terms as is seen in many spiders; the female has greater nutritional requirements than the male because of the relatively larger volume of nutrients in eggs.

Ecology

All of the preceding range of features may relate to an organisms adaptation to its environment. General roles were discussed in Chapter 1 and additional specific and general examples in later chapters. Only a few of these examples will be noted here.

The role of earthworms in soil formation was noted. In tropical

regions the termites sometimes have an important role in soil structure. Scarabaeid beetles help tremendously to remove vertebrate
wastes from the soil surface of some savannah regions.

Marine environments are sometimes structured by invertebrates.
Reef formation by corals is well known. Vermetid gastropods have a
major role in production of some very small reefs (Safriel, 1974). Tubi-
colous polychaetes can stabilize marine sediments. Some invertebrates
can be destructive of their environments, as in the case of *Acanthaster*
changing the nature of some reef environments. Some effectively penetrate carbonate substrates (Carriker et al., 1969) and modify them in
the process.

Some invertebrates survive extremes of environment by inactivity
(Bushnell et al., 1974). Many live in such rigorous environments or are
so well adapted that shifts in conditions usually result in shifts in
dominant species representing a genus, or other taxonomic unit with a
similar trophic role. Interesting examples are given by Harger (1972)
for intertidal invertebrates. *Mytilus californianus* excels in unprotected
intertidal areas of the West Coast. It is very difficult to remove from
the solid substrate. In more protected locations *M. californianus* is replaced by *M. edulis,* whose attachment is less strong (Fig. 13-2, A). *M.*

Figure 13-2 Comparison of *Mytilus californianus* and *M. edulis.* (A) Force required to dislodge a specimen *versus* its size. (B) The percent crawling to a
clear area after being covered with gravel *versus* time. (*Adapted from the data of
Harger, 1972.*)

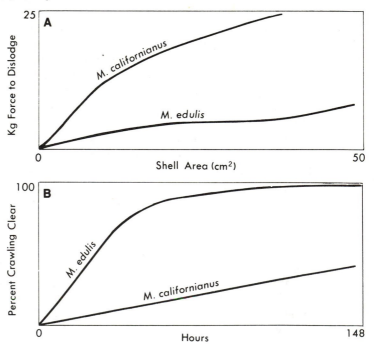

edulis can move more readily than *M. californianus* and has a better chance of escaping burial (Fig. 13-2, B) by shifting sediments in the semiprotected locations. Other species pairs noted by Harger (1972) that show adaptations of one suited for high-energy zones and the other for more protected areas include *Littorina scutulata* and *Hemigrapsus nudus* from high-energy zones, whereas *L. sitkana* and *H. oregonensis* are snails and crabs from semiprotected locations.

Sessile animals such as the lophophorates, sponges, and coelenterates are especially vulnerable to predation. Many have shells or other hard coverings. Some are toxic. Some may have relatively little nutrient reward compared to volume or the energy a predator must expend and thus be nonpreferred food. Time of larval settlement varies by species (Fig. 13-3).

Figure 13-3 Time of settlement for larvae of sessile marine epifaunal invertebrates of the Woods Hole region. The frequency of settling larvae is roughly proportional to the thickness of the lines. Numbers refer to the percent of 54 species that have some larvae settle during the month. (*Adapted from the data of Osman, 1977.*)

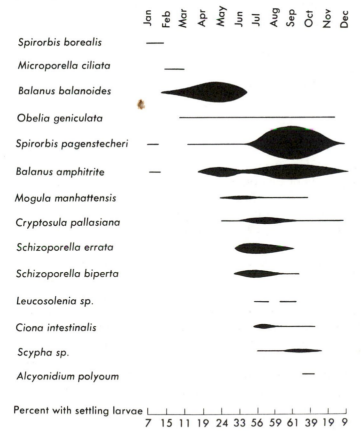

Defensive mechanisms abound. Mimicry is of use to both those with and without such defenses. An interesting case of mimicry rings was noted for insects in Chapter 11.

Development

Early cleavage of higher invertebrates is considered to belong to one of two general types, spiral (or determinate) cleavage of the Protostomia, and radial (or indeterminate) cleavage of the Deuterostomia. A trochophore larval type, unequal early cleavage into different sized cells of fixed developmental potential, entomesoderm development from the 4d cell, mouth formation from or near the blastopore, and schizocoelous coelom formation in coelomate groups—these are considered typical of the Protostomia, which includes the flatworms, the pseudocoelomate phyla, lophophorate phyla, annelids, mollusks, and arthropods. The Deuterostomia (Chaetognatha, Pogonophora, Echinodermata, Hemichordata, and Chordata) undergo early cleavages into equal-sized cells, which do not have a fixed developmental potential; their mouth does not form in the region of the blastopore; and coelom formation is enterocoelous. The range of variation within each group is greater than simple comparisons indicate when the many exceptions are not noted. (For considerable within-species variation, see Figs. 13-4 and 9-19.)

Interest in the important phenomena of development has resulted in much continuing research (Costello et al., 1976). A complete assessment seems to blur the neatness of a Protostomia–Deuterstomia dichotomy in the view of Lovtrup (1975).

Phylogeny

Speculation on the relationships of the various phyla has been based on a variety of criteria by different authors. For a discussion of some general aspects of phylogeny, see Chapter 1. Brief views of phylogeny of major groups are found in most chapters. More detailed discussion of theories may be found in Chapter 5, Volume I, of Hyman (1940); Chapter 9, Volume II, of Hyman (1951); and Chapter 23, Volume V, of Hyman (1959).

The relationship of animals to other organisms is reviewed by Whittaker (1969). Sokal (1974) points out some of the goals and pitfalls of classification. One fact of classification is that the design of a classification is determined by its use. Unfortunately, a desirable phylogenetic classification, the usual goal, is somewhat unattainable because of our present ignorance. We find the unattainable all the more desirable. Many have tried and have provided many different phylogenetic "trees." After reviewing all the phyla, it seems reasonable to join the parade and present a phylogenetic "tree" to represent one concept of phylogeny for the phyla of invertebrates. One fault of the tree as diagrammed in Figure 13-5 is the inability to show proper time

Figure 13-4 *Biomphalaria* snails from a single spring in the Mizurata area of Libya. The typical planorbid coiling is absent in the aberrant specimen. Such a gross change in a single specimen of a population indicates the possible sensitivity of morphology to slight genetic and/or environmental variables. (*Photo from specimens courtesy of Mr. Abubaker Swehli.*)

relationships. The origins of all the metazoan phyla with living representatives go back millions of years to ancestors of other phyla. Most of the phyla that provided ancestral stock for other phyla have themselves continued to radiate. So it is unlikely that any of the species alive today are unchanged descendants of the ancestral stock. By extension of some ecological principles such as Gause's, it seems likely that during adaptive radiation some of the modified forms would competitively eliminate the ancestral stock, even if it were capable of maintaining its genetic identity unchanged for eons—an unlikely event. Our ignorance of the exact relationship of the multidimensional events that occurred in evolution make it simpler to put what we do see into a two- or three-dimensional representation. If a phylogenetic tree were to be better visualized, it might be imagined as buried with tips of the various branches of all the phyla reaching the surface where the living species are now representative.

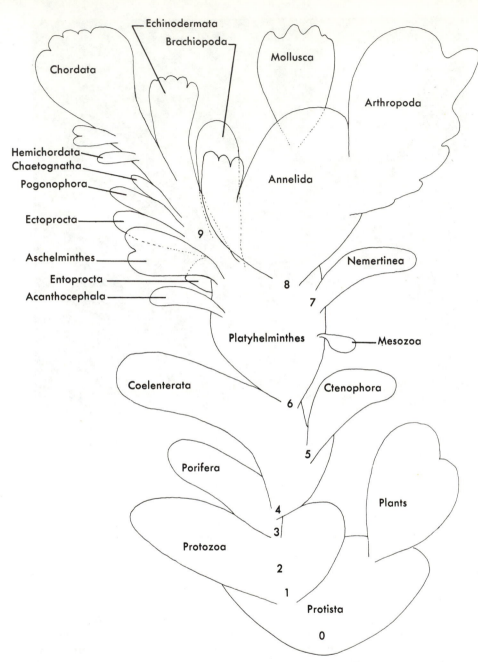

Figure 13-5 Phylogenetic tree, showing one view of suspected relationships. Numbers represent presumed point of origin of the following developments: 0, DNA-RNA. 1, Golgi apparatus, contractile vacuoles, amoeboid movement. 2, flagella and cilia. 3, collagen, spongin, cellular differentiation of multicellular organisms. 4, nerve and muscle cells. 5, biradiality. 6, bilaterality, flame cells, spiral cleavage. 7, blood vascular system. 8, coelom, coelomoducts, metamerism. 9, radial cleavage.

In the 12 years since the 1968 version of Figure 13-5 was prepared, little has occurred to warrant modification. The connection between coelenterates and poriferans seems more reasonable and has been opened wider so that nerve cells (4) can be recognized in sponges. An intuitive judgment, difficult to support concisely, resulting from additional reading and thought, is that the figure should be compressed and widened with the connections broadened so that our alternate visualization would now appear more like a stack of saucers with a central core bearing the numbers of the basic advances in close succession, the core now mostly extinct, and extant forms represented by bits of the rims in different adaptive states. Our new phylogenetic "tree" would then look more like a cabbage with the top severed and those leaves at the severed surface representing the existing classes of animals, the phyla represented by the levels of the core and the attached leaves of those levels.

Consideration of such a tree may help one to appreciate some facts of living organisms: that for all their tremendous diversity some startlingly basic themes appear and resound through succeeding groups. Study of simpler forms can show many basic facts of life in their most easily demonstrated form.

REFERENCES

BLISS, D. E., et al. 1968. Terrestrial adaptations in Crustacea [symposium]. *Am. Zoologist,* **8**:307–685.

BUSHNELL J. H., et al. 1974. Perspectives on the biology of dormancy [symposium]. *Trans. Am. Micr. Soc.,* **93**:459–631.

CARRIKER, M. R., et al. 1969. Penetration of calcium carbonate substrates by lower plants and invertebrates [symposium]. *Am. Zoologist,* **9**:629–1020.

CLARK, R. B. 1964. *Dynamics in Metazoan Evolution.* Clarendon Press, Oxford. 313 pp.

COPELAND, E. 1966. Salt transport organelle in *Artemia salenis* (brine shrimp). *Science,* **151**:470–471.

COSTELLO, D. P., et al. 1976. Spiralian development [symposium]. *Am. Zoologist,* **16**:277–626.

FITCH, W. M., and E. MARGOLIASH. 1967. Construction of phylogenetic trees. *Science,* **155**:279–284.

GARDNER, D. 1971. Bilateral symmetry and interneuronal organization in the buccal ganglia of *Aplysia. Science,* **173**:550–553.

HARGER, J. R. E. 1972. Competitive coexistence among intertidal invertebrates. *Am. Scientist,* **60**:600–607.

HEDGPETH, J. W., et al. 1977. Biology of lophophorates [symposium]. *Am. Zoologist,* **17**:1–150.

HILL, R. B., et al. 1979. Comparative physiology of invertebrate hearts [symposium]. *Am. Zoologist,* **19**:1–190.

HYMAN, L. H. 1940–1959. *The Invertebrates.* McGraw-Hill, New York.

LOVTRUP, S. 1975. Validity of the Protostomia-Deuterostomia theory. *Syst. Zool.,* **24**:96–108.

MIYAKE, A., and J. BEYER. 1974. Blepharmone: a conjugation-inducing glycoprotein in the ciliate *Blepharisma*. *Science*, **185**:621–623.

OLIVER, J. H., JR., et al. 1971. Parthenogenesis [symposium]. *Am. Zoologist*, **11**:239–398.

OSMAN, R. W. 1977. The establishment and development of a marine epifaunal community. *Ecol. Monogr.*, **47**:37–63.

PROSSER, C. L., and F. A. BROWN, JR. 1961. *Comparative Animal Physiology*. Saunders, Philadelphia. 688 pp.

RHEINS, L. A., R. D. KARP, and A. BUTZ. 1978. Humoral immunity induced in the American cockroach (*Periplaneta americana*). *Am. Zoologist*, **18**:595.

RIDDIFORD, L. M., et al. 1976. Invertebrate neurosecretion [symposium]. *Am. Zoologist*, **16**:105–175.

SAFRIEL, U. N. 1974. Vermetid gastropods and intertidal reefs in Israel and Bermuda. *Science*, **186**:1113–1115.

SANGER, J. W. 1979. Cardiac fine structure in selected arthropods and molluscs. *Am. Zoologist*, **19**:9–27.

SHERMAN, R. G., et al. 1973. Invertebrate neuromuscular systems [symposium]. *Am. Zoologist*, **13**:235–445.

SOKAL, R. R. 1974. Classification: purposes, principles, progress, prospects. *Science*, **185**:1115–1123.

THOMPSON, D'ARCY. 1917. *On Growth and Form*. An abridgement by J. T. Bonner, 1961. Cambridge Univ. Press, New York.

WAAGE, J. K. 1979. Dual function of the damselfly penis: sperm removal and transfer. *Science*, **203**:916–918.

WASSERMAN, G. S. 1973. Invertebrate color vision and the tuned-receptor paradigm. *Science*, **180**:268–275.

WHITTAKER, R. H. 1969. New concepts of kingdoms of organisms. *Science*, **163**:150–160.

WITT, P. N., et al. 1972. Invertebrate behavior [symposium]. *Am. Zoologist*, **12**:xlvi–xlvii;385–594.

Application—the Deep-Sea Invertebrates

The value of knowledge is related to its usefulness. Many scientists give accolades to pure research that has no certain value because they know it often has unforeseen applications that may even be greater in value than significant applied research.

All research has the potential of training for the teams or individuals and the accompanying development of methods and equipment.

This chapter is meant to explore one question that does not have a certain answer and is not subject to ready direct observation. The example chosen will allow us to discuss a broad range of principles of invertebrate zoology and biology in evaluating the question.

The question is—do deep-sea invertebrates reach much greater ages than their relatives from other habitats?

A question that guides research is best framed in a way that asks one specific question that can be answered by a set of observations. Such a question usually generates numerous other questions. In our case we might question our question and say—such a question is too broad; some may and some may not reach great age. The final answer will be achieved by the summation of many specific answers.

This question (Engemann, 1968) arose with the preparation of the second edition of this text. The Pogonophora had characteristics that seemed to require the implied answer. The speculative answer did not seem appropriate to a textbook without some prior exposure to criticism by the scientific community. The brief paper is reprinted here to express the first statement of the question.

POGONOPHORA: THE OLDEST LIVING ANIMALS?*

JOSEPH G. ENGEMANN
Western Michigan University

Recent studies by Webb (1963, 1964a, 1964b) on *Siboglinum fiordicum* have revealed several features of the Pogonophora previously unknown. Of interest here is the lateral branch sometimes found on the tube. Deductions from the structure and location of their tubes indicate they may reach extreme ages for animals. The tubes of *S. fiordicum* show annulations, reach a length of 243 mm, and are about one-quarter mm in diameter. They are normally without branches and are thought to be oriented perpendicularly in the bottom sediments. This assumption is supported by the rarity with which the extreme posterior end of the animal is recovered (Webb 1964a). When a lateral branch of the tube occurs, it is usually occupied by the posterior end of the animal, and the lateral branch has no annulations. These represent portions secreted when the posterior end breaks through the wall of the original tube (Webb 1964b). From this we can conclude that the normal method of tube elongation is by addition to the anterior, or upper, end.

Some other pogonophorans have flared upper openings (Ivanov 1963), and their annulations are evidently added at the anterior end. Some have annulations only in the mid-region, with the upper and lower portion of the tube flimsy and lacking annulations. Thus the arrangement of annulations and lack of structural rigidity make it probable that tubes are formed in situ and not forced downward as they form.

Thus, if tube formation begins at or near the surface of the sediments, the age of sediments at the greatest depth to which occupied tubes extend would give an indication of maximum age attained by pogonophorans. The possibility of the tube being pulled down after penetration and anchorage of the suckered and seta-bearing metasoma seems unlikely since tube length considerably exceeds body length of preserved specimens. Also, friction over such a long tube length would probably greatly exceed that on the relatively short metasoma. The short length and uniformity of the setae on the metasoma are evidence they serve as anchors in the uniform tubes. The metasoma is seldom recovered with the animal. It is likely the entire tube is seldom recovered. No evidence exists to suggest the same tube is occupied by more than one generation of adult pogonophoran. If one of the larvae replaced the parent in the tube, a discontinuity would be probable; none is noted.

Estimates of sediment age at depths reached by pogonophoran tubes give approximations of length of life reaching 25,000 years as a rather conservative estimate. Thus *Zenkevitchiana longissima* has a tube up to 150 cm long. Assuming half is embedded in the sediment and sedimentation rates of 3 cm per 1000 years (Ericson et al. 1963) 75 cm ÷ 3 cm × 1000 yrs. = 25,000 yrs. If slower sedimentation rates of 1 mm per 250 years (Ekman 1953) are assumed and a greater portion, say 100 cm, of the tube buried it gives 100 cm × 10 mm/cm × 250 yr./mm = 250,000 years.

* Reprinted by permission of The Michigan Academy of Science, Arts, and Letters. *Papers of the Michigan Academy of Science, Arts, and Letters*, **53**:105–108.

In either event it appears that the pogonophorans qualify as the oldest living animals, older by as much as one or two orders of magnitude than other long-lived animals. The sheer novelty of thinking about an animal so old makes one want to reduce it by a factor of 10 to allow for local areas of more rapid accumulation, tube not vertical or less tube embedded, and some formation beneath the surface. Even then 2500 to 25,000 years as an estimate still leaves us with the pogonophorans as the oldest animals and probably the oldest living organisms.

Supporting facts come from other aspects of the ecology of the organism. The extreme cold and uniform environment of the ocean bottom with only slight seasonal variation would be one of the best places on earth for long-lived animals. Their embedded position would confer considerable protection from predators. Their lack of a digestive system makes it likely that a part of their nutrition is saprozoic reliance on nutrients present in solution in the water of bottom sediments. Their tentacles are perhaps now primarily respiratory structures. The tentacles may have been derived from the lophophore of a filter-feeding ancestor.

If nutrition is saprozoic from fossil nutrients, radiocarbon dating of the older portions of the tube would be misleading and give older than actual ages as results. Conversely, it appears that additional layers are deposited in some tubes as the animal ages and, if food is acquired by the tentacles, it would result in younger ages by radiocarbon dating than the actual ages.

Annulations may well be related to annual increments in growth, although annual fluctuations which would produce such changes are thought to be very slight in abyssal regions. Calculations based on annulations would give ages of 1000 years and more. The extreme cold would result in slow growth rates. This could result in life cycles perhaps eight times longer at temperatures near 0°C than at surface temperatures of 30°C if the estimate of a chemical reaction being slowed by half with a drop of 10° has any relationship to enzymatic systems adapted to low temperatures. Growth and development are slowed by a scanty food supply in some invertebrates. Abyssal regions have a slow rate of input of energy available for use by animals.

Ability to survive with a slow growth rate may be the reason pogonophorans are so successful in the abyssal regions, where accumulation of sediment is so slow, and so rare in shallower seas, where sedimentation rates are normally much greater. Perhaps growth is too slow to prevent burial in the latter case.

More precise aging of pogonophorans must await more accurate findings on orientation or depth of penetration in bottom sediments where rate of accumulation is known. They will most likely fortify the hypothesis that pogonophorans are the oldest organisms alive. The fortunate discovery of several species of pogonophorans at depths of about 200 m near Florida (Nielsen 1965) may speed the answer to this fascinating question.

SUMMARY

Interpretation of research on the Pogonophora indicates they include the oldest living animals. Estimates of maximum age range from 1000 years to an unlikely 250,000 years. Various environmental and biological factors bearing on the question are briefly discussed. A slow growth rate is proposed as a reason for, and made possible by, their general abyssal distribution.

Literature Cited

EKMAN, S. 1953. Zoogeography of the Sea. London: Sidgwick and Jackson.

ERICSON, D. B., M. EWING, and G. WOLLIN. 1963. Pliocene-Pleistocene Boundary in Deep-Sea Sediments. Science **139**:727–737.

IVANOV, A. V. 1963. Pogonophora. New York: Consultants Bureau.

NIELSEN, C. 1965. Pogonophora: Living Species Found off the Coast of Florida. Science **150**:1475–1476.

WEBB, M. 1963. *Siboglinum fiordicum* sp. nov. (Pogonophora) from the Raunefjord, Western Norway. Sarsia **13**:33–44.

———. 1964a. The Posterior Extremity of *Siboglinum fiordicum* (Pogonophora). Ibid. **15**:33–36.

———. 1964b. Tube Abnormality in *Siboglinum ekmani, S. fiordicum* and *Sclerolinum brattstromi* (Pogonophora). Ibid. **15**:69–70.

In the reprinted article extreme age was inferred from the structure and placement of pogonophoran tubes. Other evidence suggested was cold temperature and low food supply.

Additional evidence will be considered, both direct and indirect, including evidence against the hypothesis.

Direct Observations in the Deep Sea

Jannasch et al. (1971) reported a remarkable instance of a very low rate of bacterial degradation of food that sank to great depths with the *Alvin*. The *Alvin* is an experimental submarine recovered after 8 months from the 1,540-meter depth of the ocean. Follow-up studies determined that the great pressure of the depth slowed bacterial action to less than 1 percent of surface rates.

The slow activity was consistent with the presumed low activity responsible for living bacteria found by Morita and ZoBell (1955) in deep-sea sediments deposited about 1 million years ago. Jannasch and Wirsen (1973) confirmed the slow growth of deep-sea microorganisms. By inoculating nutrients with bottom materials and incubating them at the great depths they found less growth per year than usually occurred in 1 month at 4°C under surface pressures.

Respiratory Rates. Smith and Teal (1973) found that the mean respiratory rates associated with an enclosed bottom sample at 1,850-meter depth was two orders of magnitude less than at shallow depths. A similar depression of respiratory rates occurs for fishes (Smith and Hessler, 1974).

Yayanos, Dietz, and VanBoxtel (1979) discovered a bacterium that grows best under pressure (Fig. 14-1). The response to pressure shows that the depression of growth is very sharp with greater increases in depth beyond 725 bars (1 bar = 1 atmosphere of pressure). A smoother curve would seem to be expected. The metabolic basis of pressure ef-

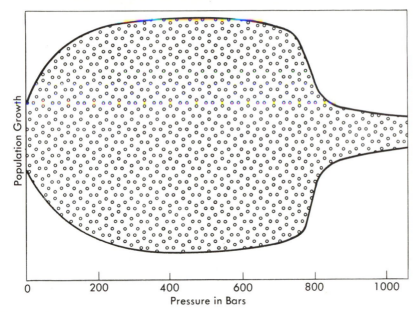

Figure 14-1 Growth relative to pressure for a barophilic bacterium. Growth is proportional to the logarithmic dimension of line thickness. (*Adapted from Yayanos, Dietz, and VanBoxtel, 1979.*)

fects is generally thought to be as stated by Low and Somero (1975): "All pressure effects, whether they be on chemical equilibria or on reaction rates, are due to volume changes which occur in the chemical process." Different substances in equilibrium reactions may have different volume changes with compression. Reactions tend toward equilibrium with products that represent reduced volume accumulating when pressure is increased.

The enzymes of deep-sea organisms seem to be of the same types as those of their shallow-water relatives. Hochachka (1971) cites studies showing that cold-water fish may be adapted to high activity at near-freezing temperatures. The adaptation to pressure would seem to occur as readily. The *in situ* measurements of Smith and Teal (1973), Jannasch and Wirsen (1973), and Smith and Hessler (1974) represent a considerable advance from the state of the art noted by Phleger (1971): "We know essentially nothing about metabolic processes of deep sea organisms."

Disturbance Levels. Wind and waves are not present in the deep sea. Tidal and other currents may be present but of low relative magnitude over the majority of the abyss. Low activity and low population levels relative to surface populations result in less disturbance. Shakers and bubblers dramatically speed growth of microorganisms. Small-scale movements by diffusion and Brownian movement may be diminished at great pressures because the asymmetry of the water molecule may interfere with molecular mobility under super compression. Hildebrand (1971) notes that a general theory is lacking, but observed cases indicate that fluidity is proportional to expansion. Chen and Millero (1976) have determined that seawater decreases about ½ percent in volume for each 100-bar increase in pressure studied at 0°C. If water behaves as other liquids, its viscosity is greater under great pressure. Either viscosity or volume changes due to pressure could affect chemical reactions of metabolism. Viscosity effects would not be subject to the evolution of adaptations for speed to the level of organisms not under such pressure.

The absence of DDT and PCB in ocean sediments (Harvey, 1974) indicates little impact from the traumatic surface events of recent years. It also substantiates the presumed extremely low level of nutrient input from surface waters, since those compounds should be abundant in feces and dead organisms from surface waters.

Clam Growth. Turekian et al. (1975) estimated the age of 8.4-mm-long *Tindaria callistiformis* to be about 98 years based on radioactive decay of incorporated radionuclides of the shell. The approximately 100 growth bands present were perhaps comforting considering the potentially large error associated with the few data points available due to the extremely difficult and time-consuming procedures involved.

Indirect Evidence

The bulk of evidence of extreme age is inferential. Some is based on deep-sea studies; some is based on application of principles learned elsewhere.

Life Histories. Grassle and Sanders (1973) suggest that small brood size and a high proportion of adult size classes indicate low growth rates as well as low rates of reproduction and mortality for the deep-sea fauna. Engemann (1978) notes that a slow growth rate would allow the pooling of data on populations gathered over a longer time span than is conventionally done for consideration as a single sample. Pooling the data from five samples gathered over a 15-month period by Rokup (1974), published for the brittle star, *Ophiacantha normani,* gives a smoothed bimodal distribution (Fig. 14-2).

Interpreting Curves. A variety of causes for mortality exist and exert their pressures, often at differing rates and different times of the life cycle. Figure 14-3 shows curve 1 for those with little natural mortality until aging processes take their toll. Curve 3 of the same figure is typical of many invertebrates that have high larval mortality due to predation, not reaching suitable habitat, or perhaps competition. Note that for all types of curves found for exclusively sexually reproducing ani-

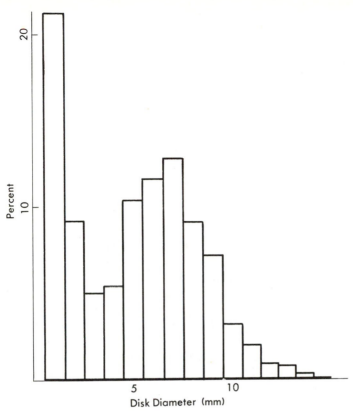

Figure 14-2 Pooled data for *Ophiacantha normani* size distribution. Five samples collected at 3-month intervals were pooled. (*Adapted from the data of Rokup, 1974.*)

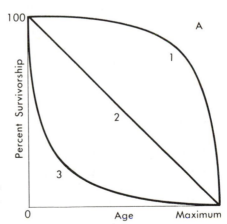

Figure 14-3 Types of curves of survivorship of a population through increasing ages.

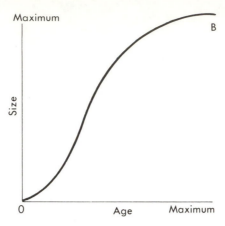

Figure 14-4 Generalized growth curve of size increasing with age.

mals there can be *no increase in numbers of an age class* beyond the starting number, only losses due to mortality.

The bimodal distribution of Figure 14-2 must then imply that another factor is involved. Several may be. One is that the feature graphed is size, not age. Most invertebrates slow in growth of linear dimensions as they age, so that size is not equivalent to age on a graph and the results tend to be graphed as a sigmoid growth curve (Fig. 14-4). The slower growth with increasing age and size is due to many factors, such as: (1) biomass increase is proportional to the cube of the linear dimensions being graphed; (2) shifting of most energy resources from growth to reproduction at maturity; (3) deterioration due to aging; (4) allometric growth, causing an increased proportion of nonfeeding structures; and (5) inability to shift to larger food or unavailability of proportionately more food since the feeding area of a benthic organism is two-dimensional whereas the organism's growth is three-dimensional.

Another possible cause of the population curve (Fig. 14-2) from the deep sea could be cyclic changes in reproductive success.

Grassle and Sanders (1973) thought that predation was not important in structuring deep-sea populations. The curve of Figure 14-2 would support that conclusion, because Estes and Palmisano (1974) found a similar curve for sea urchins in the absence of sea otter predation. The sea urchin biomass was also much higher as compared to a population subject to predation. Our interpretation is consistent with the observations of Grassle et al. (1975) from the *Alvin*. They saw a brittle star, *Ophiomusium lymani,* in high abundance with a fairly uniform distribution of individuals showing very little activity.

Thayer (1975b) shows that some of the same type of biomodal distributions for brachiopods are not artifacts of sample size or differential transport for fossil populations. Jackson, Goreau, and Hartman (1971) had found the bimodal distribution in very high density shallow-water brachiopod communities that would be expected to have a low food supply.

The best interpretation of all of the bimodal distributions appears to be that the second mode is due to the accumulation of many long-lived, slow-growing adults.

Food Supply. Input of food is extremely low for most of the ocean bottom. Under most of the ocean, sediments accumulate very slowly (Heezen and Hollister, 1971, Fig. 7-45). Large areas accumulate less than 0.5 cm per thousand years; most of the ocean accumulates less than 2 cm of sediment per thousand years. Fleischer (1979) cites some extreme reports ranging from 6 mm per 1,000 years near the East Pacific Rise to 2 mm per year in the North Sea. Kennett and Shackleton (1975) found that sedimentation rates in cores from areas of high sedimentation in the Gulf of Mexico at depths of over 2,460 m were 7 to 27 cm per 1,000 years.

Woolf (1960) notes biomasses as high as 10 to 40 grams per square meter in some deep trenches. Deep ocean water has been measured at ages of over 1,000 years. The slow replacement by polar waters and the curve for oxygen, showing that most of the reduction to the oxygen minimum beneath the ocean thermocline occurs at relatively shallow depths as water rises slowly from the deep, indicates that the vast majority of productivity from overlying waters is used before it reaches the bottom. That is also evident from the generally less than 1 percent organic content of deep sediments. An extremely slow pace of life may be necessary to account for the combination of low food supply, high oxygen levels with slow replacement, and population densities present.

We have in earlier chapters referred to the rate of development of ticks being related to feeding and the development of *Aurelia* and negative growth indicating reduced nutrition may extend life.

Gigantism. Wolff (1960) notes that gigantism is a characteristic of some deep-sea arthropods that can be opportunistic feeders on a wide range of foods. They may have the ability of some insects noted in Chapter 11 to adapt respiratory rates to activity and thus slow the use of resources between rare meals of falling fish or whales.

The gigantism is probably not due to extreme age, but to the competitive efficiency related to size, frequency, and spacing of food supplies for feeders of their type. The energy required to move identical amounts of mass through water decreases dramatically as it is put in larger packages of the same density.

The pogonophorans and other large specimens of thermal vent communities (Corliss et al., 1979) probably achieve their size as a result of unusual nutrient resources from thermal springs.

Community Structure. Connell (1978) and Osman and Whitlatch (1978) provide evidence that low diversity could be associated with stable communities. Jumars and Hessler (1976) note high standing crop and low diversity in the Aleutian Trench. Hessler and Sanders (1967) had found high diversity but had used an epibenthic sled for sampling material from over a 1,000-meter strip of bottom. The result may have given misleading diversity data if scale (Jumars, 1976) is important in the deep sea. Short lives and the resulting turnover of individuals could cause instability and higher diversity if it occurred in the deep sea. Slow recolonisation of defaunated deep-sea sediments (Grassle, 1977) indicates slow growth.

K and r Strategy. The features of low food supply, low predation, and stability in the deep sea are at the extreme of the ranges facing living organisms. We should thus expect the organisms of the deep sea to reflect the extremes of K-strategists due to K-selection.

Few young, and longer-lived individuals are thus expected to be ways of meeting the K-selective problems. This can represent a difference of several orders of magnitude in reproductive potential shifted to resources for survival of the individual.

The isopodan suborder Phreatoicoidea occurs in many freshwater habitats similar to those occupied by *Asellus* (Chapter 10). *Asellus* lives in an environment that has a greater seasonal input of nutrients and more fish and amphibian predators. A Tasmanian phreatoicid studied by Engemann (1963) took 3 years to develop to gamete production, with the female producing a maximum of 17 eggs (Fig. 14-5) per pair. Its north-temperate counterpart, *Asellus*, can complete its life cycle in from 1 year to as little as 3 months and produce in excess of 170 eggs per pair per clutch. On a yearly schedule of reproduction with no mortality *Asellus* has the potential of producing over 0.2 million offspring in the 3 years the phreatoicid, *Colubotelson*, is producing only 17 offspring. The temendous difference in biotic potential is only that—a potential. The populations of each will tend to fluctuate around the same level unless other changes occur in the environment. Reproduction is balanced by mortality in stable populations.

The *Colubotelson* egg may take 6 months to complete its embryonic development, the *Asellus* egg only a few weeks.

In the deep sea, K-selection, the low temperatures, and the slowing of respiration that appears to result from pressure could well slow egg development and other reproductive-developmental stages beyond that exhibited by *Colubotelson*. The cicadas, with sympatric populations of 2-year and 17-year life-cycle types, are an example of differences that occur without obvious variables of environmental selection.

The conclusion of year-round breeding by deep-sea forms that yield collections containing ovigerous or other index stages throughout the

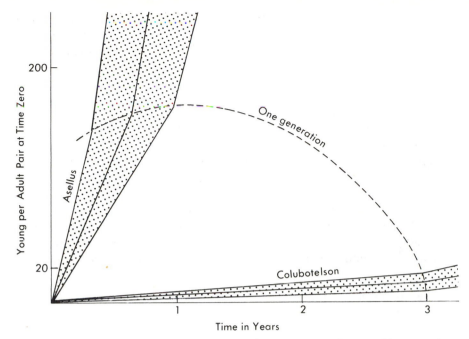

Figure 14-5 Graphic comparison of *r*- and *K*-strategists. Range of increase in one generation of two genera of isopods. (*Adapted from the data of Engemann, 1963.*)

year must be reevaluated in light of the probable slow development. It may take several years with years between successive small broods to complete one cycle of reproduction for the *K*-strategist that typically has a long reproductive life as an adult. Many have contended that there is no signal of sufficient magnitude in the deep sea to serve as a time signal for an annual breeding season (even if participation is in alternate or infrequent years). Rokup (1974) has demonstrated that some species have an annual cycle. The other species could use them as their time piece. Slight annual shifts in temperature can be significant—note the difference in water and ice at a fraction of a degree above and below freezing.

Behavior and Structure. Marine sediments seem to be quite soft in most locations. Thayer (1975a) notes that reduced growth may be a method of staying light enough to remain on the sediments.

Kitchell et al. (1978) found that meander and spiral tracks peaked at a depth of 3,000 meters. The tracks give evidence of a feeding strategy. Could the reduced evidence of deeper tracks indicate that deeper species have not had time to evolve a feeding strategy?

Back to the Drawing Board. If deep-sea animals do live long times, many things must be reevaluated. Can we determine the length of life from Figure 14-2 for the brittle star? Not with great accuracy unless more information is available. How long does it take for a new individual to grow to some small size? Does the first dip occur in 1 year, 10 years, or 100 years? That determines the scale for other projections.

A graphic solution may be attempted by partitioning the curve into compartments for each year class. Figure 14-6 illustrates the way in which the areas would be partitioned if rapid growth occurred in the deep sea. Since the valley in the bimodal distribution is clearer in the pooled data, it seems reasonable to assume that more than 1 year is needed to reach that size. Although we expect that a greater size range exists in older age classes, we can compress them to their modal values, which should become closer with age (the effect of the sigmoid-growth-curve phenomenon). If we reduce mortality projections only slightly, the area for each succeeding year class should be only slightly smaller and thus must be higher and produce the second peak that makes the bimodal curve. Since the peak is more than twice the height

Figure 14-6   Rejected projection because of excessive growth assumed. (*See text for explanation.*)

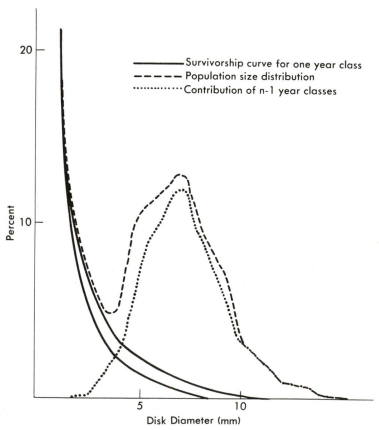

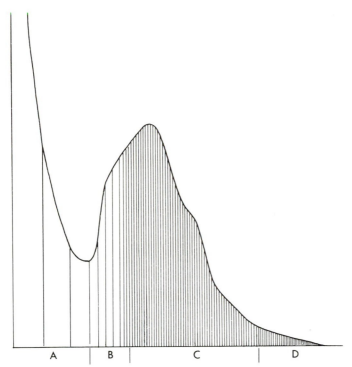

Figure 14-7 Hypothetical events of population growth curve. (A) Rapid growth, high mortality. (B) Growth and mortality slowing, onset of reproductive maturity. (C) Very low mortality of maturity. (D) Very low growth due to large size. Each vertical line indicates mean size of members of an age interval.

of the valley, the spacing at the peak should be less than half that at the valley. The continued reduction in growth can be expected and we end up with about 50-year classes, indicated in Figure 14-7. If each of the year classes were portrayed exactly, they would probably form progressively smaller and overlapping normal curves, with each next older year class peaking at a slightly larger size.

The respiration rates, two orders of magnitude lower, indicate that our divisions may not be fine enough. Increasing them 10- to 100-fold to account for the metabolic slowing would indicate that 500- to 5,000-year classes are accumulated in one population. It seems very likely that some are over 1,000 years old.

Summary

The evidence that deep-sea animals reach extreme ages is strong but largely circumstantial.

1. A direct measurement shows long life for a small clam.
2. Pogonophoran tube structure indicates long life.

3. Population structure of an ophiuroid indicates long life.
4. A long life is consistent with a slow pace of life due to observed
   a. Low food supply
   b. Low temperature
   c. Reduced respiration and metabolism due to pressure
   d. Low disturbance level
   e. *K*-selected extremes due to above and low predation
   f. Low activity levels
   g. High population relative to resources
   h. Slow sediment recolonisation

The evidence against the hypothesis is also circumstantial.

1. Rapid attraction of large scavengers to bait indicates a higher rate of activity unless sensory information can arouse them from semi-dormancy.
2. Gigantism could indicate rapid growth rather than long life, and it is certainly adaptive to deep-sea conditions for those exhibiting it.
3. Annual gametic production cycles are known for some benthic deep-sea invertebrates, albeit not the dominant ones.

The Challenge

With over half our planet covered by abyssal depths, the rate of life in such locations has considerable bearing on future potential resources. The survival of many still unknown organisms may be jeopardized by our activities. Such organisms may be relics of the past, as some deep-sea organisms already appear to be. They may answer many basic biological questions.

Knowledge and study of the details of terrestrial and shallow-water species may give hints as to ways of best studying deep-sea species.

Many questions remain to be answered for the length-of-life question alone. Some suggested questions are:

1. Are other radiological methods applicable?
2. Do cyclic phenomena express themselves with markers such as annual rings in other than shells, ovarian scars, etc.?
3. Will rates of racemization of amino acids answer questions of aging?
4. Are other functional or structural clues available?
5. Are ecological or other principles additive, synergistic, or antagonistic in concert with others having the same direction of effect?

Many times you do not find an answer until you look for it or ask a question.

REFERENCES TO LITERATURE ON THE DEEP-SEA INVERTEBRATES

CHEN, C. T., and F. J. MILLERO. 1976. The specific volume of seawater at high pressures. *Deep-Sea Res.,* **23:**595–612.

CONNELL, J. H. 1978. Diversity in tropical rain forests and coral reefs. *Science,* **199:**1302–1310.

CORLISS, J. B., et al. 1979. Submarine thermal springs on the Galapagos Rift. *Science,* **203:**1073–1083.

ENGEMANN, J. G.. 1963. A comparison of the anatomy and natural history of *Colubotelson thomsoni* Nicholls, a south temperate, fresh-water isopod and *Asellus communis* Say, a north temperate, fresh-water isopod. Ph. D. thesis, Michigan State Univ., East Lansing. 146 pp.

————. 1978. Indirect evidence shows deep-sea benthos may reach extreme ages as individuals. *Am. Zoologist,* **18:**666.

ESTES, J. A., and J. F. PALMISANO. 1974. Sea otters: their role in structuring nearshore communities. *Science,* **185:**1058–1060.

FLEISCHER, R. L. 1979. Where do nuclear tracks lead? *Am. Scientist,* **67:**194–203.

GRASSLE, J. F. 1977. Slow recolonisation of deep-sea sediment. *Nature,* **265:**618–619.

————., and H. L. Sanders. 1973. Life histories and the role of disturbance. *Deep-Sea Res.,* **20:**643–659.

————, H. L. SANDERS, R. R. HESSLER, G. T. ROWE, and T. McLELLAN. 1975. Pattern and zonation: a study of the bathyal megafauna using the research submersible *Alvin. Deep-Sea Res.,* **22:**457–481.

HARVEY, G. R. 1974. DDT and PCB in the Atlantic. *Oceanus,* **18:**18–23.

HEEZEN, B. C., and C. D. HOLLISTER. 1971. *The Face of the Deep.* Oxford Univ. Press, New York. 659 pp.

HESSLER, R. R., and H. L. SANDERS. 1967. Faunal diversity in the deep-sea. *Deep-Sea Res.,* **14:**65–78.

HILDEBRAND, J. H. 1971. Motions of molecules in liquids: viscosity and diffusivity. *Science,* **174:**490–493.

HOCHACHKA, P. W. 1971. Enzyme mechanisms in temperature and pressure adaptation of off-shore benthic organisms: the basic problem. *Am. Zoologist,* **11:**425–435.

JACKSON, J. B. C., T. F. GOREAU, and W. D. HARTMAN. 1971. Recent brachiopod-coralline sponge communities and their paleoecological significance. *Science,* **173:**623–625.

JANNASCH, H. W., and C. O. WIRSEN. 1973. Deep-sea microorganisms: in situ response to nutrient enrichment. *Science,* **180:**641–643.

————, K. EIMHJELLEN, C. O. WIRSEN, and A. FARMANFARMAIAN. 1971. Microbial degradation of organic matter in the deep sea. *Science,* **171:**672–675.

JUMARS, P. A. 1976. Deep-sea species diversity—does it have a characteristic scale? *J. Mar. Res.,* **34:**217–246.

————, and R. R. HESSLER. 1976. Hadal community structure: implications from the Aleutian Trench. *J. Mar. Res.,* **34:**547–560.

KENNETT, J. P., and N. J. SHACKLETON. 1975. Laurentide ice sheet meltwater recorded in Gulf of Mexico deep-sea cores. *Science,* **188:**147–150.

KITCHELL, J. A., J. F. KITCHELL, D. L. CLARK, and L. DANGEARD. 1978. Deep-sea foraging behavior: its bathymetric potential in the fossil record. *Science*, **200**:1289–1291.

LOW, P. S., and G. N. SOMERO. 1975. Pressure effects on enzyme structure and function *in vitro* and under simulated *in vivo* conditions. *Comp. Biochem. Physiol.*, **52B**:67–74.

MORITA, R. Y., and C. E. ZOBELL. 1955. Occurrence of bacteria in pelagic sediments collected during the Mid-Pacific Expedition. *Deep-Sea Res.*, **3**:66–73.

OSMAN, R. W., and R. B. WHITLATCH. 1978. Patterns of species diversity: fact or artifact? *Paleobiology*, **4**:41–54.

PHLEGER, C. F. 1971. Pressure effects on cholesterol and lipid synthesis by the swimbladder of an abyssal *Coryphaenoides* species. *Am. Zoologist*, **11**:559–570.

ROKOP, F. J. 1974. Reproductive patterns in the deep-sea benthos. *Science*, **186**:743–745.

SANDERS, H. L., and R. R. HESSLER. 1969. Ecology of the deep-sea benthos. *Science*, **163**:1419–1424.

SMITH, K. C., and R. R. HESSLER. 1974. Respiration of benthopelagic fishes: *in situ* measurements at 1230 meters. *Science*, **184**:72–73.

———, and J. M. TEAL. 1973. Deep-sea benthic community respiration: an *in situ* measurement at 1850 meters. *Science*, **179**:282–283.

THAYER, C. W. 1975a. Morphologic adaptations of benthic invertebrates to soft substrata. *J. Mar. Res.*, **33**:177–189.

———. 1975b. Size-frequency and population structure of brachiopods. *Palaeogeog., Palaeoclimatology, Palaeoecol.*, **17**:139–148.

TUREKIAN, K. K. et al. 1975. The slow growth rate of a deep-sea clam determined by ^{228}Ra chronology. *Proc. Nat. Acad. Sci. USA*, **72**:2829–2832.

WOLFF, T. 1960. The hadal community, an introduction. *Deep-Sea Res.*, **6**:95–124.

YAYANOS, A. A., A. S. DIETZ, and R. VANBOXTEL. 1979. Isolation of a deep-sea barophilic bacterium and some of its growth characteristics. *Science*, **205**:808–810.

Index

Numbers in *italic* type indicate illustrations. Additional illustrations may be found within multiple page citations. For terms not indexed refer to the section dealing with the appropriate taxonomic group.